Numerical Quadrature and Cubature

COMPUTATIONAL MATHEMATICS AND APPLICATIONS

Series Editor
J. R. WHITEMAN

Institute of Computational Mathematics, Brunel University, England.

E. HINTON and D. R. J. OWEN: Finite Element Programming

M. A. JASWON and G. T. SYMM: Integral Equation Methods in Potential Theory and Elastostatics

J. R. CASH: Stable Recursions: with applications to the numerical solution of stiff systems

H. ENGELS: Numerical Quadrature and Cubature

Numerical Quadrature and Cubature

H. ENGELS

Institut für Geometrie und Praktische Mathematik
Rhein.-Westf. Technische Hochschule
Aachen, West Germany

1980

ACADEMIC PRESS

A Subsidiary of Harcourt Brace Jovanovich, Publishers

London New York Toronto Sydney San Francisco

ACADEMIC PRESS INC. (LONDON) LTD
24–28 Oval Road
London NW1

US edition published by
ACADEMIC PRESS INC.
111 Fifth Avenue
New York, New York 10003

British Library Cataloguing in Publication Data

Engels, H
Numerical quadrature and cubature. – (Computational mathematics and applications).
1. Numerical Integration 2. Calculus, Integral
I. Title II. Series
515′.43 QA299.3 79–41235

ISBN 0–12–238850–X

**Printed in Great Britain by
John Wright & Sons, Ltd., at the Stonebridge Press, Bristol**

Preface

A glance through recent volumes of journals and proceedings of scientific meetings and conferences covering numerical mathematics, shows immediately that the field of numerical quadrature and cubature is now widely discussed. It seems, therefore, appropriate that a textbook should appear having strong emphasis on the applications within this field, and that it should be readily accessible to students.

The book arose from several lecture courses which I gave at the Technische Hochschule in Aachen during the period 1971–76, and includes much material based on my earlier research activities at the Central Institute of Applied Mathematics of the Kernforschungsanlage (Nuclear Research Centre) at Jülich, West Germany.

The book is directed at three groups of readers. For the graduate and undergraduate student the book has a broad approach yet includes much detail; excessive generality has been avoided. For the research worker many standard results are extended to include up-to-date research findings; this has been done, in particular, in the sections concerning cubature problems, and in the treatment of orthogonal polynomials and quadratures with derivative-free error bounds. Numerical mathematicians in practical situations will also, it is hoped, find much of interest. For all three groups of readers much auxiliary material has been included in the form of computer programs, tables, formulae and (in Chapter 9) detailed references to further sources.

It is not possible, in a volume containing some 400 pages, to cover the field completely; thus, of necessity, I have been forced to omit some important aspects and procedures. In deciding what to include and what to omit, I have been guided by considerations of applicability of the various methods. For this reason, for example, Euler Maclaurin quadrature has been omitted since, in my opinion, it is a technique of mainly theoretical interest. This decision was made easier by the fact that there exist several excellent representations of this quadrature. I also do not emphasise the use of the most general types of integral, and especially have paid no attention to the question of whether convergence is for Riemann or Lebesgue integrals. Integrals which exist in the sense of only Cauchy principle values are not treated and it has, unfortunately, been necessary also to omit detailed discussion of Sard quadratures.

By judicious choice of a sufficiently generalised definition of interpolation, of orthogonal polynomials and, as a result, of Gaussian quadrature, it has been possible to treat simultaneously several different quadrature methods. The treatment of error analysis, convergence and orthogonal polynomials in Chapters 3, 4 and 5 has been done in the language of elementary functional analysis, since this is a natural way of approaching the subject. Anyone not familiar with functional analysis may omit these chapters at a first reading or, alternatively, may just accept these results. I have, in any case, attempted to use the terms of functional analysis in such a way that the reader can treat them as "abbreviations". The only place in the book where a result of functional analysis has been proved in detail is in the Theorem 4.3.2 of Banach and Steinhaus. This has been included because of its fundamental importance, and is preferred to that of Polya and Stekloff which is concerned only with the quadrature case. In every situation, terms of functional analysis are deliberately explained as they arise rather than being collected together in a single section. The intention here is to avoid distracting the reader with time-consuming preliminaries. I have also not followed the normal practice of using the "definition – theorem – proof" sequence. It is my opinion that this is both admissable and desirable in order to help readers learn and obtain insight into mathematical thinking and research approaches.

I am greatly indebted to Professor J. R. Whiteman for the following several reasons. He not only encouraged me to write this book, but also readily accepted it for his new Academic Press series "Computational Mathematics and Applications". In addition, he worked very carefully and painstakingly through the entire manuscript improving the English and commenting on the contents. I am also grateful to Professor U. Eckhardt who read the manuscript and suggested some improvements, to Dr. I. Bausch and Dipl. Math. A. Merschen for studying the manuscript and eliminating many misprints and in addition to Dipl. Math. A. Merschen for reading the proof sheets. It is also a pleasure to acknowledge the support of several members of the Institut für Geometrie und Praktische Mathematik in Aachen and the stimulating questions and comments of my students. Finally, and most importantly, I owe most affectionate thanks to my family, who gave up so much during the period in which the book was being written.

June 1978 H. E.

Contents

List of Symbols and Abbreviations

Symbols

3.2.1	Chapter 3, Section 2.1
(7.3.4)	4th formula of Chapter 7, Section 3
P.5.2	2nd problem in the last section of Chapter 5
R.7.1	1st remark in the last section of Chapter 7
$(\cdot\,,\cdot)$	inner or scalar product
$\langle\cdot\,,\cdot\rangle$, $\langle\cdot\circ\cdot\rangle$, $\langle\cdot\mid\cdot\rangle$	different discrete inner products
$\|\cdot\|$	norm
$\vert\vert\vert\cdot\vert\vert\vert$	seminorm or pseudo-norm, not definite
$\perp$	orthogonal
$\delta_{i,k}$	$=\begin{Bmatrix} 0 & \text{for} & i\neq k \\ 1 & \text{for} & i=k \end{Bmatrix}$ Kronecker symbol
$O(\cdot)$, $o(\cdot)$	Landau growth symbols
$\mathbf{A}$	matrix
$\vert\mathbf{A}\vert=\det\mathbf{A}$	determinant of $\mathbf{A}$
$\mathbf{A}^{-1}$	inverse of $\mathbf{A}$
$\mathbf{x}$	vector
$\mathbf{x}^T, \mathbf{A}^T$	transposed vector/matrix
$f', f^{(k)}$	1st, k-th derivative of f
$'f, {}^{(k)}f$	1st, k-th integral of f
f^{-1}	inverse function
$\equiv$	defining equation
$\sim$	asymptotically equal
$\doteq$	equality occurs only if the error term is neglected
$\approx$	nearly equal
$n!$	$=1\cdot2\cdot3\cdot\ldots\cdot n$

$\binom{n}{k}$	$=\dfrac{n!}{k!\,(n-k)!}$
$\{a_n\}, \{a_n\}_1^\infty$	sequences
■	end of an example, proof, definition, corollary, lemma
2.34(−5)	$=2.34 \cdot 10^{-5}$
$f(\pm a, \pm b)$	$=f(a, b)+f(-a, b)+f(a, -b)+f(-a, -b)$
$f(\pm a, b)$	$=f(a, b)+f(-a, b)$
$\sum_{k=1}^{n} a_n$	$=a_1+a_2+\ldots+a_n$
$\sum_{k=1}^{n}{}' a_n$	$=\frac{1}{2}a_1+a_2+\ldots+a_n$
$\sum_{k=1}^{n}{}'' a_n$	$=\frac{1}{2}a_1+a_2+\ldots+a_{n-1}+\frac{1}{2}a_n$
$\prod_{k=1}^{n} a_n$	$=a_1 \cdot a_2 \cdot \ldots \cdot a_n$
$[x]$	integer part of x, e.g. $[2.5]=2$; $[3]=3$, $[-1.2]=-2$

Intervals and domains

$[a, b]$	closed non-empty finite interval
(a, b)	open non-empty finite or infinite interval
$[a, b)$, $(a, b]$	half open non-empty finite or infinite intervals
Ω	N-dimensional non-empty finite or infinite open set, simply connected; smooth or piecewise smooth boundary
$\lvert\Omega\rvert$	area of Ω
$\partial\Omega$	boundary of Ω
$\bar{\Omega}$	$=\Omega \cup \partial\Omega$
S_N	N-dimensional simplex ($N=2$: triangle; $N=3$: tetrahedron)

Operators, functionals, functions

$L_n(f, x)$	Lagrange interpolating operator with n nodes
$H_n(f, x)$	Hermite interpolating operator with n nodes
$LH_{mn}(f, x)$	mixed type interpolating operator, m nodes of Lagrange type and n nodes of Hermite type
$\mathscr{L}_{m,n}(f, x, y)$	tensor product of two Lagrange interpolating operators with m and n nodes respectively

$e_n(f, x)$	error of an interpolation or approximation
$S_n(f, x)$	interpolating spline with n nodes
$B_n(f, x)$	n-th degree Bernstein polynomial of f
$B_n(x)$	n-th degree Bernoulli polynomial
$B_n = B_n(0)$	Bernoulli numbers
$h_i(x)$	basis functions } of interpolation (usually Lagrange type interpolation)
$\lambda_i(x)$	cardinal functions } of interpolation (usually Lagrange type interpolation)
$\omega_n(x)$	nodal functions } of interpolation (usually Lagrange type interpolation)
$K(x)$, $K_n(x)$, $K_{n,m}(x)$	Peano kernel functions
$W(x)$, $W(\mathbf{x})$	weight functions
$f(x-t)_+$	$= \begin{cases} f(x-t) & \text{for} \quad x \geq t \\ 0 & \text{otherwise} \end{cases}$
sign (x)	sign function $\operatorname{sign}(x) = \begin{cases} 1 & \text{for} \quad x>0 \\ 0 & \text{for} \quad x=0 \\ -1 & \text{for} \quad x<0 \end{cases}$

$$\overset{*}{I}f = \int_0^1 W(x)f(x)\,\mathrm{d}x, \qquad If = \int_{-1}^1 W(x)f(x)\,\mathrm{d}x$$

$$\overset{*}{\underset{\infty}{I}}f = \int_0^\infty W(x)f(x)\,\mathrm{d}x, \qquad \underset{\infty}{I}f = \int_{-\infty}^\infty W(x)f(x)\,\mathrm{d}x$$

$$I_\Omega f = \int_\Omega W(\mathbf{x})f(\mathbf{x})\,\mathrm{d}\Omega$$

$\displaystyle ⨍_a^b W(x)f(x)\,\mathrm{d}x$	Cauchy principle value integral
I_k, $I_{k_1k_2\ldots k_n}$	momentum integrals
C_nf	cubature formula with n nodes
$E_nf(E_n^xf)$	quadrature or cubature error funtional (with respect to the variable x only)
Q_nf	quadrature formula with n nodes
Q_hf	quadrature formula with infinitely many nodes, every two subsequent nodes have the distance h
R_nf	special quadrature formulae with n nodes: rectangle rule
T_nf	special quadrature formulae with n nodes: trapezoidal rule
M_nf	special quadrature formulae with n nodes: midpoint rule
B_nf	special quadrature formulae with n nodes: Bernstein formula
S_nf	special quadrature formulae with n nodes: Simpson rule
$\dfrac{\mathrm{d}}{\mathrm{d}x}, \dfrac{\mathrm{d}^k}{\mathrm{d}x^k}$	1st, k-th derivation operator

$\dfrac{\partial^k}{\partial x_{k_1} \ldots \partial x_{k_n}}$, $|k| = k_1 + k_2 + \ldots + k_n$, k multi-index,
operator of the k-th derivation with respect to the variables $x_1, \ldots, x_n$

$\Delta f = f_{x_1x_1} + f_{x_2x_2} + \ldots + f_{x_nx_n}$

$\omega(f, h)$ modulus of continuity of f with increment h

Spaces

$\mathbb{N}$ natural numbers

$\mathbb{N}_0 = \mathbb{N} \cup \{0\}$

$\mathbb{R}$ real numbers

$\mathbb{R}_+(\mathbb{R}_-)$ positive (negative) real numbers

$\mathbb{R}^N$ N-dimensional Euclidian space

$\mathbb{H}$ Hilbert space

$\mathbb{H}_2$ Hardy space (for $\varepsilon = 1$)

$\tilde{\mathbb{H}}_2$ Hardy space ($\varepsilon \leqslant 1$)

$\mathscr{A}_s$ functions which are analytic in a strip

$\mathbb{E}_\rho$ functions which are analytic in the ellipse with parameter ρ

$\mathbb{P}_n$ polynomials of degree at most n

$\dot{\mathbb{P}}_n$ polynomials of *exact* degree n

$C^n[a, b]$, $C^n(\Omega)$ n times continuously differentiable functions in $[a, b]$ or Ω

$L^p[a, b]$, $L^p(\Omega)$ p-th power Lipschitz continuous functions with

$$\int_a^b W(x)\, |f(x)|^p \, dx < \infty, \qquad \int_\Omega W(\mathbf{x})\, |f(\mathbf{x})|^p \, d\Omega < \infty$$

l_p p-th power summable infinite sequence

dim ($\mathbb{D}$) dimension of space $\mathbb{D}$

1

Introduction

1.1 The necessity for numerical quadrature and cubature

By way of motivation for the whole subject we start by giving an example of an integral which cannot be evaluated exactly.

1.1.1 Motivation

Suppose that we are given the problem of determining the arc length

$$A=\int_{-a}^{a}(\mathrm{d}x^2+\mathrm{d}y^2)^{\frac{1}{2}}=\int_{-a}^{a}(1+(y')^2)^{\frac{1}{2}}\,\mathrm{d}x$$

of the half ellipse $(x/a)^2+(y/b)^2=1$, $y\geqslant 0$, $b>a$. When spherical coordinates $x=a\cos\varphi$, and $y=b\sin\varphi$, $\varphi\in[0,\pi]$, are introduced, we find that

$$A=\int_{-\pi}^{0}(a^2\sin^2\varphi+b^2\cos^2\varphi)^{\frac{1}{2}}\,\mathrm{d}\varphi,$$

from which the substitutions $k^2=1-a^2/b^2$ and $\cos^2\varphi=1-\sin^2\varphi$ lead to $A=b\int_{-\pi}^{0}(1-k^2\sin^2\varphi)^{\frac{1}{2}}\,\mathrm{d}\varphi$. This expression is called an *elliptic integral of the first kind* and cannot be calculated analytically. Thus, we are forced to develop methods which allow approximations to A to be calculated. In doing this our attention will be directed to those methods which use only a set of discrete values of the integrand and we therefore define quadrature as follows.

Definition 1.1.1 A *quadrature formula* is a numerical rule whereby the value of a definite integral is approximated by the use of information about the integrand only at a set of discrete points (where the integrand is defined). Any formula doing this is a *quadrature formula*. ■

Often the term *mechanical quadrature* is used to characterise this approach (see Klügel [3], and Vahlen [12]), which means that the required approximation is not performed analytically; for example, the integration might be done by integrating a convergent series expansion of the integrand. For multidimensional integrals the analogous problem is usually called a *cubature problem.*

In practice, because very few integrals can be evaluated exactly by analytical methods, it is necessary to have recourse to quadrature formulae and quadrature procedures. The latter are sets of quadrature formulae with certain convergence properties.

1.1.2 Integrals which cannot be evaluated analytically

There exist many famous and very simple integrals which cannot be evaluated analytically (except perhaps for special limits). Examples are:

$$\int_1^2 \frac{dx}{\ln x}, \text{ a special case of the logarithm integral } LI(z)=\int_1^z \frac{dx}{\ln x},$$

$$\int_0^\pi \frac{\sin x}{x}\,dx, \text{ a special case of the sin integral } SI(z)=\int_0^z \frac{\sin x}{x}\,dx,$$

$$\int_0^\infty \exp(-x^2)\,dx, \text{ a special case of the error function } ERF(z)=\int_0^z \exp(-x^2)\,dx,$$

$$\int_0^\pi \frac{dx}{(1-k^2\sin^2 x)^{\frac{1}{2}}}, \text{ a special case of the elliptic integral } F(k,z)=\int_0^z \frac{dx}{(1-k^2\sin^2 x)^{\frac{1}{2}}},$$

$$\int_0^\pi (1-k^2\sin^2 x)^{\frac{1}{2}}\,dx, \text{ a special case of the elliptic integral } E(k,z)=\int_0^z (1-k^2\sin^2 x)^{\frac{1}{2}}\,dx.$$

Further examples can be found in some tables mentioned in Chapter 9.

Many such functions occur frequently in mathematical physics and techniques are therefore needed to evaluate them. Many tables of numerical values are available (see Chapter 9), and these contain additional information on expansions, approximations, recurrence relations and other important background material to facilitate evaluation. The use of such tables is highly recommended.

1.1.3 Numerically inexact or inconvenient problems

Although the use of tables may facilitate the computation of integrals, we often need to decide whether to use them or whether to use a subroutine. As an example we calculate the integral

$$I=\int_{-1}^{1}\frac{2x^2+2x+13}{(x-2)(x^2+1)^2}\,dx.$$

In order to do this we split the integrand such that

$$I=\int_{-1}^{1}\left(\frac{-1}{2-x}-\frac{x+2}{1+x^2}-\frac{3x+4}{(1+x^2)^2}\right)dx$$

$$=\left[\ln(2-x)-\ln(1+x^2)^{\frac{1}{2}}-2\operatorname{arc\,tg}x+\frac{3}{2(1+x^2)}\right]_{-1}^{1}-\int_{-1}^{1}\frac{4dx}{(1+x^2)^2}.$$

We now calculate

$$\int_{-1}^{1}\frac{4\,dx}{(1+x^2)^2}-\left[\frac{2x}{1+x^2}+2\operatorname{arc\,tg}x\right]_{-1}^{1},$$

and, collecting terms, we finally obtain

$$I=\left[\tfrac{1}{2}\ln\frac{(2-x)^2}{1+x^2}-4\operatorname{arc\,tg}x+\frac{3-4x}{2(1+x^2)}\right]_{-1}^{1}=\ln\tfrac{1}{3}-8\operatorname{arc\,tg}1-2.$$

This is the exact solution but, even so, the *exact* numerical value cannot be presented as a number. Taking into account the tediousness of the above calculation it would be simpler to use a computer subroutine. Using, for example, a Gauss formula (cf. Chapter 7) which evaluates the integrand at 15 distinct points only, we find that $I=-9.381\,797\,598$, which was computed by an electronic pocket calculator in a very short time.

1.1.4 Integration of measured or tabulated functions

It is often necessary to compute the approximate value of an integral whose integrand is not known analytically but only by measured or tabled values. These could, for example, result from a computation within some program. Then there is no other alternative but to take a suitable quadrature formula which uses these discrete known values of the integrand. On the other hand, experiments should be planned such that the points of measurement can be used immediately within a good quadrature or cubature formula, but this problem of experimental design is not central to the theme of this book.

1.2 Applications of quadratures and cubatures

Although it is often necessary for a single integral to be computed numerically, the most important field of applications of numerical quadrature arises within the numerical treatment of many mathematical problems, where analytical methods are completely out of the question. We mention some of these fields here.

1.2.1 Integral transforms

In many physical and technical problems the use of integral transforms is both efficient and meaningful. Examples of such transforms are:

$$f(x)=\frac{1}{(2\pi)^{\frac{1}{2}}}\int_{-\infty}^{\infty} g(t)\exp(-itx)\,\mathrm{d}t \qquad \text{general Fourier transform,}$$

especially

$$f(x)=\left(\frac{2}{\pi}\right)^{\frac{1}{2}}\int_{0}^{\infty} g(t)\begin{pmatrix}\sin xt\\ \cos xt\end{pmatrix}\mathrm{d}t \qquad \begin{pmatrix}\sin\\ \cos\end{pmatrix}\text{-transform,}$$

$$f(x)=\int_{0}^{\infty}\exp(-xt)\,\mathrm{d}t \qquad \text{Laplace transform.}$$

Again, it is usually impossible to perform such transforms analytically. The approach is that of using suitable quadrature rules. Because of the importance of such transforms, many quadrature procedures have been developed which take into account special properties of the integrand.

1.2.2 Initial value problems in ordinary differential equations

Let us consider the initial value problem $y'+p(x)y=q(x)$, $y(x_0)=y_0$. The general solution of this is

$$y(x)=\left[\exp\left(-\int_{x_0}^{x} p(x)\,\mathrm{d}x\right)\right]\left(\int_{x_0}^{x} q(x)\left(\exp\int_{x_0}^{x} p(t)\,\mathrm{d}t\right)\mathrm{d}x+y_0\right).$$

Another initial value problem $h(x)y'+f(y)g(x)=0$, $y(x_0)=y_0$, may be handled in the form

$$\int_{y_0}^{y}\frac{\mathrm{d}y}{f(y)}+\int_{x_0}^{x}\frac{g(x)}{h(x)}\,\mathrm{d}x=0,$$

which leads to an implicit solution $F(y)+G(x)=0$.

Both the above and similar problems need quadrature formulae in order to be solved completely. Quadrature formulae are also very suitable for constructing Runge Kutta procedures (see, for example, Engels [2]).

1.2.3 Techniques used for the solution of linear boundary value problems

As another example we consider problems involving a linear operator L such as

$$Lu\equiv u''(x)+f(x)u(x), \qquad x\in[a,b],$$

or

$$Lu \equiv u(x) + \int_0^1 k(x, t)u(t)\,dt, \qquad x \in [0, 1],$$

or

$$Lu \equiv u_{xx} + u_{yy} + f(x, y)u, \qquad (x, y) \in \partial\Omega$$

where Ω is some given region with boundary $\partial\Omega$. Additional suitable boundary conditions are $y(a) = y_a$, $y(b) = y_b$ in one dimension, $u(x, y) = h(x, y)$ for $(x, y) \in \partial\Omega$ in two dimensions.

Let g be a suitable function such that the problem $Lu = g$ in which u satisfies the boundary conditions has a unique solution. Most methods for the approximate solution of this type of problem, which are not pure discretisation methods, involve the use of quadratures or cubatures. Let $v_0, v_1, \ldots, v_n$ be linearly independent, sufficiently smooth functions such that v_0 satisfies the inhomogeneous, and $v_1, \ldots, v_n$ satisfy the homogeneous boundary conditions. Then the linear combination $w \equiv v_0 + \sum_{i=1}^{n} \alpha_i v_i \equiv \sum_{i=0}^{n} \alpha_i v_i$ with $\alpha_0 = 1$ satisfies the boundary conditions. Substituting w into the operator equation $Lu = g$, we obtain an error ε unless w happens to be the exact solution: $Lw - g \equiv \varepsilon(\alpha_1, \ldots, \alpha_n)$. Because the coefficients $\alpha_1, \ldots, \alpha_n$ are still to be determined, we can require that ε approximates zero in some sense, for example in the least square norm, so that

$$Q \equiv \int \varepsilon^2 = \min. \tag{1.2.1}$$

The integrals are to be extended over the whole domain in which the solution is desired. A necessary condition for (1.2.1) to be true is that $(\partial Q/\partial \alpha_k) = 0$ for $k = 1(1)n$. This leads to $2\int \varepsilon v_k = 0$, $k = 1(1)n$, which means that $\sum_{j=1}^{n} \alpha_j (Lv_j, v_k) = (g, v_k)$, $k = 1(1)n$ with the inner product $(\cdot, \cdot)$ defined as $(Lv, w) \equiv \int w \cdot Lv$. Thus we have a system of n linear equations for the unknowns $\alpha_1, \ldots, \alpha_n$ whose coefficients (Lv_j, v_k), (g, v_k) are integrals. Again, these integrals cannot usually be evaluated in closed form and can only be approximately evaluated by quadratures or cubatures.

Another alternative, instead of minimising (1.2.1), is to require that

$$(Lw - g)v = 0, \qquad v \in \mathbb{H} \tag{1.2.2}$$

for all functions v belonging to some Hilbert space $\mathbb{H}$ with inner product

$(u, v) \equiv \int u \cdot v$. This being true, w is the solution because of the completeness of $\mathbb{H}$. In practice, we are forced to restrict the solution w of (1.2.2) to a finite-dimensional subspace $\mathbb{H}_n$ of $\mathbb{H}$. Now, let v_i, $i = 1(1)n$ be a basis of $\mathbb{H}_n$. We can then take $v = v_i$, $i = 1(1)n$, because of the linearity of $\mathbb{H}_n$ and we obtain the *Galerkin* method

$$\int (Lw - g)v_k = 0 \quad \text{for} \quad k = 1(1)n$$

or

$$\sum_{i=1}^{n} \alpha_i (Lv_i, v_k) = (g, v_k) \quad \text{for} \quad k = 1(1)n. \tag{1.2.3}$$

Another method of obtaining an approximate solution is that of *Ritz*. Here, a functional Φ is defined as

$$\Phi(v) \equiv (Lv, v) - 2(g, v) = \min, \qquad v \in \mathbb{H}. \tag{1.2.4}$$

With the Ritz technique the minimisation is with respect to all functions v belonging to some Hilbert space $\mathbb{H}$. If, in addition, L is symmetric, we can show that a solution of (1.2.4) solves the boundary value problem.

Again, we have to restrict ourselves to a finite-dimensional $\mathbb{H}_n \subset \mathbb{H}$ for practical purposes. Thus, let v_i, $i = 1(1)n$ be a basis of $\mathbb{H}_n$. Then we require that

$$\Phi(w) = \Phi(\alpha_1, \ldots, \alpha_n) = \sum_{i=1}^{n} \sum_{k=1}^{n} \alpha_i \alpha_k (Lv_i, v_k) - 2 \sum_{i=1}^{n} \alpha_i (g, v_i) = \min.$$

A necessary condition for this to be true is that $(\partial\Phi/\partial\alpha_j) = 0$, $j = 1(1)n$. We again obtain the system (1.2.3) for $\alpha_1, \ldots, \alpha_n$.

A natural requirement in order to have a linear system for $\alpha_1, \ldots, \alpha_n$ with good numerical properties is that v_i should be "nearly" orthogonal with respect to the inner product $(\cdot, \cdot)$. This is partially achieved with Finite Element Techniques where the trial functions v_i have local support so that many of the terms (Lv_j, v_k) vanish. For a detailed discussion of such problems and methods see, for example, Whiteman [13] and Mitchell and Wait [6]. In both the Ritz and the Galerkin methods, we have to approximate the integrals occurring in the system for $\alpha_1, \ldots, \alpha_n$ by quadrature or cubature methods.

Finally, we note that the numerous physical applications have led to the construction of classes of quadrature or cubature formulae which have those special features needed to achieve high accuracy in the desired approximation.

1.3 Historical remarks

A very short historical survey is now given. The evaluation of plane areas, in particular the determination of the area of the circle, is one of the oldest problems in science and seems to have been studied in every culture where men have dealt with mathematical problems.

The earliest attempts at measuring the area of a circle are thought to have been undertaken by the Babylonians and the Egyptians. The Babylonians discovered the fact that the radius of a circle is the side-length of an inscribed regular hexagon. This fact presented a possibility of characterising the arc length of a circle by the six edges, thus circumventing the problem of knowing the arc length. Four thousand years ago a remarkably good approximation of the area of a circle was discovered in Egypt. This is equivalent to the approximation $\pi \doteq (16/9)^2 = 3.1605$.

The next real progress came during the Greek period, where the question arose as to whether there exist curvilinearly-bounded regions whose area is a rational number. The ingenious Archimedes was the first person to prove that the area between a second-order parabola and one of its tangents is a rational number. However, he failed when applying his ideas to a circle. Nevertheless, he calculated the first strict bounds for the number π. For this, he computed the areas of inscribed and circumscribed regular hexagons of a circle and then successively doubled the number of edges until he had 96 (cf. Rudio [9] and Toeplitz [11]). He chose the doubling because the side s_{2n} of a regular inscribed $2n$-polygon in terms of the side of an n-polygon (radius $= 1$) is given by $s_{2n} = (2-(4-s_n^2)^{\frac{1}{2}})^{\frac{1}{2}}$ while $s_6 = 1$. In this way, Archimedes obtained the inclusion $22/7 < \pi < 223/71$.

This treatment of isolated and completely determined curves, such as parabolas or circles, is typical of the ancient way of thinking and it persisted in the following centuries. A large amount of work was done in order to find more and more complicated regions whose areas are rational numbers.

The ancient period was finally overcome by the invention of functions in the modern sense of mappings. From then on it made sense to study general properties and to treat problems such as interpolation or approximation as general problems rather than as those connected to one special curve. The idea of integration was born in order that the areas of curvilinearly-bounded regions might be measured. It was noticed that not every function could be integrable analytically, and hence the necessity for developing approximate methods arose. Newton interpolated functions at equidistant arguments and integrated the interpolants, thus becoming the father of the famous Newton–Cotes quadrature. In principle he started, like Archimedes, on approximating the arcs of the functions piecewise by secants in the most simple case. He then refined the method by replacing the secants by parabolas of increasing degree.

In the years that followed, a large number of quadratures of this type was developed. These formulae were improved by various corrections and by judicious combinations of existing formulae.

While for a long time it was regarded that the evaluation points should be preassigned equidistantly or with some other fixed distribution, Gauss was the first to notice that a suitable variation of these points led to better accuracy in general (cf. Kowalewski [4]). This fundamental result of Gauss stimulated many variations and generalisations. One of the most important of these is due to Markoff, who noticed that the Gaussian formulae can be constructed by integration of a *Hermitian interpolating polynomial.* Numerous new quadratures were subsequently found which slowly tended towards optimality properties for certain classes of functions or problems. Again and again connections between new and known methods appeared. Methods which seemed to be very far from interpolation were reconstructed by integration of suitable interpolating operators, and many more or less complete theories were developed.

We should mention that another idea used by Archimedes became important in quadrature. In calculating the area between a quadratic parabola and one of its tangents, Archimedes had used an inscribed polygon as the approximation. Subsequently he doubled the number of edges and thus was able to eliminate the error. This idea was applied by Archimedes to the circle, and it was later extended and refined in some directions by others. Snellius, Huygens and Kommerell were among the researchers who refined the idea of doubling the edges in order to compute better approximations for π. The technique was that of extrapolation to the limit zero for the length of the polygonal side. The idea was also applied more generally to raise the order of accuracy of methods. This seems first to have been noticed for quadratures (but not performed) by Klügel [3]; the actual application is due to Richardson and Gaunt [7]. In 1955 this principle was rediscovered for numerical quadrature by Romberg [8].

Today, the field of numerical quadrature is still far from being completely explored. However, in contrast to the cubature problem there exist many well-developed theories. While the quadrature problem has always profited greatly from approximation theory which has been highly developed, and from the fact that the structure of the underlying problem is simple, the cubature problem is far from having any closed theory which is comparable to that of quadrature. The difficulties are numerous; the results and formulae not only depend on the shape of the domain of integration but also on the dimension. Unlike the one-dimensional quadrature problem, many cubature problems do not have a unique solution or a real solution at all. Generalisations of one-dimensional results in general are trivial or very complicated. Nevertheless, the interest in cubature problems continues to grow rapidly and the efforts and successes are remarkable.

1.4 Concept and aims of this book

This book is intended not only to be a textbook for students at the graduate and undergraduate levels, but also to indicate some of the numerous practical applications of quadrature and cubature procedures. With the latter aim in mind, the person who is researching in this field should find much of the material appropriate.

In the writing of this book two main points have been considered to be essential. On the one hand the author has been most anxious to keep the book readable while at the same time he has sought to include as many applications as possible. Thus, many of the topics have been described initially in an elementary way and just enough detail to achieve this has been included. However, in order to keep the book to a reasonable length, it has been necessary to omit many sophisticated results. In view of the applicability of many of the procedures, stress has been put upon constructive aspects of the techniques rather than on the corresponding theoretical background. With the view to the applicability of the techniques, various programs, tables and advice concerning further material have been included.

Since much modern numerical analysis is being accomplished within the setting of functional analysis, this approach has been used here. Thus, it is assumed that the reader has some elementary knowledge of function spaces and their properties. Nevertheless, the functional analysis terms are used as far as possible in a natural way; the reader who is not familiar with these terms may consider them initially as useful abbreviations and—bearing in mind this drastic simplification—may handle them as such. It has been decided not to collect the terms of functional analysis into a separate section, but rather to define and explain them as they occur.

Each chapter of the book contains remarks, references and problems, the solution of which is left to the reader. These last are intended to facilitate learning and understanding of the material. The remarks serve to indicate certain generalisations, alternatives and related subjects which cannot be treated here. As only a very small number of books on numerical quadrature and cubature has been published to date, these are cited very frequently. As a consequence, they are listed separately at the end of this chapter and referred to simply by the author's name accompanied by the work "book".

1.5 General conventions

In this section we collect together some conventions which are commonly used and some background material which is necessary to both the theory and the applications.

1.5.1 General assumptions

We shall assume that the problems treated always make sense: that is, every integral can be shown to exist at least in the Riemann sense. Usually the integrand is assumed to be at least continuous in the domain of integration.

Results of analysis and functional analysis are, in general, used without proof; for these the reader is referred to Yosida [14].

We shall always use standard intervals or regions such as $[-1, 1]$, $[0, 1]$, $[0, \infty)$, $(-\infty, \infty)$, N-square, regular polygon, N-sphere, $\mathbb{R}^N$, spherical shell. These are listed in the list of symbols and abbreviations. In the following section transformations are given which map more general regions into standard regions.

1.5.2 Transformations

The finite interval $[a, b]$ is mapped into $[-1, 1]$ or $[0, 1]$ by

$$t \in [a, b] \rightarrow \begin{cases} x = \dfrac{2t-a-b}{b-a}, & x \in [-1, 1], \\[2ex] x = \dfrac{t-a}{b-a}, & x \in [0, 1], \end{cases}$$

$$t \in [a, \infty) \rightarrow x = t - a, \qquad x \in [0, \infty).$$

A rectangle can be mapped into a square of side-length 2 or 1 by application of linear transformations of these types to each coordinate. In the same way, an N-dimensional ellipsoid is mapped into an N-sphere. Such transformations do not affect the quality of quadrature or cubature formulae.

More generally, we consider regular mappings of the form $x=\varphi(t)$, $x\in[a, b]$, so that

$$\int_a^b f(x)\,dx=\int_{\varphi^{-1}(a)}^{\varphi^{-1}(b)} f(\varphi(t))\varphi'(t)\,dt.$$

Such a transformation often facilitates practical computation.

In N dimensions the corresponding formulae are for a transformation T

$$\mathbf{x}=T\mathbf{t}\rightarrow\int_{\bar{\Omega}} f(\mathbf{x})\,d\Omega=\int_{\tilde{\Omega}} f(T\mathbf{t})\,|\mathbf{J}|\,d\tilde{\Omega}$$

where $T^{-1}\bar{\Omega}=\tilde{\Omega}$, $\mathbf{J}$ is the Jacobian matrix of T, and $|\mathbf{J}|=\det\mathbf{J}$.

Example 1.5.1 Let T be the mapping

$$\begin{aligned} x&=r\sin\varphi\sin\vartheta, && 0\leqslant r\leqslant 1,\\ y&=r\sin\varphi\cos\vartheta, && 0\leqslant\vartheta\leqslant 2\pi,\\ z&=r\cos\varphi, && 0\leqslant\varphi\leqslant\pi, \end{aligned}$$

of a 3-sphere into a cube. The Jacobian matrix satisfies

$$|\mathbf{J}|=\begin{bmatrix} x_\varphi & y_\varphi & z_\varphi\\ x_\vartheta & y_\vartheta & z_\vartheta\\ x_r & y_r & z_r \end{bmatrix}=-r^2\sin\varphi. \quad\blacksquare$$

In N dimensions we have for $0\leqslant r\leqslant 1$, $0\leqslant\varphi_{N-k}\leqslant\pi$ with $1<k<N$, and $0\leqslant\varphi_{N-1}\leqslant 2\pi$, the mapping

$$\mathbf{x}=\begin{bmatrix} r\sin\varphi_1\sin\varphi_2\ldots\sin\varphi_{N-2}\sin\varphi_{N-1}\\ r\sin\varphi_1\sin\varphi_2\ldots\sin\varphi_{N-2}\cos\varphi_{N-1}\\ r\sin\varphi_1\sin\varphi_2\ldots\cos\varphi_{N-2}\\ \ldots\\ r\cos\varphi_1 \end{bmatrix}\equiv T\boldsymbol{\varphi},$$

with $|\mathbf{J}|=r^{N-1}\sin^{N-1}\varphi_1\sin^{N-3}\varphi_2\ldots\sin\varphi_{N-2}(-1)^N$.

The corresponding $(N+1)$-dimensional cylindrical transform is given by

$$\begin{pmatrix}\mathbf{x}\\ x_{N-1}\end{pmatrix}=\begin{pmatrix}T\boldsymbol{\varphi}\\ x_{N+1}\end{pmatrix}\quad\text{with the same } |\mathbf{J}|.$$

The most convenient transforms are, of course, the linear transforms because then $|\mathbf{J}|$ is a constant. However, there also exist simple nonlinear transformations with constant $|\mathbf{J}|$. An example is now given:

Example 1.5.2 The annulus A with inner radius ρ and outer radius 1 is mapped into the square $|r|\leq 1$, $-1<\varphi\leq 1$ by

$$x=\left(\rho^2+\frac{r+1}{2}(1-\rho^2)\right)^{\frac{1}{2}}\sin\pi\varphi, \qquad y=\left(\rho^2+\frac{r+1}{2}(1-\rho^2)\right)^{\frac{1}{2}}\cos\pi\varphi$$

with $|\mathbf{J}|=\pi(1-\rho^2)/4$. So we have that

$$\iint_A f(x,y)\,\mathrm{d}x\,\mathrm{d}y=\frac{\pi}{4}(1-\rho^2)\int_{-1}^{1}\int_{-1}^{1} f(r(x,y),\varphi(x,y))\,\mathrm{d}r\,\mathrm{d}\varphi. \quad \blacksquare$$

1.5.3 Weight functions

We often encounter weight functions $W(x)$ or $W(\mathbf{x})$ within an integral

$$If=\int_{-1}^{1} W(x)f(x)\,\mathrm{d}x \quad \text{or} \quad I_\Omega f=\int_\Omega W(\mathbf{x})f(\mathbf{x})\,\mathrm{d}\Omega.$$

For example

$$If=\int_{-1}^{1}\frac{1}{(1-x^2)^{\frac{1}{2}}}f(x)\,\mathrm{d}x \quad \text{or} \quad I_\Omega f=\int_\Omega \mathrm{e}^{-(\mathbf{x},\mathbf{x})}f(\mathbf{x})\,\mathrm{d}\Omega, \qquad \Omega=\mathbb{R}^N.$$

with $(\mathbf{x},\mathbf{x})\equiv x_1^2+\ldots+x_N^2$. Generally, we define as follows:

Definition 1.5.1 A function $W(x)$ (or $W(\mathbf{x})$) is called a *weight function* if it is a factor of $f(x)$ (or $f(\mathbf{x})$) and if the quadrature rule for the approximate calculation of $If(x)$ (or $I_\Omega f(\mathbf{x})$) uses only values of $f(x)$ (or $f(\mathbf{x})$) and none of $W(x)$ (or $(W(\mathbf{x}))$. ■

We always assume that for any weight function $W(x)$ or $W(\mathbf{x})$ the integrals exist for all admitted functions f. The use of weight functions facilitates the computation of integrals, especially in those cases where the same weight function appears repeatedly while f changes, as in the computation of integral transforms. Sometimes the Jacobian of a coordinate transform gives a weight function which facilitates the computation of the transformed integrand. The theory of orthogonal polynomials, which is fundamental to essential parts of quadrature and cubature, requires that the sign of the weight function be constant throughout the whole domain of integration.

In general, results for $W(x)\equiv 1$ or $W(\mathbf{x})\equiv 1$ remain valid for non-constant weight functions with constant sign. The changes induced by the presence of such weight functions are usually simple and straightforward. We therefore often omit the weight functions for the sake of simplicity, and mention only those cases where the results are changed significantly by the inclusion of weight functions.

1.5.4 Notation

The approximate calculation of $If \equiv \int_{-1}^{1} W(x)f(x)\,dx$ is done using information of f. Whenever f is known only in discrete form, such as from measurements or from a table, we are restricted to the use of the given values of f. If we want the formulae to remain applicable in such cases we are restricted to the evaluation of f at certain points inside or outside (or both) the interval $[-1, 1]$ (Ω resp.).

Definition 1.5.2 The discrete points of the domain of f, where the information about f is used in approximating If (or $I_\Omega f$), are called *nodes* or *points* of the quadrature formula (or cubature formula). ■

Because of the linearity of the functional If we prefer rules, which are also linear with respect to f. Thus, we immediately come to quadrature formulae of the form

$$If \doteq \sum_{i=1}^{n} A_i f(x_i) \equiv Q_n f \quad \text{or} \quad I_\Omega f \doteq \sum_{i=1}^{n} A_i f(\mathbf{x}_i) \equiv C_n f.$$

Definition 1.5.3 The numbers A_i in a quadrature formula $Q_n f$ or a cubature formula $C_n f$ are called *weights* or *coefficients*. (Not to be confused with weight functions!)

Quadrature and cubature formulae of type $Q_n f$ and $C_n f$ are sometimes called *average-formulae* because If, $I_\Omega f$ are approximated by weighted averages of some values of f. ■

The same notation and definitions are valid for

$$\overset{*}{I}f \equiv \int_0^1 W(x)f(x)\,dx, \qquad \underset{\infty}{I}f \equiv \int_{-\infty}^{\infty} W(x)f(x)\,dx, \qquad \overset{*}{\underset{\infty}{I}}f \equiv \int_0^{\infty} W(x)f(x)\,dx.$$

Clearly, we are free to choose the nodes x_i in very different ways when f is known analytically. It is also clear that the weights A_i are far from being arbitrary.

If we alter the number of nodes, we cannot expect the earlier nodes and weights still to be valid. Thus, A_i and x_i in general depend on n:

$$Q_n f \equiv \sum_{i=1}^{n} A_i^{(n)} f(x_i^{(n)}), \qquad C_n f \equiv \sum_{i=1}^{n} A_i^{(n)} f(\mathbf{x}_i^{(n)}).$$

We usually omit the upper index (n) when this does not lead to confusion.

It is often the case that $If \neq Q_n f$, and so we make the following definition:

Definition 1.5.4 The linear functional

$$E_n f \equiv If - Q_n f, \qquad E_n f \equiv I_\Omega f - C_n f$$

is called a *quadrature* or *cubature* error. ■

From this many questions immediately arise which will need to be answered in the following chapters; they are:

(i) How do we represent $E_n f$?
(ii) Does $\lim_{n\to\infty} E_n f = If$? (Convergence problem)
(iii) How do we estimate $|E_n f|$?
(iv) Do estimates of the form $|E_n f| \leq |\sigma_n| \, \|f\|$ exist for a suitable norm or pseudonorm of f?
(v) How do we minimise $|E_n f|$ for certain classes of functions?
(vi) How do we minimise $|E_n f|$ with respect to certain nodes or weights?

First, we shall answer some more elementary questions:

How can $Q_n f$ and $C_n f$ be constructed, and how can $A_i, x_i, E_n f$ be determined?

The remaining questions are treated later, particularly in Chapters 3 and 4.

1.6 Remarks, problems, references

1.6.1 Remarks

R.1.1: The development of numerical quadrature is a fascinating chapter in the history of mathematics. Very readable accounts of some aspects have been written by Cantor [1], Klügel [3], Lambert [5], Runge and Willers [10], Toeplitz [11] and Vahlen [12] (cf. Section 1.3).

R.1.2: Some very important original papers have been translated and edited by Rudio [9] and Kowalewski [4]. Archimedes' methods of measuring the area between a parabola and one of its tangents is described in detail by Toeplitz [11] (cf. Section 1.3).

R.1.3: It is a very interesting but nearly forgotten fact that some older books are concerned almost entirely with numerical quadrature. A list of these and some newer books on this topic is therefore added in Section 1.7.

R.1.4: Most of the old original papers concerning numerical quadrature and cubature up to 1910 are carefully listed and reviewed by Runge and Willers [10].

1.6.2 Problems

P.1.1: Calculate $\int_{-1}^{1} dx/(1+4x^4)$ both analytically and using a quadrature-subroutine. Compare the time needed and the reliability of the results.

P.1.2: Let E be the unit circle, s_n be the side-length and f_n the area of an inscribed regular polygon with n edges (cf. Section 1.3).

(i) Prove that $s_{2n} = (2-(4-s_n^2)^{\frac{1}{2}})^{\frac{1}{2}}$, $\quad f_{2n} = \dfrac{n}{2} s_n$.

(ii) Calculate f_4, f_8, f_{16}, etc. and construct the representation of π due to Vieta as infinite product (cf. Rudio [9]).

$$\pi = f_\infty = f_4 \cdot \left(\frac{f_8}{f_4}\right) \cdot \left(\frac{f_{16}}{f_8}\right) \cdots .$$

P.1.3: Calculate the area between the parabola $y = x^2$, $y = 0$ and $x = 1$ following Archimedes (cf. Section 1.3): $[0, 1]$ is divided into n equidistant parts, and the lower and upper rectangles are added as in Riemann sums. Show that their difference tends to zero for $n \to \infty$.

P.1.4: Calculate the area (cf. Toeplitz [11]) between the parabola $y = 1 - x^2$ and $y = 0$ following another idea of Archimedes in a modified form: $[0, 1]$ is first divided into n equidistant parts, and secondly into $2n$ parts. Thus, two inscribed polygons with n and $2n$ edges occur. Calculate their areas f_n and f_{2n}, and eliminate the error by a suitable linear combination of f_n and f_{2n} (i.e. eliminate the influence of n) (cf. Section 1.3).

P.1.5: Construct a mapping with constant Jacobian from the 3-sphere to a cube. Hint: take r^α with suitable $\alpha > 0$ instead of r, and φ instead of $\sin \varphi$ within the spherical coordinates (cf. Section 1.4).

P.1.6: For what values α and β is the weight function (due to Jacobi)

$$W(x) = (1-x)^\alpha (1+x)^\beta \quad \text{in} \quad If = \int_{-1}^{1} W(x) f(x)\, dx$$

admitted, if f belongs to the space:

(i) $C^n[-1, 1]$, $n > 0$,
(ii) of all functions satisfying $f \in C^1(-1, 1)$ and $f \in L^p[-1, 1]$
(iii) $f \in C^0(-1, 1)$ having an n-fold zero at $x = -1$ and an m-fold zero at $x = 1$ (cf. Section 1.4).

1.6.3 References

1. Cantor, M., (1907). "Vorlesungen über Geschichte der Mathematik." Volume 1. [Reprinted (1965) by Johnson Reprint Corp., New York.]
2. Engels, H. (1974). "Runge-Kutta-Verfahren auf der Basis von Quadraturformeln." ISNM, Volume 19, 83–102. Birkhäuser-Verlag, Basel.

3. Klügel, G. S. (1823). "Mathematisches Wörterbuch etc." Schwickert, Leipzig. [Being reprinted.]
4. Kowalewski, A. (1917). Newton, Cotes, Gauss, Jacobi: Vier grundlegende Abhandlungen über Interpolation und genäherte Quadratur. V. Veit, Leipzig.
5. Lambert, J. H. (1770). "Beyträge zum Gebrauch der Mathematik und deren Anwendungen." Berlin, pp. 250–313.
6. Mitchell, A. R. and Wait, R. (1977). "The Finite Element Method in Partial Differential Equations." Wiley, London.
7. Richardson, L. F. and Gaunt, J. A. (1927). The deferred approach to the limit., *Phil. Trans. Royal Soc. London, Ser. A*, **226,** 299–349.
8. Romberg, W. (1955). Vereinfachte Numerische Integration. Trondheim, *Det Kong. Norske Vidensk. Selsk. Forhandl*, **28,** 7.
9. Rudio, F. (1892) Archimedes, Huygens, Lambert, Legendre, Vier Abhandlungen über die Kreismessung. Teubner, Leipzig. [Reprinted (1971) by Sändig.]
10. Runge, C. and Willers, F. A. (1909–21). *In* "Enzyklopädie der mathematischen Wissenschaften." Volume 2, Analysis. p. 50 ff.
11. Toeplitz, O. (1949) "Die Entwicklung der Infinitesimalrechnung." Springer-Verlag, Berlin.
12. Vahlen, T. (1911). "Konstruktionen und Approximationen." Teubner, Leipzig.
13. Whiteman, J. (Ed.) (1973). "The Mathematics of Finite Elements and Applications." Academic Press, London.
14. Yosida, K. (1974). "Functional Analysis." Springer-Verlag, Berlin.

1.7 List of books concerning numerical quadrature and cubature

1. Brass, H. (1977). "Quadraturverfahren." Vandenhoeck & Ruprecht, Göttingen.
2. Davis, P. J. and Rabinowitz, P. (1967). "Numerical Integration." Blaisdell, New York.
3. Davis, P. J. and Rabinowitz, P. (1975). "Methods of Numerical Integration." Academic Press, New York. [Extended version of 2.]
4. Ghizzetti, A. and Ossicini, A. (1970). "Quadrature Formulae." ISNM, Volume 13. Birkhäuser, Basel.
5. Ionescu, D. V. (1957). "Cuadraturi Numerice." (Rumanian) Editura Technica, Bucaresti.
6. Kowalewski, G. (1932). "Interpolation und genäherte Quadratur." Teubner, Leipzig.
7. Krylov, V. I. (1962). "Approximate Calculation of Integrals." (Translation from Russian by A. Stroud.) Macmillan, New York.
8. Lobatto, R. (1852). "Lessen over de integral-rekening." den Haag.
9. Moors, B. P. (1905). "Valeur approximative d'une intégrale définie." Gauthiers-Villars, Paris.
10. Nikol'skii, S. M. (1964). "Kvadraturn'e formul." (Translation from Russian.) Hindustan Publishing Corp., Delhi.

11. Schellbach, C. H. (1884). "Mechanische Quadratur." Mayer and Müller, Berlin.
12. Sobolev, S. L. (1964). "Lectures on the Theory of Cubature Formulae I, II." (Russian) Novosibirsk.
13. Stroud, A. (1971). "Approximate Calculation of Multiple Integrals." Prentice Hall, New Jersey.
14. Stroud, A. H. and Secrest, D. (1966). "Gaussian Quadrature Formulas." Prentice Hall, New Jersey.

2

Construction Principles for Quadrature and Cubature Formulae

2.1 Geometric constructions

This section, as its title implies, is concerned with the construction of quadrature and cubature formulae following from simple geometric considerations.

2.1.1 Quadrature formulae

We recall the idea of Archimedes: approximate the circle with a polygon! Doing the same for a given function f in order to approximate If, we choose a set of distinct points $x_1, x_2, \ldots, x_n$, in $[-1, 1]$, as in Fig. 2.1.1. If we then sum the trapezia, we obtain an approximate value $\hat{T}_n f$ for If

$$\hat{T}_n f \equiv \frac{x_2 - x_1}{2}(f(x_1)+f(x_2)) + \frac{x_3 - x_2}{2}(f(x_2)+f(x_3)) + \ldots$$

$$+ \frac{x_n - x_{n-1}}{2}(f(x_n)+f(x_{n-1})). \quad \text{(cf. Fig. 2.1.1)}$$

Usually the nodes are spaced equidistantly, and the distance between any two is denoted by h, where $h = 2/(n-1)$. The above expression is called the *trapezoidal rule*, and may be written as

$$T_n f \equiv \frac{1}{n-1}\left[f(-1) + 2\sum_{i=1}^{n-2} f(-1+ih) + f(1)\right] \quad \text{with} \quad h = \frac{2}{n-1}.$$

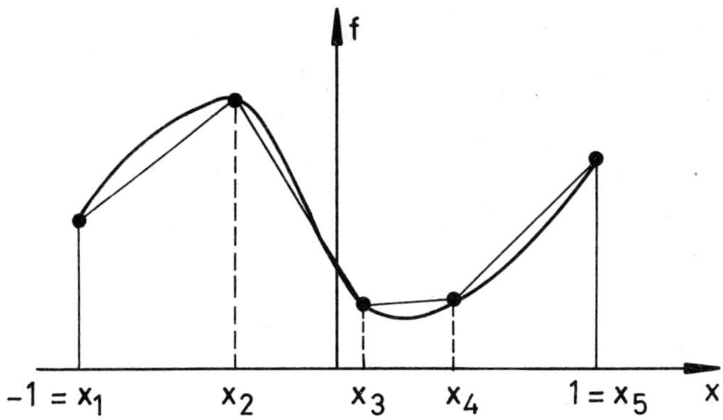

FIG. 2.1.1. Trapezoidal rule.

Of course, we can also approximate the desired area by step functions. This gives the *rectangle rule* (cf. Fig. 2.1.2)

$$\hat{R}_n f \equiv (x_2 - x_1) f(x_1) + (x_3 - x_2) f(x_2) + \ldots + (x_{n+1} - x_n) f(x_n),$$

or, for equally-spaced nodes,

$$R_n f \equiv h \sum_{i=0}^{n-1} f(-1 + ih), \quad \text{with} \quad h = 2/n.$$

Finally, we have the *midpoint rule*

$$M_n f \equiv h \sum_{i=0}^{n-1} f\left(-1 + ih + \frac{h}{2}\right), \quad \text{with} \quad h = 2/n.$$

Further, see Remark R.2.1.

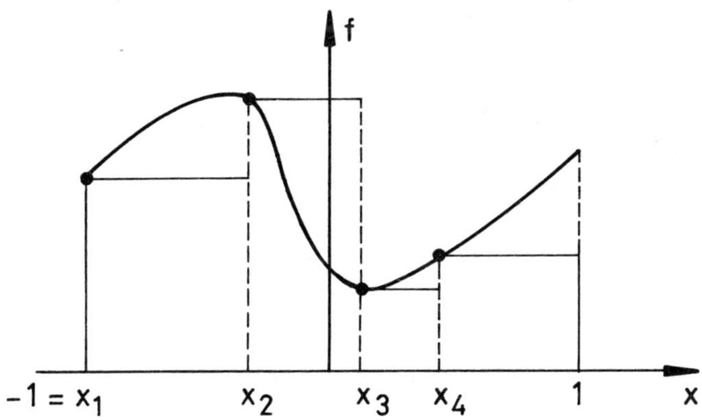

FIG. 2.1.2. Rectangle rule.

2.1.2 Cubature formulae

Simple cubature formulae can be derived in a similar manner. Thus, suppose we are given a two-dimensional domain Ω as in Fig. 2.1.3. We wish to approximate $I_\Omega f$, but there is now an additional problem: how to approximate Ω? The most natural way is to re-apply the Archimedes idea and to approximate $\partial\Omega$ piecewise by chords and $f(\mathbf{x}) = f(x, y)$ piecewise by triangles. Such a triangle within the three-dimensional space $\mathbb{R}^3$ is uniquely determined by its three corners. The orthogonal projection of such a triangle onto the x–y-plane is again a triangle. Thus, we have to triangulate Ω and calculate the volume of all the obliquely-cut prisms formed by the approximating triangles of f and their projection onto the x–y-plane. These projections or bases of the prisms have the area Δ_i respectively. We must now distinguish between different types of nodes: these are interior nodes (x_i, y_i), boundary nodes (x_b, y_b), and corner nodes (x_c, y_c). First, we sum (see again Fig. 2.1.3) the m single prisms where certain triangles meet:

$$I_\Omega f \approx \frac{1}{3} \sum_{j=1}^{m} \Delta_j [f(x_{j_1}, y_{j_1}) + f(x_{j_2}, y_{j_2}) + f(x_{j_3}, y_{j_3})]$$

where (x_{j_k}, y_{j_k}), $k = 1, 2, 3$ are the edges of the j-th triangle Δ_j. Within this formula each corner node appears at least once, each boundary node at least twice, and each interior node at least three times. In Fig. 2.1.3 the node i is five-fold, the nodes c are simple, and the nodes b are double, three-fold or five-fold, corresponding to the number of triangles meeting at the node under consideration.

For a more regularly-shaped domain Ω, such as a rectangle (Fig. 2.1.4), we may succeed in finding a triangulation consisting only of congruent triangles $\Delta_j = \Delta$, and with six triangles meeting at every interior node i,

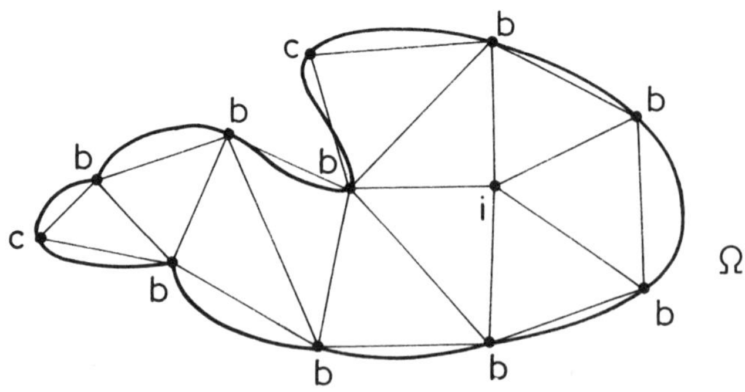

FIG. 2.1.3. Triangulated domain Ω.

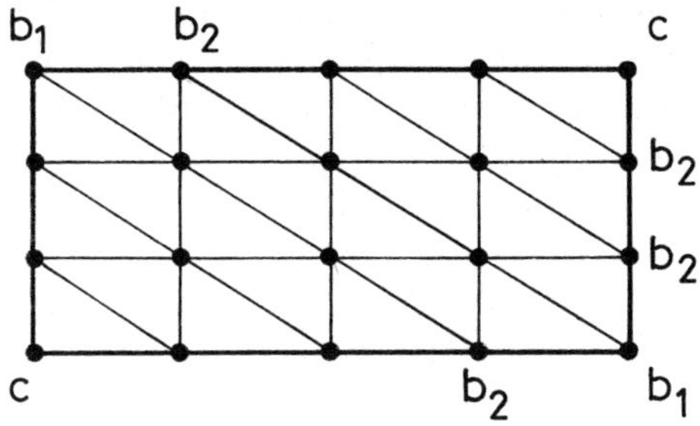

FIG. 2.1.4. Triangulated rectangle.

three at every boundary node b_2, and either one or two at a corner node c or b_1. This being true, we obtain the "trapezoidal rule":

$$T_nf \equiv \frac{\Delta}{3}\left\{\sum_c f(x_c, y_c) + 2\sum_{b_1} f(x_{b_1}, y_{b_1}) + 3\sum_{b_2} f(x_{b_2}, y_{b_2}) + 6\sum_i f(x_i, y_i)\right\}.$$

It is clear how this result can be generalised to N dimensions, but it is also clear that this method is not systematically organised, except for very well-shaped domains. However, the idea is very frequently used in Finite Element Techniques.

The simplest method for the geometric construction of a cubature formula is the "midpoint method" which uses the centre of gravity of each triangle Δ_i as node, as in Fig. 2.1.5. This leads to the midpoint rule for m triangles e_i, $i = 1(1)m$.

$$M_mf = \sum_{i=1}^{m} \Delta_i f(x_i, y_i)$$

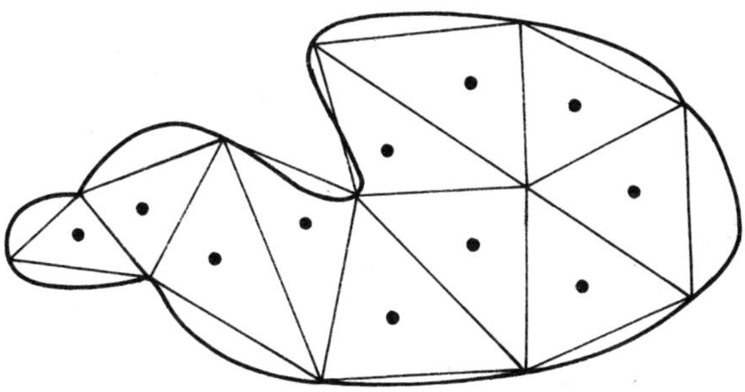

FIG. 2.1.5. Triangulated domain Ω-gravity nodes.

where now (x_i, y_i) is the centre of gravity of the triangle e_i. The generalisation of this formula to other partitions of Ω, such as quadrilaterals or hexagons, is straightforward. For dimensions higher than two we have to divide Ω not into triangles but into N-simplexes (for N dimensions) which are affine mappings of the standard simplex S_N. The N-dimensional simplex S_N is a polytope having the $N+1$ edges

$$\begin{array}{ll} E_0\colon (0,0,\ldots,0)^T & \text{origin} \\ \left.\begin{array}{l} E_1\colon (1,0,\ldots,0)^T \\ E_2\colon (0,1,\ldots,0)^T \\ \quad\cdots\cdot \\ E_N\colon (0,0,\ldots,1)^T \end{array}\right\} & \begin{array}{l}\text{unit points of each} \\ \text{coordinate axis}\end{array} \end{array}$$

The two- and three-dimensional simplexes are, respectively, the triangle and tetrahedron given in Fig. 2.1.6.

Clearly, this "method of gravity" is not restricted to a subdivision of Ω into simplexes. It also works for polytope elements such as the square and the hexagon in two dimensions (cube, octahedron, etc. in three and more dimensions) and also for curved elements as in Fig. 2.1.7. This method always works if Ω is centred.

Definition 2.1.1 The domain Ω is called *centred* if an interior point M of Ω exists such that every straight line from M has exactly one point of intersection with the boundary $\partial\Omega$ of Ω, which is not an osculating point. ■

In treating a centred region, we find first a point M of the above type and take it to be the origin of a spherical coordinate system. We then

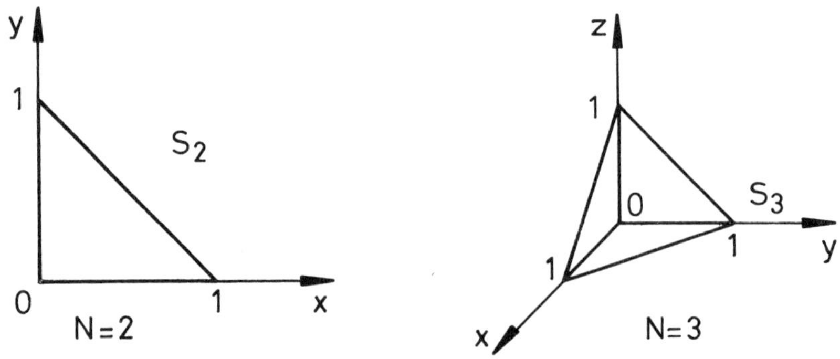

FIG. 2.1.6. N-Simplexes for $N=2$ and $N=3$.

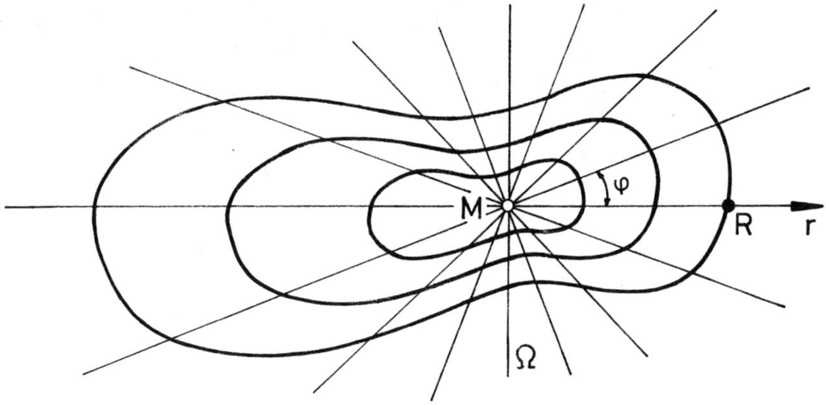

FIG. 2.1.7. Centred domain Ω with centre M.

subdivide, as in Fig. 2.1.7, and determine (approximately) the centre of gravity of the subdomains.

Geometric methods of this type can stimulate the creation of quadrature and cubature formulae, but their disadvantage is obvious: they are not systematic and they give only very weak insight into the accuracy that might be expected in applications.

2.2 The method of Taylor expansions

We now treat analytic methods, which admit the construction of more precise quadrature and cubature formulae than do the simple geometric considerations of Section 2.1.

Here, we replace the integrand f by its Taylor expansion. Then, on comparing If and Q_nf we obtain conditions for the nodes and the weights of the desired quadrature formulae (or cubature formulae analogously).

2.2.1 The one-dimensional problem

In order to shorten the following formulae we omit the weight function $W(x)$ mentioned previously. The aim of this section is the determination of nodes and weights of a quadrature formula Q_nf.

First, we consider only functions $f(x)$ from the space $C^n[-1, 1]$ so that we can apply Taylor's theorem. If both $x, a \in [-1, 1]$, we can write

$$f(x) = f(a) + (x-a)f'(a) + \ldots + \frac{(x-a)^{n-1}}{(n-1)!} f^{(n-1)}(a) + \int_a^x \frac{(x-t)^{n-1}}{(n-1)!} f^{(n)}(t)\, dt.$$

For the sake of simplicity we take $a=0$. Because of the linearity of I and Q_n both operators can be applied directly to the above formula so that

$$If = 2f(0) + \frac{2}{3} f''(0) + \ldots + \frac{1-(-1)^n}{n!} f^{(n-1)}(0) + I(e_n f)$$

$$Q_n f = f(0) \sum_{i=1}^{n} A_i + f'(0) \sum_{i=1}^{n} A_i x_i + \frac{f''(0)}{2!} \sum_{i=1}^{n} A_i x_i^2 + \ldots$$

$$+ \frac{f^{(n-1)}(0)}{(n-1)!} \sum_{i=1}^{n} A_i x_i^{n-1} + Q_n(e_n f) \quad \text{with} \quad e_n f = \int_0^x \frac{(x-t)^{n-1}}{(n-1)!} f^{(n)}(t)\, dt.$$

Neglecting $I(e_n f)$ and $Q_n(e_n f)$, we compare corresponding terms in If and $Q_n f$ and we see that equality arises for

$$\sum_{i=1}^{n} A_i x_i^k = \frac{1+(-1)^k}{k+1}, \qquad k = 0(1)n-1. \tag{2.2.1}$$

This is a linear system for the n unknown weights A_i and the equations (2.2.1) are called *momentum equations*. Now,

$$\mathbf{A} = \begin{bmatrix} 1 & 1 & \ldots & 1 \\ x_1 & x_2 & \ldots & x_n \\ x_1^2 & x_2^2 & \ldots & x_n^2 \\ \ldots\ldots \\ x_1^{n-1} & x_2^{n-1} & \ldots & x_n^{n-1} \end{bmatrix}$$

is a *Vandermonde matrix* and therefore

$$\det(\mathbf{A}) = \prod_{k=2}^{n} \prod_{j=1}^{k-1} (x_k - x_j) \neq 0$$

when the nodes x_i are distinct. It is not necessary that all $x_i \in [-1, 1]$. This means that exterior nodes are admissible provided the Taylor expansion is valid there. We thus have:

Theorem 2.2.1 *Let x_i, $i=1(1)n$, be given distinct nodes. Then the weights A_i, $i=1(1)n$ of $Q_n f$ are uniquely determined by* (2.2.1).

The first equation of (2.2.1)

$$\sum_{i=1}^{n} A_i = 2 = \int_{-1}^{1} dx \qquad \left(\text{for } W(x) \not\equiv 1\text{: } \sum_{i=1}^{n} A_i = \int_{-1}^{1} W(x)\, dx\right)$$

has a remarkable geometric meaning: the sum of the weights equals the area of the domain of integration. This is also true for cubature formulae as will be seen later, $C_n 1 = |\Omega|$.

Example 2.2.1 Determine the weights of

$$Q_3f = A_1f(-1) + A_2f(0) + A_3f(1).$$

We have the system of equations $A_1 + A_2 + A_3 = 2$, $-A_1 + A_3 = 0$, $A_1 + A_3 = \frac{2}{3}$, from which the solution is $A_1 = A_3 = \frac{1}{3}$ and $A_2 = \frac{4}{3}$. Hence, *Simpson's rule*

$$Q_3f = \tfrac{1}{3}(f(-1) + 4f(0) + f(1))$$

is obtained. This is sometimes called *Kepler's barrel rule* because it was used to calculate the volume of a barrel by rotating its defining contour. ■

2.2.2 The two-dimensional problem

The Taylor expansion approach of Section 2.2.1, for the determination of the weights of a quadrature formula, can also be used for the same problem in cubature. We omit the error term of the expansion, and restrict ourselves to two dimensions. The extension to N dimensions is straightforward:

Let $f \in C^n(\bar{\Omega})$, and $\mathbf{a}^T = (a, b) \in \bar{\Omega}$. We then have the expansion

$$\begin{aligned} f(\mathbf{x}) = f(x, y) &= f(\mathbf{a}) + (x-a)f_x(\mathbf{a}) + (y-b)f_y(\mathbf{a}) \\ &+ \tfrac{1}{2}[(x-a)^2 f_{xx}(\mathbf{a}) + 2(x-a)(y-b)f_{xy}(\mathbf{a}) + (y-b)^2 f_{yy}(\mathbf{a})] \\ &+ \ldots + \frac{1}{(n-1)!}\left[\sum_{i=0}^{n-1} \binom{n-1}{i} (x-a)^{n-i-1}(y-b)^i \frac{\partial^{n-1} f(\mathbf{a})}{\partial x^{n-i-1} \partial y^i}\right] \\ &+ \text{error}. \end{aligned}$$

Proceeding exactly as before we obtain

$$\begin{aligned} I_\Omega f &= |\Omega| f(\mathbf{a}) + f_x(\mathbf{a}) I_\Omega(x-a) + f_y(\mathbf{a}) I_\Omega(y-b) + \ldots \\ &+ \frac{1}{(n-1)!}\left[\sum_{i=0}^{n-1} \binom{n-1}{i} \frac{\partial^{n-1} f(\mathbf{a})}{\partial x^{n-i-1} \partial y^i} I_\Omega[(x-a)^{n-i-1}(y-b)^i]\right] + \text{error}, \end{aligned}$$

and

$$\begin{aligned} C_n f &= f(\mathbf{a}) \sum_{i=1}^{n} A_i + f_x(\mathbf{a}) \sum_{i=1}^{n} A_i(x_i - a) + f_y(\mathbf{a}) \sum_{i=1}^{n} A_i(y_i - b) + \ldots \\ &+ \frac{1}{(n-1)!}\left[\sum_{j=0}^{n-1} \binom{n-1}{j} \frac{\partial^{n-1} f(\mathbf{a})}{\partial x^{n-j-1} \partial y^j} \sum_{i=1}^{n} A_i(x_i - a)^{n-j-1}(y_i - b)^j\right]. \end{aligned}$$

Taking again $\mathbf{a}=\mathbf{0}$, and comparing $I_\Omega f$ with $C_n f$ we obtain the equations

$$\sum_{i=1}^{n} A_i = |\Omega|$$

$$\sum_{i=1}^{n} A_i x_i = I_\Omega x, \qquad \sum_{i=1}^{n} A_i y_i = I_\Omega y$$

$$\sum_{i=1}^{n} A_i x_i^2 = I_\Omega x^2, \qquad \sum_{i=1}^{n} A_i x_i y_i = I_\Omega xy, \qquad \sum_{i=1}^{n} A_i y_i^2 = I_\Omega y^2$$

etc.

$$\sum_{i=1}^{n} A_i x_i^{k-j} y_i^j = I_\Omega x^{k-j} y_i^j, \qquad j=0(1)k\leq n.$$

This, again, is a linear system of equations for the weights A_i, the *momentum equations* for the two-dimensional cubature problem. This system consists of $(k+2)(k+1)/2=\binom{k+2}{2}$ equations for the n weights. Hence we can obtain a unique solution only for $n=\binom{k+2}{2}$. But this is only a necessary and not a sufficient condition for the uniqueness of the desired solution. We shall show, in Chapter 5, that we can always find n nodes such that the corresponding weights are a unique solution of the above system and, in addition, that these weights are positive. If $\binom{k+2}{2}<n<\binom{k+3}{2}$ then we can choose $n-\binom{k+3}{2}$ weights arbitrarily, since the system contains more unknowns than equations.

Example 2.2.2 Let Ω be the square $|x|\leq 1$, $|y|\leq 1$. Determine the weights of the cubature formula

$$C_3 f = A_1 f(0,-1) + A_2 f(1,1) + A_3 f(1,-1).$$

The weights satisfy the equations

$$A_1+A_2+A_3=4, \qquad A_2+A_3=0, \qquad -A_1+A_2-A_3=0$$

having the solution $A_1=4$, $A_2=2$, and $A_3=-2$. ■

As an example for an N-dimensional cubature formula we treat:

Example 2.2.3 Let Ω be the hypercube $0\leq x_j\leq 1$. Determine the weights of the cubature formula

$$C_{N+1} f = A_1 f(\mathbf{0}) + \sum_{j=2}^{N+1} A_j f(\mathbf{e}_{j-1})$$

where $\mathbf{e}_j = (0, \ldots, 1, \ldots, 0)^T$ is the j-th coordinate unit vector having value unity in the j-th component and zeros else. Then the A_i satisfy

$$\begin{aligned} A_1 + A_2 + A_3 + \ldots + A_{N+1} &= 1 = |\Omega| \\ A_2 \qquad\qquad\qquad &= 1/2 \\ A_3 \qquad\quad &= 1/2 \\ \cdots\cdot \\ A_{N+1} &= 1/2, \end{aligned}$$

because $I_\Omega 1 = 1$, $I_\Omega x_j = 1/2$, $C_{N+1}1 = A_1 + A_2 + \ldots + A_{N+1}$, and $C_{N+1}x_{j-1} = A_j$, $j = 2(1)N+1$. Hence, we have the solution $A_2 = A_3 = \ldots = A_{N+1} = 1/2$, and $A_1 = 1 - N/2$, so that

$$C_{N+1}f = \left(1 - \frac{N}{2}\right) f(\mathbf{0}) + \tfrac{1}{2} \sum_{j=1}^{N} f(\mathbf{e}_j)$$

is the desired cubature formula. ■

2.2.3 Special properties and difficulties

It is possible to impose additional conditions on the weights, if we reduce the number of momentum equations in a suitable way. However, this can have the consequence that the system has no real solution at all, or no unique solution. It is usually reasonable to make some of the weights within a formula equal.

Example 2.2.4 Let Ω be the square $|x|, |y| \leq 1$. Determine the weights of the cubature formula

$$C_5 f = A_1 f(0,0) + A_2[f(1,1) + f(1,-1) + f(-1,1) + f(-1,-1)].$$

This leads to $A_1 = 8/3$, and $A_2 = 1/3$ from $A_1 + 4A_2 = 4 = |\Omega|$, $0A_2 = 0$, $4A_2 = \frac{4}{3}$. ■

How do we proceed if $\binom{k+2}{2} < n < \binom{k+3}{2}$? If this is the case we cannot take into account all monomials of the highest possible order within the expansions. Suppose that $n = 7$. This means that in the scheme

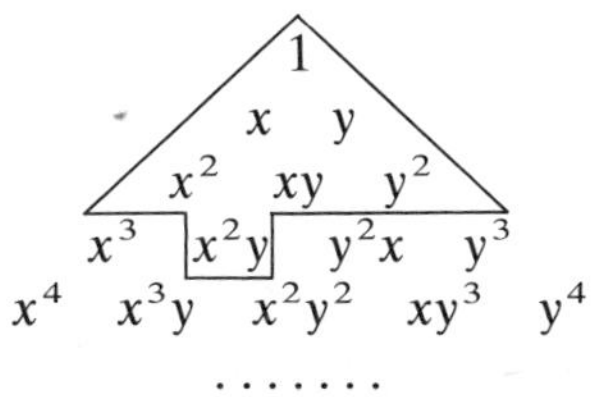

(as shown, its monomials appear in the expansions of $I_\Omega F$ and $C_n f$) we take into account the first seven monomials. But this is not a uniquely-determined problem. Uniqueness only appears for $n = 1, 3, 6, 10, \ldots$. For other values of n we have two possibilities.

Let $n_0 \leq n$ be the largest integer for which uniqueness occurs, then:

(i) Select $n - n_0$ monomials arbitrarily (as is done for $n = 7$ in the above scheme), or

(ii) preassign $n - n_0$ weights explicitly (or impose alternative $n - n_0$ conditions) and neglect the corresponding monomials within the scheme (as is done in Example 2.2.5).

Example 2.2.5 Determine the weights of

$$C_4 f = A_1 f(1, 0) + A_2 f(-1, 0) + A_3 f(0, 1) + A_4 f(0, -1),$$

where Ω is the square $|x|, |y| \leq 1$.

The system of equations for the weights here is

$$A_1 + A_2 + A_3 + A_4 = 4 = I_\Omega 1, \quad A_1 - A_2 = 0 = I_\Omega x, \quad A_3 - A_4 = 0 = I_\Omega y.$$

This system does not possess a unique solution because it has 3 equations and 4 unknowns. However, we obtain the x^2 condition by addition of $A_1 + A_2 = 4/3 = I_\Omega x^2$. Similarly, for the xy condition we do not obtain an additional equation because $0A_4 = 0 = I_\Omega xy$. Preassigning $A_1 = 1$, or requiring equal weights, we obtain the unique solution $A_1 = A_2 = A_3 = A_4 = 1$ and hence $C_4 f = f(-1, 0) + f(0, -1) + f(1, 0) + f(0, 1)$. ■

Once the linear system of equations for the weights is established, it still does not necessarily have a unique solution even in those cases where the number of nodes equals the number of monomials taken into account. This is proved by the following counterexample:

Example 2.2.6 Determine the weights of $C_3 f = \sum_{i=1}^{3} A_i f(x_i, y_i)$. We have to solve the system

$$\begin{aligned} I_\Omega 1 &= A_1 + A_2 + A_3 \\ I_\Omega x &= A_1 x_1 + A_2 x_2 + A_3 x_3 \\ I_\Omega y &= A_1 y_1 + A_2 y_2 + A_3 y_3 \end{aligned} \quad \text{with the matrix} \quad \mathbf{A} \equiv \begin{bmatrix} 1 & 1 & 1 \\ x_1 & x_2 & x_3 \\ y_1 & y_2 & y_3 \end{bmatrix}.$$

$\mathbf{A}$ is not a Vandermonde matrix, and $\det(\mathbf{A}) = 0$ if the three distinct nodes are situated on a straight line. We remember that $\det(\mathbf{A})/2$ equals the area of the triangle whose edges are the three nodes. ■

By these examples we learn whether we can or cannot produce a unique solution or, indeed, a solution at all. For further details the reader is referred to Chapter 5.

Remarks

(i) By construction we have $E_nf = If - Q_nf = 0$ for $f \in \{1, x, x^2, \ldots, x^{n-1}\}$. Hence, $E_nf = 0$ for every polynomial $f \in \mathbb{P}_{n-1}$ because every $g \in \mathbb{P}_{n-1}$ is a linear combination of the monomials x^k, $k = 0(1)n-1$. For cubature we have an analogous result concerning the difference between the number of nodes and the number of terms in the Taylor expansion.

(ii) The Taylor expansion method is applicable to N-dimensional cubature, as is demonstrated by Example 2.2.3, but the presentation is more complicated.

(iii) Instead of the Taylor expansion we can use any other suitable expansion of f, e.g. a Fourier expansion. The preference for polynomial expansions is because of the greater simplicity and the existence of an error representation which can be handled easily. This will be described in detail later.

2.2.4 Special nonlinear systems of equations–Free nodes

Until now, we have considered the nodes as fixed, preassigned numbers or vectors. There is, however, no essential reason for doing this. If the nodes are not preassigned, but are treated as n additional unknowns, we need a system of $2n$ equations for $2n$ unknowns x_i and A_i, $i = 1(1)n$. This system is the enlarged system of momentum equations

$$\sum_{i=1}^{n} A_i x_i^k = I(x^k), \qquad k = 0(1)2n-1. \tag{2.2.2}$$

Equation (2.2.2) follows from the Taylor expansion method with $f \in C^{2n}[-1, 1]$. Let us denote $I(x^k) = c_k$. The special nonlinear system (2.2.2) is solved by elimination of the linear unknowns A_i. The procedure is demonstrated for $n = 3$. The generalisation to arbitrary n is straightforward but time-consuming. The system to be solved is:

$$\begin{aligned} A_1 + A_2 + A_3 &= c_0 \\ A_1x_1 + A_2x_2 + A_3x_3 &= c_1 \\ &\cdots\cdots \\ A_1x_1^5 + A_2x_2^5 + A_3x_3^5 &= c_5. \end{aligned}$$

The weights A_1, A_2, A_3 are eliminated, using Gaussian elimination, so that firstly,

$$A_2(x_1-x_2)+A_3(x_1-x_3)=c_0x_1-c_1$$
$$A_2(x_1-x_2)x_2+A_3(x_1-x_3)x_3=c_1x_1-c_2$$
$$\cdots\cdots$$
$$A_2(x_1-x_2)x_2^4+A_3(x_1-x_3)x_3^4=c_4x_1-c_5,$$

and secondly,

$$A_3(x_1-x_3)(x_2-x_3)=c_0x_1x_2-c_1(x_1+x_2)+c_2$$
$$A_3(x_1-x_3)(x_2-x_3)x_3=c_1x_1x_2-c_2(x_1+x_2)+c_3$$
$$\cdots\cdot$$
$$A_3(x_1-x_3)(x_2-x_3)x_3^3=c_3x_1x_2-c_4(x_1+x_2)+c_5.$$

Analogously, we obtain

$$A_2(x_1-x_2)(x_3-x_2)=c_0x_1x_3-c_1(x_1+x_3)+c_2$$
$$A_1(x_2-x_1)(x_3-x_1)=c_0x_2x_3-c_1(x_2+x_3)+c_2.$$

(The weights can be obtained from these equations.)

Thirdly, for $k=0, 1, 2$

$$0=c_kx_1x_2x_3-c_{k+1}(x_1x_2+x_1x_3+x_2x_3)+c_{k+2}(x_1+x_2+x_3)-c_{k+3}.$$

This gives a linear system of equations for the Vieta coefficients

$$V_3\equiv x_1x_2x_3, \qquad V_2\equiv x_1x_2+x_1x_3+x_2x_3, \qquad V_1\equiv x_1+x_2+x_3$$

which also occur in

$$\omega_3(x)=(x-x_1)(x-x_2)(x-x_3)=x^3-V_1x^2+V_2x-V_3.$$

Thus, the zeros of the *nodal polynomial* $\omega_3(x)$ are the desired nodes. If these are distinct, the weights are given by

$$A_1=\frac{c_0x_2x_3-c_1(x_2+x_3)+c_2}{(x_2-x_1)(x_3-x_1)}, \qquad A_2=\frac{c_0x_1x_3-c_1(x_1+x_3)+c_2}{(x_1-x_2)(x_3-x_2)},$$
$$A_3=\frac{c_0x_1x_2-c_1(x_1+x_2)+c_2}{(x_1-x_3)(x_2-x_3)}.$$

(This holds also if the nodes are preassigned.)

As an example, we calculate for $W(x)\equiv 1$ the nodes and weights of a three-point formula.

Example 2.2.7 Determine for the integral If the weights A_i and nodes x_i in Q_3f. We have $c_0=2$, $c_2=2/3$, $c_4=2/5$, $c_1=c_3=c_5=0$. The Vieta coefficients are to be obtained from

$$2V_3+\tfrac{2}{3}V_1=0$$
$$-\tfrac{2}{3}V_2-\tfrac{2}{5}=0$$
$$\tfrac{2}{3}V_3+\tfrac{2}{5}V_1=0.$$

The solution is $V_1=V_3=0$, $V_2=-3/5$. Hence, $\omega_2(x)=x^3-(3/5)x$ is the nodal polynomial, and its zeros $x_1=-(3/5)^{\frac{1}{2}}$, $x_2=0$, $x_3=(3/5)^{\frac{1}{2}}$ are the desired nodes. The weights are then $A_1=A_3=5/9$, $A_2=8/9$. ■

We remark that this quadrature formula belongs to the class of Gaussian quadrature formulae, which will be treated in Chapter 7. It will be proved there, that the system (2.2.2) always has a real and unique solution, that the weights A_i are all positive, and that the nodes are distinct and in $(-1, 1)$.

The results of this section concern only quadrature and not cubature. If we turn to N-dimensional cubatures, we encounter an enormous increase of unknowns because every node vector has N unknown components.

2.2.5 Calculation of momentum integrals

If we establish a system of momentum equations, then the right-hand sides are integrated monomials, which are called *momentum integrals.* In the previous examples there has been no difficulty in calculating these integrals. Taking into account that the weight function $W(x)$ or $W(\mathbf{x})$ must satisfy only very general conditions, this task is not always a simple one. In cubature we have, in addition, the influence of the dimension N and of the shape of Ω. One of the most fortunate ideas for the calculation of the momentum integrals seems to be the creation of a functional equation which combines certain momentum integrals in such a way that recursive calculation is possible.

We now treat some simple cases in which the momentum integrals are of importance to the following sections (cf. Problems P.2.7, P.2.8, P.2.9). Let us define $I_k\equiv I(x^k)$, $I_{k,l}\equiv I_\Omega(x^k y^l)$ for the following.

Interval $(-\infty, \infty)$, $W(x)=\exp(-x^2)$

Integration by parts gives

$$\underset{\infty}{I_k}=\int_{-\infty}^{\infty}\exp(-x^2)x^k\,dx=\frac{2}{k+1}\underset{\infty}{I_{k+2}}$$

and

$$\underset{\infty}{I_0} = \sqrt{\pi}, \qquad \underset{\infty}{I_1} = 0.$$

These momentum integrals are closely related to the Gauss–Hermite quadrature (cf. Chapter 7).

Interval [0, 1], *Jacobi weight function* $W(x) = x^\alpha(1-x)^\beta$, $\alpha, \beta > -1$

The first momentum integral is

$$\overset{*}{I_0} = \frac{\Gamma(\alpha+1)\Gamma(\beta+1)}{\Gamma(\alpha+\beta+2)} \qquad \text{(cf. Problem P.2.7)},$$

and integration by parts leads to

$$\overset{*}{I_k} = \int_0^1 x^\alpha(1-x)^\beta x^k \, dx = \frac{\alpha+k}{\alpha+\beta+k+1} \overset{*}{I}_{k-1} \quad \text{for} \quad k > 1.$$

These momentum integrals are related to the Gauss–Jacobi quadrature (cf. Chapter 7).

N-dimensional hypercube

Let Ω be the N-dimensional hypercube $|x_j| \leq 1$, $j = 1(1)N$. Then

$$\begin{aligned} I_{k_1,k_2,\ldots,k_N} &= \int_{-1}^1 \int_{-1}^1 \cdots \int_{-1}^1 x_1^{k_1} x_2^{k_2} \ldots x_N^{k_N} \, dx_1 \, dx_2 \ldots dx_N \\ &= \frac{\prod_{i=1}^N [1+(-1)^{k_i}]}{(k_1+1)(k_2+1)\ldots(k_N+1)} \\ &= \frac{k_j+3}{k_j+1} I_{k_1,\ldots,k_j+2,\ldots,k_N}, \\ I_{0,0,\ldots,0} &= 2^N, \\ I_{1,0,\ldots,0} &= I_{0,1,0,\ldots,0} = \ldots = I_{0,0,\ldots,0,1} = 0. \end{aligned}$$

Piece of a parabola

We now seek $I_\Omega f$ in a very general domain Ω (cf. Filippi and Engels [10]). Let Ω be bounded by $y = 0$, $x = -h$, and by the parabola

$$P(x) \equiv y = a - \frac{x}{2} + \frac{x^2}{2h^2}(h-2a)$$

in the x–y-plane as in Fig. 2.2.1. Here, a is a parameter which can be chosen such that Ω fits a part of a larger domain with a curved boundary.

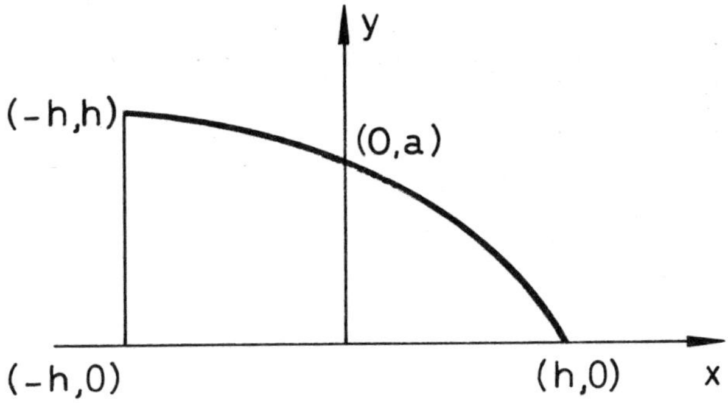

FIG. 2.2.1. Ω piece of a parabola.

Let $h=1$ for simplicity. At first we have, by integration by parts,

$$I_{k,0}=\int_{-1}^{1}\int_{0}^{P(x)} x^k\,\mathrm{d}y\,\mathrm{d}x=\frac{(-1)^k}{k+1}+\frac{1-(-1)^k}{2(k+1)(k+2)}-\frac{(1-2a)(1+(-1)^k)}{(k+1)(k+3)}.$$

In order to determine $I_{k,l}$ we integrate with respect to y

$$I_{k,l}=\int_{-1}^{1}\int_{0}^{P(x)} x^k y^l\,\mathrm{d}y\,\mathrm{d}x=\frac{1}{l+1}\int_{-1}^{1} x^k[P(x)]^{l+1}\,\mathrm{d}x.$$

Integration by parts and return to a double integral successively lead to

$$\begin{aligned}I_{k,l}&=\frac{(-1)^k}{(k+1)(l+1)}-\frac{1}{k+1}\int_{-1}^{1} x^{k+1}[P(x)]^l[P'(x)]\,\mathrm{d}x\\&=\frac{(-1)^k}{(k+1)(l+1)}+\frac{1}{2(k+1)}\int_{-1}^{1} x^{k+1}[P(x)]^l\,\mathrm{d}x\\&\qquad-\frac{1-2a}{k+1}\int_{-1}^{1} x^{k+2}[P(x)]^l\,\mathrm{d}x\\&=\frac{(-1)^k}{(k+1)(l+1)}+\frac{l}{2(k+1)}\int_{-1}^{1}\int_{0}^{P(x)} x^{k+1}y^{l-1}\,\mathrm{d}y\,\mathrm{d}x\\&\qquad-\frac{1-2a}{k+1}l\int_{-1}^{1}\int_{0}^{P(x)} x^{k+2}y^{l-1}\,\mathrm{d}y\,\mathrm{d}x\\&=\frac{(-1)^k}{(k+1)(l+1)}+\frac{l}{2(k+1)}I_{k+1,l-1}-\frac{1-2a}{k+1}l\cdot I_{k+2,l-1}.\end{aligned}$$

The same technique can be used to prove the following results.

Triangle, $W(x, y) = 1$

Let Ω be the equilateral triangle with edges $(2h, 0)$, $(-h, \pm\sqrt{3}h)$ having its centre of gravity in $(0, 0)$ (cf. Fig. 2.3.1). Then

$$I_{k,l} = \int_{-h}^{2h} \int_{(x-2h)/\sqrt{3}}^{(2h-x)/\sqrt{3}} x^k y^l \, dx \, dy,$$

and we have

$$I_{k,0} = \frac{2h^{k+2}}{(k+1)(k+2)\sqrt{3}} \{2^{k+2} + (-1)^k(3k+5)\}, \qquad I_{k,1} = 0$$

$$I_{k,l} = \frac{1}{(k+1)(k+2)3} \left\{ (-1)^k [1 + (-1)^l] h^{2+k+l} 3^{(l+1)/2} \left[3\frac{k+2}{l+1} - 1 \right] \right.$$
$$\left. + l(l-1) I_{k+2,l-2} \right\} \quad \text{for} \quad l > 1$$

(cf. Chapter 6).

Circle $x^2 + y^2 \leqslant R^2$, $W(x, y) = 1$

Let Ω be the circle $x^2 + y^2 \leqslant R^2$. For $W(x, y) = 1$ we have

$$I_{0,0} = \pi R^2, \qquad I_{1,0} = 0, \qquad I_{k,1} = 0 \quad \text{for} \quad k > 0,$$

$$I_{k,0} = \frac{k-1}{k+2} R^2 I_{k-2,0} \quad \text{for} \quad k > 1,$$

$$I_{k,l} = \frac{l-1}{k+1} I_{k+2,l-2} \quad \text{for} \quad l > 1, k > 0.$$

Hence,

$$I_{2,0} = I_{0,2} = \pi R^4/4,$$

$$I_{4,0} = I_{0,4} = \pi R^6/8, \qquad \binom{4}{2} I_{2,2} = \pi R^6/4,$$

$$I_{6,0} = I_{0,6} = 5\pi R^8/64, \qquad \binom{6}{2} I_{2,4} = \binom{6}{4} I_{4,2} = 15\pi R^8/64,$$

$$I_{8,0} = I_{0,8} = 7\pi R^{10}/128, \qquad \binom{8}{2} I_{6,2} = \binom{8}{6} I_{2,6} = 28\pi R^{10}/128,$$

$$\binom{8}{4} I_{4,4} = 42\pi R^{10}/128, \quad \text{etc.}$$

Abbreviating $\Delta f \equiv f_{xx}(\mathbf{0}) + f_{yy}(\mathbf{0})$, $\Delta^k f = \Delta(\Delta^{k-1} f)$, we obtain the following expansion of $I_\Omega f$

$$I_\Omega f = \pi R^2\left[f(\mathbf{0}) + \frac{R^2}{4\cdot 2!}\Delta f + \frac{R^4}{8\cdot 4!}\Delta^2 f + \frac{5R^6}{64\cdot 6!}\Delta^3 f + \frac{7R^8}{128\cdot 8!}\Delta^4 f + \ldots\right]$$

$$= \pi R^2\left[\Delta^0 f + \left(\frac{R}{2}\right)^2 \frac{\Delta f}{2!} + \left(\frac{R}{2}\right)^4 \frac{\Delta^2 f}{2!\,3!} + \left(\frac{R}{2}\right)^6 \frac{\Delta^3 f}{3!\,4!} + \left(\frac{R}{2}\right)^8 \frac{\Delta^4 f}{4!\,5!} + \text{etc.}\right]$$

$$= \pi R^2 \sum_{k=0}^{m} \left(\frac{R}{2}\right)^{2k} \frac{\Delta^k f}{k!\,(k+1)!} + \text{error} \quad \text{for} \quad f \in C^{2m+2}(\bar{\Omega}).$$

A general proof for this expansion has been given by Albrecht and Collatz [3].

Further spherically-shaped regions

Proceeding analogously from the case of a circle we find the following expansions.

Let Γ be the arc of the circle $x^2 + y^2 = R^2$. Then

$$I_\Gamma f = \int_\Gamma f(x, y)\,\mathrm{d}s = 2\pi R \sum_{n=0}^{k} \left(\frac{R}{2}\right)^{2n} \frac{\Delta^n f}{n!\,n!} + \text{error}$$

for $f \in C^{2k+2}(\Gamma)$. Here, s is the arc length of Γ.

Let Ω be the surface of the sphere $x^2 + y^2 + z^2 \leq R^2$, and $\mathrm{d}S$ the surface element, then

$$I_\Omega f = \int_S f(x, y, z)\,\mathrm{d}S = 4\pi R^2 \sum_{n=0}^{k} R^{2n} \frac{\Delta^n f}{(2n+1)!} + \text{error}$$

for $f \in C^{2k+2}(\Omega)$.

Let Ω be the sphere $x^2 + y^2 + z^2 \leq R^2$. Then

$$I_\Omega f = \int_\Omega f(x, y, z)\,\mathrm{d}\Omega = 4\pi R^3 \sum_{n=0}^{k} \frac{(2n+2)R^{2n}}{(2n+3)!} \Delta^n f + \text{error}$$

for $f \in C^{2k+2}(\Omega)$.

Stroud (book, pp. 220–227) gives a collection of domains for which the momentum integrals are known: N-cube, N-cubical shell, N-sphere, surface of the N-sphere, N-octahedron, N-simplex, $\mathbb{R}^N$ with $W(\mathbf{x}) = \exp(-(\mathbf{x}, \mathbf{x}))$ and $\exp(-(\mathbf{x}, \mathbf{x})^{1/2})$, and some special domains in two and three dimensions.

The existence of a recurrence relation for the momentum integrals allows their computation by a computer, especially if non-numerical compilers like FORMAC or SYMBAL are available.

FORMAC program

As an example, we present a FORMAC program for the calculation of the momentum integrals for the standard triangle given by Fig. 2.3.1. The program calculates $I_{k,l}$ for $k, l = 0$ up to $k + l = N$. $W3 \equiv \sqrt{3}$ remains as an undetermined parameter. The powers of $W3$ may be replaced by the terms 3, $3W3$, 9, $9W3$, 27, etc. using a REPLACE statement whose length depends on N.

```
MOMINT: PROC OPTIONS (MAIN);
FORMAC-OPTIONS;
N=10;
  DO K=0 TO N;
  LET (K="K");
PRINT-OUT
  (I(K, 0)=2*H**(K+2)/((K+1)*(K+2)*W3)
                         *(2**(K+2)+(-1)**K*(3*K+5));
    I(K, 1)=0);
  END;
KM1=N-1;
  DO L=2 TO N;
  LET (L="L");
  LET (Q=W3**(L+1));
  KM1=KM1-1;
    DO K=0 TO KM1;
    LET (K="K");
 PRINT-OUT
 (I(K, L)=1/((K+1)*(K+2)*3)*((-1)**K*(1+(-1)**L)
    *H**(2+K+L)*Q*(3*(K+2)/(L+1)+L*(L-1)*I(K+2, L-2))
    END;
  END;
END MOMINT;
```

2.2.6 Expansion for nodes which are regularly distributed on a circle – The general expansion for a small number of nodes

The most general form of $C_n f$ usually is not the most convenient for the determination of nodes and weights. The nodes are often assumed, therefore, to be distributed regularly on suitable subspaces of Ω, e.g. on concentric circles. For such special distributions rather far-reaching expansions are possible and indicated in order to reduce the amount of work which arises for a general distribution of nodes.

Let us assume, that we have k_τ equally-spaced nodes on the circle

$x^2+y^2=R_\tau^2$, all having the same weight A_τ. We then have

$$I_\Omega f \approx C_n f = \sum_{\tau=1}^{m} A_\tau \sum_{j=1}^{k_\tau} f\left(R_\tau \cos\left(\varphi+\frac{2\pi j}{k_\tau}\right), R_\tau \sin\left(\varphi+\frac{2\pi j}{k_\tau}\right)\right), \qquad (\varphi \equiv \varphi_\tau)$$

with $n=\sum_{\tau=1}^{m} k_\tau$. The expansions of $I_\Omega f$ and $C_n f$ are compared. Hence

$$I_\Omega 1 = \sum_{\tau=1}^{m} A_\tau k_\tau,$$

$$I_\Omega x^k y^l = \sum_{\tau=1}^{m} A_\tau R_\tau^{k+l} \sum_{j=1}^{k_\tau} \cos^k\left(\varphi+\frac{2\pi j}{k_\tau}\right)\sin^l\left(\varphi+\frac{2\pi j}{k_\tau}\right)$$

are the momentum equations of the problem. In order to simplify the sums we calculate (replacing k_τ by $\varkappa$)

$$\sum_{j=1}^{\varkappa} \cos^k\left(\varphi+\frac{2\pi j}{\varkappa}\right)\sin^l\left(\varphi+\frac{2\pi j}{\varkappa}\right).$$

This is done using complex numbers so that (i here is $\sqrt{-1}$)

$$\sum_{j=1}^{\varkappa}\left(\cos\left(\varphi+\frac{2\pi j}{\varkappa}\right)+i\sin\left(\varphi+\frac{2\pi j}{\varkappa}\right)\right)=\sum_{j=0}^{\varkappa-1}\exp\left(i\left(\varphi+\frac{2\pi j}{\varkappa}\right)\right)$$

$$=e^{i\varphi}\sum_{j=0}^{\varkappa-1}(e^{2\pi i/\varkappa})^j=\frac{1-e^{2\pi i}}{1-e^{2\pi i/\varkappa}}e^{i\varphi}=\begin{cases}\varkappa e^{i\varphi} & \text{for } \dfrac{1}{\varkappa}=\pm1,\pm2,\ldots\\ 0 & \text{otherwise.}\end{cases}$$

We separate real and imaginary parts and obtain

$$\sum_{j=1}^{\varkappa}\cos\left(\varphi+\frac{2\pi j}{\varkappa}\right)=\sum_{j=1}^{\varkappa}\sin\left(\varphi+\frac{2\pi j}{\varkappa}\right)=0 \quad \text{for} \quad \varkappa>1.$$

Analogously, we obtain

$$\sum_{j=1}^{\varkappa}\left[\cos\left(\varphi+\frac{2\pi j}{\varkappa}\right)+i\sin\left(\varphi+\frac{2\pi j}{\varkappa}\right)\right]^2=\sum_{j=0}^{\varkappa-1}e^{2i(\varphi+2\pi j/\varkappa)}$$

$$=e^{2i\varphi}\sum_{j=0}^{\varkappa-1}(e^{4\pi i/\varkappa})^j=\frac{1-e^{4\pi i}}{1-e^{4\pi i/\varkappa}}e^{2i\varphi}$$

$$=\begin{cases}\varkappa e^{2i\varphi} & \text{for } 2/\varkappa=\pm1,\pm2,\pm3,\ldots \text{ or } \varkappa=\pm2,\pm1,\pm2/3,\ldots\\ 0 & \text{otherwise.}\end{cases}$$

Thus

$$\sum_{j=1}^{\varkappa}\left[\cos^2\left(\varphi+\frac{2\pi j}{\varkappa}\right)-\sin^2\left(\varphi+\frac{2\pi j}{\varkappa}\right)\right]=0 \quad \text{for} \quad \varkappa>2,$$

and, therefore, since $\sin^2 \varphi + \cos^2 \varphi = 1$,

$$\sum_{j=1}^{\varkappa} \cos^2\left(\varphi + \frac{2\pi j}{\varkappa}\right) = \sum_{j=1}^{\varkappa} \sin^2\left(\varphi + \frac{2\pi j}{\varkappa}\right) = \varkappa/2 \quad \text{for} \quad \varkappa > 2.$$

Comparing the imaginary terms we have

$$\sum_{j=1}^{k} \cos\left(\varphi + \frac{2\pi j}{\varkappa}\right) \sin\left(\varphi + \frac{2\pi j}{\varkappa}\right) = 0 \quad \text{for} \quad \varkappa > 2.$$

This calculation may be continued, but we shall treat a special case in more detail in the following section. For the moment we state the special result for $m = \tau = 1$, $A = A_1$, $R = R_1$, $k = k_1$:

$$A \sum_{j=1}^{k} f\left(R \cos\left(\varphi + \frac{2\pi j}{k}\right), R \sin\left(\varphi + \frac{2\pi j}{k}\right)\right) = Akf(\mathbf{0}) + \frac{1}{2} AR^2 k \frac{\Delta f}{2!} + \ldots.$$

We apply this expansion in the following example.

Example 2.2.8 Derive a cubature formula $C_3 f$ for the hexagonal domain Ω given in Fig. 2.2.2 (cf. problem P.2.9).

We distribute the three nodes regularly on a circle with centre (0, 0). The momentum integrals are given by

$$I_{0,0} = \frac{3\sqrt{3}}{2} h^2 = |\Omega|, \qquad I_{1,0} = I_{0,1} = I_{1,1} = 0, \qquad I_{2,0} = I_{0,2} = \frac{5\sqrt{3}}{16} h^4,$$

and hence

$$I_\Omega f = |\Omega| f(\mathbf{0}) + \frac{5\sqrt{3}}{16} h^4 \frac{\Delta f}{2!} + \ldots.$$

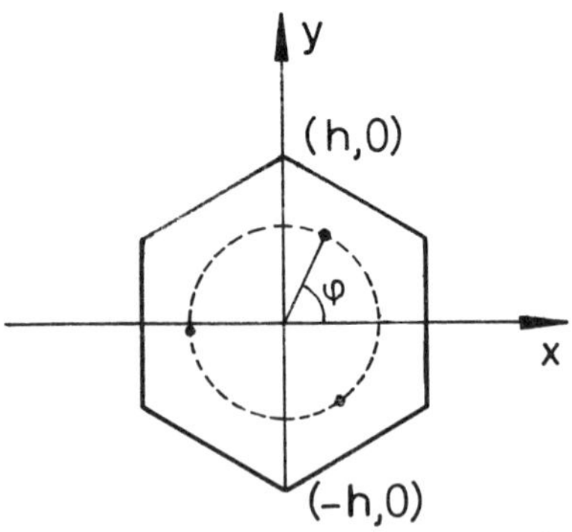

FIG. 2.2.2. Hexagonal domain Ω with three nodes.

Comparing the expansions for $n=k=3$ we obtain the equations $\frac{3\sqrt{3}}{2}h^2=3A$, $\frac{5\sqrt{3}}{16}h^4=\frac{3}{2}AR^2$ having the solution $A=\frac{1}{2}\sqrt{3}h^2$, $R=\frac{h}{2}\sqrt{\frac{5}{3}}$. Thus,

$$C_3f=\frac{\sqrt{3}}{2}h^2\sum_{j=1}^{3}f\left(\frac{h}{2}\sqrt{\frac{5}{3}}\cos\left(\varphi+\frac{2\pi j}{3}\right),\frac{h}{2}\sqrt{\frac{5}{3}}\sin\left(\varphi+\frac{2\pi j}{3}\right)\right)$$

is the desired cubature formula for the hexagon Ω. Note that the angle φ remains undetermined and hence can be chosen arbitrarily. ■

2.2.7 Expansion for nodes which are regularly spaced on a circle–The general expansion for an even number of nodes

The extension of the above expansion to a large number of nodes is very complicated. Following Albrecht and Collatz [3] (cf. Remark R.2.4) we restrict our attention to those cases where an even number of nodes is distributed regularly on the circle $x^2+y^2=R^2$ so that whenever (x_j, y_j) is a node, then $(-x_j, y_j)$, $(x_j, -y_j)$, and $(-x_j, -y_j)$ are also nodes. This implies symmetry of C_nf about both the x- and y-axes. As an immediate consequence

$$\sum_j x_j^\nu y_j^\mu=0 \quad \text{for odd} \quad \nu, \mu.$$

Let $f\in C^{2k+2}(\bar{\Omega})$. We consider the sum

$$S_{2m}(R)=\sum_{j=0}^{2m-2}f\left(R\cos\left(\varphi+\frac{\pi j}{m}\right), R\sin\left(\varphi+\frac{\pi j}{m}\right)\right),$$

where $\varphi=0$ or $\varphi=\pi/2m$ in order to preserve symmetry. Then the Taylor expansion is

$$S_{2m}(R)=\sum_{n=0}^{2k+1}\frac{R^n}{n!}\sum_{j=0}^{2m-2}\left[\cos\left(\varphi+\frac{\pi j}{m}\right)\frac{\partial}{\partial x}+\sin\left(\varphi+\frac{\pi j}{m}\right)\frac{\partial}{\partial y}\right]^n f(\mathbf{x})\Big|_{\mathbf{x}=\mathbf{0}}+\text{error}.$$

We apply the differential operator Δ to $f(\mathbf{x})$, and replace $\mathbf{x}\equiv\mathbf{0}$, $\varphi_j\equiv\varphi+\frac{\pi j}{m}$, $\cos\varphi_j\equiv(e^{i\varphi_j}+e^{-i\varphi_j})/2$, and $\sin\varphi_j\equiv(e^{i\varphi_j}-e^{-i\varphi_j})/2i$. Then, omitting the error we have

$$\begin{aligned}S_{2m}(R)&\approx\sum_{n=0}^{2k+1}\frac{R^n}{n!}\sum_{j=0}^{2m-2}\sum_{p=0}^{n}\binom{n}{p}\frac{(e^{i\varphi_j}+e^{-i\varphi_j})^{n-p}(e^{i\varphi_j}-e^{-i\varphi_j})^p}{2^n i^p}f_{n-p,p}(\mathbf{0})\\&\approx\sum_{n=0}^{2k+1}\left(\frac{R}{2}\right)^n\sum_{p=0}^{n}a_{2m,n-p,p}\,f_{n-p,p}(\mathbf{0}).\end{aligned}$$

with

$$a_{2m,n-p,p}=\frac{(-i)^p}{p!\,(n-p)!}\sum_{j=0}^{2m-2}\left[\sum_{q=0}^{n-p}\binom{n-p}{q}e^{i(n-p-2q)\varphi_j}\right]\left[\sum_{r=0}^{p}\binom{p}{r}e^{i(p-2r)\varphi_j}(-1)^r\right]$$

$$=i^p\sum_{q=0}^{n-p}\sum_{r=0}^{p}\frac{(-1)^{r+p}}{q!\,(n-p-q)!\,r!\,(p-r)!}\sigma_{2m,r,q,n},$$

when $(e^{i\varphi_j}+e^{-i\varphi_j})^{n-p}\cdot(e^{i\varphi_j}-e^{-i\varphi_j})^p$ is expanded into a sum. Now

$$\sigma_{2m,r,q,n}=\sum_{j=0}^{2m-2}e^{i(n-2q-2r)\varphi_j}=e^{i\varphi(n-2q-2r)}\sum_{j=0}^{2m-2}(e^{(\pi i/m)(n-2q-2r)})^j.$$

By the summation formula for geometric sums we obtain

$$\sigma_{2m,r,q,n}=e^{i\varphi(n-2q-2r)}\cdot\begin{cases}0 & \text{for } \dfrac{n-2q-2r}{2m}\neq l, \quad l=0,\pm1,\pm2,\ldots\\ 2m & \text{for } \dfrac{n-2q-2r}{2m}=l, \quad l=0,\pm1,\pm2,\ldots,\pm L\end{cases},$$

where L is the largest integer such that $(n-2q-2r)/2mL$ is also an integer. The numbers q and r are coupled by $2m$ and l. We have that $\sigma_{2m,r,q,n}=0$ for $2mL>n-2q-2r$.

We notice that $2r+2q=0(2)2n$ because $r=0(1)p$, and $q=0(1)n-p$. Hence $n-2q-2r=-n(2)n$, and we obtain

$$a_{2m,n-p,p}=2m\sum_{q=0}^{n-p}\sum_{r=0}^{p}\frac{(-1)^{p+r}i^p e^{2i\varphi lm}}{q!\,(n-p-q)!\,r!\,(p-r)!}.$$

As already noted above, we have $a_{2m,n-p,p}=0$ for odd p. We therefore restrict the sum to even n and p and replace n by $2n$, and p by $2p$. Then

$$S_{2m}(R)=\sum_{n=0}^{k}\left(\frac{R}{2}\right)^{2n}\sum_{p=0}^{n}a_{2m,2n-2p,2p}\,f_{2n-2p,2p}(\mathbf{0}),$$

and

$$a_{2m,2n-2p,2p}=2m\sum_{\substack{q=0\\ n-q-r=lm\\ l=-L(1)L}}^{2n-2p}\sum_{r=0}^{2p}\frac{(-1)^{2p+r}\sigma_l}{q!\,(2n-2p-q)!\,r!\,(2p-r)!}$$

where

$$\sigma_l=e^{i\varphi(n-2q-2r)}=e^{i\varphi 2lm}=\begin{cases}1 & \text{for } \varphi=0\\ (-1)^l & \text{for } \varphi=\pi/2m.\end{cases}$$

This sum is simplified by the substitution $r=p-\rho$, and $q=n-lm-p+\rho$. Thus, from $r=0(1)2p$ we obtain $\rho=-p(1)p$. Having fixed r and ρ

respectively, it follows that $q=n-lm-r=n-lm-p+\rho$, where $l=-L(1)L$. Hence

$$a_{2m,2n-2p,2p}=2m\sum_{l=-L}^{L}\sum_{\rho=-p}^{p}\frac{(-1)^{\rho}\sigma_l}{(n-lm-p+\rho)!\,(n+lm-p-\rho)!\,(p-\rho)!\,(p+\rho)!}.$$

The sum over l is split and symmetry is taken into account so that

$$\begin{aligned}a_{2m,2n-2p,2p}&=\frac{2m}{n!\,n!}\sum_{\rho=-p}^{p}(-1)^{\rho}\binom{n}{p+\rho}\binom{n}{p-\rho}\\&\quad+4m\sum_{l=1}^{L}\frac{\sigma_l}{(n+lm)!\,(n-lm)!}\sum_{\rho=-p}^{p}(-1)^{\rho}\binom{n+lm}{p+\rho}\binom{n-lm}{p-\rho}.\end{aligned}$$

When this is substituted into S_{2m} and all arguments are considered the following has been proved:

Theorem 2.2.2 (Albrecht and Collatz [3]) *Let $2m$ nodes be distributed regularly and symmetrically on the circle $x^2+y^2=R^2$. Let $\varphi=0$ or $\varphi=\pi/2m$. Then for $f\in C^{2k+2}(\bar{\Omega})$ the node distribution*

$$S_{2m}(R)=\sum_{j=0}^{2m-2}f(R\cos(\varphi+\pi j/m),\,R\sin(\varphi+\pi j/m))$$

has the expansion

$$\begin{aligned}S_{2m}(R)&=2m\sum_{n=0}^{k}\left(\frac{R}{2}\right)^{2n}\sum_{p=0}^{n}\frac{1}{n!\,n!}\sum_{\rho=-p}^{p}(-1)^{\rho}\binom{n}{p+\rho}\binom{n}{p-\rho}f_{2n-2p,2p}(\mathbf{0})\\&\quad+4m\sum_{n=0}^{k+1}\left(\frac{R}{2}\right)^{2n}\sum_{p=0}^{n}\sum_{l=1}^{L}\frac{\sigma_l}{(n+lm)!\,(n-lm)!}\sum_{\rho=-p}^{p}(-1)^{\rho}\binom{n+lm}{p+\rho}\\&\qquad\times\binom{n-lm}{p-\rho}f_{2n-2p,2p}(\mathbf{0})\\&\quad+\text{error}\end{aligned}$$

where $\sigma_l=1$ or $(-1)^l$, and L is determined by $Lm\leq n<(L+1)m$.

These results can be used to derive cubature formulae which contain few nodes by comparison with other similar formulae. This is especially so for the circle and is because the expansions of $I_\Omega f$ and $S_{2m}(R)$ are very similar. Such formulae were constructed by Albrecht [2] and later on by Albrecht and Engels [4]. They are listed in Table 2.2.1 together with some explanatory notes.

In order that the reader may become familiar with Theorem 2.2.2 the result is specialised to $m=2$. First $2L\leq n<2L+2$ leads to $L=0$ for

$n=0$ and $n=1$, and to $L=1$ for $n=2$ and $n=3$. Hence

$$S_4(R)=4\sum_{n=0}^{k}\left(\frac{R}{2}\right)^{2n}\sum_{p=0}^{n}\frac{1}{n!\,n!}\sum_{\rho=-p}^{p}\binom{n}{p+\rho}\binom{n}{p-\rho}(-1)^{\rho}f_{2n-2p,2p}(\mathbf{0})$$

$$+8\sum_{n=2}^{k}\left(\frac{R}{2}\right)^{2n}\sum_{p=0}^{n}\frac{\sigma_1}{(n+2)!\,(n-2)!}$$

$$\times\sum_{\rho=-p}^{p}\binom{n+2}{p+\rho}\binom{n-2}{p-\rho}(-1)^{\rho}f_{2n-2p,2p}(\mathbf{0})$$

$$+8\sum_{n=4}^{k}\left(\frac{R}{2}\right)^{2n}\sum_{p=0}^{n}\sum_{l=1}^{L}\cdots\cdots$$

For $n=0$ we have that $4f_{0,0}(\mathbf{0})=4f(\mathbf{0})$ and for $n=1$ we obtain

$$R^2\sum_{p=0}^{1}\sum_{\rho=-p}^{p}\binom{1}{p+\rho}\binom{1}{p-\rho}f_{2-2p,2p}(\mathbf{0})$$

$$=R^2f_{2,0}(\mathbf{0})+R^2\sum_{\rho=-1}^{1}\binom{1}{1+\rho}\binom{1}{1-\rho}f_{0,2}(\mathbf{0})=R^2[f_{2,0}(\mathbf{0})+f_{0,2}(\mathbf{0})],$$

because $\binom{1}{2}=\dfrac{1!}{2!\,(-1)!}=0$. Now for $n=2$

$$\frac{R^4}{4}\sum_{p=0}^{2}\frac{1}{4}\sum_{\rho=-p}^{p}(-1)^{\rho}\binom{2}{p+\rho}\binom{2}{p-\rho}f_{4-2p,2p}(\mathbf{0})$$

$$+\frac{R^4}{2}\sum_{p=0}^{2}\frac{\sigma_1}{24}\sum_{\rho=-p}^{p}\binom{4}{p+\rho}\binom{0}{p-\rho}(-1)^{\rho}f_{4-2p,2p}(\mathbf{0})$$

$$=\frac{R^4}{16}\left\{\left(1+\frac{\sigma_1}{3}\right)f_{4,0}(\mathbf{0})+2(1-\sigma_1)f_{2,2}(\mathbf{0})+\left(1+\frac{\sigma_1}{3}\right)f_{0,4}(\mathbf{0})\right\}.$$

Collecting the terms, we find that

$$S_4(R)=4f(\mathbf{0})+R^2[f_{2,0}(\mathbf{0})+f_{0,2}(\mathbf{0})]+$$

$$+\frac{R^4}{16}\left\{\left(1+\frac{\sigma_1}{3}\right)(f_{4,0}(\mathbf{0})+f_{0,4}(\mathbf{0}))+2(1-\sigma_1)f_{2,2}(\mathbf{0})\right\}+\cdots.$$

Further examples can be found in Problems P.2.10 and P.2.11.

2.2.8 Cubature formulae for a circle

We use Theorem 2.2.2 in order to derive cubature formulae for a circle. Following Albrecht [2], and Albrecht and Engels [4] we form

$$S_{2m}(\sigma_{2m}r)=S_{2m}(R)\quad\text{for}\quad R=\sigma_{2m}r\quad\text{and}\quad\varphi=0,$$

and

$$T_{2m}(\tau_{2m}r) = S_{2m}(R) \quad \text{for} \quad R = \tau_{2m}r \quad \text{and} \quad \varphi = -\frac{\pi}{2m}.$$

These expansions have to be displayed and are to be compared with the expansion of $I_\Omega f$. Let us denote the corresponding weights by $S_{2m,\rho}$ and $T_{2m,\rho}$ respectively. Using the abbreviations

$$\Sigma_1 \equiv 4 \sum_{\rho=1}^{N} \sigma_{4,\rho}^{2n} S_{4,\rho}, \qquad \Sigma_2 \equiv 4 \sum_{\rho=1}^{N-1} \tau_{4,\rho}^{2n} T_{4,\rho},$$

$$\Sigma_3 \equiv 8 \sum_{\rho=1}^{N-2} \tau_{8,\rho}^{2n} T_{8,\rho}, \qquad \Sigma_4 \equiv 16 \sum_{\rho=1}^{N-4} \tau_{16,\rho}^{2n} T_{16,\rho},$$

we obtain the systems of equations for nodes and weights as in Table 2.2.1.

TABLE 2.2.1

Degree 3 4 nodes, $N=1$	$\Sigma_1 = \dfrac{1}{n+1}$,	for $n = 0(1)2N-1$
Degree 7 12 nodes, $N=2$	$\Sigma_1 + \Sigma_2 = \dfrac{1}{n+1}$, $\Sigma_1 - \Sigma_2 = 0$,	for $n = 0(1)2N-1$ for $n = 2(1)2N-1$
Degree 11 28 nodes, $N=3$ Degree 15 44 nodes, $N=4$	$\Sigma_1 + \Sigma_2 + \Sigma_3 = \dfrac{1}{n+1}$, $\Sigma_1 - \Sigma_2 = 0$, $\Sigma_1 + \Sigma_2 - \Sigma_3 = 0$,	$n = 0(1)2N-1$ $n = 2(1)2N-1$ $n = 4(1)2N-1$
Degree 19 76 nodes, $N=5$ Degree 23 108 nodes, $N=6$ Degree 27 140 nodes, $N=7$ Degree 31 172 nodes, $N=8$	$\Sigma_1 + \Sigma_2 + \Sigma_3 + \Sigma_4 = \dfrac{1}{n+1}$, $\Sigma_1 - \Sigma_2 = 0$, $\Sigma_1 + \Sigma_2 - \Sigma_3 = 0$, $\Sigma_1 + \Sigma_2 + \Sigma_3 - \Sigma_4 = 0$,	$n = 0(1)2N-1$ $n = 2(1)2N-1$ $n = 4(1)2N-1$ $n = 8(1)2N-1$

etc.

These are nonlinear systems of equations for the weights and radii. They may be solved by separating the radii from the weights, and eliminating the radii first. After the remaining nonlinear system has been solved for the radii, the weights are the solutions of a linear system of equations. The radii are the zeros of a polynomial as described earlier. We now list the results:

Let $\Omega = \{(x, y) \mid x^2 + y^2 \leq r^2\}$, and

$$\frac{1}{\pi r^2} \iint_{\Omega} f(x, y)\, dx\, dy \doteq \sum_{\rho=1}^{N} S_{4,\rho} \sum_{j=1}^{4} f\left(\sigma_{4,\rho} \cos\left(\frac{\pi j}{2}\right), \sigma_{4,\rho} \sin\left(\frac{\pi j}{2}\right)\right)$$

$N = 1$

$$+ \sum_{\rho=1}^{N-1} T_{4,\rho} \sum_{j=1}^{4} f\left(\tau_{4,\rho} \cos\left(\frac{\pi j}{2} - \frac{\pi}{4}\right), \tau_{4,\rho} \sin\left(\frac{\pi j}{2} - \frac{\pi}{4}\right)\right)$$

$N = 2$

$$+ \sum_{\rho=1}^{N-2} T_{8,\rho} \sum_{j=1}^{8} f\left(\tau_{8,\rho} \cos\left(\frac{\pi j}{4} - \frac{\pi}{8}\right), \tau_{4,\rho} \sin\left(\frac{\pi j}{4} - \frac{\pi}{8}\right)\right)$$

$N = 3, 4$

$$+ \sum_{\rho=1}^{N-4} T_{16,\rho} \sum_{j=1}^{16} f\left(\tau_{16,\rho} \cos\left(\frac{\pi j}{8} - \frac{\pi}{16}\right), \tau_{16,\rho} \sin\left(\frac{\pi j}{8} - \frac{\pi}{16}\right)\right)$$

$N = 5, 6, 7, 8$

$N = 1$ $\qquad \sigma_{4,1} = \frac{1}{\sqrt{2}} \qquad S_{4,1} = \frac{1}{4}$

$N = 2$ $\qquad \tau_{4,1} = \frac{1}{2}\sqrt{3} \qquad T_{4,1} = \frac{2}{27}$

$$\sigma_{4,1} = \left(\frac{27 + 3\sqrt{29}}{52}\right)^{\frac{1}{2}} \qquad S_{4,1} = \frac{551 - 41\sqrt{29}}{6264}$$

$$\sigma_{4,2} = \left(\frac{27 - 3\sqrt{29}}{52}\right)^{\frac{1}{2}} \qquad S_{4,2} = \frac{551 + 41\sqrt{29}}{6264}$$

$N = 3$

$\sigma_{4,3} = 9.79798373636374404476(-01)$	$S_{4,3} = 1.04912371962463641123(-02)$
$\sigma_{4,2} = 7.20984642975780841580(-01)$	$S_{4,2} = 4.90399916287106928339(-02)$
$\sigma_{4,1} = 3.26655862700939836315(-01)$	$S_{4,1} = 6.55791415454133134241(-02)$
$\tau_{4,2} = 9.59818372605046949546(-01)$	$T_{4,2} = 1.41511940644713428477(-02)$
$\tau_{4,1} = 6.66984940001672175775(-01)$	$T_{4,1} = 5.88984355651582867819(-02)$
$\tau_{8,1} = 9.12870929175276855762(-01)$	$T_{8,1} = 2.59200000000000000000(-02)$

$N=4$

$\sigma_{4,4}=9.89746802511490700781(-01)$
$\sigma_{4,3}=8.73836956644881701371(-01)$
$\sigma_{4,2}=5.77728928444823464961(-01)$
$\sigma_{4,1}=2.52863797091229572651(-01)$
$\tau_{4,3}=9.74816532850201264705(-01)$
$\tau_{4,2}=8.45154254728516577510(-01)$
$\tau_{4,1}=5.30919564929349369906(-01)$
$\tau_{8,2}=9.55857425391985060322(-01)$
$\tau_{8,1}=7.65726179729415882044(-01)$

$S_{4,4}=5.31979391979622843865(-03)$
$S_{4,3}=2.10840370156484115539(-02)$
$S_{4,2}=3.48550570365141272517(-02)$
$S_{4,1}=3.98811120280412327558(-02)$
$T_{4,3}=8.30879845435517600780(-03)$
$T_{4,2}=2.10087500000000000000(-02)$
$T_{4,1}=4.05617108049040832515(-02)$
$T_{8,2}=1.35355122434771234676(-02)$
$T_{8,1}=2.59548581268932469028(-02)$

$N=5$

$\sigma_{4,5}=9.97199047393362244644(-01)$
$\sigma_{4,4}=8.75876017531398360438(-01)$
$\sigma_{4,3}=7.34821249218660899068(-01)$
$\sigma_{4,2}=4.65377356354853006817(-01)$
$\sigma_{4,1}=2.00104904292266120498(-01)$
$\tau_{4,4}=9.94076616887621244497(-01)$
$\tau_{4,3}=8.60729772357272567987(-01)$
$\tau_{4,2}=7.08384914864406499273(-01)$
$\tau_{4,1}=4.27101503406496656794(-01)$
$\tau_{8,3}=9.89062581324609777081(-01)$
$\tau_{8,2}=8.36868322058266360691(-01)$
$\tau_{8,1}=6.33612573300188710957(-01)$
$\tau_{16,1}=9.48683298050513799600(-01)$

$S_{4,5}=2.25973933290592424146(-03)$
$S_{4,4}=9.84364340406369209608(-03)$
$S_{4,3}=1.80691371313042602879(-02)$
$S_{4,2}=2.38962773639315176284(-02)$
$S_{4,1}=2.51816324964580806663(-02)$
$T_{4,4}=2.53220665348266478521(-03)$
$T_{4,3}=1.21083249537874003509(-02)$
$T_{4,2}=1.75861204519463112659(-02)$
$T_{4,1}=2.73560037442208396849(-02)$
$T_{8,3}=2.99187436435829293737(-03)$
$T_{8,2}=1.64061635788238122577(-02)$
$T_{8,1}=2.00530768425600669139(-02)$
$T_{16,1}=8.06617122410374119372(-03)$

$N=6$

$\sigma_{4,6}=9.98537750041087694197(-01)$
$\sigma_{4,5}=9.28705180368430949244(-01)$
$\sigma_{4,4}=7.60719716224967712040(-01)$
$\sigma_{4,3}=6.25374131600638069673(-01)$
$\sigma_{4,2}=3.88037396620266082814(-01)$
$\sigma_{4,1}=1.65302390505726109006(-01)$
$\tau_{4,5}=9.96994973145652888087(-01)$
$\tau_{4,4}=9.26270313010084032134(-01)$
$\tau_{4,3}=7.47372901846392235478(-01)$
$\tau_{4,2}=6.01724388894736197879(-01)$
$\tau_{4,1}=3.55778522107668142140(-01)$
$\tau_{8,4}=9.93938925750435560522(-01)$
$\tau_{8,3}=9.21243846682460433115(-01)$
$\tau_{8,2}=7.25499080495251139515(-01)$
$\tau_{8,1}=5.35430731928962873555(-01)$
$\tau_{16,2}=9.72471067475638071168(-01)$
$\tau_{16,1}=8.49097573656561364863(-01)$

$S_{4,6}=1.26479837005058384383(-03)$
$S_{4,5}=6.90583559257388670038(-03)$
$S_{4,4}=9.08071804961844617774(-03)$
$S_{4,3}=1.40875535212320291829(-02)$
$S_{4,2}=1.70676962418956173406(-02)$
$S_{4,1}=1.72570155755529147101(-02)$
$T_{4,5}=1.37901658673384064018(-03)$
$T_{4,4}=6.98484540391198785269(-03)$
$T_{4,3}=1.08746131226157802567(-02)$
$T_{4,2}=1.35983684540314590635(-02)$
$T_{4,1}=1.93780074855946817255(-02)$
$T_{8,4}=1.62740634998345280709(-03)$
$T_{8,3}=7.10061560522737619526(-03)$
$T_{8,2}=1.42327470024334650934(-02)$
$T_{8,1}=1.51079578595690371029(-02)$
$T_{16,2}=4.32361406883721038349(-03)$
$T_{16,1}=9.67240542160331714370(-03)$

$N = 7$

$\sigma_{4,7} = 9.99298013869759751650(-01)$	$S_{4,7} = 8.15082212151928122136(-04)$
$\sigma_{4,6} = 9.53845090717241175093(-01)$	$S_{4,6} = 4.56947932145008056439(-03)$
$\sigma_{4,5} = 8.44844331649548452875(-01)$	$S_{4,5} = 7.54017141152681236885(-03)$
$\sigma_{4,4} = 6.68487344060863077064(-01)$	$S_{4,4} = 7.70435450263728756205(-03)$
$\sigma_{4,3} = 5.43373096669951241108(-01)$	$S_{4,3} = 1.10616044839757304983(-02)$
$\sigma_{4,2} = 3.33225018389564246682(-01)$	$S_{4,2} = 1.27771957751626081023(-02)$
$\sigma_{4,1} = 1.41181350252440997884(-01)$	$S_{4,1} = 1.26188831850849667072(-02)$
$\tau_{4,6} = 9.98210794327259164551(-01)$	$T_{4,6} = 8.91249643330110266871(-04)$
$\tau_{4,5} = 9.52492231755159301165(-01)$	$T_{4,5} = 4.58089657978445579893(-03)$
$\tau_{4,4} = 8.42778602099899535003(-01)$	$T_{4,4} = 7.60238099421012636517(-03)$
$\tau_{4,3} = 6.56568666856091788861(-01)$	$T_{4,3} = 9.12291660782051624232(-03)$
$\tau_{4,2} = 5.22216072088580786064(-01)$	$T_{4,2} = 1.06300235434878450071(-02)$
$\tau_{4,1} = 3.05314891086880605191(-01)$	$T_{4,1} = 1.44394216499237974659(-02)$
$\tau_{8,5} = 9.95859559752949601771(-01)$	$T_{8,5} = 1.07760721496542354883(-03)$
$\tau_{8,4} = 9.49207190811658950395(-01)$	$T_{8,4} = 4.60041750383343565189(-03)$
$\tau_{8,3} = 8.38290467316815518626(-01)$	$T_{8,3} = 7.68911689633224355493(-03)$
$\tau_{8,2} = 6.36751818668873558588(-01)$	$T_{8,2} = 1.17618162931626914773(-02)$
$\tau_{8,1} = 4.63275049831781367266(-01)$	$T_{8,1} = 1.16475353004549600181(-02)$
$\tau_{16,3} = 9.82074559541336573504(-01)$	$T_{16,3} = 2.82876201937703049808(-03)$
$\tau_{16,2} = 9.04629974729031572483(-01)$	$T_{16,2} = 6.20356157958294744752(-03)$
$\tau_{16,1} = 7.57837070527892958867(-01)$	$T_{16,1} = 8.99101481902907866104(-03)$

$N = 8$

$\sigma_{4,8} = 9.99764490744675502157(-01)$	$S_{4,8} = 5.67941110155838311465(-04)$
$\sigma_{4,7} = 9.67256100033008607218(-01)$	$S_{4,7} = 3.30289515916698657603(-03)$
$\sigma_{4,6} = 8.90473272801725276940(-01)$	$S_{4,6} = 5.52079015806161839836(-03)$
$\sigma_{4,5} = 7.67529325907417756249(-01)$	$S_{4,5} = 7.10091610777722913611(-03)$
$\sigma_{4,4} = 5.94883919850400784035(-01)$	$S_{4,4} = 6.43861274578302233406(-03)$
$\sigma_{4,3} = 4.80128437817432541553(-01)$	$S_{4,3} = 8.84684511804295535797(-03)$
$\sigma_{4,2} = 2.92254872816954479513(-01)$	$S_{4,2} = 9.92112238555064595618(-03)$
$\sigma_{4.1} = 1.23391971333152181234(-01)$	$S_{4,1} = 9.65411464943065818002(-03)$
$\tau_{4,7} = 9.98869410600278636902(-01)$	$T_{4,7} = 6.28486903690406910313(-04)$
$\tau_{4,6} = 9.66240514969400640875(-01)$	$T_{4,6} = 3.30493342435324549546(-03)$
$\tau_{4,5} = 8.89289982913001897025(-01)$	$T_{4,5} = 5.52931573045200540030(-03)$
$\tau_{4,4} = 7.65695207742824259044(-01)$	$T_{4,4} = 7.15003698959518018970(-03)$
$\tau_{4,3} = 5.84139264909947412422(-01)$	$T_{4,3} = 7.57806761533915630873(-03)$
$\tau_{4,2} = 4.61081615117704509362(-01)$	$T_{4,2} = 8.47794461085056505554(-03)$
$\tau_{4,1} = 2.67649536743855674578(-01)$	$T_{4,1} = 1.11791301185964966333(-02)$
$\tau_{8,6} = 9.96866263986476252538(-01)$	$T_{8,6} = 7.86338156011656658143(-04)$
$\tau_{8,5} = 9.63570667872856569612(-01)$	$T_{8,5} = 3.30986900237964102114(-03)$
$\tau_{8,4} = 8.86299545209185581134(-01)$	$T_{8,4} = 5.53838381319441309798(-03)$
$\tau_{8,3} = 7.61624026237497729331(-01)$	$T_{8,3} = 7.21693418400025453315(-03)$
$\tau_{8,2} = 5.66152062442431667402(-01)$	$T_{8,2} = 9.68944561308410059758(-03)$
$\tau_{8,1} = 4.08241324748083122261(-01)$	$T_{8,1} = 9.21044276181345164656(-03)$
$\tau_{16,4} = 9.87174496088300047304(-01)$	$T_{16,4} = 2.03139270169502181259(-03)$
$\tau_{16,3} = 9.32294379264438078720(-01)$	$T_{16,3} = 4.48694894495153835309(-03)$
$\tau_{16,2} = 8.32405393768712739852(-01)$	$T_{16,2} = 6.42992767288510199758(-03)$
$\tau_{16,1} = 6.80745888778993107871(-01)$	$T_{16,1} = 7.87573570851507649858(-03)$

The computation of this table has been performed so that rational arithmetic has been applied as far as possible in order to reduce the influence of round-off errors. In this way all polynomials, whose zeros are the radii $\tau_{\nu\rho}$ or $\sigma_{\nu\rho}$ have been computed exactly with rational coefficients. Rounding errors entered all subsequent calculation.

Meanwhile, Ann Haegemans [14] has also calculated these formulae in her doctoral dissertation. The results differ by the multiplier π in the weights, because we have normed the integral.

2.3 The method of undetermined coefficients

We now turn to another method for the determination of the weights (and nodes) of a quadrature formula Q_nf, which is only identical with the Taylor-expansion method for certain special cases. In general this method, which is sometimes referred to as an algebraic method, gives a much simpler way of obtaining quadrature formulae with special properties.

2.3.1 Definition of the method

We choose $f \in C^k[-1, 1]$ or $C^k(\bar{\Omega})$ where k is a sufficiently large number. Then, by Taylor's theorem, we have expansions of the form

$$\begin{aligned} f(x) &= \sum_{\nu=0}^{k-1} a_\nu x^\nu + \text{error}, \\ f(x, y) &= \sum_{\nu=0}^{k-1} \sum_{i=0}^{\nu} a_{i,\nu} x^{\nu-i} y^i + \text{error}, \end{aligned} \tag{2.3.1}$$

and analogously for N dimensions. We have already mentioned that Q_nf or C_nf, derived using the Taylor-expansion method, integrates the beginning of an expansion like (2.3.1) exactly, but not the later terms. The same is true if we want *a priori* Q_nf or C_nf to integrate exactly a set of monomials such as $\{1, x, x^2, \ldots, x^\mu\}$ or $\Phi = \{1, x, y, x^2, xy, y^2, \ldots, x^\mu, x^{\mu-1}y, \ldots, y^\mu\}$ $\mu < k$. We then have

$$Q_n x^\nu = I x^\nu, \quad \text{so that} \quad E_n x^\nu = 0 \quad \text{for} \quad \nu = 0(1)\mu, \tag{2.3.2}$$

or

$$C_n \varphi_\nu = I_\Omega \varphi_\nu, \quad \text{so that} \quad E_n \varphi_\nu = 0 \quad \text{for} \quad \varphi_\nu \in \Phi. \tag{2.3.3}$$

Equations (2.3.2) and (2.3.3) are systems of equations for the determination of $\mu + 1$ or $\mu(\mu+1)/2$ unknown nodes, or weights, or both nodes and weights. Equation (2.3.3) immediately shows how to generalise the procedure, which is then defined as follows.

Definition 2.3.1 Let $\{\varphi_1, \varphi_2, \ldots, \varphi_\mu\}$ be a set of linearly independent functions which are defined and integrable in $[-1, 1]$ or Ω. Then the requirement that

$$E_n f = 0 \quad \text{for} \quad f \in \{\varphi_1, \varphi_2, \ldots, \varphi_\mu\}$$

serves to define nodes or weights or both by solving a linear or nonlinear system of equations. This procedure is called the *method of undetermined coefficients* or the *algebraic method* (cf. Remarks R.2.2 and R.2.5). ■

Example 2.3.1 Approximate $\underset{\infty}{I}\exp(-x^2) = \int_{-\infty}^{\infty} \exp(-x^2)\,dx = \sqrt{\pi}$ by a suitable quadrature formula $Q_3 f$. The integrand does not in any way possess a polynomial-like behaviour because $\underset{\infty}{I}x^k$ is senseless. We therefore choose functions with a form similar to the integrand, e.g. $\varphi_k(x) = 1/(x^2+k^2)$, $k = 1(1)\ldots$. We determine one node a and the weights of $Q_3 f = A_1 f(0) + A_2[f(a) + f(-a)]$ by requiring $E_3\varphi_k = 0$ for $k = 1, 2, 3$. Thus, the system $\pi/k = A_1/k^2 + 2A_2/(k^2+a^2)$, $k = 1, 2, 3$ has to be solved. The weights are eliminated so that $9/(9+a^2) - 8/(4+a^2) + 1/(1+a^2) = 0$ or

$$9(a^2+4)(a^2+1) - 8(a^2+9)(a^2+1) + (a^2+9)(a^2+4) = 2a(a^2-11) = 0.$$

Hence $a = 0$, which has already been used, and $a = \sqrt{11}$. We then calculate the weights A_1 and A_2 and

$$Q_3 f = \frac{6\pi}{11}[f(0) + 5(f(-\sqrt{11}) + f(\sqrt{11}))]$$

is the desired quadrature formula. Application to $\exp(-x^2)$ gives $Q_3(\exp(-x^2)) = 1.714$ while $\underset{\infty}{I}(\exp(-x^2)) = \sqrt{\pi} = 1.772$, so that $Q_3 f$ is a rather good approximation (cf. Remarks R.2.6 and R.2.7). ■

As has been said, the general method of undetermined coefficients often facilitates enormously the construction of quadrature and cubature formulae having special properties. Examples are the Ritz and Galerkin methods which involve trial functions from finite-dimensional spaces. We can use these trial functions to construct quadrature formulae which allow the trial functions or their products to be integrated exactly. In this way at least the error in the quadrature will be small. Much work remains to be done in this area of Finite Elements.

We now take $\varphi_\nu = x^{\nu-1}$, $\nu = 1(1)\ldots$ (analogously for $N > 1$ dimensions) in order to define some useful terms.

2.3.2 The leading error term—The algebraic degree of precision

Let us return to the expansions (2.3.1). Integration gives

$$E_nf = If - Q_nf = \sum_{\nu=\mu+1}^{k-1} a_\nu[Ix^\nu - Q_nx^\nu] + \text{error} = \sum_{\nu=\mu+1}^{k-1} a_\nu E_nx^\nu + \text{error}. \tag{2.3.4}$$

Definition 2.3.2 The first non-vanishing term of (2.3.4)

$$a_{\mu+1}E_n(x^{\mu+1})$$

is called the *leading error term* of the error expansion. ■

This term characterises the quadrature error of Q_nf if the coefficients a_ν decrease rapidly, which means that

$$|a_{\mu+1}| \gg |a_{\mu+2}| \gg |a_{\mu+3}| \gg \ldots, \tag{2.3.5}$$

and if the error is small enough.

Then the leading error term certainly does not provide an upper bound for $|E_nf|$ but an estimate which can be considered as an acceptable guide to the real value of $|E_nf|$.

The same is true for two-dimensional cubature if we replace $a_{\mu+1}x^{\mu+1}$ by $\sum_{i=0}^{\mu+1} a_{i,\mu+1}x^{\mu+1-i}y^i$. The extension to N dimensions is trivial. If we apply Q_nf to the interval $[0, h]$ instead of $[-1, 1]$, then the leading error term is

$$c_{\mu+1}h^{\mu+2} = O(h^{\mu+2}).$$

Thinking of h as a small positive number we sometimes say that Q_nf has the *order* $\mu+2$. This changes into $O(h^{\mu+N+1})$ for N dimensions. The dependence of the order on the dimension N disappears if we divide the leading error term by the area of the domain of integration h or $|\Omega|$. Then we obtain $O(h^{\mu+1})$. On the other hand, the method has been constructed so that polynomials of degree at most μ are integrated exactly.

Definition 2.3.3 If the quadrature error $E_nf = 0$ for $f \in \mathbb{P}_k$ and in addition there exists a polynomial $g \in \mathbb{P}_{k+1}$ with $E_ng \neq 0$, then Q_nf is said to have the *algebraic degree of precision* k. The same is valid for cubature formulae C_nf. (More concisely: *degree* or *degree of precision*.) ■

The degree of precision allows the comparison of formulae with respect to their degree. If the expansion (2.3.1) is valid for a certain class of functions, and the coefficients satisfy (2.3.5), then it is preferable to use formulae with as high a degree as possible for approximately integrating functions of this class. We often compare If with Q_nf or $I_\Omega f$ with C_nf without writing down the error term. We then write $If \doteq Q_nf$ or $I_\Omega f \doteq C_nf$ instead of $If = Q_nf + E_nf$ or $I_\Omega f = C_nf + E_nf$.

2.3.3 Determination of the leading error term

The method of undetermined coefficients is very suitable for the calculation of the leading error term of a qudrature or cubature formula.

For Q_nf having degree k it is necessary and sufficient to show that:

(i) $E_nx^\nu = 0$ for $\nu = 0(1)k$, which means that for every $f \in \mathbb{P}_k$ we have $E_nf = 0$.

(ii) $E_nx^{k+1} \neq 0$. If this were wrong, then Q_nf would have at least degree $k+1$, which would be a contradiction.

In two-dimensional cubature we have to test $x^{k+1}, x^ky, \ldots, xy^k, y^{k+1}$ instead of merely x^{k+1}. The extension to N dimensions is again trivial.

In order to see the meaning of the constants which come into the leading error term we assume that f has a convergent Taylor series expansion about the origin. Thus,

$$f(x) = \sum_{\nu=0}^{\infty} \frac{f^{(\nu)}(0)}{\nu!} x^\nu,$$

and

$$f(x, y) = \sum_{\nu=0}^{\infty} \frac{1}{\nu!} \sum_{\mu=0}^{\nu} \binom{\nu}{\mu} x^{\nu-\mu} y^\mu \frac{\partial^\nu f(\mathbf{0})}{\partial x^{\nu-\mu} \partial y^\mu}.$$

Then, for a k-th degree formula we have the leading error terms

$$\frac{f^{(k+1)}(0)}{(k+1)!} E_n(x^{k+1}) \quad \text{and} \quad \frac{1}{(k+1)!} \sum_{\mu=0}^{k+1} \binom{k+1}{\mu} f_{k+1-\mu,\mu}(\mathbf{0}) E_n(x^{k+1-\mu} y^\mu)$$

with corresponding formulae for N dimensions.

Example 2.3.2 Determine the leading error term of Simpson's rule. We apply the error functional $E_3f = If - (1/3)[f(-1) + 4f(0) + f(1)]$ to the monomials $1, x, x^2, \ldots,$. We see from symmetry that $E_3(x^{2k-1}) = 0$ for $k \in \mathbb{N}$ and, in addition, we find that $E_3(1) = E_3(x^2) = 0$, $E_3(x^4) = -4/15 \neq 0$. Hence $(-4/15)f^{(4)}(0)/4! = (-1/90)f^{(4)}(0)$ is the leading error term of Simpson's rule, and the degree of precision is 3. ■

Example 2.3.3 Let Ω be the square $|x| \leqslant h$, $|y| \leqslant h$, $r = \sqrt{2/3}\, h$.

$$C_4f = h^2 \sum_{j=0}^{3} f(r \cos(\varphi + \pi j/2), r \sin(\varphi + \pi j/2)), \quad \text{for} \quad 0 \leqslant \varphi < \pi/2$$

is a given cubature formula with four nodes. We immediately see that

$$E_4 f = 0 \quad \text{for} \quad f \in \{1, x, y, x^2, xy, y^2, x^3, x^2y, xy^2, y^3\},$$

and

$$E_4(x^4) = E_4(y^4) = \frac{4h^6}{5} - \frac{8h^6}{9}(\cos^4\varphi + \sin^4\varphi) = \frac{4h^6}{45}(-1 + 5\sin^2(2\varphi))$$

$$E_4(x^3y) = -E_4(xy^3) = \frac{8h^6}{9}(\cos^3\varphi \sin\varphi - \sin^3\varphi\cos\varphi) = \frac{2h^6}{9}\sin(4\varphi)$$

$$E_4(x^2y^2) = \frac{4}{9}h^6 - \frac{16h^6}{9}\cos^2\varphi\sin^2\varphi = \frac{4}{9}h^6\cos^2(2\varphi).$$

Hence, $C_4 f$ has degree of precision 3 because there exists no angle φ such that all five terms vanish together.

The leading error term is:

$$\frac{4h^6}{4!\,45}\left[(5\sin^2(2\varphi) - 1)(f_{4,0} + f_{0,4}) + \frac{5}{2}\sin(4\varphi)\left(\binom{4}{1}f_{3,1} - \binom{4}{3}f_{1,3}\right) + 5\cos^2(2\varphi)\binom{4}{2}f_{2,2}\right],$$

which is a function of φ that can still be minimised in some respects. ■

2.3.4 Highest possible degree of precision

The degree of precision of the Gauss quadrature formula

$$If \doteq f(1/\sqrt{3}) + f(-1/\sqrt{3}) \tag{2.3.6}$$

is easily seen to be 3. Because the corresponding error $E_2 f$ equals 0 for $f \in \{1, x, x^2, x^3, x^5, \ldots\}$ and $E_2(x^4) = 8/45$ the leading error term is $(1/135)f^{(4)}(0)$. On the other hand, $Q_2 f = f(1) + f(-1)$ has the leading error term $(-2/3)f''(0)$, so that its degree is 1. Therefore, although these two formulae have the same number of nodes, their degree differs by 2. Obviously this is due to the choice of the nodes. Hence, we can ask whether we can improve the degree of (2.3.6) by another choice of nodes. This is impossible because of:

Theorem 2.3.1 *The algebraic degree of precision of a quadrature formula with n nodes cannot exceed $2n - 1$.*

Proof Let $x_1 < x_2 < \ldots < x_n$ be the n distinct nodes of a quadrature formula $Q_n f$. Let $\omega_n(x)$ be the nodal polynomial $\omega_n(x) \equiv (x-x_1)(x-x_2)\ldots(x-x_n)$. Thus $\omega_n^2(x) \in \mathbb{P}_{2n}$. Clearly, $Q_n(\omega_n^2)=0$, but for $W(x) \geq 0$

$$I(\omega_n^2(x)) = \int_{-1}^{1} W(x)\omega_n^2(x)\,\mathrm{d}x > 0.$$

Hence $\omega_n^2(x)$ cannot be integrated exactly by $Q_n f$, so that $E_n(\omega_n^2(x) \neq 0$, and the proof is finished. ■

The Gaussian quadrature formulae reach exactly the degree $2n-1$. Further, Theorem 2.3.1 can be used to determine the nodes of Gaussian quadrature formulae (cf. Remark R.2.3). Let $Q_n f$ have the degree $2n-1$. Then $Q_n f = If$ for $f \in \mathbb{P}_{2n-1}$. Now every $\lambda_i(x) \equiv \omega_n(x)/(x-x_i) \in \dot{\mathbb{P}}_{n-1}$, and hence $\lambda_i(x)\omega_n(x) \in \dot{\mathbb{P}}_{2n-1}$. Thus,

$$E_n(\omega_n(x)\lambda_i(x)) = 0 \quad \text{or} \quad E_n(\omega_n^2/(x-x_i)) = 0, \qquad i = 1(1)n.$$

Because $Q_n(\omega_n(x)\lambda_i(x)) = 0$, $i = 1(1)n$, the nodes have to be chosen so that the system $I(\omega_n(x)\lambda_i(x)) = 0$, $i = 1(1)n$ is satisfied. This nonlinear system of equations contains only the nodes, not the weights of $Q_n f$. The problem of finding nodes and weights is thus split into two problems.

Polynomials $\omega_n(x) \in \dot{\mathbb{P}}_n$ satisfying the condition

$$\int_a^b W(x)\omega_n(x)p(x)\,\mathrm{d}x = 0$$

for every $p(x) \in \mathbb{P}_{n-1}$ are called *orthogonal* to $\mathbb{P}_{n-1}$ in $[a, b]$ with respect to $W(x)$. Comparing with the above, we see that orthogonal polynomials (or, more generally, orthogonal functions) will govern the theory of quadratures of highest possible degree of precision (cf. Chapters 5 and 7).

Theorem 2.3.1 cannot be extended in a straightforward manner to cubature problems, but the idea will be used to define orthogonal polynomials in two or more variables. This theory will be much more complicated because of the dependence on the dimension and on the shape of Ω (cf. Chapter 5). Nevertheless, we state:

Theorem 2.3.2 *Let $\mathbf{x}_1 \neq \mathbf{x}_2 \neq \ldots \neq \mathbf{x}_n$ be the n distinct nodes of a cubature formula, and let $\omega_\nu(\mathbf{x}) \in \dot{\mathbb{P}}_\nu$ be a polynomial such that $\omega_\nu(\mathbf{x}_i) = 0$ for $i = 1(1)n$. Then the degree of $C_n f$ cannot exceed $2\nu - 1$.*

The proof is analogous to that of Theorem 2.3.1. We have

$$\int_\Omega \omega_\nu^2(\mathbf{x})\,\mathrm{d}\Omega > 0 \quad \text{while} \quad C_n(\omega_\nu^2(\mathbf{x})) = 0.$$

Note that usually $\nu < n$.

Example 2.3.4 Let all n nodes of C_nf be situated on the hypersphere $(\mathbf{x}, \mathbf{x}) = r^2$. Taking $\omega_2(\mathbf{x}) \equiv (\mathbf{x}, \mathbf{x}) - r^2$ we have that $\omega_2^2 \in \dot{\mathbb{P}}_4$ and, by Theorem 2.3.2, the degree cannot exceed 3. ■

The problem of determining the minimal number n of nodes of a cubature formula C_nf having a prescribed degree is one of the most difficult problems in cubature, because of the dependence on the domain Ω and on the dimension as well as on the degree itself (cf. Chapter 5).

The simplest case is that of degree 1. We choose one node at the origin. Then $C_1f = |\Omega| f(\mathbf{0})$ is the corresponding cubature formula and it is identical with its Taylor expansion. On the other hand

$$I_\Omega f \doteq |\Omega| f(\mathbf{0}) + \sum_{i=1}^{N} \left(\int_\Omega x_i \, d\Omega \right) \frac{\partial f}{\partial x_i} \bigg|_{x_i=0}.$$

Hence, $C_1 f$ has degree 1 if and only if $\int_\Omega x_i \, d\Omega = 0$, $i = 1(1)N$, for every coordinate variable x_i. Remember now that, for $\mathbf{a}$ to be the centre of gravity of Ω, the equations $\int_\Omega (x_i - a_i) \, d\Omega = 0$, $i = 1(1)N$ must hold. Thus, when $\mathbf{a}$ is made to be the origin, the conditions are fulfilled.

The degree 2 is reached with $N+1$ suitably chosen nodes in general. For further details see Chapter 5, Stroud (book, p. 80 ff.) [12], [13], [18] and [11], where Haber and Flatto construct an N-dimensional domain with an $(N+1)$-point cubature formula of degree 3.

2.3.5 Some special results for cubatures

Let Ω be planar with centre of gravity $\mathbf{0}$ and take the nodes on the circles $(\mathbf{x}, \mathbf{x}) \equiv x^2 + y^2 = r_i^2$, $i = 1(1)k$. Then the degree cannot exceed $4k - 1$. This is seen as in the proof of Theorem 2.3.1.

Let $\omega_k(\mathbf{x}) \equiv \prod_{i=1}^{k} (x^2 + y^2 - r_i^2)$. Then $\omega_k^2(\mathbf{x}) \in \dot{\mathbb{P}}_{4k}$ and $C_n(\omega_k^2) = 0$. Hence, $E_n(\omega_k^2) = I_\Omega(\omega_k^2) \neq 0$ and the degree cannot exceed $4k - 1$.

Let $\lambda_i(\mathbf{x}) = \omega_k(\mathbf{x}) / [(\mathbf{x}, \mathbf{x}) - r_i^2]$, $i = 1(1)k$. Then $C_n(\lambda_i \omega_k) = 0$ for $i = 1(1)k$. For C_nf to have the degree $4k - 2$ the equations $E_n(\lambda_i \omega_k) = I_\Omega(\lambda_i \omega_k) = 0$, $i = 1(1)k$, have to be valid. This is a nonlinear system for the k unknown radii r_i, but we cannot say whether the system possesses a unique solution or even a real solution.

For $k = 1$ we obtain $E_n(\omega_1) = I_\Omega(x^2 + y^2 - r_1^2) = 0$. This leads to $I_\Omega(x^2) + I_\Omega(y^2) = r_1^2 I_\Omega 1 = r_1^2 |\Omega|$. Hence, $r_1 = (I_\Omega(x^2 + y^2)/|\Omega|)^{\frac{1}{2}}$. Analogously in N dimensions we have that $r_1 = (I_\Omega(x_1^2 + \ldots + x_N^2)/|\Omega|)^{\frac{1}{2}}$. Note that, although this result does not depend on the number n of nodes, n has to be so large that C_nf has the degree under discussion.

The extension of the above to N dimensions is straightforward. We thus have:

Theorem 2.3.3 *A necessary condition for C_nf, whose nodes $\mathbf{x}_j$ satisfy $(\mathbf{x}_j, \mathbf{x}_j) = r_i^2$ for $i = 1(1)k$, to have degree $4k-2$ is $I_\Omega(\lambda_i\omega_k) = 0$, $i = 1(1)k$, where $\omega_k = \prod_{i=1}^{k} [(\mathbf{x}, \mathbf{x}) - r_i^2]$ and $\lambda_i = \omega_k/[(\mathbf{x}, \mathbf{x}) - r_i^2]$, $i = 1(1)k$*

As an illustration we give some examples for $N = 2$.

Example 2.3.5

(i) Let Ω be the unit sphere $x^2 + y^2 \leq 1$. Then (cf. Problem P.2.15) $r_1 = 1/\sqrt{2}$, because $|\Omega| = \pi$ and $I_\Omega(x^2 + y^2) = \dfrac{\pi}{2}$.

(ii) Let Ω be the square $-1 \leq x$, $y \leq 1$. Then $r_1 = \sqrt{2/3}$ because $|\Omega| = 4$, and $I_\Omega(x^2 + y^2) = 8/3$.

(iii) Let Ω be the regular hexagon of Problem P.2.9 with $h = 1$. Then $r_1 = \sqrt{5/3}/2$ because $|\Omega| = 3\sqrt{3}/2$, and $I_\Omega(x^2 + y^2) = 5\sqrt{3}/8$.

(iv) Let Ω be the triangle in Section 2.2.5 with $h = 1/2$. Then $r_1 = 1/2$ because $|\Omega| = 3\sqrt{3}/4$, and $I_\Omega(x^2 + y^2) = 3\sqrt{3}/16$.

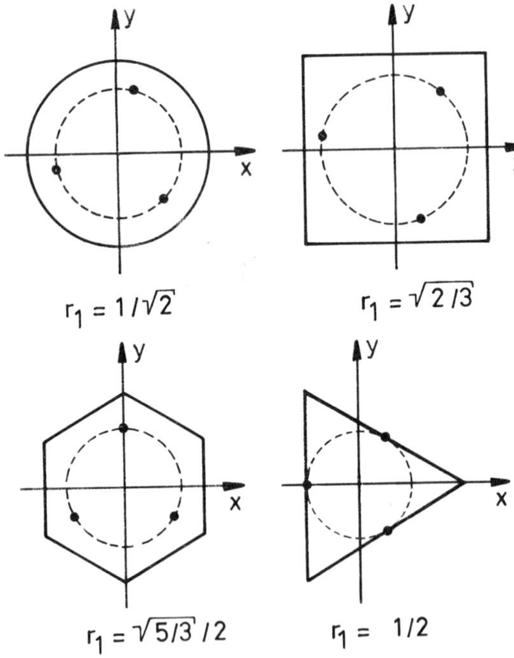

FIG. 2.3.1. Node distribution for four planar regions.

These cubature formulae

$$C_3 f = \frac{|\Omega|}{3} \sum_{j=1}^{3} f\left(r_1 \cos\left(\varphi + \frac{2\pi}{3} j\right), r_1 \sin\left(\varphi + \frac{2\pi}{3} j\right)\right)$$

with arbitrary φ have degree 2. The degree can be raised from 2 to 3 if there are four regularly-distributed nodes on the same circles for hexagon, square, and circle. We then have

$$C_4 f = \frac{|\Omega|}{4} \sum_{j=1}^{4} f(r_1 \cos(\varphi + \pi j/2), r_1 \sin(\varphi + \pi j/2))$$

with arbitrary φ having degree 3.

For the triangle the same is not true because the momentum integrals $I_{3,0}$ and $I_{1,2}$ in the expansion of $I_\Omega f$ do not vanish. We need additional nodes, in order to reach degree 3, except for $\varphi = 0$ where $C_3 f$ already has degree 3. ■

It is interesting to ask whether or not the degree changes under a variable transform. It is impossible to give an exhaustive answer to this question except for the special case of linear transformations:

Theorem 2.3.4 (Hammer and Wymore [16]) *The algebraic degree of precision of a cubature formula is invariant under linear non-singular transforms T of the independent variables.*

Proof Let

$$\left.\begin{aligned} C_m f &\equiv \sum_{i=1}^{m} A_i f(\mathbf{x}_i) \doteq I_\Omega f(\mathbf{x}) \\ \tilde{C}_m f &\equiv \sum_{i=1}^{m} B_i f(\mathbf{y}_i) \doteq I_{T\Omega} f(\mathbf{y}) \end{aligned}\right\} \quad \text{with} \quad \mathbf{y} = T\mathbf{x} = \mathbf{A}\mathbf{x} + \mathbf{b},$$

where $T:\Omega \to T\Omega$, and the inverse $\mathbf{A}^{-1}$ of $\mathbf{A}$ exists because $\mathbf{A}$ is not singular. The Jacobian matrix $\mathbf{J}$ of T has constant elements; i.e. $|\mathbf{J}| = \det \mathbf{A} = \text{const}$. Let us compare C_m with $\tilde{C}_m$:

For $p_1, p_2, \ldots, p_n \in \mathbb{N}_0$ with $\sum_{j=1}^{n} p_j \leq k$, we have that

$$P(\mathbf{y}) \equiv \prod_{j=1}^{n} y_j^{p_j} = \prod_{j=1}^{n} \left(\sum_{i=1}^{n} a_{ij} x_i + b_j\right)^{p_j} \equiv \overset{*}{P}(\mathbf{x}) \in \mathbb{P}_k$$

where a_{ij} are the elements of the matrix $\mathbf{A}$. On applying T, we obtain for the integral

$$I_{T\Omega}P(\mathbf{y}) = \int_{T\Omega} P(\mathbf{y})\, dy_1 \ldots dy_n = |\mathbf{J}| \int_{\Omega} \overset{*}{P}(\mathbf{x})\, dx_1 \ldots dx_n = |\mathbf{J}| I_\Omega \overset{*}{P}(\mathbf{x}),$$

and correspondingly for the cubature formulae

$$\tilde{C}_m P(\mathbf{y}) = \sum_{i=1}^{m} B_i P(\mathbf{y}_i) = |\mathbf{J}| \sum_{i=1}^{m} A_i \overset{*}{P}(\mathbf{x}_i) = |\mathbf{J}| C_m \overset{*}{P}(\mathbf{x}),$$

hence

$$B_i = |\mathbf{J}| A_i \quad \text{for} \quad i = 1(1)m.$$

From $I_\Omega f(\mathbf{x}) = C_m f(\mathbf{x})$ for $f \in \mathbb{P}_k$ it always follows that $I_{T\Omega} f(\mathbf{y}) = \tilde{C}_m f(\mathbf{y})$ for $f \in \mathbb{P}_k$. It can be shown analogously that there exists a $Q \in \dot{\mathbb{P}}_{k+1}$ such that $I_\Omega Q(\mathbf{x}) \neq C_m Q(\mathbf{x})$ and $I_{T\Omega} Q(\mathbf{y}) \neq \tilde{C}_m Q(\mathbf{y})$ and the proof is finished. ■

2.4 The method of generating functions

As the third and last analytic method for the determination of quadrature formulae we treat the method of generating functions, which is very closely connected with the construction of equally-weighted quadrature formulae.

2.4.1 Equally-weighted quadrature formulae

Another approach with quadrature formulae is to prescribe the weights and then to determine the corresponding nodes. Assuming that the integrand function is derived from measurements and is therefore erratic, and assuming further that the average error due to the measurement is constant, it is natural to seek formulae with constant weights. Such formulae do not diminish or increase these errors. We therefore arrive at formulae of the form $Q_n f = A \sum_{i=1}^{n} f(x_i)$ which we call *equally-weighted* quadrature formulae.

Such a formula has degree n if $E_n x^k = 0$ for $k = 0(1)n$. From $E_n 1 = 0$ we conclude that

$$nA = \int_{-1}^{1} W(x)\, dx \quad \text{or} \quad A = \frac{2}{n} \quad \text{for} \quad W(x) \equiv 1.$$

The leading error term then is $\dfrac{1}{(n+1)!} C_{n+1} f^{(n+1)}(0)$.

The system $E_n x^k = 0$, $k = 1(1)n$ can be transformed into a polynomial $\omega_n(x)$ whose zeros are the nodes, as was described in Section 2.2. There is, however, a more elegant and simple way of determining $\omega_n(x)$, which dates back to Tschebyscheff (cf. Remark R.2.12), and so the problem of constructing equally-weighted quadrature formulae is often denoted as *Tschebyscheff's quadrature problem.* We present a construction which is very similar to that of Tschebyscheff.

2.4.2 Construction of the nodal polynomial

Let us apply Q_n to the integral transform

$$g(z)=\int_{-1}^{1}\ln(1-xz)\,dx=\sum_{\nu=1}^{\infty}\frac{-z^{2\nu}}{\nu(2\nu+1)}\quad\text{for}\quad |z|<1,$$

using the series expansion of $\ln(1-t)$. We then obtain

$$g(z)\doteq\frac{2}{n}\sum_{i=1}^{n}\ln(1-x_iz)=\frac{2}{n}\ln[(1-x_1z)(1-x_2z)\dots(1-x_nz)].$$

Hence, because $\omega_n\equiv(x-x_1)\dots(x-x_n)$ we have that $g(z)\doteq\frac{2}{n}\ln\left[z^n\omega_n\left(\frac{1}{z}\right)\right]$. Taking into account that $\ln(1-xz)=-\sum_{k=1}^{\infty}\frac{1}{k}(xz)^k$ for $|xz|<1$, the quadrature error is $O(z^{n+1})$. Hence,

$$\begin{aligned}z^n\omega_n\left(\frac{1}{z}\right)&=\exp\left[\frac{n}{2}\int_{-1}^{1}\ln(1-xz)\,dx+O(z^{n+1})\right]\\&=\exp\left[-\frac{n}{2}\left[\frac{z^2}{3}+\frac{z^4}{10}+\frac{z^6}{21}+\dots\right]+O(z^{n+1})\right]\\&=\sum_{\mu=0}^{\infty}\frac{1}{\mu!}\left[-\sum_{\nu=1}^{\infty}\frac{nz^{2\nu}}{2\nu(2\nu+1)}+O(z^{n+1})\right]^{\mu}\\&=1-\frac{n}{6}z^2+\frac{5n-18}{360}nz^4+\dots+O(z^{n+1}).\end{aligned}$$

For special values of n we obtain, by comparing with $z^n\omega_n\left(\frac{1}{z}\right)=\prod_{i=1}^{n}(1-x_iz)$, Table 2.4.1.

For $n=8$ and $n>9$ the polynomials $\omega_n(x)$ have some complex nodes. The resulting quadrature formulae are rather useless in practice. Once the nodes are determined Q_nf is completely defined. We have collected some equally-weighted quadrature formulae for several weight functions in Section 2.4.3. Usually there exist only a few formulae with real nodes. For the weight function $W(x)=(1-x^2)^{-1/2}$, however, there exists a real quadrature formula Q_nf for every $n\in\mathbb{N}$, having the degree of precision $2n-1$. Therefore, these formulae belong to the class of Gaussian quadrature formulae (cf. Chapter 7).

A detailed discussion of the properties of Tschebyscheff quadrature formulae is given in Chapter 6. Further, see Remark R.2.8.

TABLE 2.4.1

n	To compare $z^n\omega_n\left(\frac{1}{z}\right)=\prod_{i=1}^{n}(1-x_iz)$	$\omega_n(x)=(x-x_1)(x-x_2)\ldots(x-x_n)$
1	$1=1-x_1z$, hence $x_1=0$	x because of $x_1=0$
2	$1-\frac{z^2}{3}=(1-x_1z)(1-x_2z)$	$x^2-\frac{1}{3}$
3	$1-\frac{z^2}{2}=(1-x_1z)\ldots(1-x_3z)$	$x^3-\frac{1}{2}x$
4	$1-\frac{2}{3}z^2+\frac{1}{45}z^4=(1-x_1z)\ldots(1-x_4z)$	$x^4-\frac{2}{3}x^2+\frac{1}{45}$
5	etc.	$x^5-\frac{5}{6}x^3+\frac{7}{72}x$
6		$x^6-x^4+\frac{1}{5}x^2-\frac{1}{105}$
7		$x^7-\frac{7}{6}x^5+\frac{119}{360}x^3-\frac{149}{6480}x$
9		$x^9-\frac{3}{2}x^7-\frac{27}{40}x^5-\frac{57}{560}x^3+\frac{53}{22400}$

2.4.3 Tables of equally-weighted quadrature formulae

TABLE 2.4.2

$$\int_{-1}^{1} f(x)\,dx \doteq \frac{2}{n}\sum_{i=1}^{n} f(x_i)$$

n	x_i
1	0
2	$\pm 1/\sqrt{3}$
3	$0;\ \pm 1/\sqrt{2}$
4	$\pm((1\pm 2/\sqrt{5})/3)^{\frac{1}{2}}$
5	$0;\ \pm((5\pm\sqrt{11})/12)^{\frac{1}{2}}$
6	±0.86624 68181 07820 59138 3598 ±0.42251 86537 61111 52911 8546 ±0.26663 54015 16704 72033 1534
7	±0.88386 17007 58049 03570 4224 ±0.52965 67752 85156 81138 5048 ±0.32391 18105 19907 63751 9673 0
9	±0.91158 93077 28434 47366 4949 ±0.60101 86553 80238 07142 8128 ±0.52876 17830 57879 99326 0181 ±0.16790 61842 14803 94306 8031 0

Complex nodes for $n=8$ and $n>9$.

TABLE 2.4.3

$$\int_{-1}^{1} (1-x^2)^{\frac{1}{2}} f(x)\,dx \doteq \frac{\pi}{2n} \sum_{i=1}^{n} f(x_i)$$

n	x_i
1	0
2	$\pm 1/2$
3	$0; \pm\frac{1}{2}\sqrt{\frac{3}{2}}$
4	$0; 0; \pm 1/\sqrt{2}$
5	$0; \pm((5\pm\sqrt{15})/16)^{\frac{1}{2}}$

Complex nodes for $n \geqslant 6$. Note the double node for $n = 4$.

TABLE 2.4.4

$$\int_{-1}^{1} x^2 f(x)\,dx \doteq \frac{2}{3n} \sum_{i=1}^{n} f(x_i)$$

n	x_i
1	0
2	$\pm\sqrt{3/5}$
3	$0; \pm 3/\sqrt{10}$
4	$\pm(\frac{3}{5}\pm\frac{2}{5}\sqrt{\frac{3}{7}})^{\frac{1}{2}}$
6	$\pm 0.94100\ 74694\ 79100\ 78740\ 6942$ $\pm 0.81334\ 08270\ 89980\ 08415\ 4968$ $\pm 0.50297\ 28038\ 10629\ 82373\ 3785$

Complex nodes for $n = 5$, $n \geqslant 7$.

2.5 Interpolatory quadratures and cubatures

The most important and systematic strategy for the construction of quadrature and cubature formulae is taken from approximation:

Approximate the integrand $f(x)$ using only functions which are analytically integrable and then integrate this approximation instead of the original $f(x)$.

We consider such approximations to $f(x)$ which are *interpolants* to $f(x)$, i.e. satisfy *exact* values at a set of points. Pure approximation, for example by Bernstein polynomials, is treated in Section 3.5.

2.5.1 General Lagrange interpolation

Let $\lambda_i(x)$, $i = 1(1)n$ be a set of functions satisfying

$$\lambda_j(x_n) = \delta_{j,n}. \tag{2.5.1}$$

Then $L_n(f, x) = \sum_{i=1}^{n} \lambda_i(x) f(x_i)$ has the *interpolating property* that $L_n(f, x_k) = f(x_k)$, $k = 1(1)n$, provided that x_i are n given distinct nodes. We now integrate $L_n(f, x)$ in order to obtain a quadrature formula $Q_n f$ and, in order to do this, we have to make the condition that the functions $\lambda_i(x)$ are integrable with respect to the weight function $W(x)$; that is

$$Q_n f = \int_{-1}^{1} W(x) L_n(f, x)\,dx = \sum_{i=1}^{n} A \cdot f(x_i),$$

with

$$A_i = \int_{-1}^{1} W(x) \lambda_i(x)\,dx, \qquad i = 1(1)n.$$

Clearly, we have to impose further conditions on the Lagrange coefficients or *cardinal functions* $\lambda_i(x)$ if we wish to prove general properties of $Q_n f$ or $L_n(f, x)$; for example, if we wish to determine the *error of interpolation* of $e_n(f, x) = f(x) - L_n(f, x)$.

A very efficient way for obtaining suitable $\lambda_i(x)$ of this type is the following.

Choose a *nodal function* $\omega_n(x)$ and n *basis functions* $h_i(x)$ which are defined at least in an open interval around every node, and which have at least a first derivative at every node, satisfying

$$\left.\begin{array}{ll} \omega_n(x_i) = 0, & 0 < |\omega_n'(x_i)| < \infty \\ h_i(x_i) = 0, & 0 < |h_i'(x_i)| < \infty \end{array}\right\} \quad \text{for} \quad i = 1(1)n$$
$$h_i(x_k) \neq 0 \quad \text{for} \quad i \neq k.$$

The Lagrange coefficients are then obtained by making

$$\lambda_i(x) = \frac{\omega_n(x) h_i'(x_i)}{\omega_n'(x_i) h_i(x)}, \qquad i = 1(1)n,$$

since these satisfy (2.5.1) as is seen by l'Hospital's theorem. If $\lambda_i(x)$ are integrable with respect to $W(x)$ in $[-1, 1]$, a quadrature formula is obtained by integration of the corresponding $L_n(f, x)$. The simplest choice of $\omega_n(x)$ is $\omega_n(x) = \prod_{i=1}^{n} h_i(x)$. The functions $\omega_n(x)$ and $h_i(x)$ are now specified in various ways.

2.5.2 Special cases

Lagrange polynomial interpolation

The oldest and perhaps the most important kind of interpolation is *polynomial interpolation.* This means that the interpolant $L_n(f, x)$ is a polynomial. The nodal function $\omega_n(x)$ is chosen to be the nodal polynomial $\omega_n(x) = \prod_{j=1}^{n} (x - x_j)$ now and $h_i(x) = x - x_i$, $i = 1(1)n$. Then

$$\lambda_i(x) = \prod_{\substack{j=1 \\ \neq i}}^{n} \frac{x - x_j}{x_i - x_j}, \qquad i = 1(1)n,$$

are the well-known Lagrange coefficients of the corresponding Lagrange interpolating *polynomial* $L_n(f, x) = \sum_{i=1}^{n} \lambda_i(x) f(x_i)$. If the nodes are chosen to be equidistant $x_i = -1 + 2(i-1)/(n-1)$, $i = 1(1)n$, then integration with $W(x) \equiv 1$ gives the famous *Newton–Cotes quadrature* formulae, which are treated in detail in Chapter 6.

Rational interpolation

Choosing n different nodes x_i with $|x_i| < 1$ and the nodal function

$$\omega_n(x) = \sum_{j=1}^{n} \frac{x - x_j}{1 - x x_j}, \qquad h_i(x) = x - x_i, \qquad i = 1(1)n$$

we obtain cardinal functions

$$\lambda_i(x) = \left(\prod_{j=1}^{n} \frac{1 - x_i x_j}{1 - x x_j} \right) \left(\prod_{\substack{j=1 \\ \neq i}}^{n} \frac{x - x_j}{x_i - x_j} \right), \qquad i = 1(1)n,$$

and the corresponding $Q_n f$ are related to Wilf's quadrature procedure, (cf. Section 7.5).

Interpolation along the real axis

Let $\{x_i\} = \{0; \pm h; \pm 2h; \ldots\}$, and

$$\omega_h(x) = \sin(\pi x / h), \qquad h_i = x - ih, \qquad i = 0(\pm 1) \pm \infty.$$

Then

$$\lambda_i(x)=\frac{(-1)^i h \sin(\pi x/h)}{\pi(x-ih)}, \qquad i=0(\pm 1)\pm\infty$$

are *congruent* cardinal functions, leading to

$$L_h(f,x)=\frac{h}{\pi}\sum_{i=-\infty}^{\infty}\frac{(-1)^i \sin(\pi x/h)}{x-ih}f(ih),$$

provided that the sum makes sense. The corresponding quadrature formula $Q_h f$ with $W(x)\equiv 1$ is a rectangle rule with infinitely many terms (cf. Section 7.6.2).

Trigonometric interpolation

Let $h_i(x)=\sin(\pi(x-x_i)/2)$, and $\omega_n(x)=\prod_{i=1}^{n} h_i(x)$, then for $|x_i|<1$ we obtain

$$\lambda_i(x)=\frac{\pi\omega_n(x)}{2\omega_n'(x_i)\sin(\pi(x-x_i)/2)}, \qquad i=1(1)n.$$

This is simplified for the case of $2n+1$ equidistant nodes $x_i=-1+2i/(2n+1)$, $i=0(1)2n$, to give

$$\lambda_i(x)=\frac{\sin\left(\frac{2n+1}{2}\pi\left(x-\frac{2n-2i+1}{2n+1}\right)\right)}{(2n+1)\sin\left(\frac{\pi}{2}\left(x-\frac{2n-2i+1}{2n+1}\right)\right)}, \qquad i=1(1)n.$$

Integration for $W(x)\equiv 1$ leads to $R_n f$ which has certain optimality properties for periodic functions (cf. Section 7.6.1). Further, see Remark R.2.11.

Polynomial spline interpolation

The Lagrange coefficients $\lambda_i(x)$ take on a most simple form which is easy to handle, if $\lambda_i(x)$ have a local support, i.e. each $\lambda_i(x)$ vanishes identically except in a "small" subinterval of $[-1, 1]$. The simplest case arises if $\lambda_i(x)$ are the roof functions

$$\lambda_i(x)=\begin{cases}\dfrac{x-x_{i-1}}{x_i-x_{i-1}} & \text{for} \quad x\in[x_{i-1},x_i]\\[2ex] \dfrac{x-x_{i+1}}{x_i-x_{i+1}} & \text{for} \quad x\in[x_i,x_{i+1}]\\[2ex] 0 & \text{otherwise}\end{cases} \qquad i=1(1)n.$$

These functions are uniquely determined if we require that $x_0<-1$, $x_1=-1$, $x_n=1$, $x_{n+1}>1$, where x_0 and x_{n+1} are unimportant fictitious nodes. Clearly, the corresponding $L_n(f, x)$ is the inscribed polygon of $f(x)$. Therefore, $L_n(f, x)\in C^0[-1, 1]$ is usually called the *linear interpolating spline* of $f(x)$.

Generally, we call $\lambda_i(x)$ a polynomial spline of degree k, if $\lambda_i(x)\in\mathbb{P}^k$ for each (x_i, x_{i+1}), $i=1(1)n-1$. Then, in general, we have that $\lambda_i(x)\notin\mathbb{P}^k[-1, 1]$ and $\lambda_i(x)\notin C^0[-1, 1]$.

We now consider cubic splines and define the restriction of $\lambda_i(x)$ to $[x_i, x_{i+1}]$ to be

$$\lambda_j(x)\,|_{[x_i,x_{i+1}]}=a_{i,j}+b_{i,j}(x-x_i)+c_{i,j}(x-x_i)^2+d_{i,j}(x-x_i)^3, \quad \begin{cases} i=1(1)n-1 \\ j=1(1)n, \end{cases}$$

where $x_1=-1$ and $x_n=1$. The natural conditions to be satisfied by the $\lambda_j(x)\in C^0[-1, 1]$ are the *interpolating conditions*

$$\begin{aligned} \lambda_j(x_i) &= \delta_{j,i}=a_{i,j} \qquad i=1(1)n-1, \\ \lambda_j(x_n) &= \delta_{j,n}=a_{n-1,j}+b_{n-1,j}(x_n-x_{n-1}) \\ &\qquad +c_{n-1,j}(x_n-x_{n-1})^2+d_{n-1,j}(x_n-x_{n-1})^3, \end{aligned} \tag{2.5.2}$$

and the *continuity conditions* for the interior nodes:

$$\lambda_j\,|_{[x_{i-1},x_i]}(x_i)=\lambda_j\,|_{[x_i,x_{i+1}]}(x_i), \qquad i=2(1)n-1,$$

which give the equation

$$a_{i-1,j}+b_{i-1,j}(x_i-x_{i-1})+c_{i-1,j}(x_i-x_{i-1})^2+d_{i-1,j}(x_i-x_{i-1})^3=a_{i,j}, \quad i=2(1)n-1. \tag{2.5.3}$$

These are $2n-2$ conditions for the $4n-4$ coefficients of $\lambda_j(x)$. There remain $2n-2$ degrees of freedom for their determination. Hence

$$\lambda_i(x)=\begin{cases} \left(\dfrac{x-x_{i-1}}{x_i-x_{i-1}}\right)^3 & \text{for } x\in[x_{i-1}, x_i] \\ \left(\dfrac{x-x_{i+1}}{x_i-x_{i+1}}\right)^3 & \text{for } x\in[x_i, x_{i+1}] \\ 0 & \text{otherwise} \end{cases} \qquad \begin{array}{l} i=1(1)n,\ x_0<-1,\ x_{n+1}>1 \\ \text{arbitrary} \end{array}$$

is one possibility for a set of $\lambda_i(x)\in C^0[-1, 1]$ having a local support. Further, there exist $\lambda_i(x)\in C^1[-1, 1]$ with local support. In order to construct such a set, we use (2.5.2) and (2.5.3) and, in addition, impose the *smoothness conditions*

$$\lambda_j'\,|_{[x_{i-1},x_i]}(x_i)=\lambda_j'\,|_{[x_i,x_{i+1}]}(x_i), \qquad i=2(1)n-1,$$

or

$$b_{i-1,j}+2c_{i-1,j}(x_i-x_{i-1})+3d_{i-1,j}(x_i-x_{i-1})^2=b_{i,j}, \qquad (2.5.4)$$
$$i=2(1)n-1.$$

We now have $3n-4$ conditions for the $4n-4$ unknown coefficients of $\lambda_j(x)$. The remaining n degrees of freedom make it possible to achieve a *local* support of $\lambda_j(x)$. For equidistant nodes $x_{i+1}=x_i+h$, $i=1(1)n-1$, we find, for example,

$$\lambda_i(x)=\begin{cases}\dfrac{1}{h^3}\{3h(x-x_{i-1})^2-2(x-x_{i-1})^3\} & \text{for} \quad x\in[x_{i-1},x_i]\\ \dfrac{1}{h^3}\{3h(x-x_{i+1})^2+2(x-x_{i+1})^3\} & \text{for} \quad x\in[x_i,x_{i+1}]\\ 0 \quad \text{otherwise} & x_0=-1-h,\ x_{n+1}=1+h\end{cases}$$

for $i=1(1)n$.

If we wish to obtain cubic splines $\lambda_j(x)\in C^2[-1,1]$ we use (2.5.2), (2.5.3) and (2.5.4) and, in addition, impose the *continuous curvature conditions*

$$\lambda_j''|_{[x_{i-1},x_i]}(x_i)=\lambda_j''|_{[x_i,x_{i+1}]}(x_i), \qquad i=2(1)n-1$$

or

$$2c_{i-1,j}+6d_{i-1,j}(x_i-x_{i-1})=2c_{i,j}, \quad i=2(1)n-1. \qquad (2.5.5)$$

Hence we have $4n-6$ conditions for the $4n-4$ unknown coefficients of $\lambda_j(x)$. The remaining two degrees of freedom in general are not sufficient to obtain $\lambda_j(x)$ with local support. Usually, these are used to impose boundary conditions like

$$\lambda_j'(x_1)=\lambda_j'(x_n), \qquad \lambda_j''(x_1)=\lambda_j''(x_n)$$

or

$$\lambda_j''(x_1)=\lambda_j''(x_n)=0 \qquad (2.5.6)$$

or

$$\lambda_j'(x_1)+\alpha_1\lambda_j''(x_1)=\beta_1, \qquad \lambda_j'(x_n)+\alpha_2\lambda_j''(x_n)=\beta_2$$

with suitable constants α_1, α_2, β_1, β_2.

It may happen, that $\lambda_j(x)\in\mathbb{P}^3$ for suitable boundary conditions which are different from (2.5.6). This is not possible for (2.5.6), because then $\lambda_j''(x)$ has two zeros, which cannot be true.

Thus, after determination of the coefficients of $\lambda_j(x)$ we have the *cardinal splines* $\lambda_j(x)$ satisfying $\lambda_j(x_k)=\delta_{j,k}$, and the corresponding $L_n(f,x)\in C^2[-1,1]$ is the cubic interpolating spline. In general, it is not possible to obtain $L_n(f, x) \in C^3[-1, 1]$. The corresponding quadrature formulae which are obtained on integration of $L_n(f, x)$ require low-order differentiability of the integrand. There exist formulae having certain optimality properties, and which are connected with the name of Sard (cf. Section 6.3). Further, see Remark R.2.9.

2.5.3 The error of Lagrange interpolation

The error of the general Lagrange interpolating operator is defined by

$$e_n(f, x) \equiv f(x) - L_n(f, x),$$

and, because of the interpolating property, vanishes at every node x_i so that $e_n(f, x_i) = 0$, $i = 1(1)n$. The error in the quadrature formula $Q_n f = \int_{-1}^{1} W(x) L_n(f, x)\, \mathrm{d}x$ is thus

$$E_n f = \int_{-1}^{1} W(x) e_n(f, x)\, \mathrm{d}x.$$

Hence, we immediately obtain an expression for $E_n f$ if we succeed in constructing a representation of $e_n(f, x)$. To do this, we introduce a further node $t = x_{n+1}$ which is considered to be a fixed but undetermined parameter satisfying $x_i \neq t$ for $i = 1(1)n$. This is done using the formula

$$L_{n+1}(f, x) = L_n(f, x) + \frac{\omega_n(x)}{\omega_n(t)} (f(t) - L_n(f, t)) \tag{2.5.7}$$

which is built like a simple linear Newton interpolation. We immediately see, that

$$L_{n+1}(f, x_i) = f(x_i) \quad \text{for} \quad i = 1(1)n+1. \tag{2.5.8}$$

Subtracting (2.5.7) from the identity $f(x) \equiv f(x)$ we find that the interpolation error $e_{n+1}(f, x)$ satisfies

$$e_{n+1}(f, x) = e_n(f, x) - \frac{\omega_n(x)}{\omega_n(t)} e_n(f, t) = \frac{1}{\omega_n(t)} \begin{vmatrix} e_n(f, x) & \omega_n(x) \\ e_n(f, t) & \omega_n(t) \end{vmatrix}.$$

The corresponding quadrature error $E_n f$ satisfies

$$E_{n+1} f = E_n f - \int_{-1}^{1} \frac{e_n(f, t)}{\omega_n(t)} W(x) \omega_n(x)\, \mathrm{d}x.$$

Because of (2.5.8) $e_{n+1}(f, x)$ has at least $n+1$ zeros in $[-1, 1]$ and hence, by Rolle's theorem, $e'_{n+1}(f, x)$ has at least n zeros in $(-1, 1)$. Further applications of Rolle's theorem show that $e^{(n)}_{n+1}(f, x)$ has at least one zero

$\xi \in (-1, 1)$. We therefore have that

$$0 = e_{n+1}^{(n)}(f, \xi) = e_n^{(n)}(f, \xi) - \frac{\omega_n^{(n)}(\xi)}{\omega_n(t)} e_n(f, t),$$

and from this

$$e_n(f, t) = \omega_n(t) \frac{e_n^{(n)}(f, \xi)}{\omega_n^{(n)}(\xi)} = \frac{\omega_n(t)}{\omega_n^{(n)}(\xi)} [f^{(n)}(\xi) - L_n^{(n)}(f, \xi)], \qquad (2.5.9)$$

Equation (2.5.9) is defined for all $t \in [-1, 1]$ because for $t = x_i$ both sides vanish.

Equation (2.5.9) makes sense if $\omega_n^{(n)}(\xi) \neq 0$, and if $e_n(f, x)$ and $\omega_n(x) \in C^n[-1, 1]$ rather than $C^{n+1}[-1, 1]$. This being true, we obtain the quadrature error by integration (integrability assumed) so that

$$E_n f = \int_{-1}^{1} \frac{W(x)\omega_n(x)}{\omega_n^{(n)}(\xi)} [f^{(n)}(\xi) - L_n^{(n)}(f, \xi)] \, dx.$$

In general, this representation is neither easy to handle nor to estimate because $\omega_n^{(n)}$ and $L_n^{(n)}(f, x)$ can be very complicated. This influences the choice of $\omega_n(x)$, which is therefore usually done so that $\omega_n(x) \in C^k[-1, 1]$, $k \geq n$, and $\omega_n^{(n)}(x) \neq 0$ for $x \in [-1, 1]$. The last condition is simplest if $\omega_n^{(n)}(x)$ is constant so that $\omega_n \in \dot{\mathbb{P}}_n$. This form appears for $\omega_n(x) = \prod_{i=1}^{n} (x - x_i)$ so that $\omega_n^{(n)} = n!$. Further, for $h_i(x) = x - x_i$, $i = 1(1)n$, we have that $\lambda_j(x) \in \mathbb{P}_{n-1}$ and thus $L_n^{(n)}(f, x) \equiv 0$. Hence, for polynomial Lagrange interpolation we obtain the error representation $e_n(f, x) = \dfrac{1}{n!} \omega_n(x) f^{(n)}(\xi)$ and the quadrature error is

$$E_n f = \int_{-1}^{1} \frac{1}{n!} \omega_n(x) f^{(n)}(\xi) \, dx \quad \text{for} \quad f \in C^n[-1, 1].$$

2.5.4 General Hermite interpolation

A more sophisticated form of interpolation is that due to Hermite in which not only function values but also derivative values are interpolated at the nodes. We limit ourselves here to Hermite interpolation with only function f and first derivative f' values. Further, see Remark R.2.10.

This means, we have to determine Hermite coefficients or cardinal functions $\mu_i(x)$ and $\nu_i(x)$ such that

$$\begin{aligned} \mu_i(x_k) &= \delta_{i,k}, & \nu_i(x_k) &= 0, \\ \mu_i'(x_k) &= 0, & \nu_i'(x_k) &= \delta_{i,k}, \end{aligned} \qquad i, k = 1(1)n.$$

If this is valid, and $\mu_i(x)$, $\nu_i(x)$ are differentiable in an open interval around each node, then the mapping

$$H_n(f, x) = \sum_{i=1}^{n} (\mu_i(x)f(x_i) + \nu_i(x)f'(x_i))$$

makes sense and $H_n(f, x)$ has the interpolating properties that

$$H_n(f, x_k) = f(x_k), \qquad H'_n(f, x_k) = f'(x_k), \qquad k = 1(1)n.$$

Let us assume that the Hermite coefficients are integrable with respect to the weight function $W(x)$. Then

$$\int_{-1}^{1} W(x)H_n(f, x)\,dx = \sum_{i=1}^{n} (A_i f(x_i) + B_i f'(x_i)) \equiv Q_n f$$

is a quadrature formula where

$$A_i = I\mu_i(x) = \int_{-1}^{1} W(x)\mu_i(x)\,dx, \qquad B_i = I\nu_i(x) = \int_{-1}^{1} W(x)\nu_i(x)\,dx.$$

It may happen, that $B_i = 0$ for $i = 1(1)n$. Then $Q_n f$ is simplified to its normal form. We shall see that this behaviour is characteristic for Gauss-type quadratures (see Chapter 7).

Let us now construct the Hermite coefficients $\mu_i(x)$, $\nu_i(x)$. Remember that the Lagrange coefficients $\lambda_i(x)$ have the property $\lambda_i(x_k) = \delta_{i,k}$; hence, also

$$\lambda_i^2(x_k) = \delta_{i,k} \quad \text{and} \quad [\lambda_i^2(x_k)]' = 2\delta_{i,k}\lambda_i'(x_k) = 2\lambda_i'(x_i)\delta_{i,k}.$$

Then $\tilde{\nu}_i(x) = h_i(x)\lambda_i^2(x)$ has the desired properties of $\nu_i(x)$ up to a constant multiplier because of the conditions

$$\begin{aligned} \tilde{\nu}_i(x_k) &= h_i(x_k)\lambda_i^2(x_k) = 0, \\ \tilde{\nu}_i'(x_k) &= h_i(x_k)2\delta_{i,k}\lambda_i'(x_k) + h_i'(x_k)\delta_{i,k} = h_i'(x_i)\delta_{i,k}, \end{aligned} \qquad i, k = 1(1)n,$$

since $h_i(x_i) = 0$. Therefore,

$$\nu_i(x) = \frac{h_i(x)}{h_i'(x_i)}\lambda_i^2(x), \qquad i = 1(1)n.$$

Again, $\lambda_i^2(x)$ has nearly all the desired properties of $\mu_i(x)$ except $\mu_i'(x_i) = 0$. This can be corrected by subtraction of a suitable term which vanishes at $x = x_i$ and preserves $\lambda_i^2(x_k) = \delta_{i,k}$.

Taking $\tilde{\mu}_i(x) = \lambda_i^2(x) - \alpha h_i(x)\lambda_i^2(x)$ we obtain

$$\tilde{\mu}_i(x_k) = \delta_{i,k},$$

and

$$\begin{aligned}\tilde{\mu}_i'(x_k) &= 2\delta_{i,k}\lambda_i'(x_k) - \alpha h_i(x_k)2\delta_{i,k}\lambda_i'(x_k) - \alpha h_i'(x_k)\delta_{i,k} \\ &= [2\lambda_i'(x_i) - \alpha h_i'(x_i)]\delta_{i,k}, \\ &= 0 \quad \text{for} \quad \alpha = 2\lambda_i'(x_i)/h_i'(x_i), \qquad i, k = 1(1)n,\end{aligned}$$

Thus

$$\mu_i(x) = \left(1 - 2\lambda_i'(x_i)\frac{h_i(x)}{h_i'(x_i)}\right)\lambda_i^2(x), \qquad \nu_i(x) = \frac{h_i(x)}{h_i'(x_i)}\lambda_i^2(x), \tag{2.5.10}$$

are cardinal functions having the desired properties. If we succeed in making $B_i = 0$, then $A_i = I\lambda_i^2(x) - 2\lambda_i'(x_i)B_i = I\lambda_i^2(x)$ which is much simpler than $I\mu_i(x)$ for $B_i \neq 0$.

The continuation to the interpolation of f'' or f''' is a matter of lengthy but elementary calculation (cf. Engels [9]).

2.5.5 The error of Hermite interpolation

The method for deriving an error expression is exactly the same as for Lagrange interpolation. Let $\omega_n(x)$, $\lambda_i(x)$, $h_i(x)$, and $f(x)$ have derivatives of sufficiently high order. We introduce the additional node $x_{n+1} = t \neq x_i$, $i = 1(1)n$, through the use of

$$\tilde{H}_n(f, x) \equiv H_n(f, x) + \omega_n^2(x)\frac{f(t) - H_n(f, t)}{\omega_n^2(t)}.$$

This operator satisfies

$$H_n(f, x_i) = f(x_i), \quad i = 1(1)n+1, \quad \tilde{H}_n'(f, x_i) = f'(x_i), \quad i = 1(1)n.$$

Subtracting $\tilde{H}_n(f, x)$ from $f(x)$ we have the error equation

$$\tilde{e}_n(f, x) = e_n(f, x) - \omega_n^2(x)\frac{e_n(f, t)}{\omega_n^2(t)}.$$

Now, $\tilde{e}_n(f, x)$ has at least one simple and n double distinct zeros, i.e. at least $2n+1$ zeros in total. We apply Rolle's theorem $2n$ times and see that $e_n^{(2n)}(f, x)$ has at least one zero, say $\xi \in (-1, 1)$. Thus

$$\tilde{e}_n^{(2n)}(f, \xi) = 0 = e_n^{(2n)}(f, \xi) - [\omega_n^2(x)]_{x=\xi}^{(2n)}\frac{e_n(f, t)}{\omega_n^2(t)},$$

and, finally, for $[\omega_n^2(x)]_{x=\xi}^{(2n)} \neq 0$ we have

$$e_n(f, t) = \omega_n^2(t)\frac{f^{(2n)}(\xi) - H_n^{(2n)}(f, \xi)}{[\omega_n^2(x)]_{x=\xi}^{(2n)}},$$

where t is not a specified point in $[-1, 1]$. This expression makes sense for every $t \in [-1, 1]$ and therefore gives the error of Hermite interpolation of f and f'.

2.5.6 Mixed interpolation

Sometimes there arises the necessity to fix or to preassign one or more nodes, while further nodes can be chosen such that Hermitian interpolation is performed there.

Let us denote the preassigned nodes by z_i, $i = 1(1)m$, and the other nodes by x_i, $i = 1(1)n$. For the nodes x_i, $i = 1(1)n$, we form $H_n(f, x)$ as before. Now the preassigned nodes z_i are introducted (in the same way as $x_{n+1} = t$ above) by a suitable Lagrangian interpolating operator $L_m(f, x)$ and then the combined operator

$$HL_{n,m}(f, x) = H_n(f, x) + \omega_n^2(x) L_m\left(\frac{f(x) - H_n(f, x)}{\omega_n^2(x)}, x\right)$$

has the desired interpolating property. This is seen as follows. The nodal multiplier $\omega_n^2(x)$ has a double zero at $x_1, x_2, \ldots, x_n$. Hence

$$HL_{n,m}^{(j)}(f, x_i) = H_n^{(j)}(f, x_i) = f^{(j)}(x_i), \qquad i = 1(1)n, \qquad j = 0, 1.$$

Remember the structure of Lagrange interpolation. Let $\tilde{\omega}_m(x)$ be the nodal function belonging to $L_m(f, x)$ and let $\tilde{h}_i(x)$ be the basis function and $\tilde{\lambda}_i(x)$ the Lagrangian coefficient. Then, because of

$$L_m(f, x) = \sum_{i=1}^{m} \tilde{\lambda}_i(x) f(z_i) \quad \text{with} \quad L_m(f, z_i) = f(z_i)$$

we obtain

$$L_m\left(\frac{f(x) - H_n(f, x)}{\omega_n^2(x)}, x\right) = \sum_{i=1}^{n} \tilde{\lambda}_i(x) \frac{f(z_i) - H_n(f, z_i)}{\omega_n^2(z_i)},$$

and

$$L_m\left(\frac{f(x) - H_n(f, x)}{\omega_n^2(x)}, z_j\right) = \frac{f(z_j) - H_n(f, z_j)}{\omega_n^2(z_j)}, \qquad j = 1(1)m.$$

Hence $HL_{n,m}(f, z_j) = f(z_j)$, $j = 1(1)m$.

Analogously with the previous chapter, we find the error expression

$$e_{n,m}(f, x) = \frac{f^{(2n+m)}(\xi) - HL_{n,m}^{(2n+m)}(f, \xi)}{[\omega_n^2(x)\tilde{\omega}_m(x)]_{x=\xi}^{(2n+m)}} \omega_n^2(x)\tilde{\omega}_m(x).$$

Again, the polynomial case gives the simplest expression for $e_{n,m}(f, x)$ because $HL_{n,m}^{(2n+m)}$ vanishes, and $(\omega_n^2(x)\tilde{\omega}_m(x))^{(2n+m)} = (2n+m)!$.

For polynomial interpolation Radau [24] uses -1 *or* 1, and Lobatto (book) -1 *and* 1 as preassigned nodes. Krylov (book) admits several "backward" preassigned abscissae $z_i \leqslant -1$.

2.5.7 Interpolation in several dimensions

The easiest construction is that of *tensor product* methods.

Let

$$L_n(f, x) = \sum_{i=1}^{n} \lambda_i(x) f(x_i, y), \qquad L_m(f, y) = \sum_{j=1}^{m} \mu_i(y) f(x, y_j)$$

be two Lagrange interpolating functions. Then

$$\mathcal{L}_{nm}(f, x, y) \equiv L_n(L_m(f, x), y) = \sum_{i=1}^{n} \sum_{j=1}^{m} \lambda_i(x)\mu_j(y) f(x_i, y_j)$$

is a two-dimensional interpolating function as is seen from $\lambda_i(x_k) = \delta_{i,k}$, $\mu_j(y_r) = \delta_{j,r}$. The extension to more dimensions is obvious.

The tensor product interpolant for cubature problems is not used extensively. Usually, the resulting formulae need too many nodes with respect to the degree of the formula. We will therefore not treat this problem further.

The problem of multidimensional interpolation is a very difficult one because of the uniqueness problem which is trivial in one-dimensional polynomial interpolation (because of the fundamental theorem of algebra).

Remember the problem of constructing a plane passing through three given distinct points in $\mathbb{R}^3$. This has a unique solution if and only if these three points are not collinear. The same in general is true for the "nodes" of this plane, i.e. the projection of the three points into the x–y-plane. Similar problems of non-uniqueness arise if we want to interpolate by a second-order parabola, i.e. a plane of second order, and the nodes are distributed on certain conic sections.

2.6 Composite rules

In practical calculations we are often forced to increase the number of nodes in order to achieve sufficient accuracy. This is most easily done by applying a quadrature or cubature formula several times *to some subintervals* of the given interval of integration. Such a procedure is now described.

2.6.1 Subdivision of the domain of integration

Let us subdivide the interval $[-1, 1]$ into M subintervals $[-1+2i/M, -1+(2i+2)/M]$ of equal length $2/M$. Let Q_nf be the quadrature formula with respect to $[-1, 1]$

$$Q_nf = \sum_{i=1}^{n} A_if(x_i) \quad \text{where} \quad \sum_{i=1}^{n} A_i = 2.$$

In addition, we apply this in each subinterval and so obtain there the composite rule

$$Q_{nM}f \equiv \sum_{k=1}^{M} \sum_{i=1}^{n} \frac{A_i}{M} f\left(-1+\frac{2k-1}{M}+\frac{x_i}{M}\right).$$

Clearly, the sum of the nM weights is again 2, as required, and $-1+(2k-1)/M$ is the midpoint of the k-th subinterval.

It is, of course, not necessary to take equal subintervals. Let us choose the set of points $t_0=-1<t_1<t_2<\cdots<t_M=1$, which generates a non-equidistant subdivision of $[-1, 1]$ into the set of intervals $[t_0, t_1]$, $[t_1, t_2], \ldots, [t_{M-1}, t_M]$ with lengths $\Delta t_1, \Delta t_2, \ldots, \Delta t_M$ respectively.

The composite quadrature rule is now

$$Q_{nM}f \equiv \sum_{k=1}^{M} \sum_{i=1}^{n} \frac{\Delta t_k}{2} A_if\left(t_k+\frac{\Delta t_k}{2}+\frac{\Delta t_k}{2} x_i\right).$$

Again, the sum of the weights equals 2 and $t_k+\dfrac{\Delta t_k}{2}$ is the midpoint of the k-th subinterval. Here Δt_k replaces $2/M$ in the previous formula.

We now generalise to $N>2$ dimensions. It turns out that the total number of nodes is reduced, if we have one or more nodes which are used twice in adjacent subdomains. We illustrate this for an N-dimensional hypercube H_N which is split into k^N hypercubes, by subdividing each side into k subintervals of equal length $2h$. The following formulae of degree 3 are easily derived by the method of undetermined coefficients. These are:

$$C_{2N+1}f = (2h)^N\left\{\left(1-\frac{N}{3}\right)f(\mathbf{0})+\frac{1}{6}\sum_{i=1}^{N} f(\pm h\mathbf{e}_i)\right\}, \tag{2.6.1}$$

where $\mathbf{e}_i$ is the i-th coordinate unit vector, e.g. $\mathbf{e}_1^T=(1, 0, \ldots, 0)$. The node distribution for $N=3$ is given in Fig. 2.6.1(a) $(1-N/3=0$ here).

If we take the 2^N corners instead of the face centres in (2.6.1) we have (see Fig. 2.6.1(b) for $N=3$)

$$C_{2N+1}f = \frac{h^N}{3}\left\{2^{N+1}f(\mathbf{0})+\sum_{i=1}^{2N} f(\mathbf{c}_i)\right\}, \tag{2.6.2}$$

where $\mathbf{c}_i$ are the 2^N corners of the hypercube.

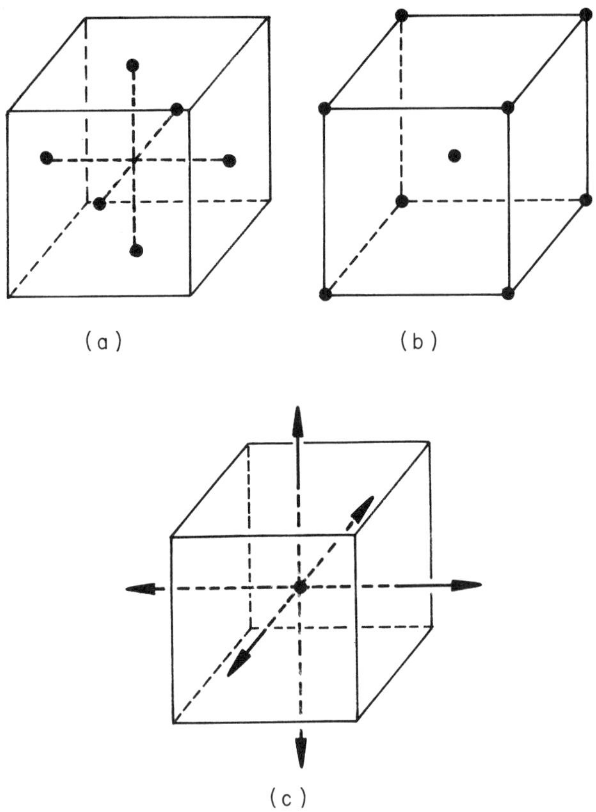

FIG. 2.6.1. Node distributions for 3-cubes.

Replacing the values of f in the surface nodes of (2.6.1) by the normal derivatives (see Fig. 2.6.1(c)) we obtain

$$C_{2N+1}f=(2h)^N\left\{f(\mathbf{0})+\frac{h}{12}\sum_{i=1}^{N}\left[\frac{\partial f}{\partial x_i}(h\mathbf{e}_i)-\frac{\partial f}{\partial x_i}(-h\mathbf{e}_i)\right]\right\}, \tag{2.6.3}$$

where $\partial f/\partial x_i$ is the partial derivative of f with respect to the i-th variable. It is identical with the normal derivative here.

We now put the surface nodes of (2.6.1) or (2.6.3) into the midpoints of the adjacent hypercubes in each coordinate direction so that

$$C_{2N+1}f=(2h)^N\left\{\left(1-\frac{N}{12}\right)f(\mathbf{0})+\frac{1}{24}\sum_{i=1}^{N}f(\pm 2h\mathbf{e}_i)\right\} \tag{2.6.4}$$

(cf. Fig. 2.6.2(a) for $N=3$). Finally, we omit the origin and choose $2N$ nodes

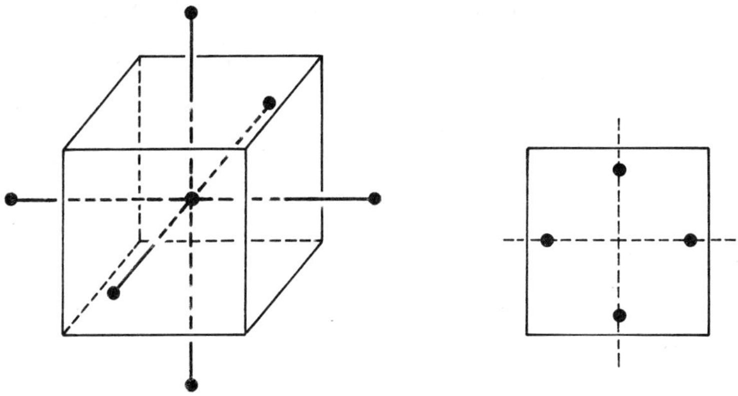

(u) (b)

FIG. 2.6.2. Node distributions for 3- and 2-cubes.

on the coordinate axis in optimal distance from the origin such that the degree is 3 (see Fig. 2.6.2(b) for $N=2$)

$$C_{2N}f = \frac{(2h)^N}{2N} \sum_{i=1}^{N} f\left(\pm\left(\frac{N}{3}\right)^{\frac{1}{2}} h\mathbf{e}_i\right). \tag{2.6.5}$$

Equation (2.6.5) is identical with (2.6.1) for $N=3$, and with (2.6.4) for $N=12$. The five formulae (2.6.1) to (2.6.5) have the same degree and differ only slightly in the number of nodes. However, we shall see immediately that these differences cause large variations in the number of function evaluations when the formulae are used in composite rules.

Let us count up the number of nodes for the hypercube H_N which is subdivided k times in each coordinate direction. The centres of the small hypercubes which correspond to the open circles in Fig. 2.6.3 occur k^N times. We now proceed by induction in order to obtain the total number of nodes:

For one dimension the interior boundary nodes of the small "hyper-cubes" are used twice and hence they finally occur $k-1$ times. We thus have a total of $k+(k+1)$ nodes for $N=1$ as is illustrated for $k=4$ in Fig. 2.6.3.

For $N=2$ we see that the grid of points for $N=1$ (cf. Fig. 2.6.3) appears k times for both dimensions. Hence we have a total of $2k[k+(k+1)]-k^2=k^2+2k(k+1)$. Since the centre points have to occur only once, we subtracted k^2.

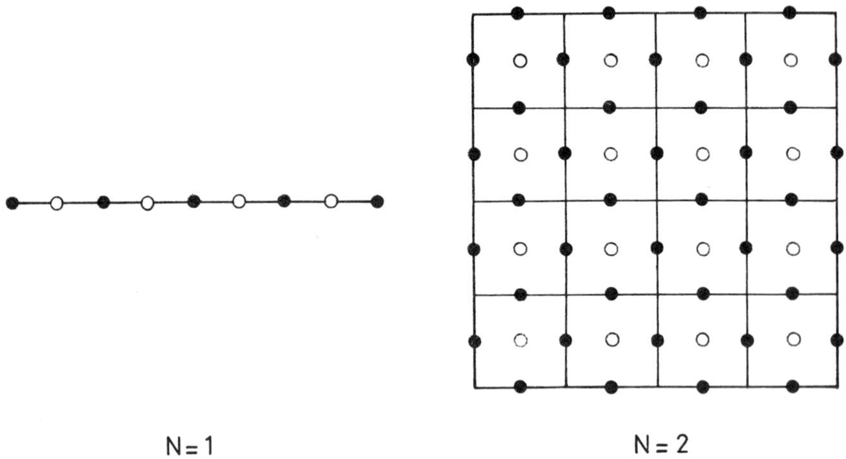

FIG. 2.6.3. Composite rules with matching nodes.

For $N=3$ the grid for $N=2$ appears k times at the levels $h, 3h, \ldots, (2k-1)h$ measured above the x–y-plane. The bold grid of $N=1$ in addition appears k^2 times in the cube-centres perpendicular to the x–y-plane. Hence we have a total of $k^3+3k^2(k+1)$ nodes for $N=3$. If we continue this process and denote by $n(N, k)$ the total number of nodes needed for H_N at k subdivisions in each coordinate, we then have $n(N, k)=k^N+Nk^{N-1}(k+1)$ nodes totally if equation (2.6.1) is applied. This has to be corrected for $N=3$ to $n(3, k)=3k^2(k+1)$ because the weight of $f(0)$ in (2.6.1) vanishes for $N=3$.

The values $n(N, k)$ of the further cubature formulae are calculated analogously, so that we finally obtain the results of Table 2.6.1. Note that the partial derivatives of (2.6.3) cancel each other in the interior of H_N. Asymptotically we have for (2.6.4) and (2.6.5) the ratio of $n(N, k)$

$$\frac{2Nk^N}{k^N+2Nk^{N-1}}=\frac{2Nk}{k+2N}\sim 2N \quad \text{for} \quad k \gg N.$$

which is a rather direct measure of the corresponding ratio of computer time needed for (2.6.5) and (2.6.4). It turns out, that it is worth developing formulae which are efficient in composite rules.

2.6.2 Product rules

Cubature formulae can be obtained by systematic application of quadrature rules, especially for irregularly-shaped domains. Assume that we

TABLE 2.6.1

Equation	$n(N,k)$	$n(4,2)$	$n(4,4)$	$n(5,50)$
(2.6.1)	$\begin{cases} N(k+1)k^{N-1}+k^N, & N\neq 3 \\ N(k+1)k^{N-1}, & N=3 \end{cases}$	112	1536	1.9×10^9
(2.6.2)	$(k+1)^N+k^N$	97	881	6.6×10^8
(2.6.3), (2.6.4)	$k^N+2N\,k^{N-1}$	80	768	3.8×10^8
(2.6.5)	$\begin{cases} 2N\,k^N, & N\neq 3, 12^{(*)} \\ N(k+1)^{N-1}, & N=3 \\ k^N+2N\,k^{N-1}, & N=12 \end{cases}$	128	2048	3.1×10^9

(*) Further special cases for $N=3m^2$, $m>2$ are neglected.

have to calculate

$$I_\Omega f=\int_{-1}^{1}\int_{u(x)}^{v(x)} f(x, y)\,dy\,dx,$$

where $u(x)$ and $v(x)$ are assumed to be continuous for $x\in[-1, 1]$. Further, we assume, without essential loss of generality, that $u(x)\leqslant v(x)$. If this is not true, the interval $[-1, 1]$ is divided into subintervals using the points of intersection of $u(x)$ and $v(x)$ to be the limits of the resulting subintervals. Let us apply the quadrature formula $Q_n f$ with respect to x:

$$Q_n^x f\equiv\sum_{i=1}^{n} A_i\int_{u_i}^{v_i} f(x_i, y)\,dy \quad \text{where} \quad v_i=v(x_i),\ u_i=u(x_i),$$

Another (or the same) rule $Q_m f$ is applied with respect to y so that the *tensor product equation*

$$Q_{n,m}f=Q_m^y[Q_n^x f]\equiv\sum_{i=1}^{n} A_i\sum_{j=1}^{m}\frac{v_i-u_i}{2}B_j f\left(x_i,\ u_i+\frac{v_i-u_i}{2}+\left(\frac{v_i-u_i}{2}\right)y_i\right)$$

is produced. We have used here

$$Q_m f=\sum_{j=1}^{m} B_j f(x_j), \quad \text{with} \quad \sum_{j=1}^{m} B_j=2.$$

Note that B_j had to be reduced so that $\sum_{j=1}^{m}\tilde{B}_j=v_i-u_i$, and that $u_i+(v_i-u_i)/2$ is the midpoint of $[u_i, v_i]$.

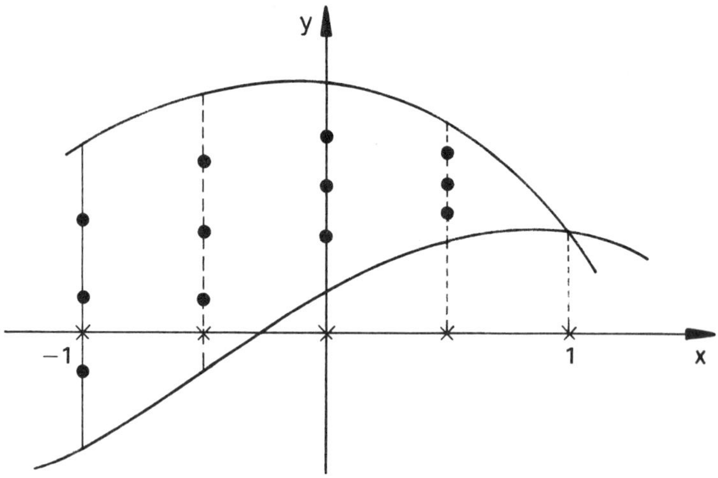

FIG. 2.6.4. Product rule.

In Fig. 2.6.4 we have $n=5$ abscissae x_i (marked by crosses) in the x-direction, and $m=3$ abscissae in the y-direction. Because $u(1)=v(1)$ the three nodes of the y-direction in $x=1$ coincide. For the same reason $u(1)-v(1)=0$ and hence their weight is zero so that they do not appear explicitly in the product cubature formula. Thus, we obtain a cubature formula whose nodes are marked by bold dots in Fig. 2.6.4. Of course we can vary the number of nodes with respect to j. This is desirable if the distance $v(x)-u(x)$ varies strongly.

As already mentioned, the tensor product technique is most simple and elegant, but usually the resulting cubature formulae contain too many nodes by comparison with other formulae of similar accuracy.

The extension of the method to N dimensions is trivial. There are many choices available:

(i) Application of various quadrature rules to the various dimensions.
(ii) Combination of quadrature and cubature rules for various dimensions.
(iii) Combination of several cubature rules to several subspaces.

2.6.3 Rule projection and rule extension

The above techniques have shown how quadrature rules may be extended to higher dimensions by the use of tensor product methods. Alternatively, it is possible to apply a cubature formula for N dimensions to $(N-k)$-dimensional problems $(k>0)$ by projecting it into the $(N-k)$-dimensional subspace. This is called *rule projection.* The following example illustrates the mechanics of a rule projection.

Example 2.6.1 Let $N=3$, and the 6-point cubature formula

$$C_6f = \tfrac{4}{3}\{f(-1,0,0)+f(1,0,0)+f(0,-1,0)+f(0,1,0)+f(0,0,-1)+f(0,0,1)\}$$

for the cube $|x_i|\leq 1$, $i=1,2,3$. The projection into the 2-space gives the 5-point cubature formula

$$C_5f = \tfrac{2}{3}\{f(-1,0)+f(1,0)+f(0,-1)+f(0,1)+2f(0,0)\}.$$

Clearly, the area of the 3-cube has to be reduced to "unit height".

A further projection into the 1-space leads to Simpson's rule (Fig. 2.6.5). ■

The opposite of a rule projection is called *rule extension.* An example of rule extension is the tensor product method. In general, the rule extension is not unique, because thinking in the opposite direction several different equations for the same Ω can have the same projection into a

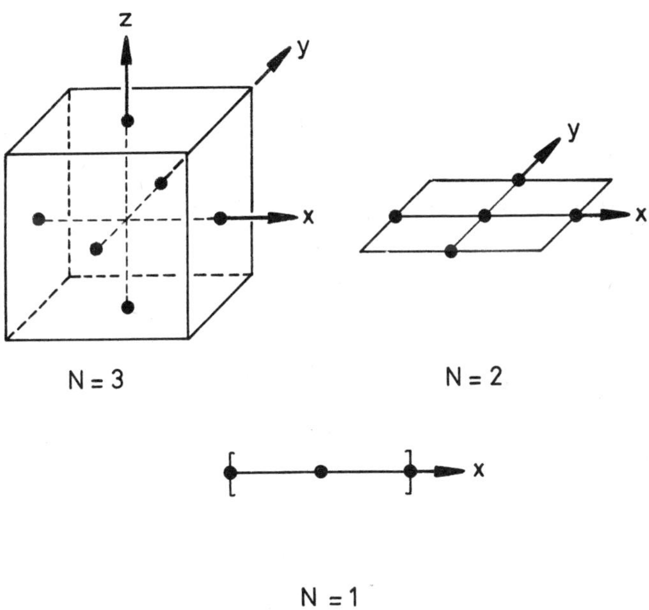

FIG. 2.6.5. Rule projection.

suitable subspace of Ω. This is demonstrated by:

Example 2.6.2 The cubature formulae

$$C_9 f = \tfrac{1}{6}\{f(1,1)+f(-1,1)+f(1,-1)+f(-1,-1)+2[f(1,0)+f(-1,0) + f(0,-1)+f(0,1)]+12f(0,0)\}$$

$$C_5 f = \tfrac{2}{3}\{f(-1,0)+f(1,0)+f(0,-1)+f(0,1)+2f(0,0)\}$$

are both projected into Simpson's rule. Hence the extension of Simpson's rule cannot be unique. ■

2.7 Monte Carlo and number theoretic methods

Somewhat exotic methods for the approximate calculation of integrals are based on statistical methods using random numbers. We mention such methods here for the sake of completeness.

Monte Carlo and quasi Monte Carlo methods have a certain importance for the approximate calculation of integrals, especially for N-dimensional integrals with large N, for very complicated domains Ω, and in those cases where only a rough approximation is required. The difference between *quasi* Monte Carlo methods and number theoretic methods is, in the first place, a geographic difference: the description "number theoretic method" is mostly used in the Russian literature. For further details the reader should study the papers of e.g. Hlawka [17], Zaremba [31], [32], [33], or Korobov [19].

2.7.1 A short outline of the method

Let us first treat the quadrature problem. Without essential loss of generality we assume that $0 \leq f(x) \leq 1$ for $x \in [0, 1]$, and we want to approximate $\overset{*}{I}f$ (see Fig. 2.7.1). We now take n pairs of uniformly-distributed random numbers (x_i, y_i), $i = 1(1)n$, satisfying $0 \leq x_i, y_i \leq 1$. There are then two possibilities for each pair:

(i) $0 \leq y_i \leq f(x_i)$, or
(ii) $f(x_i) < y_i \leq 1$.

In the first case (x_i, y_i) belongs to the area which is determined by $\overset{*}{I}f$; in the second case it does not. Let n_0 pairs satisfy condition (i). Then

$$\overset{*}{I}f = \int_0^1 f(x)\,\mathrm{d}x \doteq n_0/n.$$

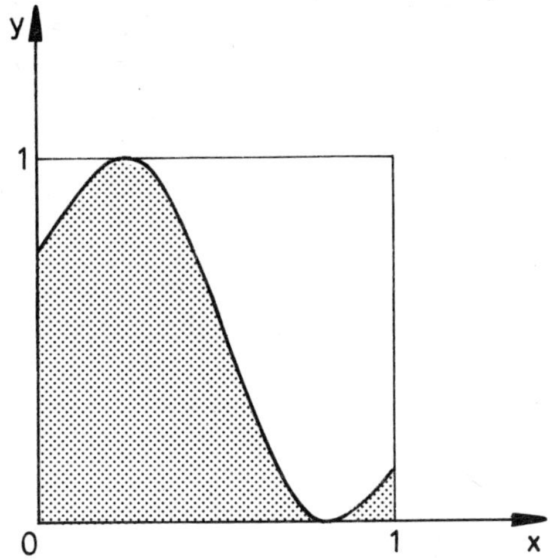

FIG. 2.7.1. Standard region for the Monte Carlo method.

The result is exact for $f \equiv 1$. For $f \not\equiv 1$, imagine that the square $0 \leqslant x, y \leqslant 1$ is subdivided into m^2 congruent subsquares. Then the assumption of uniform distribution guarantees that each subsquare contains nearly the same number of points (x_i, y_i). Hence, the ratio n_0/n is an approximation of the Riemann sum of $\overset{*}{I}f$. Obviously, the method is easily extended to N-dimensional integrals.

Let $0 \leqslant f(\mathbf{x}) \leqslant 1$ and $\Omega \subseteq S_N$, where S_N is the N-dimensional unit cube consisting of all $\mathbf{x}$ whose components x_i satisfy $0 \leqslant x_i \leqslant 1$. We now take n sets of $N+1$ uniformly-distributed random numbers from $[0, 1]$:

$$\{x_{1,i}, \ldots, x_{N,i}, y_i\}, \qquad i = 1(1)n.$$

From these, we take $2n$ vectors

$$\mathbf{x}_j^T = \{x_{1,j}, x_{2,j}, \ldots, x_{N,j}\},$$
$$\mathbf{y}_j^T = \{\mathbf{x}_j^T, y_j\}, \qquad j = 1(1)n.$$

Now, let ν vectors $\mathbf{x}_j$ belong to Ω. Then $|\Omega| \doteq \nu/n$. The corresponding ν numbers y_j satisfy

(i) $0 \leqslant y_j \leqslant f(\mathbf{x}_j)$, or
(ii) $f(\mathbf{x}_j) < y_j \leqslant 1$.

Let $n_0 \leqslant \nu$ vectors $\mathbf{y}_j$ satisfy condition (i). Then

$$I_\Omega f = \int_\Omega f(\mathbf{x})\, d\Omega \doteq \frac{n_0}{n} \doteq \frac{n_0}{\nu} |\Omega|.$$

In practice, the random numbers are usually not random numbers, but quasi random numbers, constructed by certain strategies. These numbers are the basis for lattice methods in regularly-shaped domains Ω. Certain points of a lattice then are determined to be the nodes. Then the question arises whether there exist "good lattice points" for certain function spaces from which f is taken. The following example of "good lattice points" for the space of functions with bounded variation is due to Zaremba. The lattice is given by the lines

$$x = \frac{i}{n}, \qquad y = \frac{i}{n}, \qquad i = 0(1)n = F_m,$$

and the lattice points, which are taken as nodes, are

$$\left(\frac{k}{F_m}, \left\{\frac{kF_{m-1}}{F_m}\right\}\right), \qquad k = 0(1)F_m - 1,$$

where $\{\alpha\}$ denotes the fractional part of α and F_m are the *Fibonacci*

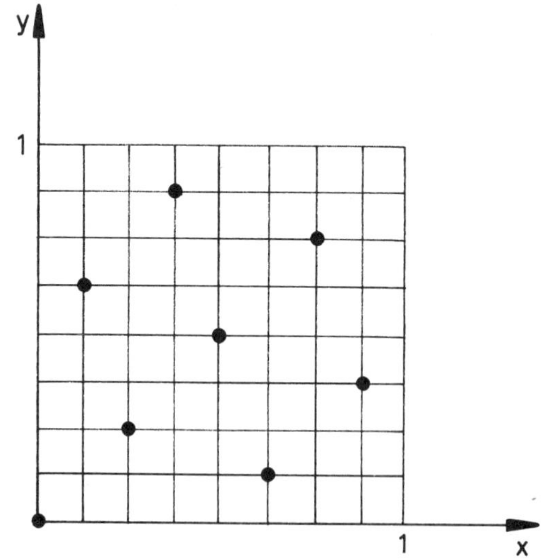

FIG. 2.7.2. Good lattice points due to Zaremba.

numbers defined by $F_1 = F_2 = 1$, $F_m = F_{m-1} + F_{m-2}$ for $m \geqslant 3$. For $n = F_6 = 8$ we obtain the nodes which are marked by bold dots in Fig. 2.7.2, and the result is an equally-weighted cubature formula $C_8 f$. This turns into an unequally-weighted formula $C_{11} f$ for periodic $f(x, y)$ (because then $f(0,0) = f(1,0) = f(0,1) = f(1,1)$) which has interesting symmetry properties. The same is true for other values of n. It remains to be determined how close the nodes are to those of comparable formulae based upon polynomial accuracy.

The main disadvantage of the Monte Carlo methods comes from their slow convergence. The error usually satisfies $O(1/\sqrt{n})$ for n random vectors $\mathbf{y}_j$. Nevertheless, good results have been obtained for a large number N of dimensions (cf. Tsuda [29]).

2.7.2 Program and examples

We give here a FORTRAN program for a Monte Carlo method.

```
      SUBROUTINE MCQ (F, N, U, O, X, G, NDGMAX, QD, QC, Q)
C     MONTE CARLO QUADRATURE PROGRAM CALCULATES
C     THE INTEGRAL OF F(X) ON AN ARBITRARY DOMAIN
C     OMEGA. THIS DOMAIN HAS TO BE CONTAINED
C     WITHIN A CUBE C. IT IS DESCRIBED BY A
C     SUBROUTINE USING THE PARAMETERS (X, N, IN).
C     IN≠0, IF X BELONGS TO OMEGA, IN=0 ELSE,
C     FOR THE N-DIMENSIONAL VECTOR X.
C     G: NAME OF THE SUBROUTINE DESCRIBING OMEGA
C     U, O: FIELDS OF LENGTH N, CONTAINING LOWER AND
C     UPPER BOUNDARIES OF C.
C     X: FIELD OF LENGTH N
C     NDGMAX: IF POSITIVE: NUMBER OF POINTS IN C
C             IF NEGATIVE: NUMBER OF POINTS IN OMEGA
C     OUTPUTPARAMETERS:
C     QD: VOLUME OF C
C     QC: VOLUME OF OMEGA
C     Q: APPROXIMATE VALUE OF THE INTEGRAL
C     RANF(X) SUBROUTINE CREATING RANDOM NUMBERS
      DIMENSION U(N), O(N), X(N)
      EXTERNAL F, G
      ND=Ø
      NG=Ø
      SUM=Ø
```

```
1 IF (NDGMAX) 3, 3, 6
6 IF (ND-NDGMAX) 4, 2, 2
3 IF(NG+NDGMAX) 4, 2, 2
4 DO 5  I=1, N
5 X(I)=(O(I)-U(I))×RANF (Ø.)+U(I)
  CALL G(X, N, IN)
  ND=ND+1
  IF (IN. EQ. Ø) GOTO 1
  NG=NG+1
  SUM=SUM+F(X)
  GOTO 1
2 QD=1
  DO 7  I=1, N
7 QD=QD×(O(I)-U(I))
  QC=NG×QD/ND
  Q=SUM×QD/ND
  RETURN
  END
```

Example 2.7.1 Let us compute

$$\int_0^1 \cdots \int_0^1 \prod_{i=1}^{N} \left(\frac{a}{a+bx_i}\right)^N \mathrm{d}x_1 \ldots \mathrm{d}x_N = \left[\frac{a(b+2a)}{2(a+b)^2}\right]^N.$$

Here $f_{\max}=1$ and $f_{\min}=(a/(a+b))^{N^2}$. For $b>a$ and large N there is a steep peak at $x=0$, and we expect a bad result. We compare the results for $a=2$, $b=1$, for $a=9$, $b=1$, and for $a=1$, $b=2$. The corresponding values of $I_\Omega f$ are $(5/9)^N$, $(171/200)^N$ and $(2/9)^N$ respectively (see Table 2.7.1).

Secondly, we compute approximate values (see Table 2.7.2) for the volume of the cone $z=1-(x^2+y^2)^{\frac{1}{2}}$, $z\geqslant 0$, by means of the integral

$$\int_{-1}^{1} \int_{-(1-x^2)^{1/2}}^{(1-x^2)^{1/2}} (1-(x^2+y^2)^{\frac{1}{2}})\,\mathrm{d}y\,\mathrm{d}x = \int_0^{2\pi}\int_0^1 (1-r)r\,\mathrm{d}r\,\mathrm{d}\varphi = \frac{\pi}{3}$$

2.8 Remarks, problems, references

2.8.1 Remarks

R.2.1: An alternative method of approximating $f(x)$ is to take circumscribed polygons of tangents instead of inscribed polygons using chords. This leads to

TABLE 2.7.1

Dimension N	Number of sample nodes	Monte Carlo quadrature result	Relative error $\lvert(I_\Omega f - C_n f)/I_\Omega f\rvert$
	$a = 2,\ b = 1$		
1	64	5.705 043(−1)	2.69(−2)
	1024	5.444 003(−1)	2.00(−2)
2	64	2.984 732(−1)	3.29(−2)
	1024	3.158 343(−1)	2.33(−2)
4	64	9.223 809(−2)	3.17(−2)
	1024	9.525 485(−2)	5.26(−5)
	4096	9.443 851(−2)	8.62(−3)
8	64	8.598 308(−3)	5.25(−2)
	1024	8.444 818(−3)	6.94(−2)
16	64	8.790 143(−5)	6.75(−2)
	1024	8.493 270(−5)	4.14(−2)
32	64	6.222 527(−9)	8.23(−2)
	1024	5.603 489(−9)	1.74(−1)
64	64	1.600 364(−17)	6.52(−1)
	1024	5.900 081(−17)	2.83(−1)
	$a = 9,\ b = 1$		
1	256	8.559 196(−1)	1.08(−3)
2		7.289 833(−1)	2.79(−3)
4		5.316 099(−1)	5.22(−3)
8		2.791 901(−1)	2.24(−2)
16		8.264 021(−2)	1.33(−2)
32		6.972 429(−3)	4.83(−2)
64		4.138 498(−5)	6.46(−2)
	$a = 1,\ b = 2$		
1	4096	2.234 060(−1)	5.33(−3)
2		4.875 944(−2)	1.26(−2)
4		2.408 443(−3)	1.24(−2)
8		5.615 385(−6)	5.58(−2)
16		1.481 895(−11)	5.81(−1)
32		1.144 390(−23)	9.91(−1)
64		4.358 208(−47)	1

TABLE 2.7.2

Number of random vectors	Approximations of $\lvert\Omega\rvert = \pi = 3.1416$	Approximations of $I_\Omega f = \pi/3 = 1.0472$
32	2.8750	1.1343
128	3.0938	0.9464
512	3.0859	1.0123
2048	3.1465	1.0584
8192	3.1553	1.0502
32768	3.1378	1.0475

equations such as:

$$If \doteq \sum_{i=0}^{n-1} \frac{h}{2}\{f(x_i)+[f(x_i)+hf'(x_i)]\}$$
$$= h\sum_{i=0}^{n-1} f(x_i)+\frac{h^2}{2}\sum_{i=0}^{n-1} f'(x_i)$$

for equidistant nodes $x_i = -1+ih$, $i = 0(1)n-1$, $h = 2/(n-1)$. Using the tangent at every node we obtain another polygon and another form of a trapezoidal rule (Fig. 2.8.1) (cf. Section 2.1.1).

R.2.2: When applying the method of undetermined coefficients we are interested in finding unique formulae. This is possible because of the following:

Theorem 2.8.1 *Let $\varphi_i(x)$, $i = 1(1)n$ be a given system of functions. These are linearly independent if and only if the* Gram matrix $\mathbf{\Delta}_n$ *satisfies* $\det(\Delta_n) \neq 0$.

The matrix elements of $\mathbf{\Delta}_n$ are

$$\Phi_{i,j} = (\varphi_i, \varphi_k) = \int_{-1}^{1} W(x)\varphi_i(x)\varphi_k(x)\,dx.$$

For a proof see Natanson, [23] p. 227 (cf. Section 2.1).

R.2.3: The Gaussian quadratures, because they have the highest possible degree of precision, are in general use in computer subroutines. For this purpose nodes and weights are floating point numbers. However, the closed form for small n is also very suitable for applications on electronic pocket computers where there is no store for long floating point numbers, but only for small integers. We

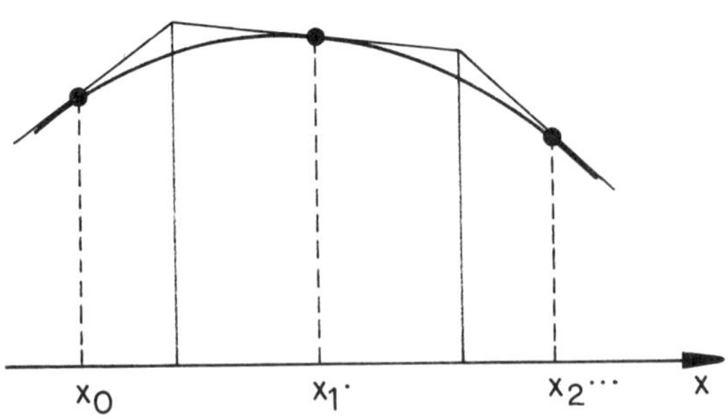

FIG. 2 8.1. Trapezoidal rule using derivatives.

therefore list here the Gaussian quadrature formulae up to $n=5$, $W(x)\equiv 1$ (cf. Sections 2.2.4, 2.3.4):

$n=1$ $\quad x_1=0, \quad A_1=2$

$n=2$ $\quad x_1=-x_2=\sqrt{1/3}, \quad A_1=A_2=1$

$n=3$ $\quad x_1=-x_3=\sqrt{3/5}, \quad x_2=0, \quad A_1=A_3=5/9, \quad A_2=8/9$

$n=4$ $\quad x_{1,2}=-x_{4,3}=((3\pm\sqrt{4.8})/7)^{\frac{1}{2}}, \quad A_{1,2}=A_{4,3}=(1\mp\sqrt{5/54})/2$

$n=5$ $\quad x_{1,2}=-x_{5,4}=(5\pm\sqrt{40/7})^{\frac{1}{2}}/3, \quad x_3=0,$
$\quad A_{1,2}=A_{5,4}=(3.22\mp 1.3\sqrt{0.7})/9, \quad A_3=64/225.$

R.2.4: The expansion of node distributions due to Albrecht and Collatz is restricted to $N=2$ dimensions. An extension to three or more dimensions seems not yet to have been done.

R.2.5: The method of Taylor expansion and the method of undetermined coefficients give identical results for $\varphi_i\in\{1, x, x^2, \ldots, x^{2n-1}\}$. This is no longer true for $\varphi_i\in\{x^{\alpha_1}, x^{\alpha_2}, \ldots, x^{\alpha_n}\}$ with $-1<\alpha_1<\alpha_2<\ldots<\alpha_n$ in general. This more general case seems to have been treated first by August [5] (cf. Sections 2.2 and 2.3).

R.2.6: In Example 2.3.1 we took $\varphi_\nu(x)=1/(x^2+\nu^2)$, $\nu=1(1)i+1$, in order to obtain nodes and weights of a quadrature formula Q_nf approximating $\underset{\infty}{If}$ by solving the corresponding nonlinear system of equations. For $2n$ symmetric nodes we always obtain an even polynomial of degree $2n$, whose zeros are the desired nodes. (The same is true for $2n-1$ nodes where 0 is a preassigned node.) It is seen by computation that for very small n there exist real distinct nodes. But there is no general knowledge about any general property (cf. Section 2.3.1).

R.2.7: The numerical solution of a nonlinear system of equations for nodes and weights is performed at most by splitting the problem into a strictly nonlinear problem for the weights by eliminating the nodes. The nonlinear system is usually solved iteratively. Practical computations show that very good starting values are needed. The larger the number of nodes, the closer the starting values have to be to the real solution. Fortunately, the nodes of two consecutive formulae usually separate each other. Hence, suitable combinations of known nodes lead to good starting values for the next set of unknown zeros.

The solution of the linear system for weights is only difficult if some nodes cluster. The weights can then differ by several powers of 10, and the system is ill-conditioned.

R.2.8: Sometimes equally-weighted quadrature formulae for odd weight functions are considered. The weights then cannot have a constant sign, e.g.

$$\int_{-1}^{1} xf(x)\,dx = A_n\sum_{i=1}^{n}(f(x_i)-f(-x_i)).$$

Here $A_n = 1\Big/\left(3\sum_{i=1}^{n} x_i\right)$. This problem seems to have been treated first by Radau [24]. Radau also considered $W(x) = x/(1-x^2)^{\frac{1}{2}}$ for $[-1, 1]$ (cf. Section 2.4).

R.2.9: As interpolating splines we have considered only those which are piecewise polynomials. Of course there are generalisations of the polynomial splines to other splines, e.g. rational or exponential splines. But we cannot go into further details about splines. The same is valid for polynomial splines of degree higher than 3. The reader may look at Ahlberg, Nilson and Walsh, [1], Böhmer [6], Böhmer, Meinardus and Schempp [7], Sard and Weintraub [25], Schoenberg [26], Schultz [27], Varga [30], and to the excellent bibliography on spline functions by Schurer and van Rooij [28] (cf. Section 2.5.2).

R.2.10: Hermitian interpolation is not necessarily restricted to the interpolating property $H_n^{(k)}(f, x_i) = f^{(k)}(x_i)$ for $i = 1(1)n$ and $k = 0(1)m$. Of course, we may cancel some of these conditions and obtain another kind of interpolation which uses derivatives. Admitting only $i = 1$, we have "Taylor-interpolation". Taking $i = 1, 2$ and $k = 0(2)2m$ we obtain an interpolation due to Lidstone. Finally, we obtain the Abel–Gontscharoff interpolation by choosing $k = i - 1$. Here we always mean polynomial interpolation (cf. Section 2.5.4).

R.2.11: The trigonometric interpolation due to $h_i(x) = \sin \pi(x - x_i)$ and $\omega_n(x) = \prod_{i=1}^{n} h_i(x)$ is not always a real interpolation. It may degenerate into an averaging or a smoothing operation for certain sets of nodes. Let us take the nodes $x_1 = -1$, $x_2 = -1/2$, $x_3 = 0$, $x_4 = 1/2$. Then

$$h_1(x) = \sin \pi(x+1) = -\sin \pi x, \qquad h_2(x) = \sin \pi(x + 1/2) = \cos \pi x,$$
$$h_3(x) = \sin \pi x, \qquad h_4(x) = -\cos \pi x,$$

and

$$\omega_4(x) = \frac{1}{4} \sin^2 (2\pi x).$$

Hence, $\omega_n(x)$ has double zeros at all nodes, which is not allowed. Nevertheless, we easily find something like cardinal functions:

$$\lambda_i(x) = \frac{\omega_4(x)}{2h_i^2(x)}, \qquad i = 1(1)4.$$

However,

$$L_4(f, -1) = 1/2(f(-1) + f(0)) = L_4(f, 0),$$

$$L_4(f, -1/2) = 1/2(f(-1/2) + f(1/2)) = L_4(f, 1/2),$$

are now no longer interpolations.

This is no disadvantage for quadrature purposes. We have

$$\begin{aligned} Q_4f &= \int_{-1}^{1} L_4(f, x)\,dx \\ &= \tfrac{1}{2}\int_{-1}^{1} [(f(-1)+f(0))\cos^2 \pi x + (f(1/2)+f(-1/2))\sin^2 \pi x]\,dx \\ &= \tfrac{1}{2}[f(-1)+f(-1/2)+f(0)+f(1/2)] = R_4f \end{aligned}$$

which is a rectangle rule.

We shall see later, that such special cases are not only due to an unfortunate choice of the nodes, but also to a general solution of $B_i = 0$, $i = 1(1)n$ (cf. Section 2.5.2).

R.2.12: In the English literature one usually writes "Chebyshev" instead of "Tschebyscheff". But this is incompatible with the custom to denote the Chebyshev polynomials by $T_n(x)$. For this reason the German translation Tschebyscheff is preferred in this book.

2.8.2 Problems

P.2.1: Instead of chordal polygons use tangential polygons in order to derive simple quadrature formulae of the form M_nf or T_nf (cf. R.2.1 and Section 2.1.1).

P.2.2: Let x_i, $i = 1(1)n$ be given real numbers. Show that

$$\det \begin{bmatrix} 1 & 1 & \cdots & 1 \\ x_1 & x_2 & \cdots & x_n \\ x_1^2 & x_2^2 & \cdots & x_n^2 \\ \vdots & \vdots & & \vdots \\ x_1^{n-1} & x_2^{n-1} & \cdots & x_n^{n-1} \end{bmatrix} = \prod_{k=2}^{n} \prod_{j=1}^{k-1} (x_k - x_j).$$

(cf. R.2.2, and Natanson p. 227, [23]; cf. Section 2.2).

P.2.3: Determine the weights of (see equation (2.2.1)):

$$Q_3f = A_1f(-1) + A_2f(-1/3) + A_3f(0) \doteq If \quad \text{(cf. Section 2.2.1).}$$

P.2.4: Determine the weights of:

$$C_4f = A_1f(1, 1) + A_2f(-1, 1) + A_3f(-1, -1) + A_4f(1, -1) \doteq I_\Omega f$$

where Ω is the square $-1 \leq x, y \leq 1$ (cf. Section 2.2).

P.2.5: Determine the weights of the cubature formula

$$C_{2N+1}f = A_1f(\mathbf{0}) + A_2 \sum_{i=1}^{N} f(\pm\mathbf{e}_i) \doteq I_\Omega f(\mathbf{x})$$

where $\mathbf{e}_i^T = (0, 0, \ldots, 0, 1, 0, \ldots, 0)$ having a 1 in the i-th component and Ω is the hypercube $-1 \leq x_1, x_2, \ldots, x_n \leq 1$ (cf. Section 2.2).

P.2.6: Calculate the leading error terms of the formulae in P.2.3, P.2.4, and P.2.5 (cf. Section 2.3.3).

P.2.7: Prove the following recursion formula:

$$\overset{*}{I}_\nu = \int_0^1 x^\alpha (1-x)^\beta x^\nu \, dx = \frac{\alpha+\nu}{\alpha+\beta+\nu+1} \overset{*}{I}_{\nu-1}, \qquad \alpha, \beta > -1.$$

Find a recursion formula for the moments

$$\underset{\infty}{\overset{*}{I}_\nu} = \int_0^\infty x^\alpha e^{-x} x^\nu \, dx, \qquad \alpha > -1$$

which are related to Gauss–Laguerre quadrature (cf. Section 2.2.5).

P.2.8: Let Ω be a circle with radius h and centre $\mathbf{0}$. Prove (cf. Engels [8]):

$$I_{0,0} = \pi h^2, \qquad I_{1,0} = I_{0,1} = I_{1,1} = 0,$$

$$I_{\nu,0} = \frac{\nu-1}{\nu+2} h^2 I_{\nu-2,0}, \qquad I_{\nu,1} = 0 \quad \text{for} \quad \nu > 1,$$

$$I_{\nu,\mu} = \frac{\mu-1}{\nu+1} I_{\nu+2,\mu-2} \quad \text{for} \quad \nu, \mu > 1 \text{ (cf. Section 2.2.5).}$$

P.2.9: Let Ω be the regular hexagon with centre $\mathbf{0}$, and diameter $2h$ on the y-axis (cf. Fig. 2.2.2). Then the side-length is h and $|x| \leqslant \sqrt{3}h/2$ for all points of Ω. Prove (cf. Engels [8] and see Section 2.2.5):

$$I_{\nu,0} = \frac{2[1+(-1)^\nu]}{(\nu+1)(\nu+2)} (\nu+3) 3^{(\nu+1)/2} \left(\frac{h}{2}\right)^{\nu+2}$$

$$I_{\nu,1} = 0 \quad \text{for} \quad \nu \geqslant 0$$

$$I_{\nu,\mu} = \frac{(\mu+\nu+3)[1+(-1)^\mu]}{2(\mu+1)(\nu+3)} \left(\frac{h}{2}\right)^\mu I_{\nu,0} + \frac{\mu(\mu-1)}{3(\nu+1)(\nu+2)} I_{\nu+2,\mu-2} \quad \text{for} \quad \mu \geqslant 2.$$

P.2.10: Calculate the first three non-vanishing terms of the expansion of (cf. Theorem 2.2.2. and Table 2.2.1, in Section 2.2.8):

$$S_6(R) = \sum_{j=1}^{6} f(R\cos(\varphi + \pi j/3), R\sin(\varphi + \pi j/3))$$

in powers of R, and construct a cubature formula

$$C_7 f = A_1 f(\mathbf{0}) + A_2 S_6(R)$$

of highest possible degree for the hexagon in Fig. 2.2.2.

P.2.11: Solve the system (cf. Table 2.2.1 in Section 2.2.8):

$$\Sigma_1 + \Sigma_2 = \frac{1}{n+1} \quad \text{for} \quad n = 0(1)3,$$

$$\Sigma_1 - \Sigma_2 = 0 \quad \text{for} \quad n = 2(1)3$$

which determines radii and weights of the cubature formula

$$C_{12}f=\sum_{\rho=1}^{2} S_{4,\rho}\sum_{j=1}^{4} f\left(\sigma_{4,\rho}\cos\left(\frac{\pi j}{2}\right), \sigma_{4,\rho}\sin\left(\frac{\pi j}{2}\right)\right)$$
$$+T_{4,1}\sum_{j=1}^{4} f\left(\tau_{4,1}\cos(2j-1)\frac{\pi}{4}, \tau_{4,1}\sin(2j-1)\frac{\pi}{4}\right)$$

for the circular domain Ω with radius 1.

P.2.12: Determine the weights of

$$Q_5f=A_1f(0)+A_2[f(-2.5)+f(2.5)]+A_3[f(-5)+f(5)]$$

for the approximate calculation of $If=\int_{-\infty}^{\infty} f(x)\,dx$ such that Q_5f is exact for

$$f_\nu(x)=1/(x^2+\nu^2), \qquad \nu=1,2,3,$$

and apply Q_5f to $f=\exp(-x^2)$ (cf. Section 2.3.1).

P.2.13: Calculate the nodes and weights of the quadrature formula:

$$Q_nf=\sum_{i=1}^{n} A_if(x_i)\doteq\int_0^{2\pi} f(x)\,dx$$

so that Q_nf is exact for $f\in\{1, \sin kx, \cos kx, k=1(1)n-1, \sin nx\}$. Hint: check whether R_nf satisfies these conditions (cf. Section 2.3.1).

P.2.14: Find nodes α, β and weights of

$$\overset{*}{I}f\doteq Q_4f=A_1f(-1)+A_2f(0)+A_3f(\alpha)+A_4f(\beta)$$

so that Q_4f has the highest possible degree. Find the leading error term (cf. Section 2.3.3).

P.2.15: Let Ω be the unit circle with centre $\mathbf{0}$. Let nodes be distributed on two circles concentric to Ω. Determine the radii such that the cubature formulae can have as high degree as possible (cf. Section 2.3.5.)

P.2.16: Apply Theorem 2.3.4 in order to obtain a cubature formula for the parallelogram with edges (0, 0), (2, 0), (1, 3), and (3, 3) from a formula for a square with four nodes and highest possible degree (cf. Section 2.3.5).

P.2.17: Determine equally-weighted quadrature formulae like

$$\int_0^1 x^\alpha f(x)\,dx\doteq A\sum_{i=1}^{n} f(x_i), \qquad \alpha>-1,$$

by calculating the first nodal polynomials and determining their zeros (cf. Section 2.4.2).

P.2.18: Find the connection between the interpolating operators:

$$L_n(f,x) \quad\text{and}\quad \Pi_n(f,x)\equiv f(x_1)^{\lambda_1(x)}f(x_2)^{\lambda_2(x)}\dots f(x_n)^{\lambda_n(x)}.$$

Compare $\Pi_2(f, x)$ with $L_2(f, x)$ for $\lambda_1(x) = \dfrac{x - x_1}{x_2 - x_1}$ and $\lambda_2(x) = \dfrac{x - x_2}{x_1 - x_2}$. What value is obtained for $x = \frac{1}{2}(x_1 + x_2)$? (cf. Section 2.5.1).

P.2.19: Construct $L_n(f, x)$ for the nodes $x_1 = -1$, $x_2 = 0$, $x_3 = 1$, and obtain the corresponding quadrature formulae for the weight functions $W(x) \equiv 1$, $W(x) = x^2$, $W(x) = 1 - x^2$, and $W(x) = (1 - x^2)^{-1/2}$ and the interval $[-1, 1]$ by integration. What are the errors? (cf. Section 2.5.2).

P.2.20: Let $x_i = 0, \pm h, \pm 2h$, etc. Replace $\sin(\pi x/h)$ by a quadratic parabola in each interval $[x_i, x_{i+1}]$. The resulting function is $\omega_h(x)$. What is the corresponding $L_h(f, x)$ for $h_i(x) = x - ih$? (cf. Section 2.5.2).

P.2.21: Show: Let $h_i(x) = \sin\left(\frac{\pi}{2}(x - x_i)\right)$, and $\omega_n = \prod_{i=1}^{n} h_i(x)$. Let

$$x_i = -1 + \frac{2i}{2n+1}, \qquad i = 0(1)2n.$$

Then

$$\lambda_i(x) = \frac{\omega_n(x) h_i'(x_i)}{\omega_n'(x_i) h_i(x)} = \frac{1}{2n+1} \frac{\sin\left(\frac{2n+1}{2}\pi(x - x_i)\right)}{\sin\left(\frac{\pi}{2}(x - x_i)\right)} \quad \text{(cf. Section 2.5.2).}$$

P.2.22: Calculate the cardinal cubic splines $\lambda_i(x)$ for the nodes $x_1 = -1$, $x_2 = -1/3$, $x_3 = 1/3$, and $x_4 = 1$ and integrate over $[-1, 1]$ (cf. Section 2.5.2).

P.2.23: Determine $H_3(f, x)$ for the nodes 0 and $\pm\alpha$ and integrate $H_3(f, x)$ over $[-1, 1]$ (cf. Section 2.5.4).

P.2.24: Calculate the mixed interpolating polynomial $HL_{1,2}(f, x)$ with $z_1 = -1$, $z_2 = 1$, and $x_1 = 0$ and integrate over $[-1, 1]$ (cf. Section 2.5.6).

P.2.25: Determine the error $e_{n,m}f$ of the mixed interpolation (cf. Section 2.5.6) in detail.

P.2.26: Calculate the expansion of the "lattice point formula"

$$C_5 F = \frac{1}{5} \sum_{k=0}^{4} f\left(\frac{k}{F_5}, \left\{\frac{kF_4}{F_5}\right\}\right)$$

(cf. Section 2.7.1). The interior nodes are on a circle round the centre!

P.2.27: Apply the Monte Carlo quadrature program to the function

$$f(\mathbf{x}) = \prod_{i=1}^{N} \sin^2(\pi x_i)$$

and compare with $I_\Omega f = 1/2^N$ for $N = 1, 2, 4, 8, 16, 32$ where Ω is the N-cube $0 \le x_i \le 1$ (cf. Section 2.7.1).

2.8.3 References

1. Ahlberg, J. H., Nilson, E. N. and Walsh, J. L. (1967). "The Theory of Splines and Their Applications." Academic Press, London.

2. Albrecht, J. (1960). Formeln zur numerischen Integration über Kreisbereiche. *Zeitschr. Angew. Math. Mech.*, **40,** 514–517.
3. Albrecht, J. and Collatz, L. (1958). Zur numerischen Auswertung mehrfacher Integrale. *Zeitschr. Angew. Math. Mech.* **38,** 1–15.
4. Albrecht, J. and Engels, H. (1977). "Cubature Formulae for the Circle." Lecture Notes, Volume 571. Springer-Verlag, Berlin, pp. 1–5.
5. August, F. (1881). Über eine Verallgemeinerung der Gauss'schen Methode der mechanischen Quadratur. *Arch. f. Mathem. u. Phys.*, **67,** 72–93.
6. Böhmer, K. (1974). "Spline-Funktionen." Teubner, Stuttgart.
7. Böhmer, K., Meinardus, G. and Schempp, W. (1974) "Spline-Funktionen." (A collection of lectures and papers concerning splines.) B-I Series, Mannheim.
8. Engels, H. (1970). Über gleichgewichtete Kubaturformeln für Kreis- und Sechseckgebiet. *Elektron. Datenverarb.*, 216–223.
9. Engels, H. (1972). Über allgemeine Gauss'sche Quadraturen. *Computing*, **10,** 83–95.
10. Filippi, S. and Engels, H. (1967). Neues über die numerische Behandlung zweidimensionaler Integrale., *Elektron. Datenverarb.*, 353–360.
11. Flatto, L. and Haber, S. (1973). A quadrature formula of degree three. *J. of Approx. Theory*, **9,** 44–52.
12. Franke, R. (1971). Obtaining cubatures for rectangles and other planar regions by using orthogonal polynomials. *Math. of. Comput.*, **25,** 803–817.
13. Franke, R. (1973). Minimal point cubatures of precision seven for symmetric planar regions., *SIAM J. Numer. Anal.*, **10,** 849–862.
14. Haegemans, A. (1975). "Constructie van Cubatuurformules voor Tweedimensionale Integralen met Behulp van Orthogonale Veeltermen." Ph.D. Thesis, Leuven.
15. Haegemans, A. (to appear). Circularly symmetrical formulas of degree $2d-1$ for two-dimensional circularly symmetrical regions. BIT.
16. Hammer, P. and Wymore, A. W. (1957). Numerical evaluation of multiple integrals. I. *MTAC*, **11,** 59–67.
17. Hlawka, E. (1962). Zur angenäherten Berechnung mehrfacher Integrale. *Monatshefte Math.*, **66,** 140–151.
18. Huelsman, C. B. III. (1973). Quadrature formulae over fully symmetric planar regions. *SIAM J. Numer. Anal.* **10,** 539–552.
19. Korobov, N. M. (1957). Approximate calculation of multiple integrals with the aid of methods in the theory of numbers. *MR*, **20,** 855#5169 (1959). [Russian: *Dokl. Akad. Nauk SSSR*, **115,** 1062–1065.
20. Korobov, N. M. (1959). On certain number theoretic methods for approximate computation of multiple integrals. (Russian) *Uspehi Math. Nauk*, **14,** 227–230.
21. Korobov, N. M. (1961). Computation of multiple integrals by the method of optimal coefficients. (Russian) *Vestnik Moskow Univ. Ser. Math. Meh. Astr. Fiz. I. Khim*, no. 4, **4,** 19–25. MDF U-227 (Engl.) *MR*, **22,** 824#4913.
22. Kress, R. (1972). On the general Hermite cardinal interpolation. *Math. of Comput.*, **26,** 925–933.
23. Natanson, I. P. (1955). "Konstruktive Funktionentheorie." Akademie-Verlag, Berlin. [English translation: (1964–65). Three volumes. Frederik Unger, New York.]
24. Radau, R. (1880). Sur les formules de quadrature à coefficients égaux. *C.R. Acad. Sci. Paris*, **90,** 520–523.

25. Sard, A. and Weintraub, S. (1971). "A Book of Splines." Wiley, New York.
26. Schoenberg, I. J. (1970). Cardinal interpolation and spline functions. *In* "Inequalities," Volume 2 (O. Shisha, Ed.). Academic Press, New York.
27. Schultz, M. H. (1973). "Spline Analysis." Prentice Hall, New Jersey.
28. Schurer, F. and van Rooij P. L. J. (1973). "A Bibliography on Spline Functions. II." (Extended version of an earlier bibliography.) Techn. Univ. Eindhoven, Rep. 73-WSK-01.
29. Tsuda, T. (1973). Numerical integration of functions of very many variables., *Numer. Math.*, **20,** 377–391.
30. Varga, R. (1971). Functional analysis and approximation theory in numerical analysis. *SIAM, Philadelphia*, **3**.
31. Zaremba, S. K. (1966). Good lattice points, discrepancy, and numerical integration., *Ann. Mat. Pura Appl.*, **73,** 293–317.
32. Zaremba, S. K. (1968). Good lattice points in the sense of Hlawka and Monte Carlo integration. *Monatshefte Math.*, **72,** 264–269.
33. Zaremba, S. K. (1968). A quasi Monte Carlo method for computing double and other multiple integrals. *Aequationes Mathematicae*, **2,** 8–19.

3

Error Analysis for Quadrature and Cubature Formulae

Throughout Chapter 2 we considered the problem of constructing quadrature and cubature formulae. There was, however, no possibility of deciding whether one formula was in any sense "better" for a given situation than another; for example, when the formulae had the same number of nodes.

In order to avoid the construction of senseless formulae, we have to state suitable criteria which should be met by all formulae. One possibility is given by the degree of precision, so that we need to determine the nodes and weights such that the resulting formula has the highest possible degree of precision when the nodes and weights are constrained to be real.

In this chapter we develop some techniques for minimising some property of a quadrature or cubature formula, thus determining the nodes, or the weights or both.

3.1 Error representations for quadratures

The most natural way of finding "good" formulae is to minimise the norm of the error in some sense. Hence we look first for suitable representations of the error. Only one- and two-dimensional cases are discussed. Although the extension to N dimensions is quite obvious, it would take up too much space here.

3.1.1 Error representation by Peano kernels

The Taylor expansion of the integrand $f(x)$ has, up to now, been used in a truncated form in which the remainder part was neglected. We now use this part in order to give a strict representation of the error of a quadrature formula Q_nf of degree k.

Let $f \in C^{k+1}[0, 1]$, $k \geq 0$. Then

$$f(x) = f(0) + \int_0^x f'(t)\,\mathrm{d}t.$$

Integration by parts leads to

$$f(x) = f(0) + xf'(x) - \int_0^x tf''(t)\,\mathrm{d}t.$$

Now

$$xf'(x) = x\left[f'(0) + \int_0^x f''(t)\,\mathrm{d}t\right],$$

and therefore

$$f(x) = f(0) + xf'(0) + \int_0^x (x-t)f''(t)\,\mathrm{d}t.$$

Proceeding in this way, after $k-1$ further integrations by parts, we finally obtain

$$f(x) = \sum_{\nu=0}^{k} \frac{f^{(\nu)}(0)}{\nu!} x^\nu + \int_0^x \frac{(x-t)^k}{k!} f^{(k+1)}(t)\,\mathrm{d}t.$$

In order to make the upper limit of the integral independent of x we define the *positive part* of the powers of $(x-t)$ by

$$(x-t)_+^m = \begin{cases} (x-t)^m & \text{for} \quad x \geq t \\ 0 & \text{for} \quad x < t \end{cases} \qquad m = 0(1)k.$$

Thus (the upper limit ∞ is also possible)

$$f(x) = \sum_{\nu=0}^{k} \frac{f^{(\nu)}(0)}{\nu!} x^\nu + \frac{1}{k!} \int_0^1 (x-t)_+^k f^{(k+1)}(t)\,\mathrm{d}t. \tag{3.1.1}$$

Note that (3.1.1) is a strict equation. In other words: the right-hand side of (3.1.1) is just a more complicated way of writing $f(x)$. We now apply the error functional E_n to (3.1.1) and so obtain

$$E_nf(x) = \sum_{\nu=0}^{k} \frac{f^{(\nu)}(0)}{\nu!} E_n(x^\nu) + \frac{1}{k!} E_n\left\{\int_0^1 (x-t)_+^k f^{(k+1)}(t)\,\mathrm{d}t\right\}.$$

Let Q_nf have the degree k. Then $E_n(x^\nu)=0$, $\nu=0(1)k$, and

$$E_nf=\frac{1}{k!}E_n\left\{\int_0^1 (x-t)_+^k f^{(k+1)}(t)\,dt\right\}.$$

Because of the continuity of the integrand and the linearity of the functionals E_nf and $\overset{*}{I}f$, we have that for $k>0$

$$E_nf=\frac{1}{k!}\int_0^1 [E_n^x(x-t)_+^k]f^{(k+1)}(t)\,dt \tag{3.1.2}$$

where the upper index x indicates that E_n^x is to apply with respect to x.

For $k=0$ the integrand is only piecewise continuous and has jumps at $x=x_i$, and at $x=t$. By subdividing $[0, 1]$ in $[0, x_1]$, $[x_1, x_2], \ldots, [x_j, t]$, $[t, x_{j+1}], \ldots, [x_n, 1]$ we can also state (3.1.2) for $k=0$.

Definition 3.1.1 The function

$$K(t)\equiv K_{n,k}(t)\equiv E_n^x(x-t)_+^k$$

is called the *Peano kernel* of the quadrature formula Q_nf of degree k. ■

If the degree of Q_nf is $m>k$ but $f\in C^{k+1}[0, 1]$, the representation of E_nf cannot be changed. However, if the degree of Q_nf is $m<k$, then we have a choice between

$$E_nf=\frac{1}{m!}\int_0^1 K_{n,m}(t)f^{(m+1)}(t)\,dt$$

and

$$E_nf=\sum_{\nu=m+1}^{k}\frac{f^{(\nu)}(0)}{\nu!}E_n(x^\nu)+\frac{1}{k!}\int_0^1 K_{n,k}(t)f^{(k+1)}(t)\,dt. \tag{3.1.3}$$

Usually, we take the first possibility and omit the subscripts of $K(t)$.

It is a matter of scaling to define E_nf for $[-1, 1]$. We collect the arguments in the following theorem:

Theorem 3.1.1 *Let $f\in C^{k+1}[-1, 1]$, and Q_nf be a quadrature formula of degree $m\geqslant k\geqslant 0$. Then the error-functional $E_nf=If-Q_nf$ has the representation*

$$E_nf=\frac{1}{k!}\int_{-1}^1 K(t)f^{(k+1)}(t)\,dt,$$

where

$$K(t) \equiv K_{n,k}(t) = E_n^x[(x-t)_+^k]$$

is the Peano kernel of $Q_n f$.

Note that $K(t)$ is a piecewise polynomial and therefore a spline. In order to illustrate the above we give some examples.

Example 3.1.1 We calculate some Peano kernels of the Radau formula of degree 2:

$$\int_0^1 f(x)\,dx \doteq \tfrac{1}{4}[3f(\tfrac{1}{3}) + f(1)] = Q_2 f \tag{3.1.4}$$

with error

$$E_2 f = \frac{1}{k!}\int_0^1 K_{2,k}(t) f^{(k+1)}(t)\,dt, \qquad k = 0(1)3.$$

Case $k = 0$:

$$\begin{aligned} K_{2,0}(t) &= E_2^x[(x-t)_+^0] \\ &= \int_0^1 (x-t)_+^0\,dx - \tfrac{1}{4}[3(\tfrac{1}{3}-t)_+^0 + (1-t)_+^0] \\ &= \int_t^1 dx - \begin{cases} \tfrac{3}{4}(\tfrac{1}{3}-t)^0 + \tfrac{1}{4}(1-t)^0 & \text{for } 0 \leqslant t \leqslant \tfrac{1}{3} \\ \tfrac{1}{4}(1-t)^0 & \text{for } \tfrac{1}{3} < t \leqslant 1 \end{cases} \\ &= \begin{cases} (1-t) - 1 = -t & \text{for } 0 \leqslant t \leqslant \tfrac{1}{3} \\ (1-t) - \tfrac{1}{4} = \tfrac{3}{4} - t & \text{for } \tfrac{1}{3} < t \leqslant 1 \end{cases} \end{aligned}$$

as shown in Fig. 3.1.1(a).

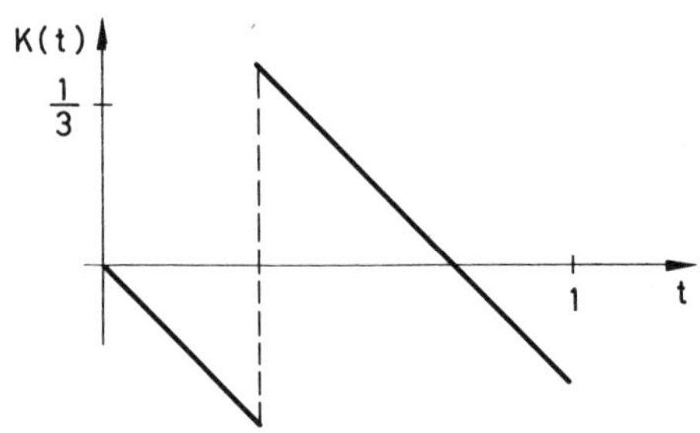

FIG. 3.1.1(a). Peano kernel of (3.1.4) for $k = 0$.

Case $k=1$:

$$K_{2,1}(t)=\begin{cases}\frac{1}{2}(1-t)^2-\frac{3}{4}(\frac{1}{3}-t)-\frac{1}{4}(1-t)=\frac{1}{2}t^2 & \text{for } 0\leqslant t\leqslant\frac{1}{3}\\ \frac{1}{2}(1-t)^2-\frac{1}{4}(1-t)=\frac{1}{4}(1-t)(1-2t) & \text{for } \frac{1}{3}<t\leqslant 1\end{cases}$$

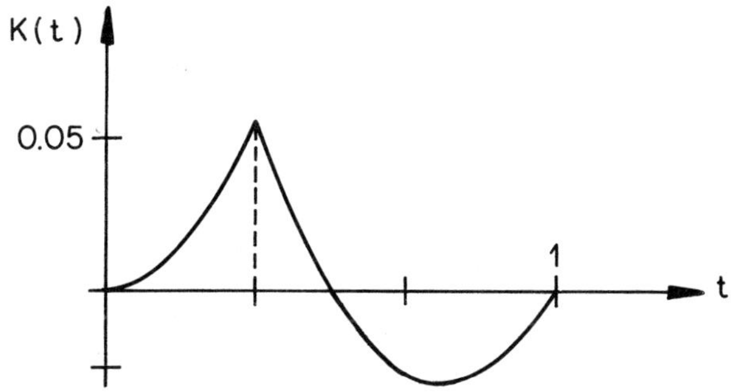

FIG. 3.1.1(b). Peano kernel of (3.1.4) for $k=1$.

Case $k=2$:

$$K_{2,2}(t)=\begin{cases}-\frac{1}{3}t^3 & \text{for } 0\leqslant t\leqslant\frac{1}{3}\\ \frac{1}{12}(1-t)^2(1-4t) & \text{for } \frac{1}{3}<t\leqslant 1\end{cases}$$

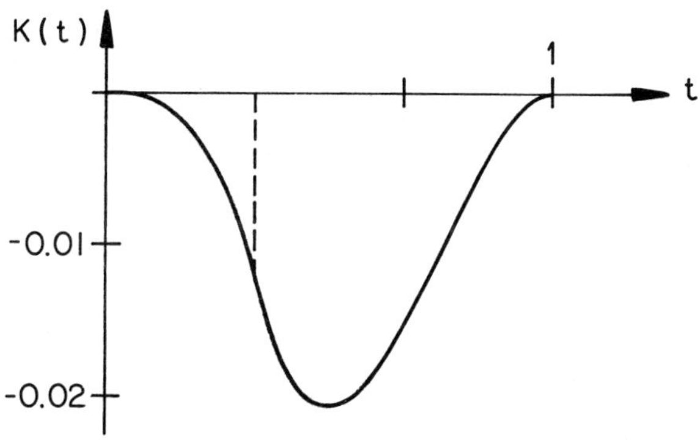

FIG. 3.1.1(c). Peano kernel of (3.1.4) for $k=2$.

Case $k=3$:

$$K_{2,3}(t)=\begin{cases}\frac{1}{4}(t^4-\frac{1}{9}) & \text{for } 0\leqslant t\leqslant\frac{1}{3}\\ -\frac{1}{4}t(1-t)^3 & \text{for } \frac{1}{3}<t\leqslant 1\end{cases}$$

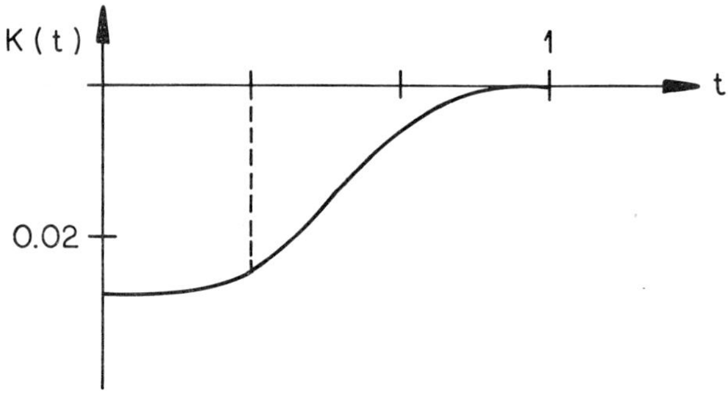

FIG. 3.1.1(d). Peano kernel of (3.1.4) for $k=3$.

The corresponding error representation is

$$E_2f=\frac{1}{3!}f'''(0)E_2x^3+\frac{1}{3!}\int_0^1 K_{2,3}(t)f^{(4)}(t)\,\mathrm{d}t. \quad \blacksquare$$

Example 3.1.2 We present the Peano kernel of

$$E_2f=\int_0^\infty \mathrm{e}^{-x}f(x)\,\mathrm{d}x-\tfrac{1}{7}[6f(0.6)+f(3.4)] \tag{3.1.5}$$

of degree 1. Here, the domain of the integral is infinite and it is easy to see that the Peano kernel representation of the error E_nf is again valid.

In the same way as in the previous example we obtain the kernel

$$K(t)=\begin{cases}\displaystyle\int_t^\infty \mathrm{e}^{-x}(x-t)\,\mathrm{d}x-\tfrac{1}{7}[6(0.6-t)+(3.4-t)] & \text{for}\quad 0\leqslant t\leqslant 0.6\\ \displaystyle\int_t^\infty \mathrm{e}^{-x}(x-t)\,\mathrm{d}x-\tfrac{1}{7}(3.4-t) & \text{for}\quad 0.6<t\leqslant 3.4\\ \displaystyle\int_t^\infty \mathrm{e}^{-x}(x-t)\,\mathrm{d}x & \text{for}\quad 3.4<t<\infty\end{cases}$$

as shown in Fig. 3.1.1(e).

More generally, we have that for $t\in(x_n,\infty)$ the Peano kernel is

$$K(t)=\frac{1}{k!}\int_t^\infty \mathrm{e}^{-x}(x-t)^k\,\mathrm{d}x$$

for a k-th degree formula $Q_nf=\sum_{i=1}^n A_if(x_i)\doteq\int_0^\infty \mathrm{e}^{-x}f(x)\,\mathrm{d}x$, where $x_1<x_2<\ldots<x_n$. $\blacksquare$

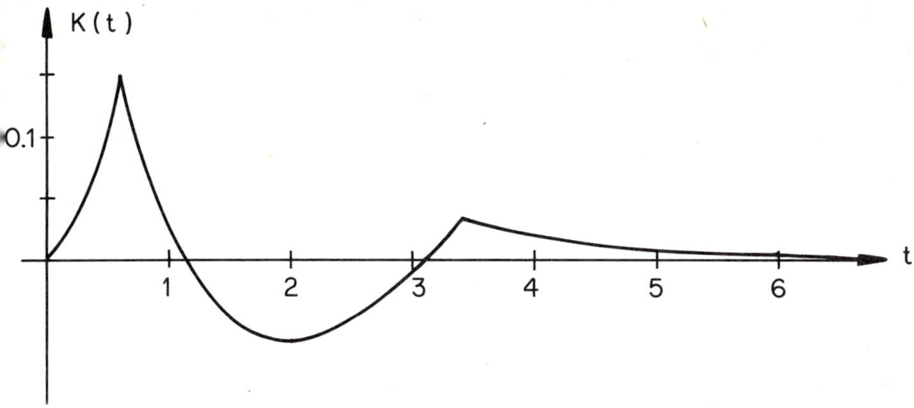

FIG. 3.1.1(e). Peano kernel of (3.1.5) for $k=1$.

3.1.2 Some properties of Peano kernels

We now state some properties of $K(t)$ which are useful for quadrature purposes. For convenience, we consider the interval $[-1, 1]$ instead of $[0, 1]$.

Theorem 3.1.2 *For given If and Q_nf the corresponding Peano kernel $K_{n,k}(t)$ for fixed k is unique.*

The proof is an immediate consequence of the method of construction of the kernel.

Example 3.1.1 suggests that the kernels $K_{n,i}(t)$ for $i<k$ are derivatives of $K_{n,k}(t)$ subject to a suitable multiplication. We compare the representations for $k=i$ and $k=i+1$, which are, respectively:

$$E_nf=\frac{1}{i!}\int_{-1}^{1} K_{n,i}(t)f^{(i+1)}(t)\,\mathrm{d}t$$

and

$$\overset{*}{E}_nf=\frac{1}{(i+1)!}\int_{-1}^{1} K_{n,i+1}(t)f^{(i+2)}(t)\,\mathrm{d}t.$$

After integration by parts,

$$\overset{*}{E}_nf=\frac{1}{(i+1)!}K_{n,i+1}(t)f^{(i+1)}(t)\,\Big|_{-1}^{1}-\frac{1}{(i+1)!}\int_{-1}^{1} K'_{n,i+1}(t)f^{(i+1)}(t)\,\mathrm{d}t$$

which is another representation for $k=i$ and has to be identical with E_nf. Now, in general, $f^{(i+1)}(\pm 1)\neq 0$. Hence

$$\left.\begin{aligned} &K_{n,i+1}(-1)=K_{n,i+1}(1)=0 \\ \text{and}\qquad & \\ &K_{n,i}(t)=-\frac{1}{i+1}K'_{n,i+1}(t)\end{aligned}\right\}\quad \text{for}\quad i=0(1)k-1,$$

must be valid. Proceeding by induction we obtain:

Theorem 3.1.3 *The Peano kernel $K_{n,k}(t)$ has zeros of multiplicity at least k at $t=\pm 1$, and satisfies*

$$(i+1)K_{n,i}(t)+K'_{n,i+1}(t)=0 \quad \textit{for} \quad i=0(1)k-1.$$

Example 3.1.1 illustrates Theorems 3.1.2 and 3.1.3 and, in addition, shows that the multiplicity $k+1$ of the zero at $x=0$ is even. Example 3.1.1 also shows that sometimes the kernels have constant sign in $[-1, 1]$. In this case we are able to apply the mean-value theorem: there exists a point $\xi\in(-1, 1)$ such that

$$E_nf=\frac{f^{(k+1)}(\xi)}{k!}\int_{-1}^{1}K(t)\,\mathrm{d}t \quad \text{for} \quad K(t)\in C^0[-1, 1]$$

is valid. Here, the integral is computable once and for all, because it no longer depends on f, a very remarkable property which can be used to produce error estimates.

Further, if Q_nf has degree k then $E_nx^{k+1}\neq 0$ and, substituting $f=x^{k+1}$, we obtain

$$E_nx^{k+1}=(k+1)\int_{-1}^{1}K_{n,k}(t)\,\mathrm{d}t.$$

Hence, if $K_{n,k}(t)\equiv K(t)$ has constant sign in $[-1, 1]$ then

$$E_nf=\frac{f^{(k+1)}(\xi)}{(k+1)!}E_nx^{k+1}$$

is valid.

3.2 Error representation for cubature formulae by Sard kernels

For higher dimensions we now proceed in the same way as for one dimension in Section 3.1. We represent the cubature error by Sard kernels, where a Sard kernel is merely a form of the Peano kernel in more than one variable. We restrict our attention to functions of only two variables having the property $f \in C^{k+1}(\bar{\Omega})$. This is the most interesting case for applications in cubature problems since the extension to $N>2$ dimensions is straightforward. Generally, the Taylor expansion can be extended as far as partial derivatives exist.

We start with the trivial one-dimensional expansion (partial derivatives are denoted by $f_{i,j}$)

$$\begin{aligned} f(x, y) &= f(x, 0) + \int_0^y f_{0,1}(x, v)\, dv \\ &= f(\mathbf{0}) + \int_0^x f_{1,0}(u, 0)\, du + \int_0^y f_{0,1}(x, v)\, dv. \end{aligned}$$

This is the expansion for $k=0$, i.e. $f \in C^1(\bar{\Omega})$.

Let $k=1$. Then both integrals are expanded to give the expansion for $k=1$:

$$\begin{aligned} f(x, y) = f(\mathbf{0}) + x f_{1,0}(\mathbf{0}) + y f_{0,1}(\mathbf{0}) + \int_0^x (x-u) f_{2,0}(u, 0)\, du \\ + \int_0^y (y-v) f_{0,2}(0, v)\, dv + \int_0^x \int_0^y f_{1,1}(u, v)\, dv\, du. \end{aligned}$$

The extension to $k=2$ is slightly more complicated. Integration by parts leads to

$$\int_0^x (x-u) f_{2,0}(u, 0)\, du = \frac{x^2}{2!} f_{2,0}(\mathbf{0}) + \int_0^x \frac{(x-u)^2}{2!} f_{3,0}(u, 0)\, du,$$

$$\int_0^y (y-v) f_{0,2}(0, v)\, dv = \frac{y^2}{2!} f_{0,2}(\mathbf{0}) + \int_0^y \frac{(y-v)^2}{2!} f_{0,3}(0, v)\, dv,$$

and

$$\int_0^x \int_0^y f_{1,1}(u, v)\, dv\, du = \int_0^x y f_{1,1}(u, y)\, du - \int_0^x \int_0^y v f_{1,2}(u, v)\, dv\, du. \qquad (3.2.1)$$

The first term is transformed into

$$\int_0^x y f_{1,1}(u, y)\, du = y \int_0^x f_{1,1}(u, 0)\, du + \int_0^x \int_0^y y f_{1,2}(u, v)\, dv\, du.$$

Another integration by parts gives

$$y \int_0^x f_{1,1}(u, 0)\, du = xy f_{1,1}(\mathbf{0}) + y \int_0^x (x-u) f_{2,1}(u, 0)\, du.$$

Hence, we finally have that

$$\int_0^x \int_0^y f_{1,1}(u, v)\, dv\, du = xyf_{1,1}(\mathbf{0}) + y\int_0^x (x-u)f_{2,1}(u, 0)\, du + \int_0^x \int_0^y (y-v)f_{1,2}(u, v)\, dv\, du.$$

In (3.2.1) the integration by parts can be performed with respect fo x instead of y. This leads to

$$\int_0^x \int_0^y f_{1,1}(u, v)\, dv\, du = xyf_{1,1}(\mathbf{0}) + x\int_0^y (y-v)f_{1,2}(0, v)\, dv + \int_0^x \int_0^y (x-u)f_{2,1}(u, v)\, dv\, du.$$

There is no essential difference between these two versions of the integral of $f_{1,1}(u, v)$, and we can average them to obtain

$$\begin{aligned}\int_0^x \int_0^y f_{1,1}(u, v)\, dv\, du &= xyf_{1,1}(\mathbf{0}) \\ &+ \frac{x}{2}\int_0^y (y-v)f_{1,2}(0, v)\, dv + \frac{y}{2}\int_0^x (x-u)f_{2,1}(u, 0)\, du \\ &+ \frac{1}{2}\int_0^x \int_0^y [(x-u)f_{2,1}(u, v) + (y-v)f_{1,2}(u, v)]\, dv\, du.\end{aligned}$$

This form for the integral seems to be the most convenient for applications in cubature problems.

Finally, for $k = 2$ we have that

$$\begin{aligned}f(x, y) &= f(\mathbf{0}) + xf_{1,0}(\mathbf{0}) + yf_{0,1}(\mathbf{0}) + \frac{1}{2!}\{x^2 f_{2,0}(\mathbf{0}) + 2xyf_{1,1}(\mathbf{0}) + y^2 f_{0,2}(\mathbf{0})\} \\ &+ \frac{1}{2!}\left\{\int_0^x (x-u)^2 f_{3,0}(u, 0)\, du + \int_0^y (y-v)^2 f_{0,3}(0, v)\, dv\right. \\ &+ y\int_0^x (x-u)f_{2,1}(u, 0)\, du + x\int_0^y (y-v)f_{1,2}(0, v)\, dv \\ &+ \left.\int_0^x \int_0^y [(x-u)f_{2,1}(u, v) + (y-v)f_{1,2}(u, v)]\, dv\, du\right\}.\end{aligned}$$

Proceeding by induction we obtain

Theorem 3.2.1 *Let $f \in C^{k+1}(\bar{\Omega})$, $k \geq 0$. Then for odd $k = 2\varkappa + 1$*

$$f(x, y) = \sum_{\nu=0}^{k} \sum_{\mu=0}^{\nu} \frac{1}{\nu!} \binom{\nu}{\mu} x^{\nu-\mu} y^{\mu} f_{\nu-\mu,\mu}(\mathbf{0})$$

$$+ \frac{1}{k!} \sum_{\mu=0}^{\varkappa} \binom{k}{\mu} \Big\{ y^{\mu} \int_0^x (x-u)^{k-\mu} f_{k+1-\mu,\mu}(u, 0)\, du$$

$$+ x^{\mu} \int_0^y (y-v)^{k-\mu} f_{\mu,k+1-\mu}(0, v)\, dv \Big\}$$

$$+ \frac{1}{\varkappa!\,\varkappa!} \int_0^x \int_0^y (x-u)^{\varkappa} (y-v)^{\varkappa} f_{\varkappa+1,\varkappa+1}(u, v)\, dv\, du,$$

and for even $k = 2\varkappa$

$$f(x, y) = \sum_{\nu=0}^{k} \sum_{\mu=0}^{\nu} \frac{1}{\nu!} \binom{\nu}{\mu} x^{\nu-\mu} y^{\mu} f_{\nu-\mu,\mu}(\mathbf{0})$$

$$+ \frac{1}{k!} \sum_{\mu=0}^{\varkappa}{}' \binom{k}{\mu} \Big\{ y^{\mu} \int_0^x (x-u)^{k-\mu} f_{k+1-\mu,\mu}(u, 0)\, du$$

$$+ x^{\mu} \int_0^y (y-v)^{k-\mu} f_{\mu,k+1-\mu}(0, v)\, dv \Big\}$$

$$+ \frac{1}{2(\varkappa-1)!\,\varkappa!} \int_0^x \int_0^y (x-u)^{\varkappa-1} (y-v)^{\varkappa-1} [(x-u) f_{\varkappa+1,\varkappa}(u, v)$$

$$+ (y-v) f_{\varkappa,\varkappa+1}(u, v)]\, dv\, du$$

with

$$\sum_{\mu=0}^{\varkappa}{}' a_{\mu} \equiv a_0 + a_1 + \ldots + a_{\varkappa-1} + \tfrac{1}{2} a_{\varkappa}.$$

Having now an expansion of $f(x, y)$ with an exact error, we can apply to this an error functional E_n. For the sake of simplicity we assume that all points $(x, y) \in \bar{\Omega}$ satisfy $x \geq 0$, $y \geq 0$, and also that $C_n f$ has degree k.

Then for $k \geqslant 0$

$$
\begin{aligned}
E_n f &= I_\Omega f - C_n f \\
&= \frac{1}{k!} \sum_{\mu=0}^{\varkappa}{}^{*} \binom{k}{\mu} E_n^{x,y} \left\{ y^\mu \int_0^x (x-u)^{k-\mu} f_{k+1-\mu,\mu}(u,0)\, du \right. \\
&\qquad \left. + x^\mu \int_0^y (y-v)^{k-\mu} f_{\mu,k+1-\mu}(0,v)\, dv \right\} \qquad (3.2.2) \\
&\quad + \begin{cases} \dfrac{1}{\varkappa!\,\varkappa!} E_n^{x,y} \displaystyle\int_0^x \int_0^y (x-u)^\varkappa (y-v)^\varkappa f_{\varkappa+1,\varkappa+1}(u,v)\, dv\, du & \text{for} \quad k = 2\varkappa+1 \\[2ex] \dfrac{1}{2(\varkappa-1)!\,\varkappa!} E_n^{x,y} \displaystyle\int_0^x \int_0^y [(x-u)^\varkappa (y-v)^{\varkappa-1} f_{\varkappa+1,\varkappa}(u,v) \\ \qquad + (x-u)^{\varkappa-1}(y-v)^\varkappa f_{\varkappa,\varkappa+1}(u,v)]\, dv\, du, & \text{for} \quad k = 2\varkappa \end{cases}
\end{aligned}
$$

with

$$
\sum{}^{*} = \begin{cases} \sum & \text{for} \quad k = 2\varkappa+1 \\ \sum' & \text{for} \quad k = 2\varkappa. \end{cases}
$$

As for $N=1$ we extend the domain of the integrands by taking their "positive parts", and can thus interchange the functional E_n and the integrals:

$$
\begin{aligned}
E_n f &= \frac{1}{k!} \sum_{\mu=0}^{\varkappa}{}^{*} \binom{k}{\mu} \left\{ \int_0^\tau E_n^{x,y}[y^\mu (x-u)_+^{k-\mu}] f_{k+1-\mu,\mu}(u,0)\, du \right. \\
&\quad \left. + \int_0^\sigma E_n^{x,y}[x^\mu (y-v)_+^{k-\mu}] f_{\mu,k+1-\mu}(0,v)\, dv \right\}, \qquad \tau = \tau(y), \qquad \sigma = \sigma(x). \\
&\quad + \begin{cases} \dfrac{1}{\varkappa!\,\varkappa!} \displaystyle\iint_\Omega [E_n^{x,y}(x-u)_+^\varkappa (y-v)_+^\varkappa] f_{\varkappa+1,\varkappa+1}(u,v)\, dv\, du & \text{for} \quad k = 2\varkappa+1 \\[2ex] \dfrac{1}{2(\varkappa-1)!\,\varkappa!} \displaystyle\iint_\Omega \{[E_n^{x,y}(x-u)_+^\varkappa (y-v)_+^{\varkappa-1}] f_{\varkappa+1,\varkappa}(u,v) \\ \qquad + [E_n^{x,y}(x-u)_+^{\varkappa-1}(y-v)_+^\varkappa] f_{\varkappa,\varkappa+1}(u,v)\}\, dv\, du & \text{for} \quad k = 2\varkappa. \end{cases}
\end{aligned}
$$

Here $\tau = \tau(y) > 0$, $\sigma = \sigma(x) > 0$ are numbers such that $(\tau, y) \in \partial\Omega$, and $(x, \sigma) \in \partial\Omega$ (cf. Fig. 3.2.1).

We define the Sard kernel functions $K_1^\mu(u)$, $K_2^\mu(v)$ and $K_{\varkappa,\lambda}(u,v)$ of $E_n f$.

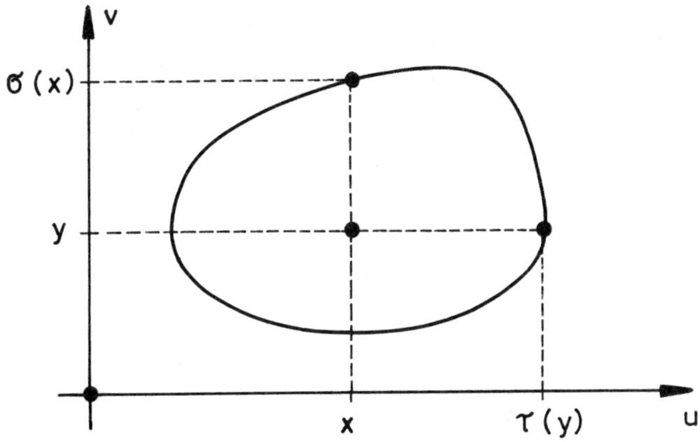

FIG. 3.2.1. Definition of σ and τ for given x, y.

Definition 3.2.1 Let $k>0$, and

$$\begin{aligned}
K_1^{\mu}(u) &\equiv E_n^{x,y}[y^{\mu}(x-u)_+^{k-\mu}], && \mu=0(1)\varkappa, && (x,y)\in\bar{\Omega}\\
K_2^{\mu}(v) &\equiv E_n^{x,y}[x^{\mu}(y-v)_+^{k-\mu}], && \mu=0(1)\varkappa, && (x,y)\in\bar{\Omega}\\
K_{\varkappa,\varkappa}(u,v) &\equiv E_n^{x,y}[(x-u)_+^{\varkappa}(y-v)_+^{\varkappa}], && k=2\varkappa+1, && (x,y)\in\bar{\Omega}\\
K_{\varkappa,\varkappa-1}(u,v) &\equiv E_n^{x,y}[(x-u)_+^{\varkappa}(y-v)_+^{\varkappa-1}], && k=2\varkappa, && (x,y)\in\bar{\Omega}
\end{aligned}$$

The functions $K_1^{\mu}(u)$, $K_2^{\mu}(v)$, and $K_{\varkappa,\lambda}(u,v)$ are called *Sard kernel functions of* E_nf. ■

Using the Sard kernel functions we have:

Theorem 3.2.2 *The functional* $E_nf=I_\Omega f-C_nf$ *has the representation*

$$E_nf=\frac{1}{k!}\sum_{\mu=0}^{\varkappa}{}^{*}\binom{k}{\mu}\left[\int_0^{\tau}K_1^{\mu}(u)f_{k+1-\mu,\mu}(u,0)\,du+\int_0^{\sigma}K_2^{\mu}(v)f_{\mu,k+1-\mu}(0,v)\,dv\right]$$

$$+\begin{cases}\dfrac{1}{\varkappa!\,\varkappa!}\displaystyle\iint_{\Omega}K_{\varkappa,\varkappa}(u,v)f_{\varkappa+1,\varkappa+1}(u,v)\,du\,dv, & k=2\varkappa+1\\[2ex]
\dfrac{1}{2(\varkappa-1)!\,\varkappa!}\left\{\displaystyle\iint_{\Omega}K_{\varkappa,\varkappa-1}(u,v)f_{\varkappa+1,\varkappa}(u,v)\,du\,dv\right.\\
\qquad\left.+\displaystyle\iint_{\Omega}K_{\varkappa-1,\varkappa}(u,v)f_{\varkappa,\varkappa+1}(u,v)\,du\,dv\right\}, & k=2\varkappa\end{cases}$$

for $f\in C^{k+1}(\bar{\Omega})$.

To illustrate the construction of Sard kernels we have:

Example 3.2.1 We calculate the error representation of the functional

$$E_1 f = \int_0^1 \int_0^1 f(x, y)\, dx\, dy - f(\tfrac{1}{2}, \tfrac{1}{2}) \tag{3.2.3}$$

for $f \in C^2(\bar{\Omega})$. Thus $k = 1$ and $\varkappa = 0$ so that

$$K_1^0(u) = E_1^{x,y}[(x-u)_+] = \int_0^1 \int_0^1 (x-u)_+\, dx\, dy - (\tfrac{1}{2} - u)_+$$

$$= \int_u^1 (x-u)\, dx - \begin{cases} \frac{1}{2} - u & \text{for} \quad 0 \leq u \leq \frac{1}{2} \\ 0 & \text{otherwise} \end{cases}$$

$$= \begin{cases} \frac{1}{2}u^2 & \text{for} \quad 0 \leq u \leq \frac{1}{2} \\ \frac{1}{2}(1-u)^2 & \text{otherwise} \end{cases}$$

Analogously,

$$K_2^0(v) = \begin{cases} \frac{1}{2}v^2 & \text{for} \quad 0 \leq v \leq \frac{1}{2} \\ \frac{1}{2}(1-v)^2 & \text{otherwise} \end{cases}$$

Further, we obtain

$$K_{0,0}(u, v) = E_1^{x,y}[(x-u)_+^0 (y-v)_+^0]$$

$$= \int_0^1 \int_0^1 (x-u)_+^0 (y-v)_+^0\, dx\, dy - (\tfrac{1}{2} - u)_+^0 (\tfrac{1}{2} - v)_+^0$$

$$= \int_u^1 \int_v^1 dy\, dx - \begin{cases} 1 & \text{for} \quad 0 \leq u, v \leq \frac{1}{2} \\ 0 & \text{otherwise.} \end{cases}$$

Hence

$$K(u, v) = \begin{cases} uv - u - v & \text{for} \quad 0 \leq u, v \leq \frac{1}{2} \\ (1-u)(1-v) & \text{otherwise} \end{cases}$$

We thus have all desired kernels. As the kernels $K_1^0(u)$ and $K_2^0(v)$ have constant sign, there exist points $(\xi_1, 0)$ and $(0, \xi_2)$ such that the error representation is

$$E_1 f = f_{2,0}(\xi_1, 0) \int_0^1 K_1^0(u)\, du + f_{0,2}(0, \xi_2) \int_0^1 K_2^0(v)\, dv$$

$$+ \int_0^1 \int_0^1 f_{1,1}(u, v) K_{0,0}(u, v)\, dv\, du$$

$$= \tfrac{1}{24}[f_{2,0}(\xi_1, 0) + f_{0,2}(0, \xi_2)] + \int_0^1 \int_0^1 f_{1,1}(u, v) K_{0,0}(u, v)\, dv\, du$$

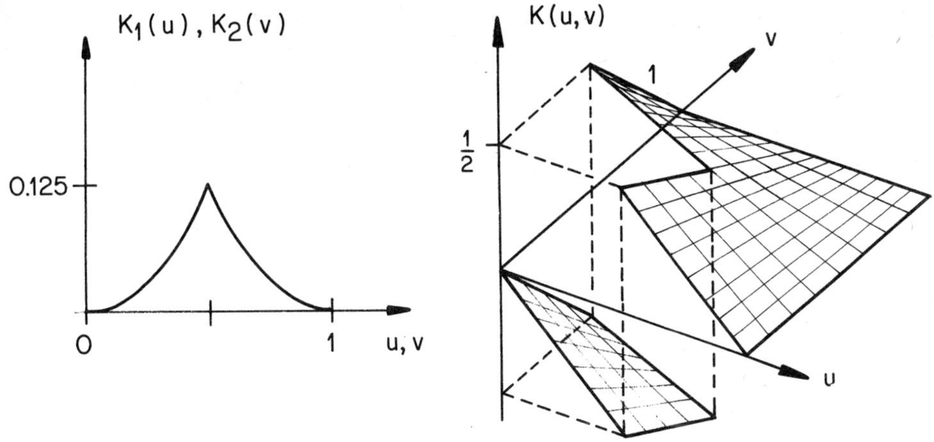

FIG. 3.2.2. Sard kernel of (3.2.3) for $k = 1$.

However, because $K(u, v)$ is not of constant sign, we cannot here apply the general mean-value theorem to the double integral. ■

In direct analogy with the one-dimensional case, the error representation is simpler if *all Sard kernels have constant sign.*

Because of the general mean-value theorem we have the existence of points $(\xi_i, 0)$, $(0, \eta_j)$ and (s, t), (s_1, t_1), $(s_2, t_2) \in \Omega$ such that

$$E_n f = \frac{1}{k!} \sum_{\mu=0}^{\varkappa}{}^{*} \binom{k}{\mu} \left[f_{k+1-\mu,\mu}(\xi_\mu, 0) \int_0^\tau K_1^\mu(u)\, du + f_{\mu,k+1-\mu}(0, \eta_\mu) \int_0^\sigma K_2^\mu(v)\, dv \right]$$

$$+ \begin{cases} \dfrac{1}{\varkappa!\,\varkappa!} f_{\varkappa+1,\varkappa+1}(s, t) \displaystyle\iint_\Omega K_{\varkappa,\varkappa}(u, v)\, du\, dv, & k = 2\varkappa + 1 \\[2ex] \dfrac{1}{2(\varkappa-1)!\,\varkappa!} \left\{ f_{\varkappa+1,\varkappa}(s_1, t_1) \displaystyle\iint_\Omega K_{\varkappa,\kappa-1}(u, v)\, du\, dv \right. & \\[2ex] \qquad \left. + f_{\varkappa,\varkappa+1}(s_2, t_2) \displaystyle\iint_\Omega K_{\varkappa-1,\varkappa}(u, v)\, du\, dv \right\} & k = 2\varkappa \end{cases}$$

for $f \in C^{k+1}(\bar{\Omega})$.

Some remarks concerning this section are presented in R.3.2, R.3.3 and R.3.4. See also the problems P.3.2 and P.3.3.

3.3 Determination and minimisation of error constants using Peano and Sard kernels

In the preceding sections error representations were developed which produce error inequalities of the form

$$|E_nf| \leqslant \sigma_n \, \|f\|,$$

where σ_n is a real number independent of f, and $\|f\|$ is a suitable norm of f. If we do not fix the nodes and weights *a priori*, then σ_n is a function of these nodes and weights and we ask whether we can minimise σ_n with respect to these free parameters. This leads to numerous versions of "optimal" or "minimal" quadratures. The problem is to separate the respective contributions of f and Q_n to E_nf.

3.3.1 Error bounds for $f \in C^0[-1, 1]$

Let us define $\|f\| \equiv \sup_{|x| \leqslant 1} |f(x)|$. Then

$$|E_nf| \leqslant \int_{-1}^{1} |f(x)| \, dx + \sum_{i=1}^{n} |A_i| \, |f(x_i)| \leqslant \left(2 + \sum_{i=1}^{n} |A_i|\right) \|f\|. \tag{3.3.1}$$

On substituting $\sigma_n \equiv 2 + \sum_{i=1}^{n} |A_i|$ we obtain

$$|E_nf| \leqslant \sigma_n \, \|f\|, \tag{3.3.2}$$

where the properties of Q_n and f are separated. Once the norm is chosen as above, the term $\|f\|$ cannot be in any way controlled. However, the size of σ_n depends on the A_i. The smaller σ_n can be made, the better the error estimate (3.3.1) for any fixed function f. For every formula of at least degree zero we have that

$$\sum_{i=1}^{n} A_i = 2, \quad \text{hence} \quad \sum_{i=1}^{n} |A_i| \geqslant 2.$$

Here, equality only occurs if $A_i \geqslant 0$ for $i = 1(1)n$. Hence the "best quadrature formulae" in the sense of (3.2.2) are those with positive weights. We shall shortly define Q_nf to be *positive* if the weights are positive. The positivity of Q_nf will be of importance for convergence.

Since positivity is a property of many quadratures we need a further requirement to reduce σ_n. The estimate (3.3.1) has the serious disadvantage that $|\sigma_n| \geqslant 4$ and we cannot conclude convergence.

The very rough estimation $|E_n f| \leqslant (2+\sum |A_i|)\,\|f\|$ can be refined in two ways:

Assume first that $Q_n f$ converges for every polynomial g if $n \to \infty$. Then

$$|E_n f| \leqslant |E_n(f-g)| + |E_n g|$$

where $\lim_{n\to\infty} |E_n g| \to 0$. If $Q_n f$ has degree $k(n)>0$, and $g \in \mathbb{P}_{k(n)}$ then, in addition, $E_n g = 0$ for $n \leqslant k(n)$. Denoting by $e_m(f, x)$ the algebraic approximation error where

$$\|e_m(f, x)\| \equiv \min_{g\in\mathbb{P}_m} \sup_{|x|\leqslant 1} |f(x) - g(x)|,$$

then

$$|E_n f| \leqslant \left(2 + \sum_{i=1}^{n} |A_i|\right) \|e_m(f, x)\| + |E_n g| = \sigma_n \, \|e_m(f, x)\| + |E_n g|,$$

and we obtain convergence if the approximating polynomials g converge uniformly to f for $n \to \infty$ and if σ_n is bounded.

Evidently, the same is true if we replace the space $\mathbb{P}$ by any other space S such that $E_n g$ converges to 0 for every $g \in S$ and which allows a uniform approximation of $f(x)$ in $[-1, 1]$. Examples of such spaces are the space of spline functions and the space of trigonometric functions.

Thus, we have turned the problem of estimating the quadrature error into that of estimating the error of uniform approximation.

The error representation (3.3.1) will, secondly, be refined to a form suitable for error estimation. This is done by splitting the integral in $E_n f$ such that

$$E_n f = \int_{-1}^{x_1} f(x)\,dx + \sum_{i=1}^{n} \left[\int_{x_i}^{x_{i+1}} f(x)\,dx - A_i f(x_i)\right],$$

with $x_{n+1} \equiv 1$. Hence

$$E_n f = \int_{-1}^{x_1} f(x)\,dx + \sum_{i=1}^{n} \left\{\int_{x_i}^{x_{i+1}} (f(x) - f(x_i))\,dx + (x_{i+1} - x_i - A_i) f(x_i)\right\}.$$

Using the modulus of continuity

$$\omega(f, h) \equiv \sup_{|x-y|\leqslant h} |f(x) - f(y)|$$

with

$$\lim_{h\to 0} \omega(f, h) = 0$$

and

$$h_n \equiv \sup_{i=1(1)n} |x_{i+1} - x_i|$$

we substitute to obtain the inequality

$$|E_n f| \leq \left[|x_1 + 1| + \sum_{i=1}^{n} |x_{i+1} - x_i - A_i| \right] \|f\| + |1 - x_1| \, \omega(f, h_n)$$

which provides a bound for the error. We thus define

$$\sigma_n \equiv |x_1 + 1| + \sum_{i=1}^{n} |x_{i+1} - x_i - A_i| + |1 - x_1| \, \omega\left(\frac{f}{\|f\|}, h_n\right)$$

and use σ_n for the estimation of $E_n f$.

This is not a good estimate in the sense of our problem, because σ_n now depends on ω and therefore on $f/\|f\|$. Nevertheless, it is a better result then (3.3.1). On the other hand, we eliminate easily the dependence of f on estimating $\omega(f/\|f\|, h_n) \leq 2$.

We now have to decide how to choose A_i, x_i in order to minimise σ_n. Note that $\lim_{h \to 0} \omega(f/\|f\|, h) = 0$ and that ω does not depend on A_i. The dependence of x_i suggests making h_n as small as possible, and this is done by taking equidistant nodes. If the nodes become non-equidistant, then ω increases.

The smallest possible value for h_n is $h_n = 2/(n+1)$, where the nodes $x_i = -1 + ih_n$, $i = 1(1)n$, are all in $(-1, 1)$. The sum in σ_n vanishes if we choose

$$A_i \equiv x_{i+1} - x_i = \frac{2}{n+1} \quad \text{for} \quad i = 1(1)n,$$

with the result that

$$\bar{\sigma}_n \equiv \frac{2}{n+1} + \left(2 - \frac{2}{n+1}\right) \omega\left(\frac{f}{\|f\|}, \frac{2}{n+1}\right) = \frac{2}{n+1}\left(1 + n\omega\left(\frac{f}{\|f\|}, \frac{2}{n+1}\right)\right).$$

If we take not the smallest possible value for h_n but one slightly larger, we can also eliminate $|x_1 + 1|$. A suitable value is $h_n = 2/n$, thus making the nodes $x_i = -1 + 2(i-1)/n$, $i = 1(1)n$. Now, $A_i = 2/n$, $i = 1(1)n$, and $\sigma_n \equiv 2\omega(f/\|f\|, 2/n) = \bar{\sigma}_n + O(1/n)$. Consequently,

$$Q_n f = \frac{2}{n} \sum_{i=1}^{n} f\left(-1 + 2\frac{i-1}{n}\right)$$

and

$$E_nf = \sum_{i=1}^{n} \int_{x_i}^{x_i+2/n} (f(x)-f(x_i))\,dx, \qquad x_i = -1+2(i-1)/n, \quad (3.3.3)$$

which even holds if $f \notin C^0[-1, 1]$ but only $f \in C^0(x_i, x_{i+1}]$, $i = 1(1)n-1$.
In the sense of the inequality $|E_nf| \leq \sigma_n \|f\|$ following from (3.3.3), we cannot find a "better" constant $\overset{*}{\sigma}_n$ with $\overset{*}{\sigma}_n < \bar{\sigma}_n$.

Theorem 3.3.1 *For* $f \in C^0[-1, 1]$ *the error constant* $\sigma_n = 2\omega(f/\|f\|, 2/n)$ *is* asymptotically *the best possible.*

Asymptotically $\bar{\sigma}_n = \sigma_n$. We therefore restrict our proof to the case of the simpler σ_n.

Proof We have to show that the error estimate is sharp, i.e. we have to find a function f such that $|E_nf| = \sigma_n \|f\|$. In order to do this we choose the step function

$$\tilde{f} = \begin{cases} -1 & \text{for} \quad x = -1 \\ -1+2(i-1)/n & \text{for} \quad x \in (x_{i-1}, x_i], \end{cases} \qquad i = 2(1)n+1, \qquad (x_{n+1} = 1).$$

Clearly, $\tilde{f} \notin C^0[-1, 1]$ but $\tilde{f}$ can be approximated arbitrarily closely by functions $f_\varepsilon(x) \in C^0[-1, 1]$ such that $\int_{-1}^{1} |\tilde{f} - f_\varepsilon|\,dx \leq \varepsilon$ for every given $\varepsilon > 0$, as shown in Fig. 3.3.1.

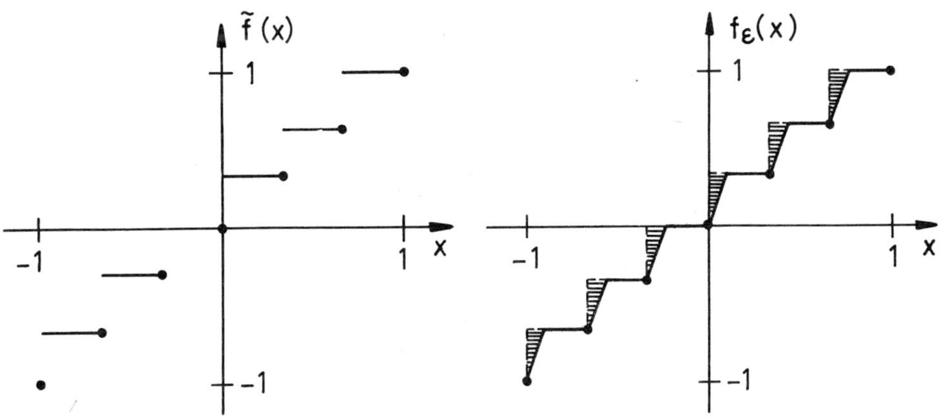

FIG. 3.3.1. $\tilde{f}(x)$ and $f_\varepsilon(x)$ for $n = 6$.

Since $Q_n\tilde{f} = Q_n f$ we have that

$$E_n\tilde{f} - E_n f_\varepsilon = E_n(\tilde{f} - f_\varepsilon) = \int_{-1}^{1} (\tilde{f} - f_\varepsilon)\,dx = \varepsilon.$$

The error coefficient σ_n has the same value both for $\tilde{f}$ and f_ε:

$$\sigma_n = 2\omega(\tilde{f}/\|\tilde{f}\|, 2/n) = 2\omega(f_\varepsilon/\|f_\varepsilon\|, 2/n) = \overline{\lim_{\varepsilon\to 0}}\, \omega(f_\varepsilon, 2/n) = 4/n.$$

On the other hand, we obtain from (3.3.3) that

$$E_n\tilde{f} = \varepsilon + E_n f_\varepsilon = \varepsilon + \sum_{i=1}^{n} \int_{x_i}^{x_i+2/n} (f_\varepsilon(x) - f_\varepsilon(x_i))\,dx$$

$$= \varepsilon + \sum_{i=1}^{n} \left(\frac{4}{n^2} - \frac{\varepsilon}{n}\right) = \frac{4}{n} = \sigma_n,$$

(cf. Fig. 3.3.1) so that actually $E_n\tilde{f} = \sigma_n \|\tilde{f}\|$ since $\|\tilde{f}\| = 1$. In addition, $\|f_\varepsilon\| = 1$ and hence, $E_n f_\varepsilon = \sigma_n \|f\| - \varepsilon$. This means that $E_n f$ comes arbitrarily close to $\sigma_n \|f\|$ in $C^0[-1, 1]$. ■

The spaces $C^k[-1, 1]$, $k = 1, 2, \ldots$, may be treated in the same way as the space $C^0[-1, 1]$. This would necessitate extended investigations for every single k in order that the error coefficients might be found and, in addition, that the nodes and weights which minimise these error coefficients might be found. A detailed discussion of such problems is beyond the scope of this book.

3.3.2 Error estimates for $f \in C^k[-1, 1]$, $f^{(k)} \in L^p[-1, 1]$, $k > 0$

When this space is considered, the Peano kernel theorem is applicable. First, let $k < degree(Q_n f)$ so that the error representation is

$$E_n f = \frac{1}{(k-1)!} \int_{-1}^{1} K(t) f^{(k)}(t)\,dt, \quad \text{where} \quad K(t) \in L^q[-1, 1].$$

In order to separate the influence of the two functions $f(t)$ and $K(t)$ we apply the Hölder inequality so that

$$|E_n f| \leq \frac{1}{(k-1)!} \left(\int_{-1}^{1} K(t)^q\,dt\right)^{1/q} \left(\int_{-1}^{1} |f^{(k)}(t)|^p\,dt\right)^{1/p} \quad \text{where} \quad \frac{1}{p} + \frac{1}{q} = 1.$$

The two most important cases are $p=q=2$, and $q=1$, $p=\infty$.

(i) $p=q=2$ gives the estimate $|E_nf| \leq \sigma_n \, |||f|||$ where

$$\sigma_n^2 \equiv \left(\frac{1}{(k-1)!}\right)^2 \int_{-1}^{1} K^2(t)\,\mathrm{d}t \quad \text{and} \quad |||f|||^2 \equiv \int_{-1}^{1} [f^{(k)}(t)]^2\,\mathrm{d}t \tag{3.3.4}$$

is a pseudonorm of f. It is not a norm since it lacks the definiteness property: $\|f\|=0$ if and only if $f\equiv 0$.

We now seek nodes and weights which minimise $Q(x_i, A_i) \equiv \int_{-1}^{1} K^2(t)\,\mathrm{d}t$; a problem which for *fixed* nodes was solved by Sard. If there exist A_i and x_i which minimise $Q(x_i, A_i)$, then these are the solutions of the system of equations

$$\partial Q/\partial x_i = 0, \qquad \partial Q/\partial A_i = 0 \quad \text{for} \quad i=1(1)n.$$

(ii) $p=\infty$, $q=1$. We now have that

$$|E_nf| \leq \frac{1}{(k-1)!} \sup_{|t|\leq 1} |f^{(k)}(t)| \int_{-1}^{1} |K(t)|\,\mathrm{d}t \tag{3.3.5}$$

and we can take

$$\sigma_n \equiv \frac{1}{(k-1)!} \int_{-1}^{1} |K(t)|\,\mathrm{d}t, \qquad |||f||| \equiv \sup_{|t|\leq 1} |f^{(k)}(t)|.$$

The problem thus becomes that of finding nodes and weights such that

$$\int_{-1}^{1} |K(t)|\,\mathrm{d}t = \min.$$

In both the general and the special cases (i) and (ii) it is very difficult to calculate the nodes and weights in general. Quadrature formulae are known for very small k and in some cases where additional restrictions are made, e.g. all nodes preassigned.

On the other hand, Peano's kernel theorem produces very efficient error estimates for *given* Q_nf. In addition to the special estimates (3.3.4) and (3.3.5) which are always applicable, we have the following very important case. This occurs when $K(t)$ has constant sign (so that the mean-value theorem is then applicable) and if $k \geq$ *degree* (Q_nf). Again, we are able to apply the Peano-kernel theorem, but there are still some derivatives at $x=0$:

$$E_nf = \sum_{r=l}^{k-1} \frac{f^{(r)}(0)}{r!} E_n x^r + \frac{1}{(k-1)!} \int_{-1}^{1} K(t) f^{(k)}(t)\,\mathrm{d}t \tag{3.3.6}$$

if degree$(Q_nf) = l-1$. (This is very advantageous if $f^{(r)}(0) = 0$, $r = l(1)k-1$, because we then have the former situation.) We apply Hölder's inequality to obtain

$$|E_nf| \leqslant \left\{ \sum_{r=l}^{k-1} (E_n x^r/r!)^q + \frac{1}{(k-1)!} \int_{-1}^{1} |K(t)|^q \, dt \right\}^{1/q} |||f|||$$

with (3.3.7)

$$|||f||| = \left\{ \sum_{r=l}^{k-1} [f^{(r)}(0)]^p \right\}^{1/p} + \left\{ \int_{-1}^{1} [f^{(k)}(t)]^p \, dt \right\}^{1/p}$$

for $1/p + 1/q = 1$. (The proof is left to the reader as a problem.) The special cases are to be treated as before.

We continue with the problem given in Example 3.1.1:

Example 3.1.1 (*contd*)

$$\begin{aligned}
k=0:&\quad |E_2f| \leqslant M_1 \overset{*}{I} |K_{2,0}(t)| = 25/144 M_1\\
k=1:&\quad |E_2f| \leqslant M_2 \overset{*}{I} |K_{2,1}(t)| = 1/48 M_2\\
k=2:&\quad E_2f = -f'''(\xi)/216 \\
k=3:&\quad \begin{cases} E_2f = f'''(0)/6 - f^{(4)}(\zeta)/405 \quad \xi, \zeta \in (-1, 1)\\ |E_2f| \leqslant M_4/405 \quad \text{for} \quad f'''(0) = 0. \end{cases}
\end{aligned}$$

These inequalities can be used to produce error bounds for special cases. Examples for sin, cos, exp (sin) are given in Table 3.3.1.

TABLE 3.3.1

<table>
<tr><th>f:</th><th>sin (πx/2)</th><th>cos (πx/2)</th><th>exp (sin x)</th></tr>
<tr><td>k</td><td></td><td></td><td></td></tr>
<tr><td>0</td><td colspan="2">0.3054</td><td>0.2836</td></tr>
<tr><td>1</td><td colspan="2">0.0514</td><td>0.02083</td></tr>
<tr><td>2</td><td colspan="2">0.0179</td><td>0.01885</td></tr>
<tr><td>3</td><td>3.8908</td><td>0.0150</td><td>0.01310</td></tr>
<tr><td>|E₂f| ≤</td><td colspan="2">0.0116</td><td>0.01093</td></tr>
</table>

We thus obtain acceptable bounds for the error $|E_nf|$. The bounds will be poor only for functions $f(x)$ for which $f'''(0) \neq 0$. This is because the bound depends very strongly on $|f'''(0)|$. Hence, the bound may become large although the integral $\overset{*}{I}f$ is approximated very well by Q_2f.

The calculation of the error bound in the L^2-norm is more complicated.

We obtain, for example,

$$|E_2 f| \leqslant 8.22(-3)\left\{\int_0^1 [f'''(x)]^2\,dx\right\}^{\frac{1}{2}} = 8.22(-3)|||f|||_2$$

For $f=\sin(\pi x/2)$ we have $|||f|||_2=\pi^3/8\sqrt{2}<2.7406$. Hence,

$$|E_2(\sin(\pi x/2))| < 0.02252 \qquad (k=2).$$

For $f=\exp(\sin x)$ the same method is not applicable. We can only bound the integral. Since $|f'''(x)|\leqslant 2$ for $|x|\leqslant 1$, we obtain

$$|E_2(\exp(\sin x))| < 1.644(-2) \qquad (k=2). \quad \blacksquare$$

For the *cubature* problem we proceed analogously, but replace Peano kernels by Sard kernels.

Let $f\in C^{k+1}(\bar{\Omega})$. Again, we apply Hölder's inequality in order to produce error bounds. To this end we restrict our attention to the form of the bound which in practice is the most important: namely, that in terms of the sup-norm $\|\cdot\|_\infty$. Let

$$M_{i,j} = \sup_{x\in\bar{\Omega}} |f_{i,j}(x, y)|.$$

Then for $p=\infty$, $q=1$ we have the error estimates

$$|E_n f| \leqslant \frac{1}{k!}\sum_{\mu=0}^{\varkappa}{}^{*}\binom{k}{\mu}\left[M_{k+1-\mu,\mu}\int_0^{\tau}|K_1^{\mu}(u)|\,du + M_{\mu,k+1-\mu}\int_0^{\sigma}|K_2^{\mu}(v)|\,dv\right]$$

$$+\begin{cases} \dfrac{M_{\varkappa+1,\varkappa+1}}{\varkappa!\,\varkappa!}\displaystyle\iint_{\Omega}|K_{\varkappa,\varkappa}(u,v)|\,du\,dv & \text{for}\quad k=2\varkappa+1 \\[2ex] \dfrac{\frac{1}{2}M_{\varkappa+1,\varkappa}}{(\varkappa-1)!\,\varkappa!}\displaystyle\iint_{\Omega}|K_{\varkappa,\varkappa-1}|\,du\,dv + \dfrac{\frac{1}{2}M_{\varkappa,\varkappa+1}}{(\varkappa-1)!\,\varkappa!}\displaystyle\iint_{\Omega}|K_{\varkappa-1,\varkappa}|\,du\,dv, & \\ & \text{for}\quad k=2\varkappa \end{cases}$$

Thus, for the midpoint rule of Example 3.2.1 we have

$$|E_1 f| \leqslant \tfrac{1}{24}(M_{2,0}+M_{0,2}) + \tfrac{9}{32}M_{1,1}.$$

Remark 1: The extension to several variables is a straightforward task, but it requires a great deal of paper!

Remark 2: We have restricted our attention to functions $f\in C^{k+1}(\bar{\Omega})$. Clearly, the derivatives which exist, especially the mixed ones, may be of different order.

The Taylor expansion can be performed in such a way, that all these derivatives which exist are taken into account (see Stroud, book).

Remark 3: One may also consider regions with curved boundaries. This, again, is straightforward but very laborious.

3.4 Further determination and minimisation of error constants

In this section we look for optimal or minimal error bounds of quadrature formulae if the integrand $f(x)$ has certain properties: the number of existing derivatives of $f(x)$ is larger than the degree of the quadrature formulae; $f(x)$ possesses a convergent power series expansion; $f(x)$ is periodic; and, finally, the values of $f(x)$ are influenced by round-off errors.

3.4.1 Error bounds and minimal error coefficients for $f \in C^k[-1, 1]$, $k \geqslant \text{degree}(Q_nf)$

We assume that the number of existing derivatives of $f(x)$ exceeds the degree of Q_nf. Now, formulae with $k = \text{degree}(Q_nf)$ are usually obtained by integration of an interpolating polynomial (cf. Section 2.3). Hence, we have the simple error representation from the error of interpolation corresponding to the Peano kernel representation. From Lagrange interpolation we have $k = n$ and from Hermite interpolation $k = 2n$.

(i) Let $k = n$: We then have, from Section 2.5.2, that

$$E_nf = \frac{1}{n!} \int_{-1}^{1} f^{(n)}(\xi)\omega_n(x)\,dx.$$

Hence

$$|E_nf| \leqslant \frac{1}{n!} \int_{-1}^{1} |f^{(n)}(\xi)|\,|\omega_n(x)|\,dx \leqslant \frac{1}{n!} M_n \int_{-1}^{1} |\omega_n(x)|\,dx$$

with

$$M_n \equiv \sup_{|x| \leqslant 1} |f^{(n)}(x)| = |||f|||.$$

Thus, $|E_nf| \leqslant \sigma_n\,|||f|||$ where $\sigma_n \equiv \dfrac{1}{n!} I\,|\omega_n(x)|$ and we see, in view of this inequality, that the best quadrature formulae are those which minimise

$I\,|\omega_n(x)|$. The general solution for the weight function $W(x)\equiv 1$ (cf. Section 1.4.3) is known because of:

Theorem 3.4.1 (Zolotareff, Korkin) *Every polynomial* $p_n(x)\in\mathbb{P}_n$ *with leading coefficient* 2^n *satisfies*

$$\int_{-1}^{1}|p_n(x)|\,dx \geqslant \int_{-1}^{1}|U_n(x)|\,dx = 2$$

where $U_n(x)=\sin(n+1)\theta/\sin\theta$ with $\theta=\arccos x$ *are the Tschebyscheff polynomials of the second kind.*

Proof We first calculate the integral

$$\int_{-1}^{1}|U_n(x)|\,dx = \int_0^{\pi}\left|\frac{\sin(n+1)\theta}{\sin\theta}\right|\sin\theta\,d\theta = \int_0^{\pi}|\sin(n+1)\theta|\,d\theta$$
$$= (n+1)\int_0^{\pi/(n+1)}\sin(n+1)\theta\,d\theta = 2.$$

Secondly, we have that

$$\int_{-1}^{1}|U_n(x)|\,dx = \int_{-1}^{1}U_n(x)\,\text{sign}\,(U_n(x))\,dx.$$

The function $U_n(x)$ is normalised such that $U_n(x)=2^nx^n+\ldots$. We consider the polynomials $p_n(x)=2^nx^n+\ldots\in\mathbb{P}_n$ (this normalisation does not affect the position of the zeros in which we are only interested) and split the integral such as

$$\int_{-1}^{1}|U_n(x)|\,dx = \int_{-1}^{1}(U_n(x)-p_n(x))\,\text{sign}\,U_n(x)\,dx + \int_{-1}^{1}p_n(x)\,\text{sign}\,U_n(x)\,dx.$$

In order to avoid theoretical difficulties due to the jumps of the integrands at the zeros of $x_1,\ldots,x_n$ of $U_n(x)$ we consider the integrals to be an abbreviation for

$$\sum_{i=0}^{n}\int_{x_i}^{x_{i+1}}(\cdot)\,dx \quad \text{with} \quad x_0=-1 \quad \text{and} \quad x_{n+1}=1.$$

We have immediately

$$\int_{-1}^{1}p_n(x)\,\text{sign}\,U_n(x)\,dx \leqslant \int_{-1}^{1}|p_n(x)|\,dx.$$

Hence, we still need

$$\int_{-1}^{1}(U_n(x)-p_n(x))\,\text{sign}\,U_n(x)\,dx = 0 \tag{3.4.1}$$

in order to complete the proof. Note that $U_n(x)-p_n(x)\in\mathbb{P}_{n-1}$ because the leading terms $2^n x^n$ cancel.

Thus (3.4.1) is an orthogonality condition which tells us that sign $U_n(x)$ is orthogonal to $\mathbb{P}_{n-1}$ because $p_n(x)-2^n x^n\in\mathbb{P}_{n-1}$ is an arbitrary polynomial. Hence, it is sufficient to show that

$$\int_{-1}^{1} x^\nu \operatorname{sign} U_n(x)\,dx=0 \quad \text{for} \quad \nu=0(1)n-1.$$

We remark that the Fourier expansion of sign $U_n(x)$ is given by (cf. Problem P.3.7)

$$\begin{aligned}\operatorname{sign} U_n(x)=\operatorname{sign}[\sin(n+1)\theta]&=\frac{4}{\pi}\sum_{k=0}^{\infty}\frac{\sin[(2k+1)(n+1)\theta]}{2k+1}, \qquad \theta\in[0,\pi]\\ &=\frac{4}{\pi}\sin\theta\sum_{k=0}^{\infty}\frac{1}{2k+1}\,U_{(2k+1)n}(x)\\ &=\frac{4}{\pi}(1-x^2)^{\frac{1}{2}}\sum_{k=0}^{\infty}\frac{1}{2k+1}\,U_{(2k+1)n}(x).\end{aligned}$$

Hence

$$\int_{-1}^{1} x^\nu(1-x^2)^{\frac{1}{2}}\{U_n(x)+\tfrac{1}{3}U_{3n}(x)+\ldots\}\,dx=0$$

must be valid, and this follows from the property that

$$\int_{-1}^{1} U_n(x)U_m(x)(1-x^2)^{\frac{1}{2}}\,dx=\int_0^{\pi}\sin(n+1)\theta\cdot\sin(m+1)\theta\,d\theta=\frac{\pi}{2}\,\delta_{n,m},$$

and the theorem is proved, since $U_m(x)$ is a linear combination of x^k, $k=0(1)m$. ■

The function $U_n(x)$ has n real distinct zeros in $[-1, 1]$. These are the nodes of the quadrature formula Q_nf minimising $I\,|\omega_n(x)|$. The result cannot easily be extended to weight functions $W(x)\neq 1$.

If we estimate E_nf in the L_2-norm instead of the sup-norm then we obtain the following result. Providing that $f^{(n)}\in L^2[-1, 1]$, we have, by the Schwarz inequality,

$$|E_nf|\leqslant\frac{1}{n!}\left(\int_{-1}^{1}[f^{(n)}(x)]^2\,dx\right)^{\frac{1}{2}}\left(\int_{-1}^{1}\omega_n^2(x)\,dx\right)^{\frac{1}{2}},$$

and the optimal quadrature formulae in the sense of this inequality are those which minimise $I\omega_n^2(x)$.

Such formulae are optimal in another sense, too, namely for $k=2n$.

(ii) Let $k=2n$: Then, from Section 2.5.5,

$$E_nf=\frac{1}{(2n)!}\int_{-1}^{1} f^{(2n)}(\xi)\omega_n^2(x)\,\mathrm{d}x=\frac{1}{(2n)!}f^{(2n)}(\zeta)\int_{-1}^{1}\omega_n^2(x)\,\mathrm{d}x$$

because of the positivity of $\omega_n^2(x)$. Hence we have

$$|E_nf|\leqslant\sigma_n\,|||f|||\quad\text{with}\quad |||f|||=\sup_{|x|\leqslant 1}|f^{(2n)}(x)|,\quad\text{and}\quad \sigma_n=\frac{1}{(2n)!}I\omega_n^2(x).$$

The nodal polynomials $\omega_n(x)$ which minimise $I\omega_n^2(x)$ are well known:

Theorem 3.4.2 *Consider the system of polynomials $\omega_n(x)\in\mathbb{P}_n$, $n=1(1)\ldots$. Then the necessary and sufficient condition for $\omega_n(x)$ to minimise $I\omega_n^2(x)$ is that the $\omega_n(x)$ be orthogonal to $\mathbb{P}_{n-1}$ with respect to the weight function $W(x)$.*

This theorem is an immediate consequence of the more general Theorem 7.1.3, and hence a proof can be found there.

3.4.2 Error bounds for functions f possessing a convergent power series expansion

Another class of optimal quadrature and cubature formulae is now obtained for such functions $f(x)$ which possess a sufficiently rapidly convergent power series expansion

$$f(x)=\sum_{i=0}^{\infty}a_i(x-x_0)^i. \tag{3.4.2}$$

Without essential loss of generality we also assume that $x_0=0$. Thus

$$|E_nf|\leqslant\sum_{i=0}^{\infty}|a_i|\,|E_nx^i| \tag{3.4.3}$$

is a bound for the quadrature error E_nf. Further, the non-zero coefficients of (3.4.2) and (3.4.3) are assumed to decrease monotonically in absolute value.

Rapidly convergent expansions

If the expansion (3.4.2) converges rapidly, we can simplify the error bound. We first define formally rapid convergence.

Definition 3.4.1 The series (3.4.2) is said to be *rapidly convergent* if the non-zero coefficients $|a_k|$ are monotonically decreasing, and if there exists a constant c independent of j such that

$$\sum_{i=j}^{\infty} |a_i| \leq |a_j|(1+c) \quad \text{for} \quad j=0(1)\ldots. \quad \blacksquare$$

As an example of such a series we mention that $a_i = 1/i!$ with $c \leq \exp(1)$. Let degree $(Q_n f) = k-1$, and let $C_k = \sup_{i \geq k} |E_n x^i|$. Then

$$|E_n f| \leq \sum_{i=k}^{\infty} |a_i|\,|E_n x^i| \leq C_k \sum_{i=k}^{\infty} |a_i| \leq C_k (1+c)\,|a_k|.$$

If, in addition, the non-zero coefficients $|a_i| \neq 0$ decrease so fast that even the sequence $\{C_i \; |a_i| \neq 0\}_{i=0}^{\infty}$ is monotonically decreasing, then the bound becomes minimal if k is as large as possible. Hence, we have:

Theorem 3.4.3 *Let $f(x)$ possess a rapidly convergent series expansion (3.4.2) such that even $\{C_i \; |a_i| \neq 0\}_{i=0}^{\infty}$ is monotonically decreasing; then the best quadrature formulae are those having the highest possible degree of precision.*

Theorem 3.4.3 exhibits the philosophy behind the common requirement that $Q_n f$ should have the highest possible degree of precision.

If we replace the monomials $(x-x_0)^i$ in (3.4.2) by suitable functions $g_i(x)$ such that $\sum b_i g_i(x)$ is another rapidly convergent series expansion of $f(x)$, then all arguments remain valid if we define the degree of precision k by requiring $E_n g_i(x) = 0$ for $i = 0(1)k$. For trigonometric functions the solution is presented in Section 3.4.3.

l_p-convergent expansions

Let degree $(Q_n f) = k-1$. Then, by Hölder's inequality, we have for $1/p + 1/q = 1$ that

$$|E_n f| \leq \left(\sum_{i=k}^{\infty} |a_i|^p\right)^{1/p} \left(\sum_{i=k}^{\infty} |E_n x^i|^q\right)^{1/q}.$$

We say that $f(x)$ is l_p-convergent if the first series converges. Let the non-zero $|a_i|$ decrease monotonically.

The special case $p=1$, $q=\infty$ has just been treated. For $p=\infty$, $q=1$ we have l_∞-convergence and hence

$$|E_n f| \leq A_k \sum_{i=k}^{\infty} |E_n x^i| \quad \text{with} \quad A_k \equiv \sup_{i \geq k} |a_i|.$$

If $a_k \neq 0$ then, clearly, $A_k = |a_k|$.

In order to obtain a minimal error bound we have to require that:

(i) k is as large as possible,

(ii) the sum $\sum_{i=k}^{\infty} |E_n x^i|$ is as small as possible.

The solution of this problem, and that of the same problem for general p and q, is unknown. If $p=q=2$ and $k=0$ then

$$|E_n f| \leq \left(\sum_{i=0}^{\infty} a_i^2\right)^{\frac{1}{2}} \cdot \left(\sum_{i=0}^{\infty} E_n^2 x^i\right)^{\frac{1}{2}}.$$

and we have the problem of finding a minimal second sum. Here, the first sum is only influenced by $f(x)$ and can be considered as a norm of $f(x)$, while the second sum depends only on the nodes and the weights of the quadrature formula. Thus, in fact, the second sum can be minimised under appropriate circumstances. This will be performed in the following Section, 3.5.4.

The cubature problem

The preceding considerations are now performed for cubature formulae. Assume that there exists a power series expansion of $f(x, y)$

$$f(x, y) = \sum_{i=0}^{\infty} \sum_{j=0}^{i} a_{ij} x^{i-j} y^j.$$

Then, for degree $(C_n f) = k-1$ we have that

$$|E_n f| \leq \sum_{i=k}^{\infty} \sum_{j=k}^{i} |a_{ij}|\, |E_n(x^{i-j} y^j)|.$$

Defining

$$A_i \equiv \sum_{j=k}^{i} a_{ij} \quad \text{and} \quad e_i \equiv \sum_{j=k}^{i} |E_n(x^{i-j} y^j)|$$

we have the same situation as before, and the corresponding extension to cubature problems is straightforward, even for $N > 2$ dimensions. Again, the formulae of the highest possible degree of precision have the same importance. The only difference appears in the additional problem of finding the smallest number of nodes for a given degree $(k-1)$.

3.4.3 The space of 2π-periodic functions

We ask for optimal quadrature formulae in the sense of (3.4.3). Let $f(x)$ and $W(x)$ be 2π-periodic functions. We then have to evaluate integrals like

$$If = \int_0^{2\pi} W(\cos \varphi) f(\cos \varphi)\, d\varphi = \int_0^{\pi} (\cdot) + \int_{\pi}^{2\pi} (\cdot).$$

The mapping $\cos \varphi = x$ leads to

$$If = 2 \int_{-1}^{1} \frac{W(x)}{(1-x^2)^{\frac{1}{2}}} f(x)\, dx.$$

Hence, we need to find optimal quadratures for a weight function $(1-x^2)^{-\frac{1}{2}} W(x)$ with respect to $[-1, 1]$. Once these formulae are known then, by $x = \cos \varphi$, we obtain the corresponding optimal formulae for periodic functions with respect to $[0, 2\pi]$. For $W(x) \equiv 1$ the trapezoidal rule $T_n f$ solves our problem. This is seen as follows.

The natural expansion of a periodic function f is the Fourier series expansion

$$f(x) = \tfrac{1}{2} b_0 + \sum_{k=1}^{\infty} (a_k \sin k\pi x + b_k \cos k\pi x).$$

From Problem P.2.13 we have: *If* $E_n g_k = 0$ is valid for

$$g_k \in \{1, \sin \pi kx, \cos \pi kx, k = 1(1)n-1, \sin n\pi x\}$$

then the corresponding quadrature formula is the trapezoidal rule. We recall that $f(0) = f(1)$ because of periodicity.

Here, the extension to periodic functions in several variables and to the corresponding cubature problem is also obvious. We only need to map each coordinate with $x_i = \cos \varphi_i$ to obtain

$$I_\Omega f = 2^N \int_{-1}^{1} \cdots \int_{-1}^{1} \frac{W(\mathbf{x}) f(\mathbf{x})}{((1-x_1^2) \ldots (1-x_N^2))^{\frac{1}{2}}}\, d\Omega$$

The one-dimensional Fourier expansion has to be replaced by the N-dimensional expansion. The corresponding cubature formulae are product type formulae if $\mathbf{W}(\mathbf{x}) = \prod_{j=1}^{N} W_j(x_j)$ is a product of one-dimensional weight functions.

3.4.4 Optimal quadrature formulae for erratic functions–Influence of round-off errors

If the values of the integrand $f(x)$ are assumed to be erratic, then the following kind of optimal formulae arises. Assume that the function values $f(x_i)$ are not known exactly but only with errors which are due to measuring or rounding off. We denote these errors by ε_i. Then

$$E_nf = If - \sum_{i=1}^{n} A_i f(x_i) - \sum_{i=1}^{n} A_i \varepsilon_i.$$

Hence

$$|E_nf| \leqslant |If - Q_nf| + \sum_{i=1}^{n} |A_i|\,|\varepsilon_i|.$$

We now look for a Q_nf where $\sum |A_i|\,|\varepsilon_i| = \min$. By the Schwarz inequality, which gives

$$\sum_{i=1}^{n} |A_i|\,|\varepsilon_i| \leqslant \left(\sum_{i=1}^{n} A_i^2\right)^{\frac{1}{2}} \cdot \left(\sum_{i=1}^{n} \varepsilon_i^2\right)^{\frac{1}{2}}, \tag{3.4.4}$$

we separate the influence of weights and errors. In general, the ε_i cannot be regulated because they are randomly determined. We therefore require $\sum_{i=1}^{n} A_i^2$ to be minimal, subject to the condition $\sum_{i=1}^{n} A_i = 2$ for $W(x) \equiv 1$, which means that Q_nf has at least the degree of precision 0.

Let λ be a Lagrangian multiplier so that the minimum of

$$B(A_1, \ldots, A_n, \lambda) \equiv \sum_{i=1}^{n} A_i^2 + \lambda\left(\sum_{i=1}^{n} A_i - 2\right)$$

is sought. A necessary condition for this is

$$\partial B/\partial A_i = 0 \quad \text{for} \quad i = 1(1)n, \qquad \partial B/\partial \lambda = 0.$$

which leads to the system of equations

$$2A_i + \lambda = 0, \qquad i = 1(1)n$$

$$A_1 + \ldots + A_n = 2.$$

This, in turn has the solution $\lambda = -4/n$ so that $A_i = 2/n$, independently of i. This solution really minimises B. In order to see this let $\overset{*}{A}_i = A_i + b_i$, $\sum b_i = 0$ in order to ensure that $\sum_{i=1}^{n} \overset{*}{A}_i = 2$. Hence, with $A_i = 2/n$

$$\sum_{i=1}^{n} (\overset{*}{A}{}_i^2 - A_i^2) = \frac{4}{n} \sum_{i=1}^{n} b_i + \sum_{i=1}^{n} b_i^2 = \sum_{i=1}^{n} b_i^2 \geqslant 0$$

with equality for $b_i = 0$, $i = 1(1)n$, and we have:

Theorem 3.4.4 *Let f be erratic and $W(x) \equiv 1$. Then $\sum |A_i|\,|\varepsilon_i|$ is minimised in the sense of* (3.4.4) *if and only if Q_nf is a quadrature formula with equal weights.*

Examples of formulae of this type with the highest possible degree of precision are the formulae of Tschebyscheff (cf. Section 2.4.1). Further examples are the formulae R_nf and B_nf.

For an optimal equally-weighted formula with $W(x)\equiv 1$ we have that

$$\lim_{n\to\infty}\left(\sum_{i=1}^{n} A_i^2\right)^{\frac{1}{2}}=\lim_{n\to\infty}(2/\sqrt{n})=0.$$

For a general quadrature or cubature formula the expression

$$V_n\equiv\left(\sum_{i=1}^{n} A_i^2\right)^{\frac{1}{2}}$$

is called the *variance* V_n of the formula Q_nf.

If Q_nf or C_nf is not equally weighted, but $V_n\to 0$ for $n\to\infty$, we call such formulae *asymptotically optimal* for erratic functions.

3.5 Error bounds without derivatives for quadratures on analytic function spaces

The previous bounds usually contain derivatives of f. This is a serious disadvantage if such derivatives are not known, do not exist or are too complicated to be handled. Hence, bounds without derivatives are very desirable and can be derived for analytic functions. The requirement of analyticity is not such a serious disadvantage as it might appear, because we only use discrete values of a function and there always exist analytic functions passing through these discrete values. If f is piecewise analytic, the bounds may be applied separately to each piece.

3.5.1 Error bounds for analytic functions by Cauchy's formulae

Assume that we are given a simply-connected region R in the complex plane such that $f(z)$ is analytic in R and $[-1, 1]$ is contained in R (Fig. 3.5.1). Let C be a simple closed contour including $[-1, 1]$, and contained completely within the interior of R. Then, by Cauchy's integral formulae we have the representation

$$f^{(n)}(z)=\frac{n!}{2\pi i}\int_C\frac{f(u)\,du}{(u-z)^{n+1}},$$

where the points u lie on C, and z belongs to the interior of C. Hence, we are able to bound the n-th derivative of f with

$$|f^{(n)}(z)|\leq\frac{n!}{2\pi}\frac{|C|\,M}{d^{n+1}},$$

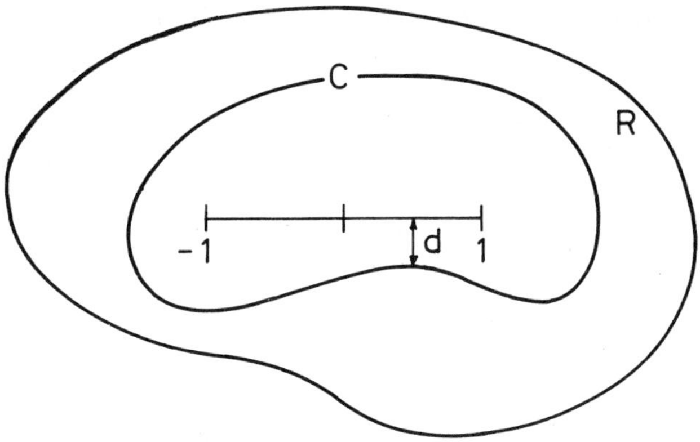

FIG. 3.5.1. Error estimation for analytic functions.

where $|C|$ is the arc length of C, $M=\max_{z\in C}|f(z)|$, and d measures the distance of C and $[-1,1]$; i.e.

$$d=\inf_{|x|\leq 1, z\in C}|x-z|.$$

This distance d is easily calculated if C is the circle $|z|=R>1$, or the boundary of the union of the set of all circles $|z-x|\leq R>1$ for every $x\in[-1,1]$.

Consider the first case so that $|z|\leq R>1$. Then the distance $d=R-1$ and $|C|=2\pi R$. Hence in this case

$$|f^{(n)}(z)|\leq\frac{n!\,RM}{(R-1)^{n+1}}.$$

Usually, M depends on R and this bound may, therefore, have a minimum with respect to R. Assume that $M=M(R)$ is differentiable. Then a necessary condition for $|f^{(n)}(z)|$ to assume its minimum is that $\frac{\mathrm{d}}{\mathrm{d}R}\left(\frac{RM(R)}{(R-1)^{n+1}}\right)=0$ from which it follows that $\frac{M'}{M}=\frac{nR+1}{R(R-1)}$. From this condition we obtain the radius R.

Example 3.5.1 For $f(z)=z\mathrm{e}^z$ we have $M=R\mathrm{e}^R$, and

$$\frac{M'}{M}=\frac{R+1}{R}.\quad \text{Hence} \quad \frac{R+1}{R}=\frac{nR+1}{R(R-1)} \quad \text{must be valid.}$$

This leads to $R=\frac{1}{2}(n+(n^2+8)^{\frac{1}{2}})$, and for $f(z)=z\mathrm{e}^z$ we have that

$$|f^{(n)}(z)|\leq\frac{n!\,R^2\mathrm{e}^R}{(R-1)^{n+1}} \quad \text{with} \quad R=\tfrac{1}{2}(n+(n^2+8)^{\frac{1}{2}}).$$

Some values of optimal radii are listed in Table 3.5.1 for $f(z)=z\exp(z)$.

TABLE 3.5.1

n	1	2	3	4	5	6	7	8	9	10
$\max\lvert f^{(n)}\rvert\leqslant$	29.6	44.1	62.3	83.3	106.8	132.5	160.2	189.8	221.2	254.2
$R(n)$	2	2.73	3.56	4.45	5.37	6.32	7.27	8.24	9.22	10.20

For Simpson's rule the error term is

$$E_3f=-\frac{f^{(4)}(t)}{90}\quad\text{with}\quad t\in(-1,1),$$

so that we obtain the optimal error bound with respect to R:

$$|E_3f|\leqslant\frac{83.3}{90}=0.926.$$

This is, of course, a poor estimate, as the actual error is 0.048. ■

Now let us consider the second case where $|z-x|\leqslant R$. We have for every $x\in[-1,1]$ that $d=R$ and hence for a circle with centre at x

$$|f^{(n)}(z)|\leqslant\frac{n!}{2\pi}\frac{2\pi RM(x)}{R^{n+1}}\quad\text{with}\quad M(x)\equiv\max_{|z-x|\leqslant R}|f(z)|\quad\text{and}\quad|x|\leqslant 1$$

$$\leqslant\frac{n!\,M}{R^n}\qquad\text{with}\quad M\equiv\max_{|x|\leqslant 1}M(x).$$

This bound assumes a minimal value for $M'/M=n/R$.

Example 3.5.1 (*contd*) We have $M=(1+R)e^{1+R}$ and

$$|f^{(n)}(z)|\leqslant\frac{n!\,(1+R)e^{1+R}}{R^n}$$

becomes minimal for $R=\frac{1}{2}(n-2+(n^2+4)^{\frac{1}{2}})$ which is slightly smaller than the previous radius.

TABLE 3.5.2

n	1	2	3	4	5	6	7	8	9	10
$\max\lvert f^{(n)}\rvert\leqslant$	13.2	27.0	44.1	64.1	86.5	111.2	138.0	166.6	197.0	229.1
$R(n)$	0.618	$\sqrt{2}$	2.30	3.24	4.19	5.16	6.14	7.12	8.11	9.10

For Simpson's rule this gives the error bound $|E_3 f| \leqslant 0.712$ which is only slightly better than the previous bound. ■

3.5.2 Hilbert spaces of analytic functions

Using Hilbert spaces of analytic functions we shall now obtain sharper error bounds than in the previous section via Cauchy's formulae. Assume that there exists a convergent expansion

$$f(x) = \sum_{k=0}^{\infty} a_k g_k(x) \tag{3.5.1}$$

where f and g_k are analytic in some subdomain of the complex plane. Because E_n is a bounded linear operator we find that

$$E_n f = \sum_{k=0}^{\infty} a_k E_n g_k(x)$$

and, by the Schwarz inequality, that

$$|E_n f| \leqslant \left(\sum_{k=0}^{\infty} a_k^2\right)^{\frac{1}{2}} \left(\sum_{k=0}^{\infty} [E_n g_k(x)]^2\right)^{\frac{1}{2}}.$$

Hence, this inequality is written in such a way that the influence of f (via a_k) and Q_n are separated. As in the previous sections, we define the error coefficient

$$\sigma_n^2 \equiv \sum_{k=0}^{\infty} [E_n g_k(x)]^2.$$

The term $(\sum a_i^2)^{\frac{1}{2}}$ is a norm of f if the set of functions $\{g_k\}$ is orthonormal with respect to a suitable inner product. This clearly takes us into the realms of inner product spaces, from which the extension to Hilbert spaces with the defining of the norm from the inner product $(\cdot, \cdot)$ is obvious.

Definition 3.5.1 A complete, linear, inner product space with norm $\|\cdot\| = (\cdot, \cdot)^{\frac{1}{2}}$ is called a *Hilbert space.* ■

Thus, let $\mathbb{H}$ be a Hilbert space with inner product $(\cdot, \cdot)$. Then, for $f, g_k \in \mathbb{H}$ we have that

$$\|f\|^2 = (f, f) = \left(\sum_{k=0}^{\infty} a_k g_k(x), \sum_{k=0}^{\infty} a_k g_k(x)\right) = \sum_{k=0}^{\infty} a_k^2$$

if and only if $(g_k, g_m) = \delta_{k,m}$, i.e. the g_i are orthonormal.

The following Theorem, 3.5.1, is most important for the construction of error bounds of quadrature formulae:

Theorem 3.5.1 (Fréchet–Riesz) *If $F(f)$ is a bounded linear functional defined on a Hilbert space $\mathbb{H}$, then there exists a unique element $g \in \mathbb{H}$ such that $F(f) = (f, g)$ for all $f \in \mathbb{H}$.*

Definition 3.5.2 The unique function $g \in \mathbb{H}$ defined in Theorem 3.5.1 is called the *representer* of the functional F. ■

We shall need the properties of the representer given in:

Theorem 3.5.2 *Assume that the functions $g_k(x) \in \mathbb{H}$ are orthonormal with respect to the inner product $(\cdot\,,\cdot)$. Then the representer is given by the series*

$$g = \sum_{k=0}^{\infty} \bar{F}(g_k) g_k.$$

Further,

$$\|g\| = \|F\|, \qquad |F(g)| \leq \|F\| \cdot \|g\| \quad \text{and} \quad \|g^2\| = (g, g) = \sum_{k=0}^{\infty} (F(g_k))^2.$$

From the completeness of $\mathbb{H}$ it follows that $g \in \mathbb{H}$ and hence there exists a function $g \in \mathbb{H}$ for which $|Fg| = \|F\|\,\|g\|$.

Theorems 3.5.1 and 3.5.2 are now applied to the error functional $E_n f$. Then the representer r_n of $E_n f$, (f, $E_n f$ real-valued for x real) is given by

$$r_n = \sum_{k=0}^{\infty} (E_n g_k) g_k \quad \text{with} \quad \|r_n\|^2 = \sum_{k=0}^{\infty} E_n^2 g_k = \|E_n\|^2.$$

Hence $\sigma_n \equiv \|r_n\|$ and, finally, from

$$E_n f = \sum_{k=0}^{\infty} (f, g_k) E_n g_k$$

and the fact that $(f, g_k) = a_k$ we obtain

$$|E_n f| \leq \left(\sum_{k=0}^{\infty} (f, g_k)^2 \right)^{\frac{1}{2}} \left(\sum_{k=0}^{\infty} E_n^2 g_k \right)^{\frac{1}{2}} = \|f\|\, \sigma_n = \|f\| \cdot \|E_n\|.$$

3.5.3 The extended Hardy space $\tilde{\mathbb{H}}_2$ – Error bounds

The series which defines the representer is summable only for simple orthonormal function systems. The summability, however, is necessary if we want to minimise the error coefficient. A sufficiently simple orthonormal system exists in $\tilde{\mathbb{H}}_2$ which we shall now define.

Let $f(z)=\sum a_k z^k$ be analytic for $|z|<R$, $\sum a_k^2<\infty$ for $|z|\leq R$ and let $1/R=\varepsilon>0$. We define an inner product by

$$(f,g)=\frac{1}{2\pi}\int_0^{2\pi} f(z)\overline{g(z)}\,\mathrm{d}\varphi \quad \text{with} \quad z=R\mathrm{e}^{i\varphi}=x+iy.$$

The particular space of functions having these properties is a Hilbert space which we call the extended *Hardy space* $\tilde{\mathbb{H}}_2$. For the ordinary $\mathbb{H}_2$ we claim $R=1$. From Cauchy's integral formula we obtain

$$f(z)=\frac{1}{2\pi i}\int_{|t|=R}\frac{f(t)}{t-z}\,\mathrm{d}t=\frac{1}{2\pi}\int_0^{2\pi}\frac{f(t)}{1-\varepsilon^2 z\bar{t}}\,\mathrm{d}\varphi$$

because of $\bar{t}=\overline{R\exp(i\varphi)}=R\exp(-i\varphi)=R^2/t=1/\varepsilon^2 t$ and $\mathrm{d}t=it\,\mathrm{d}\varphi$. Writing $\overline{K(z,\bar{t})}=1/(1-\varepsilon^2 z\bar{t})$ we have the representation

$$f(z)=\frac{1}{2\pi}\int_0^{2\pi} f(t)\overline{K(z,\bar{t})}\,\mathrm{d}\varphi=(f(t),K(z,\bar{t})).$$

Since $f(z)$ can be expressed in this way, the function $K(z,\bar{t})$ is called a *reproducing kernel* function. It follows immediately that the error functional can be written as

$$E_n f=\frac{1}{2\pi}\int_0^{2\pi} f(t)E_n\left(\frac{1}{1-\varepsilon^2 z\bar{t}}\right)\mathrm{d}\varphi=(f(t),E_n K(z,\bar{t})).$$

The powers $(\varepsilon z)^k$ are an orthonormal function system in $\tilde{\mathbb{H}}_2$ because

$$\begin{aligned}((\varepsilon z)^k,(\varepsilon z)^m)&=\frac{1}{2\pi}\int_0^{2\pi}\mathrm{e}^{i(k-m)\varphi}\,\mathrm{d}\varphi=1 \quad \text{for} \quad k=m \quad (\text{note that } \varepsilon\,|z|=1)\\ &=\frac{1}{2\pi i(k-m)}[\mathrm{e}^{2\pi i(k-m)}-1]=0 \quad \text{for} \quad k\neq m.\end{aligned}$$

Hence, the representer r_n of $E_n f$ is $r_n=\sum(E_n(\varepsilon z)^j)(\varepsilon z)^j$ with $\|r_n\|=\|E_n\|$.

Further $E_n f=\sum(f,(\varepsilon z)^j)E_n(\varepsilon z)^j$, and $|E_n f|\leq\sigma_n\,\|f\|=\|E_n\|\,\|f\|$. We now determine the respective representers $\overset{*}{r}_n$ and r_n of $E_n f$ with respect to $\overset{*}{I}f$ and If noting that they are different. Thus

$$\begin{aligned}\overset{*}{r}_n=\sum_{j=0}^{\infty}(\varepsilon z)^j(\overset{*}{I}(\varepsilon x)^j-Q_n(\varepsilon x)^j)&=\sum_{j=0}^{\infty}\varepsilon^{2j}\left[z^j/(j+1)-\sum_{l=1}^{n}A_l(x_l z)^j\right]\\ &=\frac{1}{\varepsilon^2 z}\ln\frac{1}{1-\varepsilon^2 z}-\sum_{l=1}^{n}\frac{A_l}{1-\varepsilon^2 z x_l}\end{aligned}$$

and analogously

$$r_n = \sum_{j=0}^{\infty} (\varepsilon z)^j (I(\varepsilon x)^j - Q_n(\varepsilon x)^j) = \frac{1}{\varepsilon^2 z} \ln \frac{1+\varepsilon^2 z}{1-\varepsilon^2 z} - \sum_{m=1}^{n} \frac{A_m}{1-\varepsilon^2 z x_m}.$$

It follows that

$$\|\overset{*}{r}_n\|^2 = (\overset{*}{r}_n, \overset{*}{r}_n) = E_n \overset{*}{r}_n = \overset{**}{I r}_n - Q_n \overset{*}{r}_n$$

$$= a(\varepsilon) + \frac{2}{\varepsilon^2} \sum_{m=1}^{n} \frac{A_m}{x_m} \ln(1-\varepsilon^2 x_m) + \sum_{m=1}^{n} \sum_{k=1}^{n} \frac{A_k A_m}{1-\varepsilon^2 x_k x_m}$$

with

$$a(\varepsilon) = -\frac{1}{\varepsilon^2} \int_0^1 \frac{1}{x} \ln(1-\varepsilon^2 x)\, dx = -\frac{1}{\varepsilon^2} \int_0^{-\ln(1-\varepsilon^2)} \frac{t\, dt}{1-e^t} \quad \text{for} \quad 0<\varepsilon<\infty$$

when the substitution $\ln(1-\varepsilon^2 x) = -t$ is made. For $\varepsilon = 0$ we have $a(0) = 1$. The coefficients $B_j(x)$ of the power series expansion of

$$\frac{te^{xt}}{e^t - 1} = \sum_{r=0}^{\infty} B_r(x) \frac{t^r}{r!}$$

are called *Bernoulli polynomials* and the numbers $B_r \equiv B_r(0)$ *Bernoulli numbers*. These numbers satisfy $B_r(1/2) = (2^{1-r} - 1) B_r(0)$, and it is seen that

$$B_0 = 1,\ B_1 = -1/2,\ B_2 = 1/6,\ B_4 = -1/30,\ B_6 = 1/42,\ B_8 = -1/30, \text{ etc.}$$

(cf. Abramowitz and Stegun [1], p. 803 ff.) and hence

$$a(\varepsilon) = \frac{1}{\varepsilon^2} \sum_{r=0}^{\infty} \frac{B_r}{(r+1)!} \left[\ln \frac{1}{1-\varepsilon^2} \right]^{r+1}.$$

A most interesting result occurs when $\varepsilon = 0$ where

$$\|\overset{*}{r}_n\|^2 = 1 - 2\sum_{j=1}^{n} A_j + \sum_{j=1}^{n} \sum_{k=1}^{n} A_k A_j = \left(1 - \sum_{j=1}^{n} A_j\right)^2.$$

Hence

$$\|\overset{*}{r}_n\| = \left|1 - \sum_{j=1}^{n} A_j\right|$$

which means $\|\overset{*}{r}_n\| = 0$ if $Q_n f$ has at least degree 0. For $\varepsilon = 0$ we have analyticity of f in the whole complex plane.

The result $\|\overset{*}{r}_n\| = 0$ does not imply $|E_n f| = 0$ because in general $\|f\| = \infty$ in the whole complex plane.

For $\varepsilon=1$ we have analyticity of f in the unit circle and

$$a(1)=\lim_{\varepsilon\to 1-} a(\varepsilon)=\int_0^1\left(\sum_{r=1}^{\infty}\frac{1}{r}x^{r-1}\right)\mathrm{d}x=\sum_{r=1}^{\infty}\frac{1}{r^2}=\pi^2/6.$$

Analogously, we obtain

$$\begin{aligned}\|r_n\|^2&=E_nr_n=Ir_n-Q_nr_n\\&=b(\varepsilon)+\frac{2}{\varepsilon^2}\sum_{m=1}^{n}\frac{A_m}{x_m}\ln\left(\frac{1-\varepsilon^2x_m}{1+\varepsilon^2x_m}\right)+\sum_{m=1}^{n}\sum_{k=1}^{n}\frac{A_kA_m}{1-\varepsilon^2x_kx_m}\end{aligned}$$

with

$$\begin{aligned}b(\varepsilon)&=\frac{1}{\varepsilon^2}\int_{-1}^{1}\frac{1}{x}\ln\left(\frac{1+\varepsilon^2x}{1-\varepsilon^2x}\right)\mathrm{d}x=\frac{1}{\varepsilon^2}\int_{-\tau}^{\tau}\frac{2t\mathrm{e}^t}{\mathrm{e}^{2t}-1}\mathrm{d}t\quad\text{for}\quad\tau=\ln\frac{1+\varepsilon^2}{1-\varepsilon^2}\\&=\frac{1}{\varepsilon^2}\sum_{r=0}^{\infty}B_r\left(\frac{1}{2}\right)\int_{-\tau}^{\tau}\frac{(2t)^r}{r!}\mathrm{d}t=\frac{1}{\varepsilon^2}\sum_{r=0}^{\infty}\frac{2^{r+1}B_r(0.5)}{(r+1)!}\tau^{r+1}\frac{1+(-1)^r}{2}\\&=\frac{4}{\varepsilon^2}\sum_{r=0(2)}^{\infty}(1-2^{r-1})\frac{\tau^{r+1}}{(r+1)!}B_r.\end{aligned}$$

The convergence of this series is guaranteed for $\varepsilon\leqslant\left(\dfrac{\mathrm{e}^{\pi}-1}{\mathrm{e}^{\pi}+1}\right)^{\frac{1}{2}}=0.958.$
This is seen by using asymptotic values of B_r for large r.

For $\varepsilon=0$ we obtain $b(0)=4$ and $\|r_n\|=\left|2-\sum_{j=1}^{n}A_j\right|$.

For $\varepsilon=1$ we have $b(1)=\pi^2/2$ by direct calculation.
We list some values of $a(\varepsilon)$ and $b(\varepsilon)$ in Table 3.5.3.

TABLE 3.5.3

ε	$a(\varepsilon)$	$b(\varepsilon)$
0	1	4
0.1	1.002 511 174	4.000 044 446
0.2	1.010 181 883	4.000 711 521
0.3	1.023 448 363	4.003 610 541
0.4	1.043 129 962	4.011 484 027
0.5	1.070 610 556	4.028 423 494
0.6	1.108 217 409	4.060 480 346
0.7	1.160 078 457	4.117 260 726
0.8	1.234 415 663	4.216 431 433
0.9	1.351 979 120	4.400 545 282
1.0	1.644 934 067	4.934 802 201

Table 3.5.3 now enables us to calculate $\|r_n\|$ and $\|\overset{*}{r}_n\|$ for *given* nodes and weights of Q_nf, and to estimate the error by means of $|E_nf| \leqslant \|f\|\,\|r_n\|$ or $|E_nf| \leqslant \|f\|\,\|\overset{*}{r}_n\|$. For this bound to be satisfied, $\|f\|$ must be calculated. However, such a calculation is usually very complicated and often impossible. We therefore take the upper bound $\max_{|z|\leqslant R} |f(z)|$ which appears on the boundary $|z| = R$ because $f(z)$ is assumed to be analytic for $|z| < R$. Hence $\|f\| \leqslant \max_{|z|=R} |f(z)|$.

Example 3.5.2 Estimate the error of $Q_2f = \frac{1}{4}[f(0) + 3f(\frac{2}{3})] \doteq \overset{*}{I}f$. In this case

$$\|\overset{*}{r}_2\| = a(\varepsilon) + \frac{1}{2\varepsilon^2}\Bigg[\underbrace{\lim_{x\to 0} \frac{\ln(1-\varepsilon^2 x)}{x}}_{=-\varepsilon^2} + \frac{9}{2}\left(1 - \frac{2\varepsilon^2}{3}\right)\Bigg] + \frac{1}{16}\left[7 + \frac{81}{9-4\varepsilon^2}\right].$$

The power series expansion is more convenient for small ε^2

$$\|\overset{*}{r}_2\|^2 = \sum_{k=3}^{\infty} (\varepsilon^2)^k \left(\frac{1}{k+1} - \frac{3}{4}\left(\frac{2}{3}\right)^k\right)^2.$$

From this and from Table 3.5.3 for $a(\varepsilon)$ we obtain Table 3.5.4.

TABLE 3.5.4

ε	0	0.1	0.2	0.4	0.6	0.8	1
$\|\overset{*}{r}_2\|$	0	2.83(−5)	2.38(−4)	2.35(−3)	1.14(−2)	4.80(−2)	3.51(−1)

The rate of change in the norms $\|\overset{*}{r}_2\|$ is decreasing with decreasing ε, but $\|f\|$ is growing for decreasing ε. Hence, there usually exists an ε for which $|E_2f| = \|f\|\,\|\overset{*}{r}_2\|$ obtains a minimal value. Usually it is difficult to calculate this point exactly, but this is not too detrimental to the estimate because in general the curve $(E_2f)(\varepsilon)$ is very flat near the minimum (see Table 3.5.5). Consider the case $f(x) = \exp(x)$. We have $\|f\| < \exp(R) = \exp(1/\varepsilon)$.

TABLE 3.5.5

ε	0.1	0.2	0.4	0.6	0.8	1
$\|E_2f\| <$	0.623	0.0354	0.0286	0.0604	0.168	0.954

■

Example 3.5.3 We apply the estimation to Simpson's rule

$$Q_3 f = \tfrac{1}{3}[f(-1) + 4f(0) + f(1)] \doteq If$$

and we obtain

$$\|r_3\|^2 = b(\varepsilon) + \frac{2}{3\varepsilon^2}\left[2\ln\frac{1-\varepsilon^2}{1+\varepsilon^2} + \underbrace{4\lim_{x\to 0}\frac{1}{x}\ln\frac{1-\varepsilon^2 x}{1+\varepsilon^2 x}}_{=-2\varepsilon^2}\right] + \frac{1}{9}\left[32 + \frac{2}{1-\varepsilon^2} + \frac{2}{1+\varepsilon^2}\right]$$

or

$$\|r_3\|^2 = 4\sum_{k=5(2)}^{\infty} (\varepsilon^2)^{k-1}\left(\frac{1}{k} - \frac{1}{3}\right)^2.$$

The estimates in this case are as in Table 3.5.6.

TABLE 3.5.6

ε	0	0.1	0.2	0.3	0.4	0.6	0.8	0.9	0.99
$\|r_3\|$	0	2.67(−5)	4.27(−4)	2.18(−3)	7.01(−3)	3.97(−2)	1.80(−1)	4.52(−1)	2.87

Here we cannot take $\varepsilon = 1$ because $\ln(1-\varepsilon^2)$ and $1/(1-\varepsilon^2)$ are not bounded for $\varepsilon^2 \to 1$ and they do not cancel.

For $f(x) = \exp(x)$ we have $\|f\| < \exp(1/\varepsilon)$ and thus the values as in Table 3.5.7.

TABLE 3.5.7

ε	0.1	0.2	0.3	0.4	0.6	0.8	0.9	0.99
$\|E_3 f\| <$	0.588	0.0634	0.0611	0.0854	0.210	0.628	1.37	7.87

■

Remark: We did not take into account the degree of $Q_n f$ within the error estimates. This is automatically done within the sums containing the error functional $E_n f$.

3.5.4 Minimal norm quadrature formulae

Until now we have assumed that the nodes and weights of $Q_n f$ are given. In this section we shall determine them in such a way that the *error coefficient is minimised.* As the simplest case let us consider the formula

$$Q_1 f = Af(\alpha) \doteq \overset{*}{I}f$$

with one node and one weight.

We can easily derive the expressions

$$\|E_1\|^2 = \|\overset{*}{r}_1\|^2 = a(\varepsilon) + \frac{2}{\varepsilon^2}\frac{A}{\alpha}\ln(1-\varepsilon^2\alpha) + \frac{A^2}{1-\varepsilon^2\alpha^2} = \sum_{k=0}^{\infty} \varepsilon^{2k}\left(\frac{1}{k+1} - A\alpha^k\right)^2.$$

Hence $\|\overset{*}{r}_1\|^2$ is a function of A and α, and we ask whether there exist A and α such that $\|\overset{*}{r}_1\|^2$ attains its smallest possible value. For this to be true the necessary conditions

$$\frac{\partial\|\overset{*}{r}_1\|^2}{\partial A} = 0 = 2\left[\frac{1}{\varepsilon^2\alpha}\ln(1-\varepsilon^2\alpha) + \frac{A}{1-\varepsilon^2\alpha^2}\right]$$

$$\frac{\partial\|\overset{*}{r}_1\|^2}{\partial \alpha} = 0 = 2A\left[-\frac{1}{\alpha(1-\varepsilon^2\alpha)} + \frac{1}{\varepsilon^2\alpha^2}\ln\frac{1}{1-\varepsilon^2\alpha} + \frac{A\varepsilon^2\alpha}{(1-\varepsilon^2\alpha^2)^2}\right]$$

must be satisfied.

From the first of these two conditions it follows that

$$A = \frac{1-\varepsilon^2\alpha^2}{\varepsilon^2\alpha}\ln\frac{1}{1-\varepsilon^2\alpha},$$

and, by replacing A in the second one, we obtain

$$\alpha\varepsilon^2(1-\varepsilon^2\alpha^2) + (1-\varepsilon^2\alpha)\ln(1-\varepsilon^2\alpha) = 0.$$

This is a nonlinear equation for $\alpha = \alpha(\varepsilon)$, while $A = A(\varepsilon)$ can be evaluated immediately as soon as $\alpha(\varepsilon)$ is known.

We find that the values are as in Table 3.5.8.

TABLE 3.5.8

ε	$\alpha(\varepsilon)$	$A(\varepsilon)$
1	0.683 803	0.896 480
0.9	0.612 471	0.961 757
0.8	0.576 132	0.982 551
0.7	0.552 436	0.991 738
0.6	0.535 650	0.996 184
0.5	0.523 356	0.998 363
0.4	0.514 309	0.999 386
0.3	0.507 796	0.999 818
0.2	0.503 390	0.999 966
0.1	0.500 837	0.999 998
0	0.5	1

Hence we have a family of quadrature formulae depending on ε and containing the midpoint rule for $\varepsilon = 0$. For $\varepsilon = 1$ such formulae were discussed originally by Davis [9], Eckhardt [10] and Wilf [26]. The corresponding norms of the representers $\|\overset{*}{r}_1\|$ of $Q_1 f$ and $\|\overset{*}{r}_1\|_M$ of the midpoint rule for some values of ε are as in Table 3.5.9.

TABLE 3.5.9

ε	0	0.1	0.2	0.4	0.6	0.8	1
$\|\overset{*}{r}_1\|$	0	8.39(−4)	3.42(−3)	1.48(−2)	3.90(−2)	9.28(−2)	3.68(−1)
$\|\overset{*}{r}_1\|_M$	0	8.43(−4)	3.49(−3)	1.60(−2)	4.59(−2)	1.20(−1)	4.53(−1)

Comparing these norms we see that the difference $\|\overset{*}{r}_1\|_M - \|\overset{*}{r}_1\| > 0$ is growing for increasing ε. This effect is enlarged enormously if the number of nodes grows.

An interesting connection with the Gaussian quadrature formulae follows from the expansion of $\|\overset{*}{r}_n\|^2$, $\|r_n\|^2$ in powers of ε^2. We obtain

$$\|E_n\|^2 = \|\overset{*}{r}_n\|^2 = \sum_{k=0}^{\infty} \varepsilon^{2k}\left(\frac{1}{k+1} - \sum_{j=1}^{n} A_j x_j^k\right)^2 = \sum_{k=0}^{\infty} \varepsilon^{2k} E_n^2 x^k. \qquad (3.5.2)$$

This representation has the consequence:

Theorem 3.5.3 *For $\varepsilon \to 0$ the minimal norm quadrature coincides with the n-point Gauss quadrature formula having the highest possible degree of precision.*

Proof The Gaussian quadrature formulae of highest possible degree $(2n-1)$ have the error norm

$$\|E_n^G\|^2 = \sum_{k=2n}^{\infty} \varepsilon^{2k}(E_n^G x^k)^2.$$

A fortiori the quadratures with minimal $\|E_n\|^2$ satisfy

$$\|E_n^G\|^2 \geqslant \|E_n\|^2 = \sum_{k=0}^{\infty} \varepsilon^{2k}(E_n x^k)^2.$$

We then have the inclusion

$$0 \leqslant \varepsilon^{2j}(E_n x^j)^2 \leqslant \sum_{k=0}^{\infty} \varepsilon^{2k}(E_n x^k)^2 \leqslant \sum_{k=2n}^{\infty} \varepsilon^{2k}(E_n^G x^k)^2$$

for $j = 0(1)2n-1$. Hence, as $\varepsilon \to 0$ we conclude that

$$E_n x^j = 0 \quad \text{for} \quad j = 0(1)2n-1,$$

and the proof is finished. ■

The expression (3.5.2) is a weighted sum of quadrature errors for *all* monomials. The only change for $\|r_n\|$ is the replacement of $\overset{*}{I}x^{k-1}=1/k$ by $Ix^{k-1}=(1-(-1)^k)/k$. In order to minimise $\|\overset{*}{r}_n\|^2$ or $\|r_n\|^2$ we have to solve the variational problem

$$\|\overset{*}{r}_n\|^2=\sum_{k=0}^{\infty}\varepsilon^{2k}(E_n x^k)^2 \quad \text{to be minimal.} \tag{3.5.3}$$

This is a function of nodes and weights and leads to the necessary conditions

$$\frac{\partial\|\overset{*}{r}_n\|^2}{\partial x_j}=0 \quad \text{and} \quad \frac{\partial\|\overset{*}{r}_n\|^2}{\partial A_j}=0 \quad \text{for} \quad j=1(1)n$$

for this minimum. From the representation of $\|\overset{*}{r}_n\|^2$ and $\|r_n\|^2$ we obtain the coupled nonlinear systems of equations:

$$\left.\begin{aligned}\frac{1}{\varepsilon^2 x_j}\ln\frac{1}{1-\varepsilon^2 x_j}&=\sum_{m=1}^{n}\frac{A_m}{1-\varepsilon^2 x_m x_j} &\quad (3.5.4\text{a})\\ \frac{1}{\varepsilon^2 x_j(1-\varepsilon^2 x^j)}-\frac{1}{\varepsilon^4 x_j^2}\ln\frac{1}{1-\varepsilon^2 x_j}&=\sum_{m=1}^{n}\frac{A_m x_m}{(1-\varepsilon^2 x_j x_m)^2} &\quad (3.5.4\text{b})\end{aligned}\right\} j=1(1)n$$

and

$$\left.\begin{aligned}\frac{1}{\varepsilon^2 x_j}\ln\frac{1+\varepsilon^2 x_j}{1-\varepsilon^2 x_j}&=\sum_{m=1}^{n}\frac{A_m}{1-\varepsilon^2 x_m x_j} &\quad (3.5.5\text{a})\\ \frac{2}{\varepsilon^2 x_j(1-\varepsilon^4 x_j^2)}-\frac{1}{\varepsilon^4 x_j^2}\ln\frac{1+\varepsilon^2 x_j}{1-\varepsilon^2 x_j}&=\sum_{m=1}^{n}\frac{A_m x_m}{(1-\varepsilon^2 x_m x_j)^2} &\quad (3.5.5\text{b})\end{aligned}\right\} j=1(1)n$$

respectively. Once these systems are solved, we obtain simpler representations for $\|\overset{*}{r}_n\|$ and $\|r_n\|$. As we have seen in Section 3.5.3, the representation

$$\|\overset{*}{r}_n\|^2=a(\varepsilon)+\frac{2}{\varepsilon^2}\sum_{m=1}^{n}\frac{A_m}{x_m}\ln(1-\varepsilon^2 x_m)+\sum_{m=1}^{n}\sum_{k=1}^{n}\frac{A_k A_m}{1-\varepsilon^2 x_k x_m}$$

(and for $\|r_n\|^2$ with $b(\varepsilon)$ instead of $a(\varepsilon)$) is valid. Rearranging the terms we see that

$$\|\overset{*}{r}_n\|^2 = a(\varepsilon) + \frac{1}{\varepsilon^2}\sum_{m=1}^{n}\frac{A_m}{x_m}\ln(1-\varepsilon^2 x_m)$$

$$+\sum_{m=1}^{n} A_m\left\{\frac{-1}{\varepsilon^2 x_m}\ln\frac{1}{1-\varepsilon^2 x_m} + \sum_{k=1}^{n}\frac{A_k}{1-\varepsilon^2 x_k x_m}\right\}$$

$$= a(\varepsilon) - \frac{1}{\varepsilon^2}\sum_{m=1}^{n}\frac{A_m}{x_m}\ln\frac{1}{1-\varepsilon^2 x_m}$$

because of (3.5.4a). Analogously, we have that

$$\|r_n\|^2 = b(\varepsilon) - \frac{1}{\varepsilon^2}\sum_{m=1}^{n}\frac{A_m}{x_m}\ln\frac{1+\varepsilon^2 x_m}{1-\varepsilon^2 x_m}.$$

The expansion (3.5.3) was the starting point for Wilf, who used the condition $\|\overset{*}{r}_n\|^2$ to be minimum for $\varepsilon = 1$ to derive quadrature formulae for $\overset{*}{I}f$. He did not take account of $\varepsilon < 1$ and symmetric formulae for the interval $[-1, 1]$.

We will not go into further details here as we shall be deriving these quadrature formulae later by integration of a rational interpolant of f and by the use of orthogonal polynomials.

The equations (3.5.4a,b) and (3.5.5a,b) have additional interesting meanings:

Theorem 3.5.4 *The collocation requirements* $\overset{*}{r}_n(x_j) = \overset{*}{r}{}'_n(x_j) = 0$ *for* $j = 1(1)n$ *are satisfied by* (3.5.4a,b) *and the analogous requirements* $r_n(x_j) = r'_n(x_j) = 0$ *for* $j = 1(1)n$ *are satisfied by* (3.5.4a,b).

The proof can be easily performed by rewriting the representers.

Theorem 3.5.5 *Every* $Q_n f$ *satisfying* (3.5.4a,b) *or* (3.5.5a,b) *integrates* exactly *the* $2n$ *rational functions*

$$\frac{1}{1-\varepsilon^2 x x_j} \quad \textit{and} \quad \frac{x}{(1-\varepsilon^2 x x_j)^2} \qquad j = 1(1)n.$$

Proof We have

$$\int_0^1 \frac{dx}{1-\varepsilon^2 x x_j} = \frac{1}{\varepsilon^2 x_j} \ln \frac{1}{1-\varepsilon^2 x^j} = \sum_{m=1}^{n} \frac{A_m}{1-\varepsilon^2 x_m x_j} = Q_n \frac{1}{1-\varepsilon^2 x x_j}$$

$$\int_{-1}^1 \frac{dx}{1-\varepsilon^2 x x_j} = \frac{1}{\varepsilon^2 x_j} \ln \frac{1+\varepsilon^2 x_j}{1-\varepsilon^2 x_j} = Q_n \frac{1}{1-\varepsilon^2 x x_j}$$

$$\int_0^1 \frac{x\, dx}{(1-\varepsilon^2 x x_j)^2} = \frac{1}{\varepsilon^2 x_j(1-\varepsilon^2 x_j)} - \frac{1}{\varepsilon^4 x_j^2} \ln \frac{1}{1-\varepsilon^2 x_j} = Q_n \frac{x}{(1-\varepsilon^2 x x_j)^2}$$

$$\int_{-1}^1 \frac{x\, dx}{(1-\varepsilon^2 x x_j)^2} = \frac{2}{\varepsilon^2 x_j(1-\varepsilon^4 x_j)} - \frac{1}{\varepsilon^4 x_j^2} \ln \frac{1+\varepsilon^2 x_j}{1-\varepsilon^2 x_j} = Q_n \frac{x}{(1-\varepsilon^2 x x_j)^2}$$

for $j = 1(1)n$. ■

This property will be important for the construction of a Hermitian interpolating operator $H_n(f, x)$ having exactly the functions of Theorem 3.5.5 as fixed elements (g is a fixed element of $H_n(f, x)$ if $H(g, x) \equiv g$).

If the inequality $|E_n f| \leq \|E_n\| \, \|f\|$ is to be used to produce a practical error bound we need to calculate $\|f\|$, which is determined by $\sqrt{(f, f)}$. Usually this integral cannot be evaluated exactly. We therefore majorise $\|f\|$ by $\|f\| \leq M$ where $M = \max |f(z)|$ for $|z| = R$. It will be seen that for every $n \in \mathbb{N}$ such formulae exist in $[0, 1]$ and $[-1, 1]$ with positive weights.

If the nodes are known, then (3.5.4a) or (3.5.5a) are linear systems of equations for the weights. It is possible to invert the matrix explicitly as is shown by Engels [14].

A table of such optimal formulae is given in Chapter 7: Table 7.5.1.

3.5.5 Error bounds without derivatives for cubature formulae

We now turn to cubature formulae in the same way as in the preceding sections by using cartesian products of Hilbert space $\tilde{\mathbb{H}}_2$.

Let us consider the two-dimensional case. (The extension to more dimensions will be obvious.) The Hilbert space which now corresponds to the space $\tilde{\mathbb{H}}_2$ in one dimension is the cartesian product space $\tilde{\mathbb{H}}_2^2 \equiv \tilde{\mathbb{H}}_2 \times \tilde{\mathbb{H}}_2$ with the inner product

$$(f, g) = \frac{1}{4\pi^2} \int_0^{2\pi} \int_0^{2\pi} f(z, w)\overline{g(z, w)}\, d\varphi\, d\psi.$$

The functions $f(z, w)$ of this space $\tilde{\mathbb{H}}_2^2$ depend on the two complex variables $z = x + iy = |z| \exp(i\varphi)$, and $w = u + iv = |w| \exp(i\psi)$. We assume that the domain of integration $\bar{\Omega}$ is contained in the region $\{(z, w) \,|\, |z| \leq$

$R_1=1/\varepsilon,\ |w|\leqslant R_2=1/\eta\}$. In $\tilde{\mathbb{H}}_2^2$ there exists the orthonormal function system

$\{(\varepsilon z)^\nu(\eta w)^\mu\},\ \nu=0(1)\ldots,\ \mu=0(1)\ldots,$ with $\varepsilon=1/R_1$ and $\eta=1/R_2$.

This is true because of

$$((\varepsilon z)^\nu(\eta w)^\mu,(\varepsilon z)^\sigma(\eta w)^\tau)=\frac{1}{4\pi^2}\int_0^{2\pi}\int_0^{2\pi}e^{i(\nu-\sigma)\varphi+i(\mu-\tau)\psi}\,d\varphi\,d\psi$$

$$=\left(\frac{1}{2\pi}\int_0^{2\pi}e^{i(\nu-\sigma)\varphi}\,d\varphi\right)\left(\frac{1}{2\pi}\int_0^{2\pi}e^{i(\mu-\tau)\psi}\,d\psi\right)=\delta_{\nu,\sigma}\,\delta_{\mu,\tau}.$$

The representer r_n of the functional $E_nf=I_\Omega f-C_nf$ is

$$r_n(z,w)=\sum_{l=0}^{\infty}\sum_{m=0}^{\infty}\varepsilon^{2l}\eta^{2m}[E_n(x^lu^m)]z^lw^m.$$

Hence

$$\|E_n\|^2=\sum_{l=0}^{\infty}\sum_{m=0}^{\infty}\varepsilon^{2l}\eta^{2m}E_n^2(x^lu^m)=E_nr_n.$$

We illustrate the results by the following:

Example 3.5.4 Calculate the error norm $\|E_n\|$ for

$$C_4f=Af(\pm\alpha,\pm\beta)=A[f(\alpha,\beta)+f(\alpha,-\beta)+f(-\alpha,\beta)+f(-\alpha,-\beta)]$$

where Ω is the square $|x|,|u|\leqslant 1$.

The representer r_4 is given for $(0<\varepsilon,\eta<1)$ by

$$r_4(z,w)=\sum_{l=0}^{\infty}\sum_{m=0}^{\infty}\varepsilon^{2l}\eta^{2m}\left[\frac{(1+(-1)^l)(1+(-1)^m)}{(l+1)(m+1)}\right.$$
$$\left.-A\alpha^l(1+(-1)^l)\beta^m(1+(-1)^m)\right]z^lw^m$$

$$=4\sum_{l=0(2)}^{\infty}\sum_{m=0(2)}^{\infty}\left[\frac{(\varepsilon^2z)^l(\eta^2w)^m}{(l+1)(m+1)}-A(\alpha\varepsilon^2z)^l(\beta\eta^2w)^m\right]$$

$$=2\sum_{l=0(2)}^{\infty}\left[\frac{(\varepsilon^2z)^l}{l+1}2\vartheta\sum_{m=0(2)}^{\infty}\frac{(\eta^2w)^m}{m+1}\right]-$$
$$4A\sum_{l=0(2)}^{\infty}\left[(\alpha\varepsilon^2z)^l\sum_{m=0(2)}^{\infty}(\beta\eta^2w)^m\right]$$

$$=2\sum_{l=0(2)}^{\infty}\frac{(\varepsilon^2z)^l}{l+1}\cdot\frac{1}{\eta^2w}\ln\frac{1+\eta^2w}{1-\eta^2w}-4A\sum_{l=0(2)}^{\infty}\frac{(\alpha\varepsilon^2z)^l}{1-\beta^2\eta^4w^2}$$

$$=\frac{1}{\varepsilon^2z\eta^2w}\ln\frac{1+\varepsilon^2z}{1-\varepsilon^2z}\ln\frac{1+\eta^2w}{1-\eta^2w}-\frac{4A}{(1-\alpha^2\varepsilon^4z^2)(1-\beta^2\eta^4w^2)}.$$

Hence

$$\|E_4\|^2 = E_4 r_4(x, u) = c(\varepsilon, \eta) - \frac{8A}{\varepsilon^2 \alpha \eta^2 \beta} \ln \frac{1+\varepsilon^2\alpha}{1-\varepsilon^2\alpha} \ln \frac{1+\eta^2\beta}{1-\eta^2\beta} + \frac{16A^2}{(1-\alpha^4\varepsilon^4)(1-\beta^4\eta^4)}$$

with

$$c(\varepsilon, \eta) = \int_{-1}^{1} \int_{-1}^{1} \frac{1}{\varepsilon^2 x \eta^2 u} \ln \frac{1+\varepsilon^2 x}{1-\varepsilon^2 x} \ln \frac{1+\eta^2 u}{1-\eta^2 u} \, dx \, du = b(\varepsilon) b(\eta).$$

Again we have to look for special cases.

For $\varepsilon = 0$, $0 < \eta < 1$ we have that

$$\|E_4\|^2 = 4b(\eta) - \frac{16A}{\eta^2\beta} \ln \frac{1+\eta^2\beta}{1-\eta^2\beta} + \frac{16A^2}{1-\beta^4\eta^4}$$

and, analogously, that

$$\|E_4\|^2 = 4b(\varepsilon) - \frac{16A}{\varepsilon^2\alpha} \ln \frac{1+\varepsilon^2\alpha}{1-\varepsilon^2\beta} + \frac{16A^2}{1-\alpha^4\varepsilon^4} \quad \text{for} \quad 0 < \varepsilon < 1 \quad \text{and} \quad \eta = 0.$$

Finally,

$$\|E_4\|^2 = 16(1-A)^2 \quad \text{for} \quad \varepsilon = \eta = 0.$$

For $\varepsilon = 1$, $0 < \eta < 1$ we obtain

$$\|E_4\|^2 = \frac{\pi^2}{2} b(\eta) - \frac{8A}{\alpha\eta^2\beta} \ln \frac{1+\alpha}{1-\alpha} \ln \frac{1+\eta^2\beta}{1-\eta^2\beta} + \frac{16A^2}{(1-\alpha^4)(1-\eta^4\beta^4)}$$

and for $\eta = 1$, $0 < \varepsilon < 1$

$$\|E_4\|^2 = \frac{\pi^2}{2} b(\varepsilon) - \frac{8A}{\varepsilon^2\alpha\beta} \ln \frac{1+\varepsilon^2\alpha}{1-\varepsilon^2\alpha} \ln \frac{1+\beta}{1-\beta} + \frac{16A^2}{(1-\varepsilon^4\alpha^4)(1-\beta^4)}$$

and, finally, for $\varepsilon = \eta = 1$ we obtain

$$\|E_4\|^2 = \frac{\pi^4}{4} - \frac{8A}{\alpha\beta} \ln \frac{1+\alpha}{1-\alpha} \ln \frac{1+\beta}{1-\beta} + \frac{16A^2}{(1-\alpha^4)(1-\beta^4)}.$$

For arbitrary ε, η we have the expansion

$$\|E_4\|^2 = \sum_{l=0(2)}^{\infty} \sum_{m=0(2)}^{\infty} \varepsilon^{2l} \eta^{2m} E_4^2(x^l u^m)$$

which reveals the same principle as for the case of quadrature. A minimisation of $\|E_4\|^2$ requires necessarily that

$$\frac{\partial\|E_4\|^2}{\partial A}=\frac{\partial\|E_4\|^2}{\partial\alpha}=\frac{\partial\|E_4\|^2}{\partial\beta}=0.$$

From the first of these three equations we obtain

$$A=\frac{(1-\alpha^4)(1-\beta^4)}{4\alpha\beta}\left(\ln\frac{1+\alpha}{1-\alpha}\right)\left(\ln\frac{1+\beta}{1-\beta}\right).$$

Substituting this into the second and third equations we see that the coupling between α and β disappears. Thus, α satisfies

$$\frac{1+\alpha^4}{1+\alpha^2}\ln\frac{1+\alpha}{1-\alpha}=2,$$

and β satisfies the same equation if α is replaced by β. Hence $\alpha=\beta=0.665\,742\,696\,0$, $A=0.939\,548\,796\,2$ and $\|E_4\|=1.574\,385$.

If we compare the result with the first quadrature formula of Table 7.5.1 (Chapter 7) we see that the nodes $\pm\alpha$, $\pm\beta$ are identical and that the weight is squared. Hence, C_4f is a product rule (cf. Section 2.6.2).

For $\varepsilon=\eta\to 0$ the cubature formula C_4f turns into an optimal cubature formula of degree 3, whose nodes and weight satisfy

$$E_4f=0 \quad \text{for} \quad f\in\{1, x, u, x^2, xu, u^2, x^3, x^2u, xu^2, u^3\}$$

Hence $\alpha=\beta=1/\sqrt{3}$, and $A=1$.

For $0<\varepsilon$, $\eta<1$ we obtain intermediate values $\alpha(\varepsilon,\eta)$, $\beta(\varepsilon,\eta)$ and $A(\varepsilon,\eta)$. ■

The results of this example can be extended immediately to product cubature formulae for the square. The extension to a non-product domain Ω is not yet explored.

3.5.6 The space $\mathbb{E}_\rho$

The error estimates of the preceding section were connected with the circle $\varepsilon\,|z|\leq 1$ of the complex domain. If we want to integrate real-valued functions this is perhaps a disadvantage because the influence of the whole area outside $[-1, 1]$ cannot be eliminated. For this reason Davis and other authors consider the space $\mathbb{E}_\rho$ instead of $\tilde{\mathbb{H}}_2$: Let E_ρ denote the set of confocal ellipses with major semiaxis a, minor semiaxis b and foci ± 1. Then $\rho=(a+b)^2$ is the parameter on which the ellipses depend and for $\rho\to 1$ they collapse into $[-1, 1]$. Now consider E_ρ as a given domain in the complex plane. The space of functions $f(z)$ which are analytic in E_ρ

is then the Hilbert space $\mathbb{E}_\rho$ with the inner product

$$(f, g) = \iint_{E_\rho} f(z)\overline{g(z)}\, dx\, dy \quad \text{with} \quad z = x + iy$$

and norm $\|f\|^2 = (f, f)$. This space contains a complete orthonormal system of polynomials (the orthonormal polynomials are, however, no longer monomials), and a reproducing kernel function, called *Bergmann kernel.* Hence, there exists a theory similar to that of the preceding section. The error bounds of this theory seem to be better than those given above for some *fixed* $Q_n f$ such as Simpson's rule.

Nevertheless, for the following reasons we shall not go into details:

(i) The theory is more complicated for $\mathbb{E}_\rho$ than for $\tilde{\mathbb{H}}_2$.
(ii) Using the theory for $\tilde{\mathbb{H}}_2$ we can treat functions f having an integrable singularity at ± 1. This is impossible with the theory for $\mathbb{E}_\rho$ where $[-1, 1]$ belongs to the *interior* of E_ρ.
(iii) The construction of optimal formulae which minimise $\|E_n\|$ is very difficult because $\|E_n\|$ does not explictly depend on the zeros x_i.
(iv) The calculation of $\|f\|$ is usually replaced by taking $\sqrt{\pi ab} \max_{|z|=a} |f|$ which corresponds to Section 3.5.4 for $a = 1/\varepsilon$.
(v) The optimal formulae coincide for $\rho \to \infty$ with those for $\tilde{\mathbb{H}}_2$ and hence with Gaussian quadrature formulae.
(vi) The extension to two and higher dimensions, which was first performed by Barnhill [5], is analogous to that in $\tilde{\mathbb{H}}_2$. Instead of $\tilde{\mathbb{H}}_2 \times \tilde{\mathbb{H}}_2$ we have $\mathbb{E}_\rho \times \mathbb{E}_\rho$.

For details the reader is referred to the papers of Davis [9] and, concerning (vi), to Barnhill [5] and Stroud (book, p. 180 ff.). A similar theory is associated with the inner product

$$(f, g) = \int_{\mathrm{Cont}(E_\rho)} f(z)\overline{g(z)}\, ds, \qquad s \text{ arc length of } E_\rho.$$

3.6 Error bounds without derivatives for quadratures and cubatures on spaces of functions which are analytic in a strip

3.6.1 Quadrature error bounds for functions which are analytic in a strip

As already seen in Section 2.5.2, the rectangle rule is a most important quadrature rule for periodic functions. We now study a method due to

Martensen for approximating $\underset{\infty}{I}f$ by the rectangle rule and desire error bounds without derivatives.

Assume that $f(z)$, $z = x + iy$, is analytic in the strip $-\infty < x < \infty$, $0 \leq y \leq s$, real-valued for $y = 0$, uniformly convergent for every fixed $y \in [0, s]$ as $x \to \pm\infty$ and that

$$\int_{-\infty+is}^{\infty+is} |f(z)|\, dx \quad \text{exists.}$$

Because of these properties $f(z)$ can be continued analytically into the strip $-\infty < x < \infty$, $-s \leq y \leq 0$ by means of $f(\bar{z}) = \overline{f(z)}$, and because $\underset{\infty}{I}f$ exists and the "rectangle sum"

$$Q_h f \equiv h \sum_{j=-\infty}^{\infty} f(jh)$$

is convergent. We denote the set of functions $f(z)$ of this type by $\mathscr{A}_s$ and on $\mathscr{A}_s$ we prove the existence of error estimates without derivatives

$$|E_h f| = |\underset{\infty}{I}f - Q_h f| \leq \sigma_n(s)\, \|f\|.$$

Theorem 3.6.1 (Martensen) *Let $f(z) \in \mathscr{A}_s$ with $s > 0$. Then the error $E_h f$ satisfies*

$$E_h f = \operatorname{Re} \int_{-\infty+is}^{\infty+is} \left(1 - i \cot \frac{\pi z}{h}\right) f(z)\, dz,$$

$$|E_h f| \leq \frac{2}{e^{2\pi s/h} - 1} \int_{-\infty+is}^{\infty+is} |f(z)|\, dx.$$

The proof essentially uses the well-known residue theorem of complex analysis.

Proof We look for a function $g(z)$ having the residue $hf(jh)$ at $x = jh$, $j = -\infty(1)\infty$, and no further residue, in order to obtain an integral representation of $Q_h f$.

For such a function $g(z)$ we apply the residue theorem for the strip $-\infty < x < \infty$, $-s \leq y \leq s$ with the directed boundary C_∞ such that

$$\frac{1}{2\pi i} \int_{C_\infty} g(z)\, dz = ch \sum_{j=-\infty}^{\infty} f(jh) \quad \text{with suitable constant c.}$$

The simplest choice for $g(z)$ is a function which has a periodic factor $m(z) = m(z + h)$ with period h and the Laurent expansion

$$\frac{a_{-1}}{z} + \sum_{k=0}^{\infty} a_k z^k \quad \text{at} \quad x = z = 0.$$

In this case the residue of $g(z)$ at $z=0$ is ca_{-1} with constant c. If we denote $g(z)=f(z)m(z)$, then the residue is $f(0)a_{-1}$ at $z=0$ and hence $f(jh)a_{-1}$ at $z=x=jh$, $j=-\infty(1)\infty$ (since $m(z)$ has period h). An example for $m(z)$ is $\cot(\pi z/h)$ with residue h/π. Now, $\cot \bar{z} = \overline{\cot z}$, since

$$\cot(x \pm iy) = \frac{\sin x \cos x \mp i \sinh y \cosh y}{\sin^2 x \cosh^2 y + \cos^2 x \sinh^2 y}$$

and hence $g(z)=f(z)\cot z$ is a meromorphic function in the interior of the strip.

As is usual in applications of the residue theorem, we first treat a finite strip

$$S_n \equiv \{z = x+iy \mid -(n+1/2)h \leq x \leq (n+1/2)h \equiv a_n, |y| \leq s\}.$$

Hence

$$\frac{1}{2\pi i}\int_{C_n} f(z)\cot\frac{\pi z}{h}\,dz$$

$$= \frac{1}{2\pi i}\left\{\int_{-a_n-is}^{a_n-is}\cdots + \int_{a_n-is}^{a_n+is}\cdots + \int_{a_n+is}^{-a_n+is}\cdots + \int_{-a_n+is}^{-a_n-is}\cdots\right\} = \frac{h}{\pi}\sum_{j=-n}^{n} f(jh),$$

where C_n is the directed boundary of S_n. Let

$$g(z) = f(z)\cot(z/h) = u(x, y) + iv(x, y).$$

Since $g(\bar{z}) = \overline{g(z)}$, we have $u(x, -y)+iv(x, -y) = u(x, y) - iv(x, y)$ and, with this decomposition, the sum of integrals becomes

$$\int_{-a_n}^{a_n}(u(x, -s)+iv(x, -s))\,dx + \left(\int_{-s}^{0} + \int_{0}^{s}\right)(u(a_n, y)+iv(a_n, y))i\,dy$$

$$+ \int_{a_n}^{-a_n}(u(x, s)+iv(x, s))\,dx + \left(\int_{s}^{0} + \int_{0}^{-s}\right)(u(-a_n, y)+iv(-a_n, y))i\,dy.$$

Within the integrals with the limits $-s$ and 0 we replace y by $-y$, and, in addition, we always apply $g(\bar{z}) = \overline{g(z)}$. We then have

$$\int_{-a_n}^{a_n}(u(x, s)-iv(x, s))\,dx + i\int_0^s (u(a_n, y)-iv(a_n, y))\,dy$$

$$+ i\int_0^s (u(a_n, y)+iv(a_n, y))\,dy - \int_{-a_n}^{a_n}(u(x, s)+iv(x, s))\,dx$$

$$- i\int_0^s (u(-a_n, y)+iv(-a_n, y))\,dy - i\int_0^s (u(-a_n, y)-iv(-a_n, y))\,dy$$

$$= -2i\int_{-a_n}^{a_n} v(x, s)\,dx + 2i\int_0^s u(a_n, y)\,dy - 2i\int_0^s u(-a_n, y)\,dy$$

$$= -2i\int_{-a_n+is}^{a_n+is} v(x, y)\,dz + 2\int_{a_n}^{a_n+is} u(x, y)\,dz - 2\int_{-a_n}^{-a_n+is} u(x, y)\,dz$$

$$= -2i\int_{-a_n+is}^{a_n+is} (\operatorname{Im} g(z))\,dz + 2\int_{a_n}^{a_n+is} (\operatorname{Re} g(z))\,dz - 2\int_{-a_n}^{-a_n+is} (\operatorname{Re} g(z))\,dz$$

Finally, with respect to

$$\int_{-a_n+is}^{a_n+is} (\operatorname{Im} g(z))\,dz = \operatorname{Im}\int_{-a_n+is}^{a_n+is} g(z)\,dz$$

(the two integrals are real-valued), we have that

$$h\sum_{j=-n}^{n} f(jh) = -\operatorname{Im}\int_{-a_n+is}^{a_n+is} g(z)\,dz + \frac{1}{i}\int_{a_n}^{a_n+is} \operatorname{Re} g(z)\,dz - \frac{1}{i}\int_{-a_n}^{-a_n+is} \operatorname{Re} g(z)\,dz. \tag{3.6.1}$$

As $n \to \infty$ the last two integrals vanish. This is seen as follows. From $|\sin z| = (\sinh^2 y + \sin^2 x)^{\frac{1}{2}}$ and $|\cos z| = (\cosh^2 y - \sin^2 x)^{\frac{1}{2}}$ we have for $x = a_n = (n + 1/2)h$

$$\left|\cot\frac{\pi z}{h}\right| \leqslant \left(\frac{\cosh^2(\pi y/h) - 1}{\sinh^2(\pi y/h) + \sin^2 x}\right)^{\frac{1}{2}} \leqslant \left(\frac{\sinh^2(\pi y/h) + 1 - 1}{\sinh^2(\pi y/h)}\right)^{\frac{1}{2}} = 1.$$

Hence

$$\left|\int_{a_n}^{a_n+is} f(z)\cot\frac{\pi z}{h}\,dz\right| \leqslant \int_{a_n}^{a_n+is} |f(z)|\,dx \leqslant s \max_{0\leqslant y\leqslant s} |f(a_n + is)|,$$

which tends to zero as $n \to \infty$. The same is true for the third integral. Thus we have the desired integral representation

$$Q_h f = h\sum_{j=-\infty}^{\infty} f(jh) = -\operatorname{Im}\int_{-\infty+is}^{\infty+is} f(z)\cot\frac{\pi z}{h}\,dz$$

$$= \operatorname{Re} i\int_{-\infty+is}^{\infty+is} f(z)\cot\frac{\pi z}{h}\,dz, \tag{3.6.2}$$

since, in general, $-\operatorname{Im} z = -y = \operatorname{Re}(iz)$.

In order to obtain a comparable representation for $\underset{\infty}{I} f$ we apply Cauchy's integral theorem to $f(z)$ for the directed contours of the two

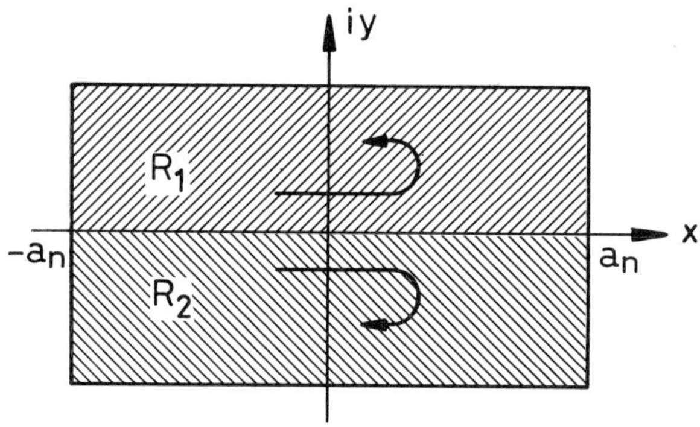

FIG. 3.6.1. Application of Cauchy's theorem to R_1 and R_2.

rectangles R_1 and R_2 as indicated in Fig. 3.6.1. Thus

$$0=\frac{1}{2}\left(\int_{-a_n}^{a_n}+\int_{a_n}^{a_n+is}+\int_{a_n+is}^{-a_n+is}+\int_{-a_n+is}^{-a_n}\right)f(z)\,dz$$

$$=\frac{1}{2}\left(\int_{-a_n}^{a_n}+\int_{a_n}^{a_n-is}+\int_{a_n-is}^{-a_n-is}+\int_{-a_n-is}^{-a_n}\right)f(z)\,dz.$$

Writing $f(z)=u(x, y)+iv(x, y)$ we have first that

$$0=\frac{1}{2}\int_{-a_n}^{a_n} f(x)\,dx+\frac{i}{2}\int_0^s (u(a_n, y)+iv(a_n, y))\,dy-$$
$$\frac{1}{2}\int_{-a_n}^{a_n} (u(x, s)+iv(x, s))\,dx$$
$$-\frac{i}{2}\int_0^s (u(-a_n, y)-iv(-a_n, y))\,dy.$$

The addition of the two results for R_1 and R_2 gives

$$0=\int_{-a_n}^{a_n} f(x)\,dx-\int_{-a_n}^{a_n} u(x, s)\,dx+\int_0^s iv(a_n, y)\,dy-i\int_0^s iv(-a_n, y)\,dy$$

$$=\int_{-a_n}^{a_n} f(x)\,dx-\text{Re}\int_{-a_n+is}^{a_n+is} f(z)\,dz-\int_{a_n}^{a_n+is}(\text{Im}\,f(z))\,dz+\int_{-a_n}^{-a_n+is}(\text{Im}\,f(z))\,dz.$$

Here, the last two integrals vanish as before, and hence

$$\underset{\infty}{I}f=\int_{-\infty}^{\infty} f(x)\,dx=\text{Re}\int_{-\infty+is}^{\infty+is} f(z)\,dz.$$

We subtract (3.6.2) from this equation to obtain

$$E_hf=\int_{-\infty}^{\infty} f(x)\,dx-h\sum_{j=-\infty}^{\infty} f(jh)=\text{Re}\int_{-\infty+is}^{\infty+is} f(z)\left(1-i\cot\frac{\pi z}{h}\right)dz.$$

From this representation we find the bound

$$|E_h f| \leqslant \sup_{|z|=|x+is|} \left|1 - i \cot \frac{\pi z}{h}\right| \int_{-\infty+is}^{\infty+is} |f(z)| \, dx$$

and note that

$$\left|1 - i \cot \frac{\pi z}{h}\right| = \left|\frac{\sin(\pi z/h) - i \cos(\pi z/h)}{\sin(\pi z/h)}\right| = \frac{|e^{-i\pi z/h}|}{|\sin(\pi z/h)|}$$

$$= \frac{e^{-\pi s/h}}{\sinh(\pi s/h)} = \frac{2}{\exp(2\pi s/h) - 1}$$

for $|z| = |x + iy|$. Hence, finally,

$$|E_h f| \leqslant \frac{e^{-\pi s/h}}{\sinh(\pi s/h)} \int_{-\infty+is}^{\infty+is} |f(z)| \, dx = \frac{2}{e^{2\pi s/h} - 1} \int_{-\infty+is}^{\infty+is} |f(z)| \, dx$$

which proves the theorem. ■

The following example shows that the error bound is extremely sharp.

Example 3.6.1 The evaluation of $\underset{\infty}{I}(\exp(-x^2)) = \sqrt{\pi}$ for some h gives us Table 3.6.1.

TABLE 3.6.1

h	$Q_h \exp(-x^2)$	Actual |error|	Estimated |error| for $s = \pi/h$
2	2.073 263 006	3.008 (−1)	3.028 (−1)
1	1.772 637 205	1.833 (−4)	1.834 (−4)
0.9	1.772 471 957	1.811 (−5)	1.811 (−5)
0.8	1.772 454 562	7.116 (−7)	7.116 (−7)
0.5	1.772 453 851	1.0 (−9) due to round-off	2.537 (−17)

In order to find the results of the last column we estimate the error using Theorem 3.6.1:

$$|\exp(-z^2)| = |\exp(-x^2 + s^2 - 2ixs)| = \exp(s^2 - x^2)$$

and hence

$$|E_h(e^{-x^2})| \leqslant \frac{2e^{s^2}}{e^{2\pi s/h} - 1} \int_{-\infty}^{\infty} e^{-x^2} \, dx = \frac{2\sqrt{\pi} e^{s^2}}{e^{2\pi s/h} - 1}.$$

This, of course, is not a very suitable error estimate because the unknown value $\underset{\infty}{I}(\exp(-x^2))$ appears as a multiplier in E_h. But we can circumvent this difficulty by taking a rough estimate of $\underset{\infty}{I}(\exp(-x^2))$ *a priori* or the computed value $Q_h(\exp(-x^2))$ *a posteriori*.

The error bound E_h still depends on s so that we can seek a minimum of $|E_hf|(s)$. The condition that the first derivative with respect to s vanishes leads to

$$s = \frac{\pi/h}{1-\exp(-2\pi s/h)} \tag{3.6.3}$$

for such an optimal value of s. For $h \ll 1$ we have the asymptotic value $s_{\text{opt}} = \pi/h$.

The solution of (3.6.3) for $h = 2$, and with $s = \pi/h$ is otherwise used to calculate the bounds in the last column of Table 3.6.1. ■

Asymptotically, the error bound is simplified to

$$|E_hf|_{s=\pi/h} = \frac{2\sqrt{\pi}\exp(\pi^2/h^2)}{\exp(2\pi^2/h^2)-1} = \frac{\sqrt{\pi}}{\sinh(\pi^2/h^2)},$$

which indicates an exponential convergence of Q_h for $h \to 0$.

Because $Q_h(\exp(-x^2)) \geq \underset{\infty}{I}(\exp(-x^2))$, we have the interesting inequality and series representation of $\sqrt{\pi}$:

$$h\sum_{j=-\infty}^{\infty} e^{-(jh)^2} - \sqrt{\pi} \leq \frac{\sqrt{\pi}}{\sinh(\pi^2/h^2)}$$

with equality for $h = 0$.

It should be noted that for this method E_hf depends on an integral with limits $\pm\infty$. Clearly, this is a serious disadvantage in view of automatic integration. However, this integral may itself be approximated by Q_hf and, when this is done, the method is again applicable.

3.6.2 Quadrature and cubature error bounds without derivatives for periodic, analytic functions

We extend the results of the preceding section, 3.6.1, to functions which are periodic, but for which the limits of integration are now assumed to be finite.

Let $f(z)$, $z = x + iy$ be an analytic periodic function with period 2π in the strip $-\infty < x < \infty$, $|y| \leq s > 0$. The calculations of the preceding section remain valid for $a_n = \pi$, $h = 2\pi/n$. The only difficulty which arises is that the n-th node is situated on the boundary of the finite strip S_n. This is eliminated by replacing Q_nf by $Q_{n,\varepsilon}f$ with the nodes $-\pi + jh - \varepsilon$ for $0 < \varepsilon < h$ and, finally, performing $\varepsilon \to 0$. This is only a technical and not an essential difficulty.

Instead of (3.6.2) we now have

$$Q_nf = h\sum_{j=1}^{n} f(-\pi + jh) = -\text{Im}\int_{-\pi+is}^{\pi+is} g(z)\,dz$$

with $g(z) = f(z)\cot(\pi z/h) = u(x, y) + iv(x, y)$ since

$$2i\int_0^s u(\pi, y)\,dy - 2i\int_0^s u(-\pi, y)\,dy = 0$$

which follows from $u(x, y) = u(x + 2\pi, y)$. Analogously,

$$0 = \int_{-\pi}^{\pi} f(x)\,dx - h\sum_{j=1}^{n} f(-\pi + jh) = \text{Re}\int_{-\pi+is}^{\pi+is} f(z)\left(1 - i\cot\frac{nz}{2}\right)dz.$$

Let $M = \max_{y=s}|f(z)|$, then (as before)

$$|E_nf| \leq 2\pi M\left|1 - i\cot\frac{nz}{2}\right| \leq \frac{2\pi M\exp(-sn/2)}{\sinh(sn/2)} = \frac{4\pi M}{\exp(ns) - 1} \tag{3.6.4}$$

$$\frac{4\pi M}{\exp(ns) - 1} \equiv (\sigma_n)_D \cdot M = \sigma_n(s)M.$$

This result is due to Davis [8]. The error coefficients $(\sigma_n)_D$ are not the best possible but, following Kress [22], we can obtain slightly better results by again using the Hilbert space theory.

Let $\mathbb{H}_G$ and $\mathbb{H}_H$ be the Hilbert spaces of *periodic* functions $f \in \mathscr{A}_s$ with, respectively, the inner products

$$(f, g)_G = \iint_G f(z)\overline{g(z)}\,dx\,dy \quad \text{and} \quad (f, g)_H = \int_h f(z)\overline{g(z)}\,dx$$

where

$$G = \{z = x + iy \mid |x| \leq \pi, |y| \leq s\}, \text{ resp. } H = \{z \mid |x| \leq \pi, |y| = s\}.$$

These spaces contain the complete orthonormal function system $g_m(z) = a_m e^{imz}$, $m = -\infty(1)\infty$, with

$$a_m = (m/(2\pi\sinh(2ms)))^{\frac{1}{2}}, \text{ resp. } a_m = 1/(4\pi\cosh(2ms))^{\frac{1}{2}}.$$

The representer of

$$E_nf=\int_{-\pi}^{\pi} f(x)\,dx-h\sum_{j=1}^{n} f(-\pi+jh)$$

is

$$r_nf=2\pi\sum_{k=-\infty}^{\infty} a_{kn}(-1)^{nk+1}g_{kn}(z).$$

Further

$$\|E_n\|^2=(2\pi)^2\sum_{k=-\infty}^{\infty} a_{kn}^2=2\sum_{k=1}^{\infty}(2\pi a_{kn})^2$$

from which

$$\|E_n\|^2=4\pi n\sum_{k=1}^{\infty}\frac{k}{\sinh(2kns)},\quad \text{resp.}\quad \|E_n\|^2=2\pi\sum_{k=1}^{\infty}\frac{1}{\cosh(2kns)},$$

follows. These series are not very practical. We therefore majorise them by (see Problem P.3.17)

$$\frac{\pi n e^{2ns}}{\sinh(2ns)\sinh^2(ns)}\quad\text{and}\quad\frac{2\pi e^{-ns}}{\sinh(ns)}=\frac{4\pi}{e^{2ns}-1}.$$

We thus have, respectively, the error bounds

$$|E_nf|<\frac{e^{ns}}{\sinh(ns)}\left(\frac{\pi n}{\sinh(2ns)}\right)^{\frac{1}{2}}\|f\|_G\quad\text{and}\quad|E_nf|<2\left(\frac{\pi}{e^{2ns}-1}\right)^{\frac{1}{2}}\|f\|_H.$$

The norms, which usually cannot be computed exactly, are estimated by

$$\|f\|_G\leqslant 2\sqrt{\pi s}M_1\quad\text{and}\quad\|f\|_H=2\sqrt{\pi}M$$

where $M_1=\max_{z\in G}|f(z)|$, and $M=\max_{z\in H}|f(z)|$, so that

$$|E_nf|<\sigma_n(s)M_1\equiv(\sigma_n)_G M_1\quad\text{and}\quad|E_nf|<\sigma_n(s)M\equiv(\sigma_n)_H M$$

with

$$(\sigma_n)_G=\frac{2\pi e^{ns}}{\sinh(ns)}\left(\frac{sn}{\sinh(2ns)}\right)^{\frac{1}{2}}\quad\text{and}\quad(\sigma_n)_H=\frac{4\pi}{(\exp(2ns)-1)^{\frac{1}{2}}}$$

The first error constant $(\sigma_n)_D$ and the third $(\sigma_n)_H$ are easily compared. We have that

$$(\sigma_n)_H=\frac{4\pi}{(e^{2ns}-1)^{\frac{1}{2}}}\leqslant\frac{4\pi}{e^{ns}-1}=(\sigma_n)_D.$$

There is no such simple relation between $(\sigma_n)_G$ and $(\sigma_n)_D$, or between $(\sigma_n)_G$ and $(\sigma_n)_H$. It may happen that for $M = M_1$ we have $(E_nf)_G \leqslant (E_nf)_H \leqslant (E_nf)_D$ for certain values of s but, possibly, $(E_nf)_H \leqslant (E_nf)_D \leqslant (E_nf)_G$ for other values of s. Hence, we cannot predict, which bound will be smaller.

Example 3.6.2 We apply the rectangle rule to $f(x) = \sin^{2k} x$ and bound the error. For $k = 8$ we obtain the values as in Table 3.6.2.

TABLE 3.6.2

n	Q_nf	$\lvert\text{error}\rvert$
3	0.419 351 99	8.1454 (−1)
5	1.126 517 14	1.0738 (−1)
7	1.230 827 84	3.0680 (−3)
9	1.233 895 79	1.0 (−8)

For $n \geqslant 9$ we have only the influence of round-off errors and, hence, the exact result if we neglect the round-off errors. We compare with the error bounds below which are calculated for each of the three possibilities D, H, G, always taking that value of s for which $\|E_n\|$ is minimal. These assure the vanishing of the error bound only for $n \geqslant 17$ (see Table 3.6.3).

TABLE 3.6.3

	$n=3$		$n=5$		$n=7$		$n=16$		$n=17$	
	s	$\|E\|$	s	$\|E\|$	s	$\|E\|$	s	$\|E\|$	s	$\|E\|$
D	0.32	17.5	0.38	6.9	0.49	2.7	10	1.9 (−4)	∞	0
H	0.25	11.0	0.34	5.8	0.47	2.6	7	1.9 (−4)	∞	0
G	0.23	15.5	0.25	10.3	0.38	6.3	7	2.9 (−3)	∞	0

Hence, for $n \leqslant 16$ these bounds are not small enough. For $n \geqslant 17$ we find the result as follows

$$|E_nf|_D \leqslant \frac{4\pi}{e^{ns}-1} \cosh^{2k} s \frac{4\pi e^{2ks}}{e^{ns}-1}$$

since $|\sin^{2k} z| = (\sin^2 x + \sinh^2 y)^k \leqslant (1 + \sinh^2 s)^k = \cosh^{2k} s$. But s is not bounded and so for $n > 2k$ we can choose s arbitrarily large so that E_nf vanishes for $f(x) = \sin^{2k} x$. ■

We extend the results to $N>1$ dimensions:

Let $f(\mathbf{x})=f(x_1, x_2, \ldots, x_N)$ be 2π periodic in all N dimensions, and let $f(\mathbf{z})$, $z_j=u_j+iv_j$, $j=1(1)N$, be analytic in the hyperstrip D_s^N which is determined by $-\infty<x_j<\infty$, $|y_j|\leq s$. We approximate the integral

$$\int_{-\pi}^{\pi} \cdots \int_{-\pi}^{\pi} f(\mathbf{x})\, \mathrm{d}\Omega$$

by the N-dimensional product rectangle rule

$$C_{n^N}f=h^N \sum_{j_1, j_2, \ldots, j_N=1}^{n} f(t_{j_1}, t_{j_2}, \ldots, t_{j_N}), \qquad t_{j_k}=-\pi+j_k h.$$

We then have the following:

Theorem 3.6.2 (Martensen) *Let $f(z)$ be analytic in D_s^N and* $M=\max\limits_{\mathbf{z}\in D_s^N} f(\mathbf{z})$. *Then*

$$|E_{n^N}f|=|I_\Omega f-C_{n^N}f|\leq(2\pi)^{N-1}N\sigma_n(s)M.$$

Proof We proceed by induction. For $N=1$ we have the well-known case of quadrature. Assume that the theorem is valid for $N=k>1$ dimensions. Then, for every fixed number x_{k+1} with $|x_{k+1}|\leq\pi$

$$\int_{-\pi}^{\pi} \cdots \int_{-\pi}^{\pi} f(x_1, \ldots x_k, x_{k+1})\, \mathrm{d}x_1 \ldots \mathrm{d}x_k$$
$$=h^k \sum_{j_1, \ldots, j_k=1}^{n} f(t_{j_1}, t_{j_2}, \ldots t_{j_k}, x_{k+1})+E_{n^k}(x_{k+1})$$

and $|E_{n^k}(x_{k+1})|\leq(2\pi)^{k-1}k\sigma_n(s)M$. Hence $E_{n^k}(x_{k+1})$ is bounded and integrable. We integrate once more

$$\int_{-\pi}^{\pi} f(t_{j_1}, \ldots t_{j_k}, x_{k+1})\, \mathrm{d}x_{k+1}=h \sum_{j_{k+1}=1}^{n} f(t_{j_1}, \ldots, t_{j_{k+1}})+E_n(j_1, \ldots, j_k).$$

The error satisfies $|E_n(j_1, \ldots j_k)|\leq\sigma_n(s)M$. Hence

$$\int_{-\pi}^{\pi} \cdots \int_{-\pi}^{\pi} f(x_1, \ldots, x_{k+1})\, \mathrm{d}x_1 \ldots \mathrm{d}x_{k+1}$$
$$=h^k \sum_{j_1, \ldots, j_k=1}^{n} \int_{-\pi}^{\pi} f(t_{j_1}, \ldots, t_{j_k}, x_{k+1})\, \mathrm{d}x_{k+1}+\int_{-\pi}^{\pi} E_{n^k}(x_{k+1})\, \mathrm{d}x_{k+1}$$
$$=h^{k+1} \sum_{j_1, \ldots, j_{k+1}=1}^{n} f(t_{j_1}, \ldots, t_{j_{k+1}})+h^k \sum_{j_1, \ldots, j_k=1}^{n} E_n(j_1, \ldots, j_k)$$
$$+\int_{-\pi}^{\pi} E_{n^k}(x_{k+1})\, \mathrm{d}x_{k+1}$$

The total error is estimated as follows:

$$|E_{n^{k+1}}| \leq h^k \sum_{j_1,\ldots,j_k=1}^{n} E_n(j_1, \ldots, j_k) + \int_{-\pi}^{\pi} E_{n^k}(x_{k+1})\,dx_{k+1}$$

$$\leq h^k n^k \sigma_n(s) M + 2\pi k (2\pi)^{k-1} \sigma_n(s) M = (2\pi)^k (1+k) \sigma_n(s) M$$

since $hn = 2\pi$, and thus the proof is completed. ■

If we apply this result to the N-fold integral

$$\int_{-\pi}^{\pi} \cdots \int_{-\pi}^{\pi} (\sin x_1 \sin x_2 \ldots \sin x_N)^{2k}\,dx_1 \ldots dx_N,$$

generalising the last Example 3.6.2, we obtain the estimate

$$|E_{n^N} f| \leq (2\pi)^{N-1} N (e^{2ns} - 1)^{-\frac{1}{2}} \cosh^{2k} s.$$

3.7 Reducing errors

In formulating quadrature and cubature formulae we have seen how to choose the nodes and weights so that the corresponding quadrature formulae or cubature formulae are optimal in the sense that the error constants are in some way minimised. There exist many further formulae of this type such as those derived using geometrical considerations or those which follow from practical considerations such as composite rules.

Optimal quadrature and cubature formulae, however, generally have one essential practical disadvantage: if for a certain number n of nodes the necessary accuracy has not been achieved, we must replace n by $m > n$. Now, in general, $Q_n f$ and $Q_m f$ are completely different for $n \neq m$. Hence, we cannot use a quadrature subroutine of such formulae by simply changing the input parameter n, because a subroutine usually contains quadrature formulae of a given type (like Gauss quadratures) for only a very few values of $n \leq n_0$. Hence, this subroutine fails for $m > n_0$. We now look for methods for reducing the error of *one* (or a small number of) given $Q_n f$ by changing only n.

3.7.1 Subdivision of the domain of integration

In Section 2.6 we mentioned composite quadrature rules. These rules retain the shape of the $Q_n f$ used, and change only n. Let us study the behaviour of the corresponding errors.

Using the error representation defined from Peano kernels we have that

$$E_nf = \overset{*}{I}f - Q_nf = \frac{1}{(k-1)!}\int_0^1 K(t)f^{(k)}(t)\,dt \quad \text{with} \quad K(t) = E_n^x[(x-t)_+^{k-1}].$$

for a given Q_nf. We choose $m+1$ points t_j: $-1 = t_0 < t_1 < \ldots < t_{m-1} < t_m = 1$. These points subdivide the interval $[-1, 1]$ into the m subintervals $[t_{j-1}, t_j]$, $j = 1(1)m$ of length Δt_j. We apply Q_nf to each subinterval and add, thus obtaining the composite rule

$$Q_{nm}f = \sum_{j=1}^{m} \Delta t_j \sum_{i=1}^{n} A_i f(t_{j-1} + \Delta t_j x_i) \doteq \overset{*}{I}f.$$

Let us choose the interval $[t_{j-1}, t_j]$ where the error is

$$(E_nf)_j = \int_{t_{j-1}}^{t_j} f(x)\,dx - \Delta t_j \sum_{i=1}^{n} A_i f(t_{j-1} + \Delta t_j x_i)$$

$$= \frac{1}{(k-1)!}\int_{t_{j-1}}^{t_j} K_j(t)f^{(k)}(t)\,dt \quad \text{with} \quad K_j(t) = (E_n^x[(x-t)_+^{k-1}])_j.$$

We substitute $t = t_{j-1} + \Delta t_j u$, and $x = t_{j-1} + \Delta t_j v$. Then

$$(E_nf)_j = \frac{\Delta t_j}{(k-1)!}\int_0^1 K_j(u)f^{(k)}(u)\,du \quad \text{with} \quad K_j(u) = (\Delta t_j)^k (E_n^v[(v-u)_+^{k-1}]).$$

Hence

$$(E_nf)_j = (\Delta t_j)^{k+1} \frac{1}{(k-1)!}\int_0^1 E_n^v[(v-u)_+^{k-1}] f^{(k)}(t_{j-1} + \Delta t_j u)\,du$$

and

$$E_{nm}f = \frac{1}{(k-1)!}\sum_{j=1}^{m} (\Delta t_j)^{k+1} \int_0^1 E_n^v[(v-u)_+^{k-1}] f^{(k)}(t_{j-1} + \Delta t_j u)\,du.$$

This result becomes simpler if the kernels have constant sign. Let $K_j(t) \geq 0$. Then there exists $\xi_j \in [t_{j-1}, t_j]$ such that

$$\int_0^1 E_n^v[(v-u)_+^{k-1}] f^{(k)}(t_{j-1} + \Delta t_j u)\,du = f^{(k)}(\xi_j)\int_0^1 E_n^v[(v-u)_+^{k-1}]\,du$$

and

$$E_{nm}f = \left(\frac{1}{(k-1)!}\int_0^1 E_n^x[(x-t)_+^{k-1}]\,dt\right)\left(\sum_{j=1}^{m} (\Delta t_j)^{k+1} f^{(k)}(\xi_j)\right).$$

For equal subintervals $\Delta t_j = 1/m$ and with $M_k = \sup_{x \in [0,1]} |f^{(k)}(x)|$ we obtain the error estimate

$$|E_{nm}f| \leq \frac{M_k}{m^k} \frac{1}{(k-1)!} \int_0^1 E_n^x[(x-t)_+^{k-1}]\,dt$$

which differs from the estimate for $|E_nf|$ by the multiplier m^{-k}. Hence, the error is reduced by a large amount, the larger m and k become. It turns out that it is more economical to take large k than large m.

Example 3.7.1 In Section 3.3.2 we calculated $E_2f = -f'''(\xi)/216$ for the Radau formula $Q_2f = [3f(1/3) + f(1)]/4$. Hence, by subdividing into m equal subintervals we have that

$$E_{2m}f = -\frac{1}{216m^4} \sum_{j=1}^{m} f'''(\xi_j) \quad \text{and} \quad |E_{2m}f| \leq \frac{1}{216m^3} M_3.$$

For $f = 1/(1+x^2)$ we find $M_3 < 4.564$. Hence $|E_{2m}f| < 0.02113/m^3$.

To obtain $|E_{2m}f| < 10^{-6}$ we need at most $m = 28$. However, we have the values given in Table 3.7.1.

TABLE 3.7.1

m	$Q_{2m}f$	$E_{2m}f$
1	0.8	1.46 (−2)
2	0.786 980 249	1.58 (−3)
4	0.785 582 537	1.84 (−4)
8	0.785 420 876	2.27 (−5)
16	0.785 400 993	2.83 (−6)
23	0.785 399 115	0.95 (−6)
32	0.785 398 517	3.53 (−7)

Hence, we need only $m = 23$ instead of the estimated $m \leq 28$. This overestimation is, of course, a consequence of the variation of f''' in $[0, 1]$. The majorisation of $f^{(k)}(\xi_j)$ by M_k gives good results only if $f^{(k)}$ does not have steep peaks or oscillations. The upper bound for m has a practical meaning as a stopping condition for a subroutine, which automatically increases n by subdividing the interval of integration.

For $f \in C^2[0, 1]$ we have had the error estimate $|E_2f| \leq M_2/48$. For this we have the worse result $|E_{2m}f| < 1(24\,m^2)$. ■

The extension to cubature problems is straightforward, if the domain Ω can be subdivided into similar subdomains such that the same C_nf is applicable to every subdomain, e.g. rectangle, triangle, hexagon or an L-shaped region.

3.7.2 Raising the degree of precision

The calculations for the last example show that an equal subdivision is not necessarily good in view of estimating the number of subintervals required in order to guarantee a prescribed accuracy. Nevertheless, an equidistant subdivision may raise, at least in a generalised sense, the degree of precision.

Remember that the rectangle rule

$$R_n f = \frac{1}{n} \sum_{i=1}^{n} f\left(\frac{i}{n}\right) \doteq \overset{*}{I} f$$

has the property $R_n f = \overset{*}{I} f$ for $f \in \{1, \sin k\pi x, \cos k\pi x, k = 1(1)n-1, \sin n\pi x\}$. Equal subdivision gives

$$R_{nm} f = \sum_{j=1}^{m} \frac{1}{m} \sum_{i=1}^{n} \frac{1}{n} f\left(\frac{j-1}{m} + \frac{1}{m}\frac{i}{n}\right) = \frac{1}{nm} \sum_{i=1}^{nm} f\left(\frac{i}{nm}\right).$$

This, again, is a rectangle rule but now $\overset{*}{I} f = R_{nm} f$ for $f \in \{1, \sin k\pi x, \cos k\pi x, k = 1(1)nm-1, \sin nm\pi x\}$.

Let F be the space of functions f satisfying $E_l f = 0$ for every $f \in F$. Then the dimension of F for $R_l f$ is $2l$. Because of this property, together with the fact that $1, \sin k\pi x, \cos k\pi x, k = 1(1)l-1, \sin l\pi x$, are a basis of F we state the following result:

Theorem 3.7.1 *For periodic functions the best possible subdivision of the interval for raising the degree of precision is that in which it is divided into equal subintervals.*

Unfortunately, this result does not remain valid in general if $Q_n f$ is based upon polynomial accuracy. This means that F is the space of monomials x^k, $k = 0(1)\text{degree}(Q_n f)$. Nevertheless, we use the idea in order to choose the subdividing points t_k such that the degree of $Q_{nm} f$ is larger than that of $Q_n f$.

Addition of nodes

Assume that $Q_n f$ is obtained by integration of a Lagrange interpolating polynomial $L_n(f, x)$:

$$Q_n f = \sum_{i=1}^{n} f(x_i) \int_{-1}^{1} \lambda_i(x)\, dx$$

which has the degree of precision of at least $n-1$ and at most $2n-1$. If we want to add further nodes to those of Q_nf, we construct a mixed interpolating operator based on the nodes x_i, $i=1(1)n$, and the m further nodes t_j, $j=1(1)m$. There are several possibilities for performing this mixed interpolation. The operator

$$LH_{n,m}(f, x)=L_n(f, x)+\omega_n(x)H_m\left(\frac{f(x)-L_n(f, x)}{\omega_n(x)}, x\right) \tag{3.7.1}$$

interpolates f at $n+m$ nodes and f' at least at m nodes. Here

$$H_m(f, x)=\sum_{j=1}^{m}\left\{\left[1-\Lambda_j'(t_j)\frac{k_j(x)}{k_j'(t_j)}\right]f(t_j)+\frac{k_j(x)}{k_j'(t_j)}f'(t_j)\right\}\Lambda_j^2(x)$$

with the basic functions $k_j(x)$, the cardinal functions $\Lambda_j(x)$, and the nodal polynomial $\Omega_m(x)$. We integrate the operator $LH_{nm}(f, x)$:

$$Q_{n+m}f\equiv\int_{-1}^{1}LH_{n,m}(f, x)\,dx=Q_nf+\int_{-1}^{1}\omega_n(x)H_m\left(\frac{f(x)-L_n(f, x)}{\omega_n(x)}, x\right)dx.$$

The calculation of the last integral is facilitated enormously if we demand that for $j=1(1)m$

$$\int_{-1}^{1}\omega_n(x)\Lambda_j^2(x)k_j(x)\,dx=0 \quad\text{or}\quad \int_{-1}^{1}\omega_n(x)\Omega_m(x)\Lambda_j(x)\,dx=0 \tag{3.7.2}$$

is valid. This means that the nodal functions $\Omega_m(x)$ are an orthogonal system of functions with respect to the weight function $\omega_n(x)$. With respect to (3.7.2) we have

$$\begin{aligned}Q_{n+m}f&=Q_nf+\sum_{j=1}^{m}\left(\int_{-1}^{1}\omega_n(x)\Lambda_j^2(x)\,dx\right)\frac{f(t_j)-L_n(f, t_j)}{\omega_n(t_j)}\\&=\sum_{i=1}^{n}\sum_{j=1}^{m}\left(\int_{-1}^{1}\left[\lambda_i(x)-\frac{\omega_n(x)\Lambda_j^2(x)\lambda_i(t_j)}{\omega_n(t_j)}\right]dx\right)f(x_i)\\&\quad+\sum_{j=1}^{m}\left(\int_{-1}^{1}\frac{\omega_n(x)\Lambda_j^2(x)}{\omega_n(t_j)}\,dx\right)f(t_j).\end{aligned} \tag{3.7.3}$$

In general $Q_{n+m}f$ has degree $n+2m-1$. If (3.7.2) is not valid we have that with the Hermite cardinal polynomials $\nu_j(x)$ and $\mu_j(x)$

$$\begin{aligned}Q_{n+m}f=Q_nf+\sum_{j=1}^{m}\Bigg[&\int_{-1}^{1}\omega_n(x)\nu_j(x)\frac{f(t_j)-L_n(f, t_j)}{\omega_n(t_j)}\,dx\\&+\int_{-1}^{1}\omega_n(x)\mu_j(x)\frac{d}{dt}\left(\frac{f(t)-L_n(f, t)}{\omega_n(t)}\right)_{t=t_j}dx\Bigg].\end{aligned}$$

Hence, the derivatives $f'(t_j)$ additionally come into $Q_{n+m}f$. We then have to evaluate f

at $n+m$ nodes and f' at m nodes. The amount of work then is not less than that for a $Q_{n+2m}f$ *without* derivatives (and degree $n+2m-1$ in general). Hence, we do not determine such formulae but only those for which (3.7.2) is valid.

Example 3.7.2 We raise the degree of the Radau formula

$$Q_2 f = (1/4)[f(0)+3f(2/3)] \doteq \overset{*}{I}f$$

in order to produce a formula of type (3.7.3).

Here $x_1=0$, $x_2=2/3$, $h_1(x)=x$, $h_2(x)=x-2/3$, $\omega_2=x(x-2/3)$, and $\lambda_1=-3/2(x-2/3)$, $\lambda_2=(3/2)x$. We need to determine $\Omega_m(x)$ which satisfies (3.7.2). For this purpose we need the momentum integrals

$$\int_0^1 \omega_2(x)x^k \, dx = \int_0^1 \left(x-\frac{2}{3}\right)x^{k+1} \, dx = \frac{1}{k+3} - \frac{2}{3(k+2)}.$$

Let $m=1$ and $\Omega_1(x)=x+a$. This leads to the contradiction

$$0 = \int_0^1 x(x-2/3)(x+a) \, dx = 1/36.$$

Hence, it is impossible to have $m=1$ and we try $m=2$. Here, let $\Omega_2(x)=x^2+ax+b$, so that

$$0 = \int_0^1 x(x-2/3)(x^2+ax+b)x^k \, dx \quad \text{for} \quad k=0, 1.$$

The corresponding system $1/30+(1/36)a=0$ and $1/30+(1/30)a+(1/36)b=0$ has the solution $a=-6/5$ and $b=6/25$. Hence,

$$\Omega_2 = x^2-(6/5)x+6/25 \quad \text{with zeros} \quad t_{1,2}=(1/5)(3\mp\sqrt{3}) = \begin{cases} 0.94641 \\ 0.25359 \end{cases}$$

and the cardinal functions are

$$\Lambda_1 = \frac{-5x+3+\sqrt{3}}{2\sqrt{3}} \quad \text{and} \quad \Lambda_2 = \frac{5x-3+\sqrt{3}}{2\sqrt{3}}.$$

Further,

$$\lambda_1(t_1)=(1+3\sqrt{3})/10 \qquad \lambda_2(t_2)=3(3-\sqrt{3})/10$$

$$\lambda_1(t_2)=(1-3\sqrt{3})/10 \qquad \lambda_2(t_2)=3(3+\sqrt{3})/10$$

$$\omega_2(t_1)=2(3-4\sqrt{3})/75 \qquad \omega_2(t_2)=2(3+4\sqrt{3})/75$$

$$L_2(f, t_1) = [(1+3\sqrt{3})f(0)]/10 + 3[(3-\sqrt{3})f(2/3)]/10$$

$$L_2(f, t_2) = [(1-3\sqrt{3})f(0)]/10 + 3[(3+\sqrt{3})f(2/3)]/10$$

$$\int_0^1 \omega_2(x)\Lambda_1^2(x)\,dx = -\int_0^1 \omega_2(x)\Lambda_2^2(x)\,dx = -5\sqrt{3}/216.$$

Hence

$$Q_4 f = Q_2 f - \frac{25}{144} f(0) - \frac{75}{208} f\left(\frac{2}{3}\right) + \frac{125\sqrt{3}}{5616}[(3+4\sqrt{3})f(t_1) - (3-4\sqrt{3})f(t_2)]$$

$$= \frac{11}{144} f(0) + \frac{81}{208} f\left(\frac{2}{3}\right) + \frac{125\sqrt{3}}{5616}[(3+4\sqrt{3})f(t_1) - (3-4\sqrt{3})f(t_2)]$$

which has degree 4 whereas $Q_2 f$ has degree 2. ■

For quadrature formulae which are improved in this way there exists a very simple and practicable error control without taking a completely new quadrature formula of higher degree: if we compute $\overset{*}{I}f$ stepwise by subdividing the domain of integration, we always compute $Q_n f$ and $Q_{n+m} f$ for each subinterval. If the two values do not differ, then the interval is too small. If they differ too much, then the interval is too large. This simple test allows a good choice of the step-length, i.e. the length of the subinterval under consideration. Such a technique is particularly necessary if the interval of integration is not finite, because then an equal choice of subintervals is generally out of the question.

Example 3.7.3 The step-length control is demonstrated for the midpoint rule $Q_1 f = 2f(0) \doteq If$ of degree 1, which is improved by $Q_3 f = (5/9)f(-\sqrt{3/5}) + (8/9)f(0) + (5/9)f(\sqrt{3/5})$ having degree 5. We compute $\underset{\infty}{\overset{*}{I}} 1/(1+x^2) = \pi/2$ and $\underset{\infty}{\overset{*}{I}} \exp(-x^2)$, always computing $Q_1 f$ and $Q_3 f$ for the same subinterval. If the values have more than four or five decimal figures in common, we change to a subdivision with larger intervals. Starting with $h = 1$ we successively reduce the length h of this first subinterval to 0.1 in order to obtain the desired accuracy. Values are presented in Table 3.7.2. Between two subsequent values of x in the table, the step-size h has a constant value, which is indicated in the first column; n is the number of steps from 0 to x.

TABLE 3.7.2

Approximation to $\int_0^\infty \frac{1}{1+x^2}\,dx = \frac{\pi}{2} = 1.570\,796\,327$

h	x	n	Q_3f	Q_1f
0.1	0.1	1	0.099 668 7	0.099 750 6
0.1	0.6	6	0.540 419 5	0.540 423 0
0.2	3.6	21	1.229 649 5	1.299 839 2
0.4	6	27	1.405 647 7	1.405 634 8
1	15	36	1.504 228 2	1.504 222 6
5	50	43	1.550 790 0	1.550 792 8
25	300	53	1.567 462 9	1.567 462 3
1. (+2)	1. (+3)	60	1.569 796 2	1.569 795 9
1. (+3)	1. (+4)	69	1.570 696 1	1.570 696 1
1. (+4)	1. (+5)	78	1.570 786 1	1.570 786 1
1. (+5)	1. (+6)	87	1.570 795 1	1.570 795 1
1. (+6)	1. (+7)	96	1.570 796 0	1.570 796 0
1. (+7)	1. (+8)	105	1.570 796 1	1.570 796 1

Approximation to $\int_0^\infty \exp(-x^2)\,dx = 0.886\,226\,926$

h	x	n	Q_3f	Q_1f
0.1	0.1	1	0.099 667 7	0.099 750 3
0.1	1	10	0.746 824 1	0.461 321 5
0.5	4	16	0.886 227 0	0.886 227 0

■

Boundary corrections

For equidistant subdivisions of the interval of integration there is another possibility for raising the degree of precision: the *method of boundary corrections.*

The integral is split so that

$$\overset{*}{I}f = \sum_{i=0}^{n-1} \int_{ih}^{(i+1)h} f(x)\,dx \tag{3.7.4}$$

and a composite quadrature rule is obtained from the application of some elementary quadrature rule to each of the individual integrals in the sum (3.7.4). By application of the trapezoidal rule to each integral we have that

$$T_{n+1}f = \frac{h}{2}\left[f_0 + 2\sum_{i=1}^{n-1} f_i + f_n\right] \equiv \frac{h}{2}\,[\![1, 2, \ldots / \ldots]\!]f$$

with $f_i = f(ih)$ and $nh = 1$, having the degree 1 and leading error term $-(1/12)f''(0)/n^2$.

We cannot raise the degree by taking any other elementary quadrature rule containing only points of the subdivision. We thus include exterior nodes for each subinterval and construct appropriate formulae; e.g.:

$$Q_4 f = \frac{1}{24}[9f_0 + 19f_1 - 5f_2 + f_3] \doteq \overset{*}{I}f$$

$$Q_4 f = \frac{1}{24}[-f_{-1} + 13f_0 + 13f_1 - f_2] \doteq \overset{*}{I}f$$

$$Q_4 f = \frac{1}{24}[f_{-2} - 5f_{-1} + 19f_0 + 9f_1] \doteq \overset{*}{I}f.$$

If we apply the first equation to $[0, h]$, the third to $[1-h, 1]$, and the second to the interior intervals and then add all three together, we do not need any node outside $[0, 1]$ which is different from ih, $i = 0(1)n$; thus

$$\hat{T}_{n+1}f = \frac{h}{24}\left\{8(f_0 + f_n) + 31(f_1 + f_{n-1}) + 20(f_2 + f_{n-2}) + 25(f_3 + f_{n-3}) + 24\sum_{i=4}^{n-4} f_i\right\}$$

which is now written as

$$\hat{T}_{n+1}f \equiv \frac{h}{24}[\![8, 31, 20, 25, 24, \ldots / \ldots]\!]f$$

$$= T_{n+1}f + \frac{h}{24}[\![-4, 7, -4, 1, 0, \ldots / \ldots]\!]f \quad \text{for} \quad n \geqslant 8.$$

This formula has degree 3 and leading error term

$$\frac{1}{6!\, n^5}(11n - 60)f^{(4)}(0).$$

The corresponding formula of degree 5 for $n \geqslant 12$ is

$$\hat{\hat{T}}_{n+1}f \equiv \frac{h}{1440}[\![459, 1982, 944, 1746, 1333, 1456, 1440, \ldots / \ldots]\!]f$$

$$\hat{T}_{n+1}f + \frac{h}{1440}[\![-21, 122, -256, 246, -107, 16, 0, \ldots / \ldots]\!]f$$

having the leading error term

$$-\frac{6}{9!\, n^7}(191n + 420)f^{(6)}(0).$$

Finally, for $n \geqslant 16$ we note that

$$\check{T}_{n+1} \equiv \frac{h}{120960} [\![35584, 185153, 29336, 220509, 46912, 156451, 111080, 122175, 120960, \ldots / \ldots]\!] f$$

$$= \hat{\hat{T}}_{n+1} f + \frac{h}{120960} [\![-2972, 18665, 79296, 73845, -65060, 34147, -9880, 1215, 0, \ldots / \ldots]\!] f,$$

having degree 7 and the leading error term

$$\frac{1}{10!\, n^9} (2497n - 74760) f^{(8)}(0).$$

This method of constructing rules $T_{n+1} f$ with boundary corrections was described by Engels [11].

Example 3.7.4 For $\overset{*}{I}(\ln(1+x)) = \ln 4 - 1 = 0.386\,294\,3610$ we obtain

$h = 0.05$:

	$T_{21} f = 0.386\,190\,210$
1st correction	$0.000\,104\,249$ (+)
	$\hat{T}_{21} f = 0.386\,294\,459$
2nd correction	$-0.000\,000\,098$ (+)
	$\hat{\hat{T}}_{21} f = 0.386\,294\,361$ ■

Naturally, other methods for constructing boundary corrections do exist. The best-known formulae are those due to Euler–Maclaurin and Gregory. Although these formulae are very efficient if the integrand is tabulated at equidistant points so that hand calculations are possible, they are not well-suited to automatic computation on a computer. Other procedures, such as the Gauss quadrature or Romberg quadrature, are simpler and generally have a higher degree of precision.

Elimination of leading error terms

If the degree of precision of a quadrature or cubature formula is less than the order of differentiability of the integrand, we can expand the error of $Q_n f$ as far as possible. Assume, for simplicity, that $f \in C^\infty[-1, 1]$.

Then, for example, we have the expansions

$$If - 2f(0) = \frac{1}{3} f''(0) + \frac{1}{60} f^{(4)}(0) + \frac{2}{7!} f^{(6)}(0) + \ldots$$

$$If - (f(-1) + f(1)) = -\frac{2}{3} f''(0) - \frac{1}{15} f^{(4)}(0) - \frac{12}{7!} f^{(6)}(0) + \ldots .$$

Hence, $f''(0)$ can be eliminated so that

$$2(If - 2f(0)) + (If - (f(-1) + f(1))) = -\frac{1}{30} f^{(4)}(0) - \frac{8}{7!} f^{(6)}(0) + \ldots$$

or

$$E_3^S f = If - \frac{1}{3}(f(-1) + 4f(0) + f(1)) = -\frac{1}{90} f^{(4)}(0) - \frac{8}{3 \cdot 7!} f^{(6)}(0) + \ldots$$

which is exactly Simpson's rule of degree 3. Another formula of degree 3 is the Gauss two-point formula, having the error expansion

$$E_2^G f = If - \left(f\left(-\frac{1}{\sqrt{3}}\right) + f\left(\frac{1}{\sqrt{3}}\right)\right) = \frac{1}{135} f^{(4)}(0) + \frac{20}{27 \cdot 7!} f^{(6)}(0) + \ldots .$$

Hence

$$\begin{aligned}\frac{2E_3^S f + 3E_2^G f}{5} &= If - \frac{1}{15}\left[2f(-1) + 9f\left(-\frac{1}{\sqrt{3}}\right) + 8f(0) + 9f\left(\frac{1}{\sqrt{3}}\right) + 2f(1)\right] \\ &= -\frac{28}{45 \cdot 7!} f^{(6)}(0) - \ldots .\end{aligned}$$

Thus we eliminate successively the leading error terms of two quadrature formulae having the same degree.

For a long time following Newton and Cotes, great efforts were made to raise the degree of the known quadrature formulae by such eliminations. However, such formulae are seldom used today because their degree is low in comparison to Gauss formulae for the same number of nodes.

However, we need not restrict our efforts to combining *different formulae* of the same degree. The same formula can be used with *different subdivisions* of the interval of integration, and these can again be combined to give formulae of the same degree. For example, consider

$$If-(f(-1)+f(1))=-\frac{2}{3}f''(0)-\frac{1}{15}f^{(4)}(0)-\frac{12}{7!}f^{(6)}(0)-\dots$$

$$If-\frac{1}{2}(f(-1)+2f(0)+f(1))=-\frac{1}{6}f''(0)-\frac{3}{5!}f^{(4)}(0)-\frac{5}{7!}f^{(6)}(0)-\dots$$

$$If-\frac{1}{3}\left(f(-1)+2f\left(-\frac{1}{3}\right)+2f\left(\frac{1}{3}\right)+f(1)\right)$$

$$=-\frac{2}{27}f''(0)-\frac{43}{3645}f^{(4)}(0)-\frac{5860}{7!\,3^7}f^{(6)}(0)-\dots.$$

It is now possible to eliminate $f''(0)$ and $f^{(4)}(0)$. This technique leads to quadrature formulae which in general are not interpolating quadratures, but which have other advantageous properties. For a special sequence of subdivisions they are known as *Romberg quadrature formulae*, and have proved to be important in practical applications. They will be considered in detail in Chapter 8.

3.8 Remarks, problems, references

3.8.1 Remarks

R.3.1: The sign of the Peano kernel (analogously of the Sard kernel) plays an important part in error estimates. If the sign is constant, error representations and bounds can be easily derived using the mean-value theorem. Hence, it would be desirable to know at least sufficient conditions for which the Peano kernel is of constant sign. Such conditions are given by Ferguson [15] but unfortunately are rather complicated (cf. Section 3.1.2).

R.3.2: We have developed the expansion of the integrand $f(x, y)$ into powers of x and y with integral error terms for $N=2$ dimensions only. The extension to $N>2$ dimensions is straightforward, but it is obvious that the essential difficulty comes from writing down the corresponding expressions (cf. Section 3.2).

R.3.3: Our attention has been restricted to functions $f\in C^{k+1}(\bar{\Omega})$ in order to obtain expansions in powers of x, y of f. Clearly, there may exist mixed (or pure) derivatives of f of order $>k+1$. In general the expansion can be continued as long as it contains only those continuous partial derivatives of f which exist (cf. Stroud, book). This is not of great general importance for cubature purposes and such expansions have been ignored (cf. Section 3.2).

R.3.4: In order to obtain representations of cubature error functionals one can consider fairly general domains Ω with curved boundaries. The performance is straightforward but very laborious (cf. Section 3.2).

R.3.5: Hämmerlin [19] enlarged the region of analyticity of the integrand function to a circle with radius $R>1$. He used the inner product

$$(f, g)=\frac{1}{2\pi}\int_0^{2\pi} f(z)g(z)\,ds$$

where s is the arc length of the circle $|z|=R$. We replaced ds by $d\varphi$ and hence avoided the factor r in $ds=r\,d\varphi$ (cf. Section 3.5.3).

R.3.6: The type of region in which $f(z)$ should be holomorphic is not restricted to circles, ellipses and strips. Other domains are treated in the literature. For example, Hetherington [20] takes the rectangle $|x|\leqslant 1$, $|y|\leqslant r$ $(z=x+iy)$ together with the two semicircles

$$(x-1)^2+y^2\leqslant r^2, x\geqslant 1, \quad \text{and} \quad (x+1)^2+y^2\leqslant r^2, x\leqslant -1.$$

It is very difficult to find quadratures with *minimal* error bounds for such regions (cf. Section 3.5.4).

R.3.7: Instead of choosing the inner product round a contour

$$(f, g)=\frac{1}{2\pi}\int_0^{2\pi} f(z)g(z)\,d\varphi,$$

it is possible to take an area integral for the interior of this contour, e.g. for the circle

$$(f, g)=\frac{1}{2\pi}\int_0^{2\pi}\int_0^{2\pi} f(z)g(z)\,dx\,dy, \qquad z=x+iy=re^{i\varphi}.$$

Here, $dx\,dy=r\,dr\,d\varphi$. Neither this nor any similar inner product seems yet to have been used for the determination of error norms or optimal quadrature formulae in the sense of this norm (cf. Sections 3.5.3 and 3.5.4).

R.3.8: Tables of quadrature formulae with minimal error coefficients for the Hardy space $\mathbb{H}_2$ were originally calculated by Wilf [26] and Eckhardt [10]. However, only small numbers of nodes have been used and the accuracy achieved is poor. The first extended table of nodes and weights and of the corresponding error constants was presented by Engels and Eckhardt [12]. We shall list some of these formulae in Chapter 7 (cf. Section 3.5.4).

R.3.9: The construction of error bounds for cubature formulae and analytic integrands was mainly performed by Barnhill in several papers. He used product-type Hilbert spaces of analytic functions of the form $\mathbb{E}_\rho\times\mathbb{E}_\rho\times\ldots\times\mathbb{E}_\rho$. Further, he observed connections between product-type cubature formulae with minimal error constants for these Hilbert spaces and Gaussian-type product cubature formulae for functions which are analytic in the whole n-dimensional complex plane.

R.3.10: Although not necessary, it is convenient to use Hilbert space techniques on account of the representation Theorem 3.5.1 for linear functionals. In non-Hilbert spaces the theory becomes more complicated. Nevertheless, there may exist formulae with minimal quadrature error. See, for example, Barrar, Loeb and Werner [7] who also obtained better rates of convergence than those of Section 4.2.5, Theorem 4.2.8.)

R.3.11: For minimal quadratures for analytic functions, complex nodes can be taken into account in order to improve the error norms. Barrar, Loeb, and Werner [7] showed that the optimal formulae cannot have complex nodes (cf. Section 3.5.4).

3.8.2 Problems

P.3.1: Determine the Peano kernels of Simpson's rule for $f \in C^{k+1}[-1, 1]$ for $k = 0, 1, 2, 3$ (cf. Section 3.1.1).

P.3.2: Let $f(x, y, z) \in C^2(\bar{\Omega})$. Calculate the expansion of f in powers of x, y, z and the integral leading error term (cf. Section 3.2).

P.3.3: Determine the Sard kernel of the cubature formula

$$C_4 f = \tfrac{1}{4}[f(0.5, 0) + f(0.5, 1) + f(0, 0.5) + f(1, 0.5)] \doteq I_\Omega f$$

where Ω is the square $0 \leq x, y \leq 1$ (cf. Section 3.2).

P.3.4: Seek the "best" error constant σ_n for $f \in C^1[-1, 1]$ by changing from the modulus of continuity $\omega(f, h_n)$ to the first derivative $f'(x)$ in Section 3.3.1.

P.3.5: For given nodes $x_1 = -1$, $x_2 = 0$, $x_3 = 1$ minimise

$$Q_3(A_1, A_2, A_3) = \int_{-1}^{1} K^2(t)\, dt$$

by choosing suitable weights A_1, A_2, A_3. Assume $A_1 = A_3$ (why?), and require

$$\frac{\partial Q_3}{\partial A_i} = 0, \qquad i = 1, 2 \quad \text{(cf. Section 3.3.2).}$$

P.3.6: Estimate the quadrature error for $f(x) = \ln(x+2)$, $x \in [-1, 1]$ for $k = 2, 3, 4$ using Simpson's rule (cf. Section 3.3.2).

P.3.7: Calculate the Fourier series expansion of the function

$$\operatorname{sign}(U_n(x)) = \operatorname{sign}(\sin(n+1)\theta) \quad \text{(cf. Section 3.4.1).}$$

P.3.8: Are the power series expansions of $\sin x$, $\ln(1+x)$, and $J_0(x)$ (=0-th Bessel function) rapidly convergent (cf. Section 3.4.2)?

P.3.9: Use the second degree formula

$$Q_3 f = f(-1) - 3f(-1/3) + 4f(0) \doteq If$$

in order to compute Ie^x. The necessary function values are to be rounded off to two decimal places. Compare with the third degree formula

$$Q_2 = f(1/\sqrt{3}) + f(-1/\sqrt{3}) \quad \text{(cf. Section 3.4.4).}$$

P.3.10: Estimate the error in Section 3.5.1 for functions f which are analytic in the domain

$$\{z \mid |z - x| \leq R \quad \text{for} \quad |x| \leq 1\}.$$

What is changed?

P.3.11: C is the boundary of the ellipse E with major semiaxis a and minor semiaxis b. Assume that b is chosen such that the distance d from $[-1, 1]$ to C satisfies $d > a-1$. Let $f(z)$ be holomorphic in a region which contains C completely within its interior. Estimate $f^{(n)}(z)$ by Cauchy's formulae and apply the result to Simpson's rule for $f(z) = ze^z$ (cf. Section 3.5.1).

P.3.12: Calculate the error norm $\|E_n\| = \|r_n\|$ for the Gauss quadrature formula $Q_2f = f(-1/\sqrt{3}) + f(1/\sqrt{3})$ depending on ε. Determine approximately that value of ε for which $\|E_2\| = \|r_2\|$ becomes minimal (cf. Section 3.5.3).

P.3.13: Determine the error norm $\|E_4\|$ of the cubature formula

$$C_4f = f(\pm 1/\sqrt{3}, \pm 1/\sqrt{3}) = I_\Omega f.$$

Ω is the square $|x|, |y| \leq 1$. Let $\varepsilon = \eta$ in $\tilde{\mathbb{H}}_2 \times \tilde{\mathbb{H}}_2$ (cf. Section 3.5.5).

P.3.14: Determine A, B and α in $Q_3f = A[f(\alpha) + f(-\alpha)] + Bf(0)$ depending on ε such that $\|E_3\| = \|r_3\|$ becomes minimal (cf. Section 3.5.4).

P.3.15: Compute approximately (cf. Section 3.6.1)

$$\int_{-\infty}^{\infty} \frac{dx}{1 + \exp(x^2)}$$

using the trapezoidal rule Q_hf for different values of h. What can be said about bounds for the quadrature error?

P.3.16: Prove: Let s be (cf. Section 3.6.1) the asymptotically optimal strip width. Then the sequence of trapezoidal sums Q_hf converges exponentially towards If for $h \to 0$, where $f = \exp(-x^2)$. From this

$$\lim_{h\to 0} \left(h \sum_{j=-\infty}^{\infty} \exp(-j^2h^2) - \sqrt{\pi} \right) \to 0 \quad \text{exponentially.}$$

P.3.17: Show that (cf. Section 3.6.2)

$$4\pi n \sum_{k=1}^{\infty} \frac{k}{\sinh(2kns)} \leq \frac{\pi n e^{2ns}}{\sinh(2ns)\sinh^2(ns)}$$

and

$$2\pi \sum_{k=1}^{\infty} \frac{1}{\cosh(2kns)} \leq \frac{2\pi e^{-ns}}{\sinh(ns)} = \frac{4\pi}{e^{2ns} - 1}.$$

P.3.18: The interval $[0, 1]$ is subdivided equidistantly into m parts. Apply Simpson's rule for every part. Estimate the error (cf. Section 3.7.2). Use this method of subdivisions in order to compute $\int_0^1 e^{-0.5x}\, dx$ for $m = 1, 2, 4, 8$, and 16.

P.3.19: Add the nodes $-s$ and s to Simpson's rule and determine s and the weights A, B, C such that (cf. Section 3.7.2)

$$Q_5f = A(f(-1)+f(1)) + B(f(-s)+f(s)) + Cf(0) \doteq If$$

has a higher degree of precision than Simpson's rule.

P.3.20: Determine suitable boundary corrections for the trapezoidal rule such that a formula of degree 5 arises (for at least 12 partial intervals) (cf. Section 3.7.2).

P.3.21: Eliminate the leading error term for the two Radau quadrature formulae of degree 2

$$Q_2f = (1/4)(f(0)+3f(2/3)) \quad \text{and} \quad Q_2f = (1/4)(3f(1/3)+f(1))$$

approximating $\overset{*}{I}f$ (cf. Section 3.7.2).

3.8.3 References

1. Abramowitz, M. and Stegun, I. A. (1965). "Handbook of Mathematical Functions." Dover, New York.
2. Barnhill, R. E. (1965). Complex quadratures with remainders of minimum norm. *Numer. Math.*, **7,** 384–390.
3. Barnhill, R. E. (1967). Optimal quadratures in $L^2(E_\rho)$. I. *SIAM J. Numer. Anal.*, **4,** 390–399; and II, *ibid.*, 534–541.
4. Barnhill, R. E. (1968). Asymptotic properties of minimum norm and optimal quadratures. *Numer. Math.*, **12,** 384–393.
5. Barnhill, R. E. (1968). An error analysis for numerical multiple integration. I. *Math. Comput.*, **22,** 98–109; and II, *ibid.*, 286–292.
6. Barnhill, R. E. (1969). The convergence of complex cubatures. *SIAM J. Numer. Anal.*, **6,** 82–89.
7. Barrar, R. B., Loeb, H. L. and Werner, H. (1974). On the existence of optimal integration formulas for analytic functions. *Numer. Math.*, **23,** 105–117.
8. Davis, P. J. (1959). On the numerical integration of periodic analytic functions. *In* "On Numerical Approximation" (R. E. Langer, Ed.). pp. 45–60. The University of Wisconsin Press, Madison.
9. Davis, P. J. (1962). Errors of numerical approximation for analytic functions. *In* "Survey of Numerical Analysis" (J. Todd, Ed.). pp. 468–484. McGraw-Hill, New York.
10. Eckhardt, U. (1968). Einige Eigenschaften Wilf'scher Quadraturformeln. *Numer. Math.*, **12,** 1–7.
11. Engels, H. (1969). Über eine Klasse von Quadraturformeln mit äquidistanten Stützstellen für große Intervalle. *ZAMM*, **49,** 751–753.
12. Engels, H. and Eckhardt, U. (1974). "Symmetrische Wilf'sche Quadraturformeln." Report KFA Jülich, Jül-1094-MA.
13. Engels, H. and Eckhardt, U. (1979). The determination of nodes and weights in Wilf quadrature formulas. *Abhdl. d. Math. Sem. d. Univ. Hamburg.*
14. Engels, H. (1972). Ein numerisches Differentiationsverfahren mit ableitungsfreien Fehlerschranken. *ZAMM*, **52,** 145–154.

15. Ferguson, D. (1973). Sufficient conditions for Peano's kernel to be of one sign. *SIAM J. Numer. Anal.* **10,** 1047–1054.
16. Hämmerlin, G. (1963). Über ableitungsfreie Fehlerschranken für Quadraturfehler. *Numer. Math.*, **5,** 226–233.
17. Hämmerlin, G. (1965). Über ableitungsfreie Fehlerschranken für Quadraturfehler. II. Ergänzungen und Möglichkeiten zur Verbesserung. *Numer. Math.*, **7,** 232–237.
18. Hämmerlin, G. (1966). Zur Abschätzung von Quadraturfehlern für analytische Funktionen. *Numer. Math.*, **8,** 334–344.
19. Hämmerlin, G. (1972). Fehlerabschätzung bei numerischer Integration nach Gauss. *Meth. u. Verf. d. Math. Phys.*, **6,** 153–163. Bibl. Inst., Mannheim–Wien–Zürich.
20. Hetherington, J. H. (1972). An error bound for quadrature. *Math. Comput.*, **29,** 695–698.
21. Kress, R. (1970). Interpolation auf einem unendlichen Intervall. *Computing,* **6,** 274–288.
22. Kress, R. (1972). Zur numerischen Integration periodischer Funktionen nach der Rechteckregel. *Numer. Math.*, **20,** 87–92.
23. Martensen, E. (1966). Verallgemeinerung der Quadraturfehlerschranken von Davis auf mehrfache Integrale. *ZAMP,* **17,** 633–635.
24. Martensen, E. (1968). Zur numerischen Auswertung uneigentlicher Integrale. *ZAMM,* **48,** T83-T85.
25. Sard, A. (1963). "Linear Approximation." AMS 190 Hope Street, Providence, Rhode Island.
26. Wilf, H. S. (1964). Exactness conditions in numerical quadrature. *Numer. Math.*, **6,** 315–319.

4

Convergence of Quadrature and Cubature Procedures

We have so far considered how to construct error bounds for quadrature and cubature formulae. A concept which is closely allied to that of the error bound is that of *convergence* of the quadrature or cubature to the integral in the sense that

$$\lim_{n\to\infty} (Q_n f) = If, \quad \text{or} \quad \lim_{n\to\infty} (C_n f) = I_\Omega f$$

for all elements f of a suitable space of functions. If convergence takes place, the error $E_n f$ must vanish as $n \to \infty$. We shall establish some convergence conditions for fairly general sets of quadrature and cubature formulae. In order to do this we need some knowledge concerning the method by which $Q_n f$ and $C_n f$ can be extended as $n \to \infty$. These methods of extension will be discussed first.

4.1 Definitions

Before treating the convergence of quadratures and cubatures in detail, we define the term *procedure* for quadrature and cubature problems.

Definition 4.1.1 Let $\nu(n): \mathbb{N} \to \mathbb{N}$ be a strictly increasing function, such that for every $\nu(n)$ there is defined a quadrature or cubature rule having exactly $\nu(n)$ distinct nodes. The set of all such rules for $n = 1(1)\infty$ is called a *quadrature procedure* (or a *cubature procedure*, as appropriate). ■

Note that nodes and weights are not necessarily real numbers. We denote such quadrature and cubature procedures respectively by

$$Q_{\nu(n)}f = \sum_{i=1}^{\nu(n)} A_i^{(\nu)} f(x_i^{(\nu)}) \quad \text{and} \quad C_{\nu(n)}f = \sum_{i=1}^{\nu(n)} A_i^{(\nu)} f(\mathbf{x}_i^{(\nu)}); \qquad n = 1(1)\infty,$$

where we have written ν instead of $\nu(n)$ in the nodes and weights.

Very often, but by no means always, we have $\nu(n) = n$ as, for example, in $T_n f$, $R_n f$, $M_n f$. As an illustration we apply the trapezoidal rule and systematically halve the interval. Thus

$$\begin{aligned} Q_2 f &= f(-1) + f(1) && \text{for } n = 1, \\ Q_3 f &= \tfrac{1}{2}(f(-1) + 2f(0) + f(1)) && n = 2, \\ Q_5 f &= \tfrac{1}{4}(f(-1) + 2f(-\tfrac{1}{2}) + 2f(0) + 2f(\tfrac{1}{2}) + f(1)) && n = 3, \end{aligned}$$

etc.

$$Q_{2^k+1} f = \frac{1}{2^k}\left(f(-1) + f(1) + 2\sum_{i=1}^{2^k-1} f\left(-1 + \frac{2i}{2^k}\right)\right) \qquad n = k+1,$$

and so

$$\nu(n) = 2^{n-1} + 1$$

In Definition 4.1.1 we have not taken into account derivatives of f. If these are present, we define further sets of weights corresponding to the orders of the derivatives. The procedures are then called *Hermitian quadrature procedures* (or *Hermitian cubature procedures* as appropriate), and are written as

$$Q_{\nu(n),k} f = \sum_{i=1}^{\nu(n)} \sum_{j=0}^{k} A_{i,j}^{(\nu)} f^{(j)}(x_i^{(\nu)}), \qquad n = 0(1)\infty,$$

where the function f possesses derivatives up to and including those of order k in an interval which includes all the nodes; (k is always finite). Clearly, we can admit *exterior nodes* outside the domain of integration but in the domain of definition of f.

We now define *convergence* of quadrature and cubature procedures.

Definition 4.1.2 The quadrature procedure $Q_{\nu(n),k}f$ or the cubature procedure $C_{\nu(n),k}f$ are said to be *convergent* if for every $\varepsilon > 0$ there exists a number $\bar{\nu}(\varepsilon)$ such that for $\nu > \bar{\nu}(\varepsilon)$

$$|E_\nu f| = |If - Q_{\nu(n),k} f| < \varepsilon \quad \text{or} \quad |E_\nu f| = |I_\Omega f - C_{\nu(n),k} f| < \varepsilon. \quad \blacksquare$$

In what follows we assume for the sake of simplicity that any $Q_{\nu(n)}f$ or

$C_{\nu(n)}f$ can be embedded in a natural way into the corresponding procedures Q_nf or C_nf, such that $\{Q_{\nu(n)}f\}_{n=1}^{\infty}$ and $\{C_{\nu(n)}f\}_{n=1}^{\infty}$ are subsequences of $\{Q_nf\}_{n=1}^{\infty}$ and $\{C_nf\}_{n=1}^{\infty}$ respectively. We can then replace $\nu(n)$ by n.

4.2 How to prove convergence of quadrature and cubature procedures

One of the most important ways of constructing a quadrature or cubature procedure is by integrating an approximation to f or a function which interpolates to f. If such an approximation or interpolation converges uniformly to f in the max-norm, the convergence of the corresponding quadrature or cubature procedure is immediately assured. As an example we treat the Bernstein polynomials of f and the corresponding quadrature procedure $B_{n+1}f$.

4.2.1 Bernstein polynomials and convergence of some quadratures

The general Bernstein polynomial of a function $f(x)$ within the interval $[0, 1]$ is given by

$$B_n(f, x) \equiv \sum_{k=0}^{n} \binom{n}{k} x^k (1-x)^{n-k} f\left(\frac{k}{n}\right).$$

Integration of this leads to

$$\overset{*}{I}f \doteq B_{n+1}f \equiv \frac{1}{n+1} \sum_{k=0}^{n} f\left(\frac{k}{n}\right),$$

which is a quadrature formula with equal weights. We now use some of the well-known convergence properties of $B_n(f, x)$ in order to show the convergence of both $B_{n+1}f$ and some related simple quadrature formulae. From the Weierstrass approximation theorem (cf. [14]), we know that for every given $\varepsilon > 0$ there exists an $n_0 = n_0(\varepsilon)$ such that for $f \in C^0[0, 1]$ $\sup_{x \in [0,1]} |f(x) - B_n(f, x)| < \varepsilon$ for $n > n_0$. We immediately conclude that for $f \in C^0[0, 1]$

$$|\overset{*}{I}f - B_{n+1}f| \leqslant \int_0^1 |f(x) - B_n(f, x)| \, dx < \varepsilon,$$

and hence have

Theorem 4.2.1 *If* $f \in C^0[0, 1]$, *then* $\lim_{n \to \infty} B_nf = \overset{*}{I}f = \int_0^1 f(x)\, dx$.

This theorem is now used to derive further results.

Definition 4.2.1 The quadrature formula $Q_{n+k}f$ with $k=0(1)\ldots$ is said to be *consistent* with $B_{n+1}f$ if

$$Q_{n+k}f - B_{n+1}f = O\left(\frac{1}{n^{\alpha}}\right) \quad \text{for} \quad \alpha > 0. \quad \blacksquare$$

Several useful results are now proved.

Theorem 4.2.2 *For $f \in C^0[0, 1]$ every $Q_{n+k}f$ which is consistent with $B_{n+1}f$ is convergent.*

Proof Using the triangle inequality we find that

$$|\overset{*}{I}f - Q_{n+k}f| \leqslant |\overset{*}{I}f - B_{n+1}f| + |B_{n+1}f - Q_{n+k}f| < \varepsilon.$$

From Theorem 4.2.1 there exists an $n_1(\varepsilon)$ such that

$$|\overset{*}{I}f - B_{n+1}f| < \varepsilon/2 \quad \text{for} \quad n > n_1(\varepsilon).$$

From Definition 4.2.1 there exists an $n_2(\varepsilon)$ such that

$$|B_{n+1}f - Q_{n+k}f| < \varepsilon/2 \quad \text{for} \quad n > n_2(\varepsilon).$$

Then, for $n_0 = \max(n_1, n_2)$ the theorem is proved. $\blacksquare$

As an immediate consequence we have:

Theorem 4.2.3 *The quadrature rules $T_{n+1}f$, R_nf, $\overline{R}_nf$ and M_nf (see Section 2.1) converge for $f \in C^0[0, 1]$.*

Proof Convergence of $T_{n+1}f$: we have that

$$\begin{aligned} T_{n+1}f - B_{n+1}f &= \frac{1}{2n}\left\{f(0) + 2\sum_{k=1}^{n-1} f\left(\frac{k}{n}\right) + f(1)\right\} - B_{n+1}f \\ &= \frac{1}{n(n+1)} B_{n+1}f - \frac{1}{2n}[f(0) + f(1)] \end{aligned}$$

and, since $f(x)$ and $B_{n+1}f$ are bounded, $T_{n+1}f - B_{n+1}f = O(1/n)$. Therefore, $T_{n+1}f$ is consistent with $B_{n+1}f$ and, from Theorem 4.2.2, convergence follows.

Analogously, we conclude convergence of $R_n f$ and $\bar{R}_n f$ since

$$R_n f - B_{n+1} f = \frac{1}{n(n+1)} B_{n+1} f - \frac{1}{n} f(1) = O\left(\frac{1}{n}\right),$$

$$\bar{R}_n f - B_{n+1} f = \frac{1}{n(n+1)} B_{n+1} f - \frac{1}{n} f(0) = O\left(\frac{1}{n}\right).$$

Finally, for $M_n f$ we have that

$$|M_n f - B_{n+1} f| \leqslant \left| \sum_{k=1}^{n} \frac{1}{n} f\left(\frac{2k-1}{2n}\right) - \frac{1}{n+1} f\left(\frac{k}{n}\right) \right| + \frac{1}{n+1} |f(0)|$$

$$\leqslant \frac{1}{n+1} \sum_{k=1}^{n} \left\{ \left| f\left(\frac{2k-1}{2n}\right) - f\left(\frac{k}{n}\right) \right| + \frac{1}{n} \left| f\left(\frac{2k-1}{2n}\right) \right| \right\} + \frac{1}{n+1} |f(0)|$$

$$\leqslant \frac{n}{n+1} \omega\left(f, \frac{1}{2n}\right) + \frac{2}{n+1} \|f\|,$$

where $\omega(f, h)$ is the modulus of continuity with increment h, and $\|f\|$ is the maximum norm. Since $f \in C^0[0, 1]$, there exists an $\alpha > 0$ such that $\omega(f, 1/2n) = O(1/n^\alpha)$. Hence we have consistency of $M_n f$ with $B_{n+1} f$ and the convergence of $M_n f$ is proved. ■

4.2.2 Jackson conditions and convergence

Another useful way of proving convergence of quadrature and cubature procedures is through the use of *Jackson conditions*. Conditions of this type are treated in the book [18] and in this section repeated reference is made to this.

Theorem 4.2.4 *Let $f \in C^0[-1, 1]$ with modulus of continuity $\omega(f, h)$. Then*

$$|f(x) - B_n(f, x)| \leqslant \tfrac{5}{4} \omega(f, 1/\sqrt{n}).$$

For proof see [18] p. 37.

Now, let the quadrature $Q_n f$ satisfy $E_n f = 0$ for polynomials $f \in \mathbb{P}_{k(n)}$ with $k(n) \to \infty$ as $n \to \infty$. In this case

$$E_n B_{k(n)}(f, x) = 0, \qquad n = 1(1) \ldots,$$

and, writing k for $k(n)$, we have that

$$|E_nf|=|E_n(f(x)-B_k(f,x))|\leqslant|I(f(x)-B_k(f,x))|+|Q_n(f(x)-B_k(f,x))|$$
$$\leqslant \sup_{|x|\leqslant 1}|f(x)-B_k(f,x)|\left(\int_{-1}^{1} W(x)\,dx+\sum_{i=1}^{n}|A_i|\right)$$
$$\leqslant \tfrac{5}{4}\omega(f,1/\sqrt{k})\left(\int_{-1}^{1} W(x)\,dx+\sum_{i=1}^{n}|A_i|\right).$$

Thus, if $\sum_{i=1}^{n}|A_i|<k$ uniformly with respect to n we have convergence, since $\omega(f,1/\sqrt{k})\to 0$ as $k\to\infty$.

4.2.3 Further use of uniform polynomial approximation for $f\in\tilde{\mathbb{H}}_2$

We now turn our attention to the Hilbert space of analytic functions $\tilde{\mathbb{H}}_2$ and show how convergence, based on the stronger properties of this space, can be proved.

Assume that we are given a function space S such that every $g\in S$ is integrated exactly by Q_nf. Then $E_ng=0$, and

$$|E_nf|=|E_n(f-g)|\leqslant\|E_n\|\inf_{g\in S}\sup_{|x|\leqslant 1}|f-g|$$

If $S=\mathbb{P}_{n-1}$, we have that

$$|E_nf|\leqslant\|E_n\|\cdot\|e_n(f,x)\| \tag{4.2.1}$$

where $e_n(f,x)$ is the error arising from approximating (or interpolating) f uniformly by polynomials which belong to $\mathbb{P}_{n-1}$. There sometimes exist convenient estimates for $e_n(f,x)$, (cf. [12] and [13]).

Assume, for example, that $f\in\tilde{\mathbb{H}}_2$ (cf. Section 3.5.3) with $R\geqslant R_0>1$. There then exists the expansion $f(z)=\sum_{k=0}^{\infty}a_kz^k$. From Cauchy's integral formula

$$f^{(k)}(z)=\frac{k!}{2\pi i}\int_{|z|=R}\frac{f(u)}{(u-z)^{k+1}}\,du$$

we obtain

$$a_k=\frac{1}{k!}f^{(k)}(0)=\frac{1}{2\pi i}\int_{|z|=R}\frac{f(u)}{u^{k+1}}\,du$$

and, hence, $|a_k|\leqslant\dfrac{M}{R^k}$ with $M=\max_{|z|\leqslant R}|f(z)|$. Let us denote by $p_n(f,x)\in\mathbb{P}_{n-1}$

the polynomial of best uniform approximation to f in $[-1, 1]$. Then the approximation error $e_n(f, x)$ of $p_n(f, x)$ can be estimated as follows. In the sup-norm we have for $|x| \leqslant 1$, and $R > 1$

$$\|e_n(f, x)\| = \|f(x) - p_n(f, x)\| \leqslant \left\| f(x) - \sum_{k=0}^{n-1} a_k x^k \right\| = \left\| \sum_{k=n}^{\infty} a_k x^k \right\|$$
$$\leqslant \sum_{k=n}^{\infty} |a_k| \leqslant M \sum_{k=n}^{\infty} R^{-k} = \frac{M}{R^{n-1}(R-1)}.$$

Hence we have:

Lemma 4.2.1 Let $f \in \tilde{\mathbb{H}}_2$, with $R \geqslant R_0 > 1$. Then for $|x| \leqslant 1$ the error of uniform approximation of f by polynomials of $\mathbb{P}_{n-1}$ satisfies

$$\|e_n(f, x)\| \leqslant \frac{M}{R^{n-1}(R-1)} \quad \text{where} \quad M = \sup_{|z| \leqslant R} |f(z)|. \quad \blacksquare$$

Using Lemma 4.2.1 we immediately obtain:

Theorem 4.2.5 *If $Q_n f$ has degree $n-1$, then $Q_n f$ converges for every $f \in \tilde{\mathbb{H}}_2$ with $R \geqslant R_0 > 1$ if $\lim_{n\to\infty} R^{-n} \|Q_n\| = 0$.*

Proof Since $Q_n f$ has degree $n-1$ we have that

$$|E_n f| \leqslant \|E_n\| \cdot \|e_n(f, x)\| \leqslant (\|Q_n\| + \|I\|) \|e_n(f, x)\|$$
$$\leqslant \left(\|Q_n\| + \int_{-1}^{1} W(x)\,\mathrm{d}x \right) \frac{MR}{R^n(R-1)}$$

for the sup-norm $\|\cdot\|$. Hence

$$\lim_{n\to\infty} |E_n f| \leqslant \lim_{n\to\infty} \left\{ \left(\frac{\|Q_n\|}{R^n} + \frac{1}{R^n} \int_{-1}^{1} W(x)\,\mathrm{d}x \right) \frac{MR}{R-1} \right\} = 0,$$

which completes the proof. Note that $\|Q_n\|$ must not be bounded. $\blacksquare$

4.2.4 Interpolation of the integrand

In this chapter we have so far used only properties of the function space. We now replace the integrand by an interpolant and integrate this.

If $Q_n f$ is an interpolatory quadrature formula, then

$$|E_n f| = |If - Q_n f| = |I(f - T_n(f, x))|$$
$$\leqslant \|I\| \cdot \|e_n(f, x)\| = \|e_n(f, x)\| \int_{-1}^{1} W(x)\,\mathrm{d}x,$$

where $T_n(f, x)$ interpolates f and $e_n(f, x)$ is the corresponding error of interpolation. Note that we cannot apply the above estimates as we do not know whether the error of polynomial interpolation satisfies

$$\|e_n(f, x)\| \leq \left\| f(x) - \sum_{k=1}^{n-1} a_k x^k \right\|.$$

Nevertheless, we obtain convergence if we can show that for the error of interpolation $\lim_{n \to \infty} \|e_n(f, x)\| = 0$, $|x| \leq 1$, is valid.

There are particularly weak conditions for the uniform convergence of *spline* interpolants; cf. [1] for cubic splines. These conditions can be exploited by the integration of an interpolating spline to produce a convergent quadrature procedure.

Polynomial Lagrange interpolation does not give similar results, and in this case the situation is rather disappointing. We have:

Theorem 4.2.6 (Faber) *Let $x_i^{(n)} \in [-1, 1]$ be the nodes of a quadrature procedure. Then there always exists a function $f \in C^0[-1, 1]$ such that the Lagrange interpolating polynomials do not converge uniformly to f.*

Wendroff [18] (p. 45) shows that the function $1/(1+25x^2)$, $x \in [-1, 1]$ is such a function if equidistant nodes are used.

In general, we cannot prove convergence of quadratures which are constructed by integration of the polynomials $L_n(f, x)$. The crucial point is that $\|L_n\| = \sup_{|x| \leq 1} \sum_{i=1}^{n} |\lambda_i(x)|$ is usually unbounded for increasing n.

The situation is better for Hermite polynomial interpolation $H_n(f, x)$. In this case, cf. [14], we have:

Theorem 4.2.7 *Let $f \in C^1[-1, 1]$ and $[1 - 2\lambda_i'(x_i)(x - x_i)] \geq a > 0$, $i = 1(1)n$, for $x \in [-1-a, 1+a]$. Then $H_n(f, x)$ converges uniformly to f in $[-1, 1]$ where the λ_i are as in Section 2.5.2.*

This is an encouraging result. However, Lagrange interpolation need not be based on polynomials and we therefore need more general conditions to prove convergence of a given quadrature procedure.

4.2.5 Convergence of $\|E_n\|$

The error estimates for $\|E_n\|$ derived in Section 3.5 are now used to prove convergence of the corresponding quadrature procedures.

Very important results come from representing the error functional in a suitable Hilbert space (i.e. in a complete, linear, inner product space). In this case we calculate the norm and then estimate $|E_nf| \leq \|E_n\| \cdot \|f\|$ as was done for the space $\tilde{\mathbb{H}}_2$ in Section 3.5.3. From such an error estimate we obtain convergence if $\|E_n\| \cdot \|f\| \to 0$ for $n \to \infty$. For convenience it is necessary that $\lim\limits_{n\to\infty} \|E_n\| = 0$. If this cannot be shown immediately it may be possible to use a modified $\tilde{E}_nf$ for which $\lim\limits_{n\to\infty} \|E_n\| \leq \lim\limits_{n\to\infty} \|\tilde{E}_n\| = 0$. As an example we treat the extended Hardy space $\mathbb{H}_2$.

The precise calculation of E_n for a minimal quadrature formula Q_nf is very difficult. We therefore majorise this $\|E_n\|$ using the norm $\|\tilde{E}_n\|$ of the quadrature formula

$$Q_{2n}f = \frac{1}{n} \sum_{\substack{i=-n \\ \neq 0}}^{n} f\left(\frac{i}{n+1}\right).$$

which, because of the simple weights and nodes, is very convenient for this purpose. Then

$$\sigma_{2n} \equiv \|\tilde{E}_{2n}\| \geq \|E_{2n}\|.$$

We have seen in Section 3.5.3 that

$$\sigma_{2n}^2 = b(\varepsilon) + \frac{2}{\varepsilon^2} \sum_{m=1}^{2n} \frac{A_m}{x_m} \ln \frac{1-\varepsilon^2 x_m}{1+\varepsilon^2 x_m} + \sum_{m=1}^{2n} \sum_{k=1}^{2n} \frac{A_m A_k}{1-\varepsilon^2 x_m x_k}$$

and hence we obtain

$$\sigma_{2n}^2 = b(\varepsilon) - \frac{2}{n} \sum_{\substack{i=-n \\ \neq 0}}^{n} \frac{n+1}{i\varepsilon^2} \ln \frac{1+i\varepsilon^2/(n+1)}{1-i\varepsilon^2/(n+1)} + \left(\frac{1}{n}\right)^2 \sum_{\substack{k=-n \\ \neq 0}}^{n} \sum_{\substack{i=-n \\ \neq 0}}^{n} \frac{1}{1-ik\varepsilon^2/(n+1)^2},$$

or, using symmetry,

$$\sigma_{2n}^2 = b(\varepsilon) - \frac{4}{n} \sum_{i=1}^{n} \frac{n+1}{i\varepsilon^2} \ln \frac{1+i\varepsilon^2/(n+1)}{1-i\varepsilon^2/(n+1)} + \frac{1}{n^2} \sum_{\substack{k=-n \\ \neq 0}}^{n} \sum_{\substack{i=-n \\ \neq 0}}^{n} \frac{1}{1+ik\varepsilon^2/(n+1)^2}.$$

We have that

$$\int_0^n \frac{n+1}{\varepsilon^2 x} \ln \frac{1+\varepsilon^2 x/(n+1)}{1-\varepsilon^2 x/(n+1)} \mathrm{d}x \leq \sum_{i=1}^{n} \frac{n+1}{i\varepsilon^2} \ln \frac{1+i\varepsilon^2/(n+1)}{1-i\varepsilon^2/(n+1)}$$

since the integrand is a strictly monotonically-increasing positive function in $[0, n]$ and hence the sum is an upper Riemann sum of the integral.

Further, the two functions

$$\frac{1}{1-xy\varepsilon^2/(n+1)^2} \quad \text{and} \quad \frac{1}{1+xy\varepsilon^2/(n+1)^2}$$

are, respectively, strictly monotonically-increasing and -decreasing positive functions in $[0, n+1]\times[0, n+1]$. Hence

$$\sum_{i=1}^{n}\sum_{k=1}^{n}\frac{1}{1-ik\varepsilon^2/(n+1)^2} \leqslant \int_1^{n+1}\int_1^{n+1}\frac{dx\,dy}{1-xy\varepsilon^2/(n+1)^2} < \int_0^{n+1}\int_0^{n+1}\frac{dx\,dy}{1-xy\varepsilon^2/(n+1)^2}$$

and

$$\sum_{i=1}^{n}\sum_{k=1}^{n}\frac{1}{1+ik\varepsilon^2/(n+1)^2} \leqslant \int_0^{n}\int_0^{n}\frac{dx\,dy}{1+xy\varepsilon^2/(n+1)^2} < \int_0^{n+1}\int_0^{n+1}\frac{dx\,dy}{1+xy\varepsilon^2/(n+1)^2}.$$

The sums are lower Riemann sums for the integrals. We therefore have that

$$\sigma_{2n}^2 < b(\varepsilon) - \frac{4}{n}\int_0^n \frac{n+1}{\varepsilon^2 x}\ln\frac{1+\varepsilon^2 x/(n+1)}{1-\varepsilon^2 x/(n+1)}\,dx + \frac{2}{n^2}\left[\int_0^{n+1}\int_0^{n+1}\frac{dx\,dy}{1-xy\varepsilon^2/(n+1)^2} + \int_0^{n+1}\int_0^{n+1}\frac{dx\,dy}{1+xy\varepsilon^2/(n+1)^2}\right].$$

We substitute $x=(n+1)u$, $y=(n+1)v$ and integrate with respect to v in the last two integrals. Then

$$\sigma_{2n}^2 < b(\varepsilon) - \frac{4}{n}(n+1)\int_0^a g(u)\,du + \frac{2}{n^2}(n+1)^2\int_0^1 g(u)\,du \tag{4.2.2}$$

where $a = n/(n+1)$ and $g(u) = 1/(\varepsilon^2 u)\ln((1+\varepsilon^2 u)/(1-\varepsilon^2 u))$.

Recalling that $b(\varepsilon) = 2\int_0^1 g(u)\,du$, we split the integrals in (4.2.2) and cancel suitable terms so that (cf. p. 131 for $b(\varepsilon)$ and $b(1) = \pi^2/2$)

$$\begin{aligned}\sigma_{2n}^2 &< \left(2+2+\frac{4}{n}+\frac{2}{n^2}\right)\int_0^1 g(u)\,du - \left(4+\frac{4}{n}\right)\int_0^a g(u)\,du \\ &= \frac{2}{n^2}\int_0^1 g(u)\,du + \left(4+\frac{4}{n}\right)\int_a^1 g(u)\,du = \frac{b(\varepsilon)}{n^2} + \frac{4}{a}\int_a^1 g(u)\,du \\ &< \frac{b(1)}{n^2} + 4\frac{1-a}{a}\sup_{u\in[a,1]} g(u) = \frac{1}{2}\left(\frac{\pi}{n}\right)^2 + \frac{4}{n}g(1) \quad \text{for} \quad \varepsilon<1.\end{aligned}$$

Hence, for $\varepsilon<1$, $\sigma_{2n}^2 < \frac{1}{2}\left(\frac{\pi}{n}\right)^2 + \frac{4}{n}\frac{1}{\varepsilon^2}\ln\frac{1+\varepsilon^2}{1-\varepsilon^2}$.

In the case $\varepsilon=1$ we have that

$$\sigma_{2n}^2<\frac{\pi^2}{2n^2}+4\frac{n+1}{n}\int_{n/n+1}^{1}\frac{1}{u}\ln\frac{1+u}{1-u}\,du,$$

where the last integral is $O((\ln 2n)/2n)$. Replacing $2n$ by n we have that

$$\sigma_n=\begin{cases} O\left(\dfrac{\ln n}{n}\right)^{\frac{1}{2}} & \text{for}\quad \varepsilon=1,\\[2ex] O\left(\dfrac{1}{n}\right)^{\frac{1}{2}} & \text{for}\quad 0\leqslant\varepsilon<1.\end{cases}$$

These asymptotic expressions are certainly not the best possible ones (cf. Remark R.3.10, Chapter 3); firstly, because the Q_nf are not known exactly for all n, and secondly because of the rough estimations of all quantities in the calculations. If we estimate more precisely, we obtain $\sigma_n=O(\sqrt{1/n})$ also for $\varepsilon=1$ (see [7]).

Theorem 4.2.8 *The error norm of minimal quadratures in the sense of the estimate $|E_nf|\leqslant\|E_n\|\cdot\|f\|$ for the space $\tilde{\mathbb{H}}_2$ satisfies*

$$\|E_n\|\leqslant\begin{cases} C_1\left(\dfrac{\ln n}{n}\right)^{\frac{1}{2}} & \textit{for}\quad \varepsilon=1,\\[2ex] C_2\left(\dfrac{1}{n}\right)^{\frac{1}{2}} & \textit{for}\quad 0\leqslant\varepsilon<1,\end{cases}$$

where C_1 and C_2 are constants independent of n.

Although this is a good result we need conditions for the more general situation when $f\in C^0[-1,1]$ only. In the following we shall, for the sake of simplicity, restrict our attention to the interval $[-1,1]$, but the results remain valid for every finite interval and for suitable weight functions guaranteeing the existence of all the required integrals for even infinite intervals of integration.

For a survey of convergence problems for quadrature see, for example, [5].

4.3 Convergence conditions for general quadrature procedures

In order to test whether a given quadrature procedure converges we need to produce a set of convergence criteria or convergence conditions which will be stated as theorems. We generally provide that the nodes are arranged in the form $-1\leqslant x_1<x_2<\ldots<x_n\leqslant 1$.

The first condition is a necessary one and answers the question whether there may remain gaps between the nodes as $n\to\infty$.

Theorem 4.3.1 *A necessary condition for Q_nf to converge is that the nodes become dense in $[-1, 1]$ as $n\to\infty$.*

Proof Assume the contrary, that there remains an interval $(a, b)\subseteq [-1, 1]$, $a<b$ such that all $x_i\notin(a, b)$ for every n. Consider the function $f\in C^{k-1}[-1, 1]$, which is defined by

$$\tilde{f}\equiv\begin{cases}(x-a)^k(b-x)^k & \text{for} \quad x\in(a, b),\\ 0 & \text{otherwise.}\end{cases}$$

Clearly $Q_{n,k}\tilde{f}=0$, but $I\tilde{f}>0$ and hence convergence is impossible for $\tilde{f}$. ■

Note that this result allows gaps only between *exterior* nodes $x_i\notin[-1, 1]$. In order to obtain *sufficient* convergence conditions we can estimate the error functional as was done in Theorem 4.2.8. However, the most interesting conditions are those which are both necessary *and* sufficient. These are to be obtained from representations of the error. The following theorem is fundamental to our theory.

4.3.1 The theorem of Banach and Steinhaus

The following conditions are necessary *and* sufficient.

Theorem 4.3.2 (Banach and Steinhaus) *Let T_n be a sequence of continuous linear operators which map a Banach space R_1 into a Banach space R_2. Then T_n converges to the linear operator T if and only if*

(1) *there exists K such that $\|T_n\|\leq K$ for all n,*
(2) *$T_nf\to Tf$ for every $f\in X\subset R_1$, X dense in R_1.*

Throughout this chapter the repeated and corresponding necessary and sufficient conditions are, for clarity and easy reference, assigned the numbers (1) and (2).

Proof

Sufficiency:

Assume (1) and (2) are true. For any $f\in R_1$ and every $\varepsilon>0$ there exists $f_\varepsilon\in X$ such that $\|f-f_\varepsilon\|<\varepsilon$, and there exists an integer $n_0(\varepsilon)$ such that $\|T_n f_\varepsilon - T_m f_\varepsilon\| < \varepsilon$ for $n, m > n_0(\varepsilon)$ and thus, using (1),

$$\begin{aligned}\|T_n f - T_m f\| &\leqslant \|T_n f - T_n f_\varepsilon\| + \|T_n f_\varepsilon - T_m f_\varepsilon\| + T_m f_\varepsilon - T_m f\| \\ &\leqslant \|T_n\| \cdot \|f - f_\varepsilon\| + \varepsilon + \|T_m\| \cdot \|f - f_\varepsilon\| \\ &< (\|T_n\| + 1 + \|T_m\|)\varepsilon < (1 + 2K)\varepsilon,\end{aligned}$$

from which convergence of the sequence $\{T_n\}$ follows.

Necessity:

Assume $\{T_n\}$ is a convergent sequence of linear operators. We have to demonstrate (1) as condition (2) is trivially satisfied. Take any function $\tilde{f} \in R_1$ and consider the "ball" $S(\tilde{f}, r)$ containing every $f \in R_1$ satisfying $|f - \tilde{f}| \leqslant r$. Assume $\|T_n f\|$ is bounded for $f \in S(\tilde{f}, r)$. For every $g \in R_1$ with $g \neq 0$ the element $g_\varepsilon \equiv \tilde{f} + \frac{\varepsilon}{\|g\|} g \in S(\tilde{f}, \varepsilon)$. Then

$$\|T_n g_\varepsilon\| = \left\| T_n \tilde{f} + \frac{\varepsilon}{\|g\|} T_n g \right\| \leqslant \tilde{K}.$$

Since

$$\tilde{K} \geqslant \left\| T_n \tilde{f} + \frac{\varepsilon}{\|g\|} T_n g \right\| \geqslant \left| \|T_n \tilde{f}\| - \frac{\varepsilon}{\|g\|} \|T_n g\| \right|,$$

we have that

$$-\tilde{K} \leqslant \|T_n \tilde{f}\| - \frac{\varepsilon}{\|g\|} \|T_n g\| \leqslant \tilde{K}$$

and thus that $\|T_n g\| \leqslant \frac{\|g\|}{\varepsilon} (\tilde{K} + \|T_n \tilde{f}\|)$.

Because of the convergence of $T_n \tilde{f}$ there exists a constant K such that $\|T_n g\| \leqslant K \|g\|$ and hence $\|T_n\| \leqslant K$, where g is not necessarily inside the ball $S(\tilde{f}, \varepsilon)$.

Assume that there does not exist a ball $S(\tilde{f}, \varepsilon)$ in which $T_n f$ is bounded. Then, for any such $S(\tilde{f}, \varepsilon)$ there exists an $f_1 \in S(\tilde{f}, \varepsilon)$ such that $\|T_{n_1} f_1\| > 1$. Because of the continuity of T_{n_1} we have a ball $S_1(f_1, \varepsilon_1) \subseteq S(\tilde{f}, \varepsilon)$ such that $\|T_{n_1} f\| > 1$ for $f \in S_1(f_1, \varepsilon_1)$. We repeat in this way and proceed by induction. Thus we have a subsequence T_{n_k}, elements f_k, and balls

$$S_k(f_k, \varepsilon_k) \subseteq S_{k-1}(f_{k-1}, \varepsilon_{k-1}) \quad \text{such that} \quad \|T_{n_k} f\| \geqslant k \quad \text{for} \quad f \in S_k(f_k, \varepsilon_k).$$

If $\varepsilon_k \not\to 0$ as $k \to \infty$ we take $\tilde{\varepsilon}_k \leqslant \varepsilon_k$ with $\tilde{\varepsilon}_k \to 0$. Then $f_k \to f \in R_1$ because of the completeness of R_1. But $\|T_{n_k} f_k\| \to \infty$ for $k \to \infty$ which is a contradiction, and hence the proof is finished. ■

4.3.2 Necessary and sufficient convergence conditions

We now apply this very general Theorem 4.3.2 to quadrature procedures.

The functionals If and $Q_{n,k}f$ are bounded continuous linear operators which map functions $f \in C^0[-1, 1]$ into the real numbers $\mathbb{R}^1$, where $\mathbb{R}^1$ is a Banach space in the max-norm. Thus, we choose the Banach space R_1 and then apply Theorem 4.3.2. The simplest case is that in which $R_1 = C^0[-1, 1]$ with $\|f\| = \max_{|x| \leq 1} |f(x)|$ for all $f \in R_1$. Because of the Weierstrass approximation theorem the space of all polynomials is dense in $C^0[-1, 1]$. Hence we have:

Corollary 4.3.1 (Polya and Stekloff) The quadrature procedure Q_nf converges to If for every $f \in C^0[-1, 1]$ if and only if

(i) $\sum_{i=1}^{n} |A_i^{(n)}| \leq K$ for every n,

(ii) $Q_nf \to If$ for every polynomial f. ■

The sufficiency of the corollary was proved by Stekloff and the necessity by Polya, both independently of Banach and Steinhaus.

In order to apply Theorem 4.3.2 to quadratures we have to specify the norm of Q_n: $\|Q_n\| = \sum_{i=1}^{n} |A_i^{(n)}|$.

Now, $|Q_nf| \leq \|f\| \sum_{i=1}^{n} |A_i^{(n)}|$ and equality arises for $f \equiv \varphi$ with

$$\varphi(x) = \begin{cases} \text{sign } (A_i^{(n)}) \quad \text{for} \quad x = x_i^{(n)}, & i = 1(1)n, \\ \text{linear in} \quad [x_i^{(n)}, x_{i+1}^{(n)}], & i = 1(1)n-1, \\ \text{constant in} \quad [-1, x_1^{(n)}] \quad \text{and} \quad [x_n^{(n)}, 1]. \end{cases}$$

Note that $\varphi(x)$ is a linear spline with $\|\varphi\| = 1$ in the sup-norm.

Corollary 4.3.1 does not take into account derivatives of f. (The extension to cover these was done by several authors; cf. for example [10], [2], and [5].) In order to perform this extension we have to define appropriate norms. The space $C^k[-1, 1]$ is a Banach space with the norm

$$\|f\| = |f(t_0)| + |f'(t_1)| + \ldots + |f^{(k-1)}(t_{k-1})| + \sup_{|x| \leq 1} |f^{(k)}(x)|, \; t_i \in [-1, 1]. \tag{4.3.1}$$

Clearly this norm is homogeneous and satisfies $\|f + g\| \leq \|f\| + \|g\|$.

If $f \equiv 0$ then, trivially, $\|f\| = 0$. Correspondingly, if $\|f\| = 0$ then

$$f(t_0) = f'(t_1) = \ldots = f^{(k-1)}(t_{k-1}) = 0 \quad \text{and} \quad f^{(k)}(x) \equiv 0.$$

From the last condition f must be a polynomial which we write as $f = a_0 + a_1(x - t_1) + \ldots + a_{k-1}(x - t_{k-1})^{k-1}$. Further, $f^{(q)}(t_q) = 0$, $q = 0(1)k-1$, so that $f^{(q)}(t_q)/q! = a_q = 0$, $q = k-1(-1)0$ and hence $a_{k-1} = \ldots = a_0 = 0$ which implies that $f \equiv 0$.

Clearly, the points t_i need not be distinct. We shall choose $t_i = -1$. We now proceed to find the corresponding norm $\|Q_{n,k}\|$, $k \neq 0$. Let us first restrict our attention to the case $k = 1$. Then (suppressing the upper index (n))

$$\begin{aligned} Q_{n,1}f &= \sum_{i=1}^{n} A_{i,0} f(x_i) + \sum_{i=1}^{n} A_{i,1} f'(x_i) \\ &= \sum_{i=1}^{n} A_{i,0} \left\{ \int_{-1}^{x_i} f'(x)\,\mathrm{d}x + f(-1) \right\} + \sum_{i=1}^{n} A_{i,1} f'(x_i) \\ &= f(-1) Q_{n,1}(1) + \sum_{i=1}^{n} A_{i,0} \sum_{j=0}^{i-1} \int_{x_j}^{x_{j+1}} f'(x)\,\mathrm{d}x + \sum_{i=1}^{n} A_{i,1} f'(x_i) \end{aligned}$$

where $x_0 \equiv -1$. This calculation was done in order to estimate $Q_{n,1}f$ using $\sup_{|x| \leq 1} |f'(x)|$. Writing $c_0 \equiv Q_{n,1}(1)$ and $\bar{f}(x) \equiv f(x) - f(-1)$ and rearranging the integrals we have that

$$Q_{n,1}f = f(-1)c_0 + \sum_{j=0}^{n-1} \int_{x_j}^{x_{j+1}} \sum_{i=j+1}^{n} A_{i,0} f'(x)\,\mathrm{d}x + \sum_{i=1}^{n} A_{i,1} f'(x_i) \qquad (4.3.2)$$

and hence that

$$\begin{aligned} |Q_{n,1}\bar{f}| &\leq \sum_{j=0}^{n-1} \left| \sum_{i=j+1}^{n} A_{i,0} \right| (x_{j+1} - x_j) \sup_{|x| \leq 1} |\bar{f}'(x)| + \sum_{i=1}^{n} |A_{i,0}| \sup_{|x| \leq 1} |\bar{f}'(x)| \\ &\leq \left(\sum_{j=0}^{n-1} \left| \sum_{i=j+1}^{n} A_{i,0} \right| (x_{j+1} - x_j) + \sum_{j=1}^{n} |A_{j,1}| \right) \|\bar{f}\| \quad \text{where} \quad \|\bar{f}\| = \|f\|. \end{aligned}$$

Thus, for $\bar{f} \in C^1[-1, 1] \cap \{f \mid f(-1) = 0\}$ $\left(\text{note that now } \sup_{|x| \leq 1} |f'(x)| \text{ is a norm}\right)$

$$\|Q_{n,1}\| = \sum_{j=0}^{n-1} \left| \sum_{i=j+1}^{n} A_{i,0} \right| (x_{j+1} - x_j) + \sum_{j=1}^{n} |A_{j,1}| \qquad (4.3.3)$$

(Returning to $C^1[-1, 1]$ the constant $C_0 \equiv |c_0|$ has to be added.) $\|Q_{n,1}\|$ is in fact a norm if we can find a function φ with $\varphi(-1) = 0$ and with $\|\varphi\| = 1$ such that $Q_{n,1}\varphi = \|Q_{n,1}\|$. We now produce such a function φ as follows: Let the derivative φ' of the function with $\|\varphi\| = 1$ be such that $\varphi'(x_i) =$

sign $A_{i,1}$, $i=1(1)n$, and $|\varphi'(x)|=1$ in (x_i, x_{i+1}) such that $\varphi'(x)=\operatorname{sign}\left(\sum_{i=j+1}^{n} A_{i,0}\right)$ for $x\in(x_j, x_{j+1})$ and $j=0(1)n-1$. Then $|\varphi'(x)|\leqslant 1$ for $|x|\leqslant 1$ and $\varphi'(x)$ is a step function (with isolated points) as in Fig. 4.3.1. The function φ itself is obtained on integration. When all these conditions are satisfied then $\|\varphi\|=|\varphi(-1)|+\sup_{|x|\leqslant 1}|\varphi'(x)|=\sup_{|x|\leqslant 1}|\varphi'(x)|=1$ and $Q_{n,1}\varphi=\|Q_{n,1}\|$. This follows immediately from the representation (4.3.2) of $Q_{n,1}f$. However, $\varphi(x)\notin C^1[-1,1]$. In order to eliminate this difficulty we approximate $\varphi'(x)$ by a continuous function $\varphi'_\varepsilon(x)$ of form as in Fig. 4.3.1,

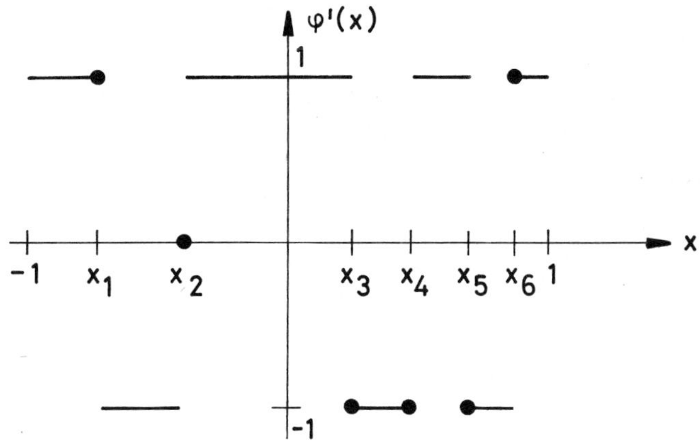

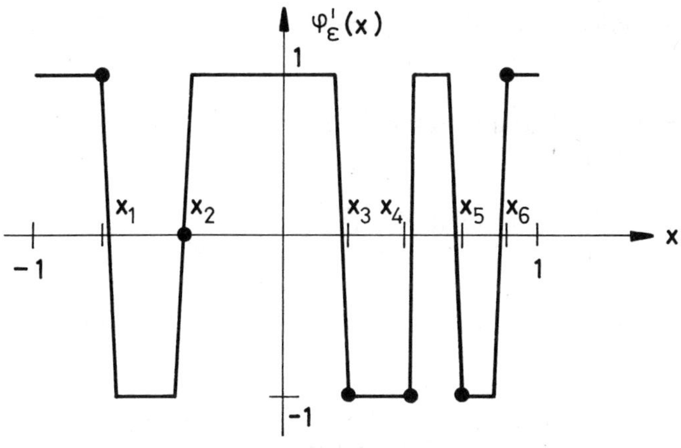

FIG. 4.3.1. $\varphi'(x)$ and $\varphi'_\varepsilon(x)$ for $n=6$.

where the slopes of the additional non-horizontal lines are $\pm 1/\varepsilon$. Then $\|\varphi_\varepsilon\| = 1$ and $Q_{n,1}\varphi_\varepsilon = Q_{n,1}\varphi + O(\varepsilon)$ as a consequence of (4.3.3). Hence we can always find $\varepsilon > 0$ such that $\varphi_\varepsilon \in C^1[-1, 1]$, and $Q_{n,1}\varphi_\varepsilon$ gets arbitrarily close to $\|Q_{n,1}\|$.

Repeating such calculations for $k = 2$ we obtain with $c_0 \equiv Q_{n,2}(1)$, $c_1 \equiv Q_{n,2}(x+1)$, and $\bar{f}(x) = f(x) - f(-1) - (x+1)f'(-1)$

$$Q_{n,2}\bar{f} = \sum_{j=0}^{n-1} \int_{x_j}^{x_{j+1}} \sum_{i=j+1}^{n} (A_{i,0}(x_i - x) + A_{i,1})\bar{f}''(x)\,dx + \sum_{i=1}^{n} A_{i,2}\bar{f}''(x_i),$$

with $x_0 = -1$. We now proceed as for $k = 1$ so that

$$\|Q_{n,2}\| = \sum_{j=0}^{n-1} \int_{x_j}^{x_{j+1}} \left| \sum_{i=j+1}^{n} A_{i,0}(x_i - x) + A_{i,1} \right| dx + \sum_{i=1}^{n} |A_{i,2}|$$

for the space $C^2[-1, 1] \cap \{f \mid f(-1) = f'(-1) = 0\}$ with norm $\|f\| = \sup_{|x| \le 1} |f''(x)|$. Returning to $C^2[-1, 1]$ the constant $C_2 \equiv |c_0| + |c_1|$ has to be added. Finally, for $\bar{f} \in C^k[-1, 1] \cap \{f \mid f(-1) = \ldots = f^{(k-1)}(-1) = 0\}$ we have that

$$\|Q_{n,k}\| = \sum_{j=0}^{n-1} \int_{x_j}^{x_{j+1}} \left| \sum_{r=0}^{k-1} \sum_{i=j+1}^{n} A_{i,r} \frac{(x_i - x)^{k-r-1}}{(k-r-1)!} \right| dx + \sum_{i=1}^{n} |A_{i,k}|.$$

Returning from $C^k[-1, 1] \cap \{f \mid f^{(j)}(-1) = 0,\ j = 0(1)k-1\}$ to $C^k[-1, 1]$ the constant

$$C_k \equiv \sum_{j=0}^{k-1} |c_j| = \sum_{m=1}^{k} \left| \sum_{r=0}^{k-m} \sum_{i=1}^{n} A_{i,r} \frac{(x_i + 1)^{k-r-1}}{(k-r-1)!} \right|$$

has to be added. With these appropriate norms the Theorem 4.3.2 of Banach and Steinhaus can again be applied so that we obtain:

Corollary 4.3.2 The quadrature procedure $Q_{n,k}f$ for $f \in C^k[-1, 1]$ converges if and only if

(i) there exists a constant K such that $\|Q_{n,k}\| \le K$,
(ii) $Q_{n,k}f$ converges for every polynomial. ■

These conditions seem originally to have been given in this context in [10].

Remark: In condition (i) of Corollary 4.3.2 we can omit C_k because this is bounded; were it not bounded, this would mean that $Q_{n,k}f$ would not converge for every polynomial.

For $k=1$ we write with $A_i = A_{i,0}$, the norm of $Q_{n,1}$ explicitly as

$$\text{(i)}\quad |A_1+\ldots+A_n|(x_1+1)+|A_2+\ldots+A_n|(x_2-x_1)+ \\ +|A_3+\ldots+A_n|(x_3-x_2)+\ldots+|A_n|(x_n-x_{n-1})+\sum_{j=1}^{n}|A_{j,1}|\leqslant K. \quad (4.3.4)$$

If the calculations had been done for the reverse situation that $x_{n+1}\equiv 1$ instead of $x_0 \equiv -1$, then we would have produced instead of (4.3.4) the inequality

$$\text{(i)}\quad |A_1|(x_2-x_1)+|A_1+A_2|(x_3-x_2)+\ldots \\ +|A_1+\ldots+A_n|(1-x_n)+\sum_{j=1}^{n}|A_{j,1}|\leqslant K \quad (4.3.5)$$

given by Krylov (book, p. 268) which is the condition, in terms of the weights and nodes, for the quadrature $Q_{n,1}f$ to be bounded.

Some important quadrature procedures are those with $A_{i,k}=0$ throughout for $i=1(1)n$, which means that $f^{(k)}(x_i)$ does not appear in $Q_{n,k}f$. In each case we can either cancel the weights $A_{i,k}$ in conditions (1) or we can recalculate $\|Q_{n,k}\|$ by expanding only up to $f^{(k-1)}$, instead of $f^{(k)}$. We then have fewer difficulties because we can choose the function $\varphi^{(k)}$ *a priori* continuously. In this case we find:

Corollary 4.3.3 The quadrature procedure $Q_{n,k}f$ for $f\in C^k[-1, 1]$ with $A_{i,k}=0$, $i=1(1)n$, converges if and only if

(i) there exists a constant K such that

$$\sum_{j=0}^{n-1}\int_{x_j}^{x_{j+1}}\left|\sum_{r=0}^{k-2}\sum_{i=j+1}^{n}A_{i,r}\frac{(x_i-x)^{k-r-1}}{(k-r-1)!}\right|dx+\sum_{i=1}^{n}|A_{i,k-1}|\leqslant K,$$

and

(ii) $Q_{n,k}f$ converges for every polynomial. ■

For $k=1$ Corollary 4.3.3 simply means that we have:

Corollary 4.3.4 The quadrature procedure $Q_{n,1}f$ for $f\in C^1[-1, 1]$ with $A_{i,1}=0$, $i=1(1)n$ converges if and only if

(i) there exists a constant K such that $\sum_{i=1}^{n}|A_{i,0}|<K$,

(ii) $Q_{n,1}f$ converges for every polynomial. ■

Corollary 4.3.4 governs the theory of Gauss quadratures of Chapter 7. The space $C^k[-1, 1]$ is not only a Banach space under the norm (4.3.1) but also under the norm

$$\|f\|_1 \equiv \max_{i=0(1)k} \left(\sup_{|x| \leq 1} |f^{(i)}(x)| \right).$$

Clearly, $\|\cdot\|_1$ is homogeneous, definite, satisfies $\|f+g\|_1 \leq \|f\|_1 + \|g\|_1$, and is equivalent to $\|\cdot\|$. The equivalence follows because

$$\|f\| = |f(t_0)| + \ldots + |f^{(k-1)}(t_{k-1})| + \sup_{|x| \leq 1} |f^{(k)}(x)| \leq (k+1) \|f\|_1,$$

and there exists a finite positive constant c such that

$$\|f\|_1 = \max_i \left(\sup_{|x| \leq 1} |f^{(i)}(x)| \right) \leq \mathrm{c} \sup_{|x| \leq 1} |f^{(k)}(x)|$$

$$\leq \mathrm{c} \left(|f(t_0)| + \ldots + |f^{(k-1)}(t_{k-1})| + \sup_{|x| \leq 1} |f^{(k)}(x)| \right)$$

so that $\|f\|_1 \leq \mathrm{c}\,\|f\|$. The reader will recall the usefulness of equivalence for estimating one norm by means of another. We immediately estimate $|Q_{n,k} f| \leq \left(\sum_{j=1}^{n} \sum_{i=0}^{k} |A_{j,i}| \right) \|f\|_1$. However, in general $\sum_{j=1}^{n} \sum_{i=0}^{k} |A_{j,i}| > \|Q_{n,k}\|_1$. Using the Banach–Steinhaus Theorem 4.3.2 and the norm equivalence of $\|f\|_1$ and $\|f\|$ we *a fortiori* obtain *sufficient* conditions for convergence which we state as:

Corollary 4.3.5 (Bandemer)

(i) If there exists a constant K such that $\sum_{j=1}^{n} \sum_{i=0}^{k} |A_{j,i}| \leq \mathrm{K}$, and

(ii) $Q_{n,k} f$ converges for every polynomial,

then the convergence of $Q_{n,k} f$ for $f \in C^k[-1, 1]$ is guaranteed. ■

Keda and Janovic [8] have given similar conditions for the space $C^k_{2\pi}$ of 2π-periodic functions.

4.3.3 Convergence conditions for special quadrature procedures

We now consider quadratures which exhibit certain features. If special additional properties are imposed on the quadrature procedures, these naturally change the convergence conditions. One important property is that of positiveness.

Definition 4.3.1 A quadrature or cubature procedure is said to be *positive*, if and only if all weights are non-negative. ■

The most important positive quadrature procedures are Hermite procedures ($k=1$) in which the weights of the derivatives $f'(x_i)$ are zero (see Chapter 7). If all weights in $Q_{n,k}f$ are non-negative, we immediately see that

$$\|Q_{n,k}\| = Q_{n,k}\left(1+(x+1)+\frac{1}{2}(x+1)^2+\ldots+\frac{1}{k!}(x+1)^k\right).$$

If this were not uniformly bounded, we would have a contradiction to the convergence for every polynomial. Hence we can omit conditions (i) for positive quadratures:

Corollary 4.3.6 A positive quadrature $Q_{n,k}f$ for $f \in C^k[-1, 1]$ converges if and only if $Q_{n,k}f$ converges for every polynomial. ■

For positive quadratures, Bandemer's convergence condition is also a necessary condition.

Until now we have not required $Q_{n,k}f$ to integrate a polynomial *exactly*. However, in practice many quadrature procedures do this and we therefore look for convergence conditions for this important type of procedure.

Definition 4.3.2 A quadrature or cubature procedure $Q_{n,k}f$ or $C_{n,k}f$ is said to have *polynomial accuracy* if $E_nf=0$ for every given polynomial provided that n is large enough. ■

Hence, the convergence of procedures having polynomial accuracy is trivial and we have

Corollary 4.3.7 The quadrature procedure $Q_{n,k}$ for $f \in C^k[-1, 1]$ having polynomial accuracy converges if and only if there exists a constant K such that $\|Q_{n,k}\| \leq K$. ■

Corollaries 4.3.6 and 4.3.7 can be combined to produce a special corollary. This is:

Corollary 4.3.8 A positive quadrature procedure $Q_{n,k}f$ for $f \in C^k[-1, 1]$ having polynomial accuracy is *always* convergent. ■

The polynomial has played a particularly important role in all the above convergence conditions. This is because polynomials are dense in

$C^k[-1, 1]$ and hence, by the Weierstrass approximation theorem, allow a uniform approximation of every $f \in C^k[-1, 1]$, especially for the case $k = 0$. Trigonometric polynomials play the same important part for the spaces of periodic functions. In fact, we can replace the polynomials by any other system of functions which produces a uniform approximation in $C^k[-1, 1]$, especially in $C^0[-1, 1]$. However, the polynomials are the simplest and most convenient system of such functions to use and they are therefore preferred to other systems, such as rational or exponential functions.

Some quadrature procedures consist of composite rules (cf. Section 2.6). Convergence conditions for such procedures will be obtained later in this chapter (see Theorem 4.5.3).

4.4 Convergence conditions for cubature procedures

We now turn to cubature procedures and consider convergence criteria for these. The methods for constructing convergence conditions for cubature procedures are naturally similar to those for quadratures. However, much work is still being done in this area, so the subject is far from being closed.

We again restrict our attention mainly to dimension $N = 2$. First, we give a necessary condition for convergence.

Theorem 4.4.1 *Let $C_n f$ be a convergent cubature procedure for the domain $\bar{\Omega}$. Then as $n \to \infty$ the nodes become dense in $\bar{\Omega}$.*

Proof Assume this is not true. Then there exists a ball $S \subset \Omega$ such that $\mathbf{x}_i^{(n)} \notin S$ for every n and i. Define

$$\varphi(\mathbf{x}) = \begin{cases} ((\mathbf{x}, \mathbf{x}_0) - r^2)^2 & \text{for} \quad \mathbf{x} \in \bar{S}, \\ 0 & \text{otherwise}, \end{cases}$$

where $\mathbf{x}_0$ is the centre of S, and r is the radius. Thus, $I_\Omega \varphi > 0$ which contradicts $C_n \varphi = 0$, and hence the original assumption is false. ∎

In order to obtain necessary and sufficient conditions we again apply Theorem 4.3.2, of Banach and Steinhaus. With the norm defined as $\|f\| = \sup_{x \in \Omega} |f(\mathbf{x})|$ the space $C^0(\bar{\Omega})$ of functions $f(\mathbf{x})$, which are continuous in

$\bar{\Omega}$, is a Banach space. The polynomials are a dense subset of $C^0(\bar{\Omega})$. Hence we have:

Corollary 4.4.1 The cubature procedure $C_n f$ for $f \in C^0(\bar{\Omega})$ converges if and only if

(i) there exists a constant K such that $\sum_{i=1}^{n} |A_i| \leq K$,

(ii) $C_n f$ converges for every polynomial. ■

In order see that $\|C_n\| = \sum_{i=1}^{n} |A_i|$, we define the function $\varphi(\mathbf{x})$ as $\varphi(\mathbf{x}) =$ sign A_i for $\mathbf{x} = \mathbf{x}_i$. Here $\varphi(\mathbf{x})$ is a circular cone with vertical axis and top at $\mathbf{x}_i$, cut off by the plane $z = 0$. The intersections of these cones with the plane $z = 0$ are circles (see Fig. 4.4.1). We assume further that the radii

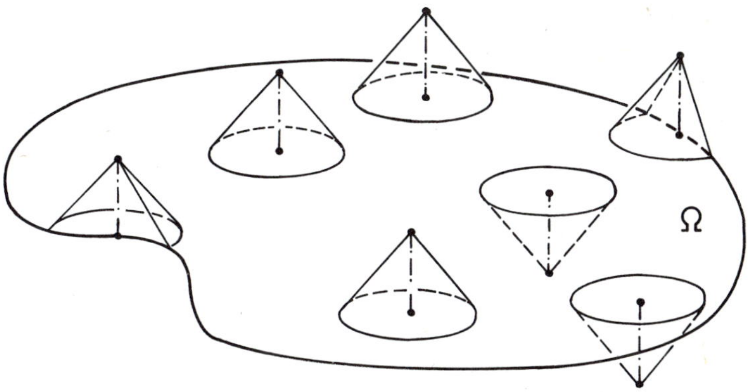

FIG. 4.4.1. Function $\varphi(\mathbf{x}) \in C^0(\bar{\Omega})$.

$r_i > 0$ of these circles are so small that there is no intersection of neighbouring circles. Outside the union of these circles we require that $\varphi(\mathbf{x}) \equiv 0$. Hence $\varphi(\mathbf{x}) \in C^0(\bar{\Omega})$, and $C_n \varphi = \sum_{i=1}^{n} |A_i|$.

The situation for cubatures having polynomial accuracy is similar to that for quadrature procedures; thus, we have:

Corollary 4.4.2 A cubature procedure $C_n f$ for $f \in C^0(\bar{\Omega})$ having polynomial accuracy converges if and only if there exists a constant K such that $\sum |A_i| < K$. ■

For cubatures with positive weights $A_i > 0$ such a constant K exists always. Usually $|\Omega| = \sum A_i$ for $W(\mathbf{x}) \equiv 1$. We have $\|C_n\| = \sum A_i$. Hence:

Corollary 4.4.3 A positive cubature procedure $C_n f$ for $f \in C^0(\bar{\Omega})$ converges if and only if $C_n f$ converges for every polynomial. ■

We can again combine positivity and the property of polynomial accuracy to give:

Corollary 4.4.4 A positive cubature procedure $C_n f$ for $f \in C^0(\bar{\Omega})$ having polynomial accuracy is always convergent. ■

First convergence conditions for cubature procedures seem to be due to Janovič [6], and to Krylov and Janovič [9]. The existence of positive cubatures was shown by Tschakaloff [17] for N dimensions. Davis [4] gave a constructive proof for weight function $W(\mathbf{x}) \equiv 1$ (see Section 6.3.3) from which Wilson [20] produced an algorithm.

Considering Hermitian cubature procedures we limit ourselves to generalising the sufficient conditions due to Bandemer [2]. Thus:

Corollary 4.4.5 A Hermitian cubature procedure $C_n f$ converges if

(i) there exists a constant K such that $\sum_{i=1}^{n} |A_i| < K$, where the sum is taken over all weights contained in $C_n f$.

(ii) $C_n f$ converges for every polynomial. ■

The validity of this corollary is due to the fact that

$$\max\left\{\sup_{\mathbf{x}\in\Omega} |f|, \sup_{\mathbf{x}\in\Omega} |f_{x_i}|, \ldots \sup_{\mathbf{x}\in\Omega} |f_{x_i\,x_j\,\ldots\,x_m}|\right\}$$

is a norm $\|f\|$ in the space $C^k(\bar{\Omega})$.

Further results can be obtained for product-type cubature formulae (see Section 2.6.2). Let $Q_{n_i} f$, $i = 1(1)N$, be N given *quadrature* procedures. The cartesian product formula of these procedures is denoted by

$$C_{n_1 \ldots n_N} f \equiv Q_{n_1}(Q_{n_2} \ldots (Q_{n_N} f(x_1, \ldots, x_N)) \ldots) = Q_{n_1} \ldots Q_{n_N} f.$$

Now, the number of nodes $n_1 \cdot \ldots \cdot n_N \to \infty$ if only one of the $n_i \to \infty$. This precludes convergence in the form so far considered and thus we are forced to resort to another definition for convergence. This is:

Definition 4.4.1 The *product-type cubature procedure* $C_{n_1 \ldots n_N} f =$

$Q_{n_1} \ldots Q_{n_N} f$ is said to be *convergent* if and only if

$$\lim_{\min_i (n_i) \to \infty} (C_{n_1 \ldots n_N} f) = I_\Omega f \quad \text{for every} \quad f \in C^0(\bar{\Omega}). \quad \blacksquare$$

With this definition, convergence of product-type cubature procedures can be discussed, and we have the following theorem:

Theorem 4.4.2 *Let $C_{n_1 \ldots n_N} f = Q_{n_1} Q_{n_2} \ldots Q_{n_N} f$ be a given product-type cubature formula. Then $C_{n_1 \ldots n_N} f$ converges if and only if all Q_{n_i} converge, $i = 1(1)N$.*

Proof Assume, for example, that $C_{n_1 \ldots n_N} f$ is convergent for every $f \in C^0(\bar{\Omega})$. Then

$$\lim_{n_i \to \infty} |I_\Omega f(\mathbf{x}_i) - C_{n_1 \ldots n_N} g(\mathbf{x}_i)| = 0, \qquad i = 1(1)N,$$
$$f = f(x_1, \ldots, x_N) \in C^0(\bar{\Omega}).$$

Assume that $Q_{n_1} f$ is not convergent. Then there exists a function of one variable $g(x_1)$ such that in

$$|I_\Omega g(x_1) - C_{n_1 \ldots n_N} g(x_1)| =$$
$$\left| \left(\int g(x_1)\, dx_1 \right) \int dx_2 \ldots dx_N - Q_{n_1} g(x_1) Q_{n_2}(1) \ldots Q_{n_N}(1) \right|$$

the multiplier $|Q_{n_1} g(x_1)|$ becomes arbitrarily large while the other terms are all bounded. The right-hand side is therefore not bounded for this $g(x_1)$, which contradicts the convergence assumption. Further, if $Q_{n_1} f$ does not converge, then $\sum |A_{i_1}|$ is unbounded. Hence

$$\|C_{n_1 \ldots n_N}\| = \prod_{i=1}^{N} \sum_{j=1}^{n_i} |A_{ij}| = \left(\sum_{j=1}^{n_1} |A_{1j}| \right) \prod_{i=2}^{N} \|Q_{n_i}\|$$

is unbounded and $C_{n_1 \ldots n_N} f$ cannot converge for every $f \in C^0(\bar{\Omega})$.

For cubature procedures, convergence can also be proved via estimates of the type $|E_n f| \leq \|E_n\| \cdot \|f\|$ by showing that $\|E_n\| \to 0$ as $n \to \infty$ for suitable function spaces, for example, those in which the functions are analytic in $\tilde{\mathbb{H}}_2 \times \tilde{\mathbb{H}}_2 \times \ldots \times \tilde{\mathbb{H}}_2$. Further, the error $e_n(f, \mathbf{x})$ obtained in the uniform approximation of f, can be used to prove convergence of the cubature procedure.

For cubatures of interpolatory type or of polynomial accuracy, respectively, we have (cf. Section 4.2.3)

$$|E_n f| \leq \|I_\Omega\| \cdot \|f - g\| \quad \text{or} \quad |E_n f| = |E_n(f - g)| \leq \|E_n\| \cdot \|f - g\|, \qquad g \in \mathbb{P}_k.$$

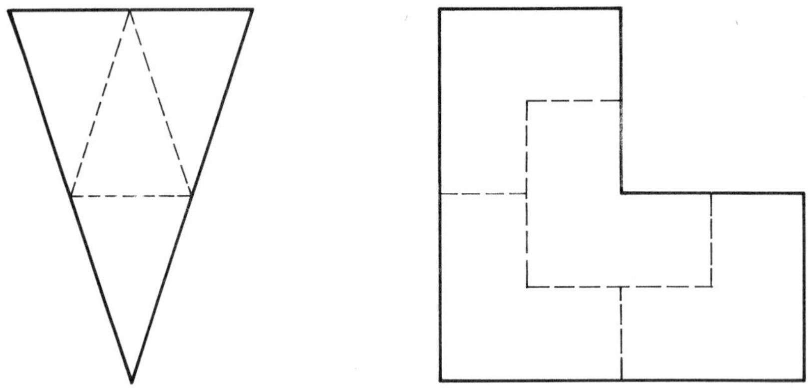

FIG. 4.4.2. Examples of subdivisible domains for $N = 2$.

A very general convergence result is now obtained for those composite cubature procedures which are generated by dividing Ω into successive subdivisions and by applying cubature formulae to each subinterval (see Fig. 4.4.2).

4.5 Convergence for procedures consisting of composite rules

A general domain Ω cannot usually be subdivided in subdomains which are either congruent or affine. Thus, in order to produce acceptable results, we make the assumption that Ω can be subdivided into 2^N congruent subdomains. This is possible if, for example, Ω is an N-cube or an N-simplex. We therefore define:

Definition 4.5.1 The domain Ω is said to be *subdivisible* if Ω can be subdivided into 2^N congruent non-empty subdomains Ω_i, $i = 1(1)2^N$, such that for each Ω_i the diameter $d(\Omega_i)$ is such that $d(\Omega_i) = d(\Omega)/2$.

Immediately, we make the two trivial conclusions:

Lemma 4.5.1 Any subdivisible domain Ω can be subdivided arbitrarily often. ■

Lemma 4.5.2 The area of the subdomain Ω_i of a subdivided domain Ω is $|\Omega|/2^N$. ■

We now take as a cubature formula over Ω

$$C_n f \equiv \sum_{i=1}^{n} A_i f(\mathbf{x}_i) \equiv \sum_{i=1}^{n} A_i f(\mathbf{t} + \tilde{\mathbf{x}}_i)$$

where $\mathbf{t}$ is the centre of gravity (or another significant point) of Ω. This cubature formula can be applied to each subdomain. The weights then become $A_i/2^N$ and hence with matrices $\mathbf{T}_i$ which produce a suitable rotation or reflexion (cf. Fig. 4.4.2) and in most cases are identity matrices

$$C_{n2^N} f = \frac{1}{2^N} \sum_{j=1}^{2^N} \sum_{i=1}^{n} A_i f(\mathbf{x}_{i,j}) = \frac{1}{2^N} \sum_{j=1}^{2N} \sum_{i=1}^{n} A_i f\left(\mathbf{t}_j + \frac{1}{2^N} \mathbf{T}_i \tilde{\mathbf{x}}_i\right)$$

where $\mathbf{t}_j$ is the centre of gravity of the j-th subdomain Ω_j. When $\|f\| = \sup_{x \in \bar{\Omega}} |f(\mathbf{x})|$ we have that

$$\|C_{n2^N}\| \leq \frac{1}{2^N} \sum_{j=1}^{2N} \sum_{i=1}^{n} |A_i| = \|C_n\|.$$

The norm of the resulting cubature formula does not grow with further subdivision of Ω. Hence, by induction, we have:

Theorem 4.5.1 *Let $C_n f$ be a cubature formula for the subdivisible domain Ω. The region Ω is then subdivided successively and the cubature formula is applied to each subdivision. The cubature procedure generated in this way has bounded norm.*

Applying Theorem 4.3.2 of Banach and Steinhaus we have convergence if and only if the procedure converges for every polynomial. In order to eliminate the dependence of convergence on polynomials, further restrictions are necessary.

Theorem 4.5.2 *Assume we are given a cubature formula $C_n f$ for $f \in C^1(\bar{\Omega})$, having degree at least zero. If this is applied to any subdomain of a subdivisible domain Ω, then the resulting cubature procedure is convergent.*

Proof The first subdivision gives 2^N congruent subdomains Ω_i. For every Ω_i we apply a mean-value theorem: there exists a point $\boldsymbol{\xi}_i \in \Omega_i$ such that $\int_{\Omega_i} f(\mathbf{x})\, d\Omega_i = |\Omega_i|\, f(\boldsymbol{\xi}_i)$. Let us expand $f(\mathbf{x})$ about $\boldsymbol{\xi}_i$: $f(\mathbf{x}) = f(\boldsymbol{\xi}_i) + (\mathbf{x} - \boldsymbol{\xi}_i)^T \operatorname{grad} f(\boldsymbol{\zeta}_i)$ where $\boldsymbol{\zeta}_i \in \Omega_i$ is a suitable point. Hence, we obtain

$$\left|\int_{\Omega_i} f(\mathbf{x})\, d\Omega_i - \sum_{j=1}^{n} \frac{A_j}{2^N} f(\mathbf{x}_{ij})\right|$$

$$= \left| |\Omega_i|\, f(\boldsymbol{\xi}_i) - \sum_{j=1}^{n} \frac{A_j}{2^N} f(\boldsymbol{\xi}_i) - \sum_{j=1}^{n} \frac{A_j}{2^N} (\mathbf{x}_{ij} - \boldsymbol{\xi}_i)^T \operatorname{grad} f(\boldsymbol{\zeta}_{ij}) \right|$$

$$\leqslant \left| |\Omega_i| - \sum_{j=1}^{n} A_j 2^{-N} \right| \cdot \|f\| + \sum_{j=1}^{n} |A_j|\, 2^{-N} d(\Omega_i) \sup_{\mathbf{x}\in\Omega} |\operatorname{grad} f|$$

$$= 2^{-N} \tfrac{1}{2} d(\Omega) \sup_{\mathbf{x}\in\Omega} |\operatorname{grad} f| \sum_{j=1}^{n} |A_j|$$

since the degree of the cubature formula is at least 0 and $d(\Omega_i) = \frac{1}{2} d(\Omega)$. Therefore, considering the whole domain Ω, we have that

$$\left|\int_{\Omega} f(\mathbf{x})\, d\Omega - C_{n2^N} f\right| \leqslant \sum_{i=1}^{2N} \left|\int_{\Omega_i} f(\mathbf{x})\, d\Omega_i - \sum_{j=1}^{n} \frac{A_j}{2^N} f(\mathbf{x}_{ij})\right|$$

$$\leqslant \tfrac{1}{2} d(\Omega) \sup_{\mathbf{x}\in\Omega} |\operatorname{grad} f| \sum_{j=1}^{n} |A_j|.$$

Repeating this process k times, we have that

$$\left|\int_{\Omega} f(\mathbf{x})\, d\Omega - C_{n2^{kN}} f\right| \leqslant 2^{-k} \cdot d(\Omega) \sup_{\mathbf{x}\in\Omega} |\operatorname{grad} f| \sum_{j=1}^{n} |A_j|$$

in which $d(\Omega)$, $\left|\underset{\mathbf{x}\in\Omega}{\operatorname{grad}} f\right|$, and $\sum_{j=1}^{n} |A_j|$ are bounded. Hence, from the factor 2^{-k} we obtain convergence. ■

This result is a *quadrature* convergence condition for $N = 1$. For $N = 1$ we can even prove the weaker result:

Theorem 4.5.3 *Assume we are given a quadrature $Q_n f$ for $f \in C^0[-1, 1]$. We halve $[-1, 1]$ succcesively and apply $Q_n f$ to every subinterval. Then the corresponding composite quadrature procedure converges if it converges for every polynomial.*

Proof The quadrature procedure contains the quadrature formulae

$$Q_{2^k n} f = \sum_{j=1}^{2^k} \frac{1}{2^{k-1}} \sum_{i=1}^{n} A_i f\left(-1 + \frac{2j-1}{2^k} + \frac{1}{2^k} x_i\right) \quad \text{for} \quad k = 0(1) \ldots.$$

Hence $Q_{2^k n}$ has bounded norm, and the conditions of the Banach–Steinhaus Theorem 4.3.2 are thus fulfilled.

4.6 Remarks, references

4.6.1 Remarks

R.4.1: The numerical effectiveness of a quadrature procedure is closely connected with the questions: How slowly or how fast can quadrature procedures possibly converge? If the velocity of convergence is too slow, then in practice convergence cannot be obtained as we can only compute with a relatively small n. This question is considered by Lipow and Stenger [11].

R.4.2: Locher [12] (and Locher and Zeller [13]) have used the algebraic approximation error in order to obtain error norms without derivatives. Hence the strong requirement of analyticity is circumvented but is replaced by the need to have a rather sharp bound for the approximation error (cf. Section 4.2.3).

4.6.2 References

1. Ahlberg, J. H., Nilson, E. N., and Walsh, J. L. (1967). "The Theory of Splines and their Applications." Academic Press, New York and London.
2. Bandemer, H. (1967). Über verallgemeinerte Interpolationsquadratur. *Math. Nachrichten*, **34,** 379–387.
3. Chawla, M. (1971). Convergence of Newton–Cotes quadratures for analytic functions. *BIT*, **11,** 159–167.
4. Davis, PH. J. (1967). A construction of nonnegative approximate quadratures. *Math. Comput.*, **21,** 578–582.
5. Filippi, S. and Esser, H. (1969). Eine Reihe neuer Sätze und Ergebnisse zur numerischen Quadratur., *Elektronische datenverarbeitung*, 166–180.
6. Janovič, L. A. (1962). The convergence of cubature procedures for absolutely continuous functions. [Russian] *Dokl. Akad. Nauk BSSR*, **6,** 77–78.
7. Johnson, L. W., and Riess, R. D. (1971). Minimal quadratures of low order continuity. *Math. Comput.*, **25,** 832–835.
8. Keda, N. P. and Janovič, L. A. (1967). Convergence of a quadrature procedure for derivable periodic functions. [Russian] *Dokl. Akad. Nauk BSSR*, **11,** 293–295.
9. Krylov, V. I. and Janovič, L. A. (1961). Conditions for the convergence of a cubature procedure for continuously differentiable functions. [Russian] *Dokl. Akad. Nauk BSSR*, **5,** 486–488.
10. Krylov, V. I. and Arljuk, T. K. (1967). On the convergence of quadrature procedures containing values of the derivatives of the integrated function. [Russian] *Dokl. Akad. Nauk BSSR*, **11,** 293–295.
11. Lipow, R. and Stenger, F. (1972). How slowly can quadrature formulas converge? *Math. Comput.*, **26,** 917–922.
12. Locher, F. (1968). "Approximationsverfahren zur Gewinnung von ableitungsfreien Fehlerschranken für Quadraturfehler." Thesis, Tübingen.
13. Locher, F. and Zeller, K. (1968). Approximationsgüte und numerische Integration., *Math. Zeitschrift*, **104,** 249–251.
14. Natanson, P. I. (1964/65). "Constructive Theory of Functions." Unger, New York.
15. Polya, G. (1933). Über Konvergenz von Quadraturverfahren. *Math. Zeitschrift*, **37,** 264–286.

16. Steklov, W. (1916). On approximate evaluation of definite integrals by means of formulas of mechanical quadrature. *Bull. de l'Académie Impériale des Sciences, Petrograd,* **10** (6), 169–186.
17. Tschakaloff, V. M., (1957). Formules de cubature méchanique à coefficients non negatifs. *Bull. Sci. Math.,* **81,** 123–134.
18. Wendroff, B. (1966). "Theoretical Numerical Analysis." Academic Press, New York and London.
19. Wilf, H. S. (1964). Exactness conditions in numerical quadrature. *Num. Math.,* **6,** 315–319.
20. Wilson, M. W. (1969). A general algorithm for nonnegative quadrature formulas. *Math. Comput.,* **23,** 253–258.

5

Orthogonal Polynomials

In the previous chapters we have already met orthogonal polynomials; these polynomials are of basic importance for many quadrature and cubature procedures. We shall see that the zeros of such polynomials are usually real, distinct and contained in a particular interval. As a result, these systems of zeros can be chosen as systems of nodes of quadrature procedures. However, this is not the only useful aspect of such polynomials. The quadratures based upon these systems of nodes usually have additional important properties, such as positivity or minimality of the quadrature error in some sense. We treat this important class of polynomials separately and produce general results which will be applied mainly in Chapter 7.

5.1 Definitions and properties

5.1.1 Preliminaries

Assume that we are given an inner product space, with inner product $(\cdot\,,\cdot)$. We then define the orthogonality of any two elements f and g of this space as follows:

Definition 5.1.1 Two elements f and g of an inner product space are said to be orthogonal: $f \perp g$, if their inner product $(f, g) = 0$ for $f, g \neq 0$. ■

As seen above, the Hardy-space $\tilde{\mathbb{H}}_2$ (Section 3.5.3) is an inner product space with

$$(f, g)_{\tilde{\mathbb{H}}_2} = \frac{1}{2\pi}\int_0^{2\pi} f(z)\,\overline{g(z)}\,\mathrm{d}\varphi, \qquad z = |z|\mathrm{e}^{i\varphi}, \tag{5.1.1}$$

where f and g are analytic for $|z| \leq R$ with $R > 1$.

The space of continuous functions $C^0[a, b]$, or in more than one dimension $C^0(\Omega)$, is an inner product space with

$$(f, g) \equiv \int_a^b W(x)f(x)g(x)\,dx \quad \text{or} \quad (f, g) \equiv \int_\Omega W(\mathbf{x})f(\mathbf{x})g(\mathbf{x})\,d\Omega, \tag{5.1.2}$$

where $W \not\equiv 0$ is a non-negative integrable weight function. The domain of integration is not necessarily finite.

The space of grid functions defined on a grid of nodes (i.e. where all functions are defined in terms of grid values), is an inner product space with *discrete* inner product

$$\langle f, g\rangle \equiv \sum_{i=0}^{n} W_i f(x_i)g(x_i), \tag{5.1.3}$$

where $W_i > 0$ are real numbers.

In these contexts we usually consider *systems* of orthogonal functions:

Definition 5.1.2 The system of functions $(f_1, \ldots, f_n)$, $f_i \neq 0$, is said to be an *orthogonal system* if

$$(f_i, f_j) = 0 \quad \text{for} \quad i \neq j \quad \text{and} \quad (f_i, f_i) \neq 0 \quad \text{for} \quad i, j = 1(1)n. \quad \blacksquare$$

It is convenient to normalise the elements of an orthogonal set of functions by $\|f_j\| \equiv (f_j, f_j)^{\frac{1}{2}} = 1$.

Definition 5.1.3 The orthogonal functions f_j are said to be *orthonormal* if $(f_j, f_j) = 1$, $j = 1(1)\ldots$ $\blacksquare$

If sets of orthogonal or orthonormal functions consist only of polynomials we call them *orthogonal or orthonormal polynomial systems*, respectively. As seen in Section 3.5.3 the monomials z^ν, $\nu = 0(1)\ldots$ are an orthonormal polynomial system for the inner product (5.1.1).

We now turn to the one-dimensional inner product (5.1.2). We shall show that a set of orthonormal polynomials exists for every non-negative weight function $W(x)$. The existence condition will be a sufficient one. Hence it may happen that we have orthonormal polynomial systems for weight functions which change sign within $[-1, 1]$ (cf. Theorem 7.4.4). Although we restrict our attention to the interval $[-1, 1]$ the results remain valid for other intervals. This is true even for non-finite intervals, provided that we have a suitable weight function.

Let $\{f_i\}$ be any set of functions from an inner product space. Then:

Definition 5.1.4 The set of functions $\{f_i,\ i=1(1)n\}$ is said to be *linearly independent* if $\sum_{i=1}^{n} a_i f_i = 0$ is only possible for $a_i = 0,\ i = 1(1)n$. Any set of functions which is not linearly independent is *linearly dependent.* ∎

We now have

Lemma 5.1.1 The monomials x^i, $i = 0(1)n$ are linearly independent.

Proof Assume this is not true. Then the system

$$
\begin{aligned}
(P_n(x), 1) &= (a_0 + a_1 x + \cdots + a_n x^n, 1) = 0 \\
(P_n(x), x) &= (a_0 + a_1 x + \cdots + a_n x^n, x) = 0 \qquad (5.1.4)\\
\cdots\cdot &\qquad \cdots\cdots\cdot \qquad \cdots \\
(P_n(x), x^n) &= (a_0 + a_1 x + \cdots + a_n x^n, x^n) = 0
\end{aligned}
$$

has non-trivial solutions. The system (5.1.4) is equivalent with

$$
\begin{aligned}
a_0(1, 1) + a_1(x, 1) + \cdots + a_n(x^n, 1) &= 0 \\
a_0(1, x) + a_1(x, x) + \cdots + a_n(x^n, x) &= 0 \\
\cdots\cdot \qquad \cdots\cdots\cdot \qquad &\cdots \\
a_0(1, x^n) + a_1(x, x^n) + \cdots + a_n(x^n, x^n) &= 0.
\end{aligned}
$$

In turn, we multiply the i-th equation by a_{i-1}, $i = 1(1)n+1$, and add all the resulting equations. Then

$$
0 = \sum_{i=1}^{n+1} a_{i-1} \sum_{j=0}^{n} a_j (x^{i-1}, x^j) = \sum_{i=0}^{n} \sum_{j=0}^{n} (a_i x^i, a_j x^j) = (P_n(x), P_n(x)) = \|P_n\|^2 > 0,
$$

which is a contradiction, thus proving the lemma. ∎

All the results of this section also hold for the discrete inner product $\langle f, g\rangle$. In particular,

$$
\langle P_n, P_n \rangle \equiv \sum_{i=0}^{n} W_i P_n^2(x_i) > 0, \qquad P_n \neq 0,
$$

because a polynomial of degree n cannot have $n+1$ zeros.

5.1.2 Existence and uniqueness

First, we prove the existence of systems of orthogonal polynomials.

Theorem 5.1.1 (Existence theorem) *For every admissible weight function there exists a system of orthogonal polynomials with respect to the inner products* $(\cdot\,,\cdot)$ *and* $\langle\cdot\,,\cdot\rangle$.

Proof The proof is constructive and follows that used in the $(n+1)$-dimensional Euclidian vector space R^{n+1}. We denote by $\omega_n(x)$ the polynomials that we seek. Then $\omega_0 \equiv \alpha_{0,0}$ since $(\omega_0, \omega_0) = \alpha_{0,0}^2(1, 1)$. Hence, for $(\omega_0, \omega_0) = 1$ we have that $|\alpha_{0,0}| = 1/\sqrt{(1, 1)}$.

Let $\omega_1(x) = \alpha_{0,1}\omega_0(x) + \alpha_{1,1}x$. Then, from $\omega_1 \perp \omega_0$ we obtain

$$(\omega_1, \omega_0) = \alpha_{0,1}(\omega_0, \omega_0) + \alpha_{1,1}(x, \omega_0) = 0.$$

Hence

$$\alpha_{0,1} = -\alpha_{1,1}\frac{(x, \omega_0)}{(\omega_0, \omega_0)} \quad \text{and} \quad \omega_1 = \alpha_{1,1}\left(-\frac{(x, \omega_0)}{(\omega_0, \omega_0)}\,\omega_0 + x\right).$$

Further, for $(\omega_1, \omega_1) = 1$ we have that

$$(\omega_1, \omega_1) = \alpha_{1,1}^2\left((x, x) + \frac{(x, \omega_0)^2}{(\omega_0, \omega_0)^2}(\omega_0, \omega_0) - 2\frac{(x, \omega_0)}{(\omega_0, \omega_0)}(x, \omega_0)\right) = 1,$$

and hence, if $(\omega_0, \omega_0) = 1$, that

$$|\alpha_{1,1}| = ((x, x) - (x, \omega_0)^2/(\omega_0, \omega_0))^{-\frac{1}{2}} = ((x, x) - (x, \omega_0)^2)^{-\frac{1}{2}}.$$

Similary, from $\omega_2 = \alpha_{0,2}\omega_0 + \alpha_{1,2}\omega_1 + \alpha_{2,2}x^2$ and $\omega_2 \perp \omega_0$, $\omega_2 \perp \omega_1$ we obtain

$$\begin{aligned}(\omega_2, \omega_0) &= \alpha_{0,2}(\omega_0, \omega_0) + \alpha_{1,2}(\omega_1, \omega_0) + \alpha_{2,2}(x^2, \omega_0)\\ &= \alpha_{0,2}(\omega_0, \omega_0) + \alpha_{2,2}(x^2, \omega_0) = 0\\ (\omega_2, \omega_1) &= \alpha_{0,2}(\omega_0, \omega_1) + \alpha_{1,2}(\omega_1, \omega_1) + \alpha_{2,2}(x^2, \omega_1)\\ &= \alpha_{1,2}(\omega_1, \omega_1) + \alpha_{2,2}(x^2, \omega_1) = 0\end{aligned}$$

and hence

$$\alpha_{0,2} = -\alpha_{2,2}\frac{(x^2, \omega_0)}{(\omega_0, \omega_0)}, \qquad \alpha_{1,2} = -\alpha_{2,2}\frac{(x^2, \omega_1)}{(\omega_1, \omega_1)}$$

so that

$$\omega_2 = \alpha_{2,2}\left(-\frac{(x^2, \omega_0)}{(\omega_0, \omega_0)}\,\omega_0 - \frac{(x^2, \omega_1)}{(\omega_1, \omega_1)}\,\omega_1 + x^2\right)$$

and finally $\alpha_{2,2}$ is obtained from

$$(\omega_2, \omega_2) = \alpha_{2,2}^2\left((x^2, x^2) - \frac{(x^2, \omega_0)^2}{(\omega_0, \omega_0)} - \frac{(x^2, \omega_1)^2}{(\omega_1, \omega_1)}\right) = 1.$$

Proceeding by induction we have that

$$\omega_n = \alpha_{n,n}\left(-\frac{(x^n, \omega_0)}{(\omega_0, \omega_0)}\omega_0 - \frac{(x^n, \omega_1)}{(\omega_1, \omega_1)}\omega_1 \cdots - \frac{(x^n, \omega_{n-1})}{(\omega_{n-1}, \omega_{n-1})}\omega_{n-1} + x^n\right)$$

and $\alpha_{n,n}$ is obtained from

$$(\omega_n, \omega_n) = \alpha_{n,n}^2\left((x^n, x^n) - \frac{(x^n, \omega_0)^2}{(\omega_0, \omega_0)} - \cdots - \frac{(x^n, \omega_{n-1})^2}{(\omega_{n-1}, \omega_{n-1})}\right) = 1.$$

This well-known method is the Gram–Schmidt orthogonalisation process. We thus construct the sequence of orthogonal polynomials $\omega_n(x)$. ■

From the manner of construction we see that, except from the multiplier $\alpha_{i,i}$, the polynomials are uniquely determined. This multiplier can be determined by normalising ω_n such that $(\omega_n, \omega_n) = 1$ is valid. Then

$$|\alpha_{n,n}| = ((x^n, x^n) - (x^n, \omega_0)^2 - \cdots - (x^n, \omega_{n-1})^2)^{-\frac{1}{2}}.$$

Now $\omega_n(x)$ is unique if we choose $\alpha_{n,n} > 0$, in which case

$$\omega_n = \alpha_{n,n}\left(x^n - (x^n, \omega_0)\omega_0 - \cdots - (x^n, \omega_{n-1})\omega_{n-1}\right).$$

Hence we have:

Theorem 5.1.2 (Uniqueness theorem) *If the leading coefficient of ω_n is normalised so that*

$$\alpha_{n,n} = \left((x^n, x^n) - \frac{(x^n, \omega_0)^2}{(\omega_0, \omega_0)} - \cdots - \frac{(x^n, \omega_{n-1})^2}{(\omega_{n-1}, \omega_{n-1})}\right)^{-\frac{1}{2}}$$

or

$$\alpha_{n,n} = \left(\langle x^n, x^n\rangle - \frac{\langle x^n, \omega_0\rangle^2}{\langle \omega_0, \omega_0\rangle} - \cdots - \frac{\langle x^n, \omega_{n-1}\rangle^2}{\langle \omega_{n-1}, \omega_{n-1}\rangle}\right)^{-\frac{1}{2}},$$

then $\omega_n(x)$ is uniquely determined.

This result can be used to expand a given polynomial in terms of a given set of orthogonal polynomials by replacing each of the single powers x^n by

$$x^n = \frac{(x^n, \omega_0)}{(\omega_0, \omega_0)}\omega_0(x) + \frac{(x^n, \omega_1)}{(\omega_1, \omega_1)}\omega_1(x) + \cdots + \frac{(x^n, \omega_{n-1})}{(\omega_{n-1}, \omega_{n-1})}\omega_{n-1}(x) + \frac{1}{\alpha_{n,n}}\omega_n(x), \qquad n = 0(1)\cdots. \quad (5.1.5)$$

From the above calculations we immediatly conclude:

Lemma 5.1.2 Let $\omega_j(x)$, $j=0(1)\ldots$ be a given orthogonal polynomial system. Then $\omega_n \perp \omega_j$, $j<n$ is equivalent with $\omega_n \perp x^j$, $j<n$ or with $\omega_n \perp \mathbb{P}_{n-1}$. ■

5.1.3. Special properties of orthogonal polynomials

The explicit determination of any set of orthogonal polynomials is facilitated enormously by the existence of a three-term recurrence relationship. In order to derive this we expand $x\omega_{n-1}(x) \in \dot{\mathbb{P}}_n$ in terms of $\omega_0, \omega_1, \ldots, \omega_n$: $x\omega_{n-1}(x) = \sum_{k=0}^{n} \alpha_k \omega_k(x)$. The inner product of this equation with $\omega_j(x)$ gives

$$(x\omega_{n-1}, \omega_j) = \sum_{k=0}^{n} \alpha_k (\omega_k, \omega_j) = \alpha_j (\omega_j, \omega_j), \qquad j=0(1)n.$$

On the other hand, from the definition of the inner product we have that $(x\omega_{n-1}, \omega_j) = (\omega_{n-1}, x\omega_j)$. This vanishes for $j=0(1)n\text{-}3$ because $x\omega_j(x) \in \dot{\mathbb{P}}_{j+1}$. Hence, $\alpha_j = 0$ for $j=0(1)n\text{-}3$. Assume, in addition, that the ω_j are orthonormal, $(\omega_j, \omega_j) = 1$, for $j=0(1)n$. Then $\alpha_j = (x\omega_{n-1}, \omega_j)$ for $j = n-2, n-1, n$, and

$$x\omega_{n-1} = (x\omega_{n-1}, \omega_n)\omega_n + (x\omega_{n-1}, \omega_{n-1})\omega_{n-1} + (x\omega_{n-1}, \omega_{n-2})\omega_{n-2}$$

or

$$\omega_n(x) = \frac{x - (x\omega_{n-1}, \omega_{n-1})}{(x\omega_{n-1}, \omega_n)} \omega_{n-1}(x) - \frac{(x\omega_{n-1}, \omega_{n-2})}{(x\omega_{n-1}, \omega_n)} \omega_{n-2}(x). \quad (5.1.6)$$

Suppose now that the polynomials $\omega_n(x)$ have the explicit form

$$\omega_n(x) = a_n x^n + b_n x^{n-1} + c_n x^{n-2} + \cdots,$$

$$\omega_{n-1}(x) = a_{n-1} x^{n-1} + b_{n-1} x^{n-2} + \cdots.$$

Comparing the coefficients of x^n, x^{n-1} we obtain that

$$a_n = \frac{1}{(x\omega_{n-1}, \omega_n)} a_{n-1} \qquad b_n = \frac{b_{n-1} - (x\omega_{n-1}, \omega_{n-1}) a_{n-1}}{(x\omega_{n-1}, \omega_n)}.$$

Hence $(x\omega_{n-1}, \omega_n) = a_{n-1}/a_n$ from which it follows that $(x\omega_{n-2}, \omega_{n-1}) = (x\omega_{n-1}, \omega_{n-2}) = a_{n-2}/a_{n-1}$, and that

$$(x\omega_{n-1}, \omega_{n-1}) = \frac{b_{n-1}}{a_{n-1}} - \frac{b_n}{a_n}.$$

Finally, for *orthonormal* polynomials we have that

$$\omega_n(x)=\frac{a_n}{a_{n-1}}\left(x-\frac{b_{n-1}}{a_{n-1}}+\frac{b_n}{a_n}\right)\omega_{n-1}(x)-\frac{a_n a_{n-2}}{a_{n-1}^2}\,\omega_{n-2}(x). \tag{5.1.7}$$

This is valid for $n>1$. For $n\leqslant 1$ we have the orthonormal polynomials

$$\omega_0(x)=\frac{1}{(1,1)^{\frac{1}{2}}}, \qquad \omega_1(x)=\frac{x-(x,1)/(1,1)^{\frac{1}{2}}}{((x,x)-(x,1)^2/(1,1))^{\frac{1}{2}}}.$$

These results again hold for discrete inner products.

The $\omega_2, \omega_3 \ldots$ can now be calculated recursively. This is done automatically by non-numerical compilers such as SYMBAL or FORMAC if the numbers $a_n, a_{n-1}, b_n, b_{n-1}$ are known explicitly. A program for the special system of Jacobi polynomials is given in Section 5.2.3.

A most important property of orthogonal polynomials in view of quadrature is the following. For several quadrature procedures the nodes turn out to be the zeros of orthogonal polynomials. Hence it is necessary to know whether these zeros are real, distinct and in the interval $[-1, 1]$, because otherwise we can not obtain a reasonable quadrature procedure. The orthogonal polynomials do, indeed, have this property.

Theorem 5.1.3 *The polynomial $\omega_n(x)\perp \mathbb{P}_{n-1}$ with respect to a non-negative weight function $W(x)$ has n real distinct zeros in $(-1, 1)$.*

Proof The proof is performed indirectly. Assume ω_n has exactly $m<n$ distinct real roots t_i of odd multiplicity in $(-1, 1)$. Let $p_m(x)\equiv (x-t_1)\cdots(x-t_m)\in\dot{\mathbb{P}}_m$. Then the product $\omega_n\cdot p_m$ does not change sign in $(-1, 1)$, because both $\omega_n(x)$ and $p_m(x)$ change their sign exactly at the points $t_1, \ldots, t_m$. Hence, $(p_m, \omega_n)\neq 0$, which contradicts the orthogonality condition, $\omega_n \perp p_m$ for $m<n$. The discrepancy only disappears for $m=n$ so that in fact ω_n has n real distinct roots in $(-1, 1)$. ■

Note that for concluding $(p_m, \omega_n)\neq 0$ in the proof it is essential that $W(x)$ does not change sign in $[-1, 1]$. If $W(x)$ does change sign, then the existence of n distinct real zeros in $(-1, 1)$ depends on further properties of $W(x)$ (see, for example, Section 7.4.6).

We now take a second look at the construction of orthogonal polynomials which comes from the property $\omega_n \perp x^\nu$, $\nu=0(1)n-1$, and which, as will be seen in the next section, sometimes facilitates the calculation of the zeros. Let us define

$$'\omega_n(x) \equiv \int_{-1}^{x} W(t)\omega_n(t)\,\mathrm{d}t,$$

so that $'\omega_n(-1)=0$. Further, since $\omega_n \perp 1$, we have that $'\omega_n(1)=0$. Similarly, for

$$''\omega_n(x) \equiv \int_{-1}^{x} {'\omega_n(t)}\,\mathrm{d}t$$

we have that $''\omega_n(-1)=0$, and integration by parts gives

$$''\omega_n(1) = x \cdot {'\omega_n(x)}\Big|_{-1}^{1} - \int_{-1}^{1} x\left(\frac{\mathrm{d}}{\mathrm{d}x}\,{'\omega_n(x)}\right)\mathrm{d}x = -\int_{-1}^{1} x\, W(x)\omega_n(x)\,\mathrm{d}x = 0.$$

Proceeding in this way we find that for $k=2(1)n$

$$^{(k)}\omega_n(x) \equiv \int_{-1}^{x} {^{(k-1)}\omega_n(t)}\,\mathrm{d}t, \qquad {^{(k)}\omega_n(\pm 1)} = 0.$$

Assume that $W(x)\omega_n(x)$ is integrable in $[-1, 1]$. Then $^{(k)}\omega_n(1)=0$, $k=1(1)n$, is a system of n non-linear equations for the n zeros of $\omega_n(x)$.

This process is simplest if it produces polynomials because then we can obtain a polynomial of degree $2n$ with the two n-fold and hence the only zeros ± 1. From Rolle's theorem we conclude that the first derivative has a zero $t_{1,1}$ in $(-1, 1)$, the second derivative has two zeros $t_{2,1}$, $t_{2,2}$, $t_{2,1} \in (-1, t_{1,1})$ and $t_{2,2} \in (t_{1,1}, 1)$ etc. Hence, by n-fold application of Rolle's theorem we have the existence of n distinct real roots in $(-1, 1)$. Let us now look for the corresponding weight functions. Clearly, $^{(n)}\omega_n$ has n derivatives so that

$$\frac{\mathrm{d}^n}{\mathrm{d}x^n}\left({^{(n)}\omega_n(x)}\right) = W(x)\omega_n(x)$$

and hence that

$$\frac{1}{W(x)}\frac{\mathrm{d}^n}{\mathrm{d}x^n}\left({^{(n)}\omega_n(x)}\right) = \omega_n(x) \in \dot{\mathbb{P}}_n,$$

while, in general, $^{(n)}\omega(x)/W(x) \notin \dot{\mathbb{P}}_{2n}$. However, if $^{(n)}\omega_n(x)/W(x) \in \dot{\mathbb{P}}_{2n}$, then we have:

Theorem 5.1.4 *If the n-fold integral $^{(n)}\omega_n(x)$ of $W(x)\omega_n(x)$ is decomposable into a multiplier $W(x)$ and a polynomial of degree $2n$ having the n-fold zeros ± 1, then*

$$\omega_n(x) = \frac{A_n}{W(x)}\frac{\mathrm{d}^n}{\mathrm{d}x^n}\left({^{(n)}\omega_n(x)}\right) \tag{5.1.8a}$$

has n distinct real roots in $(-1, 1)$, when A_n is a constant.

Sometimes we have an even simpler formula, for example,

$$\omega_n(x) = \frac{A_n}{W(x)} \frac{d^n}{dx^n} W(x) \in \dot{\mathbb{P}}_n \tag{5.1.8b}$$

with $\omega_n(x) \perp \mathbb{P}_{n-1}$ as for the Hermite polynomials in Section 5.2.4.

Formulae such as (5.1.8) are called *Rodriguez formulae*, examples of which occur for the weight functions $(1-x)^\alpha \cdot (1+x)^\beta$, $x^\alpha e^{-x}$, e^{-x^2}, with $\alpha, \beta > -1$. The practical calculation of the nodes of $\omega_n(x)$ is facilitated by the following property of orthogonal polynomials.

Theorem 5.1.5 *The zeros of two consecutive orthogonal polynomials with respect to $(\cdot\,,\cdot)$ seperate each other. The same holds with $\langle\cdot\,,\cdot\rangle$.*

Proof From the three-term recurrence relationship we conclude that ω_n and ω_{n-1} cannot have a common zero. If this were true then ω_{n-2} would also have this zero, as would ω_{n-3} etc. up to ω_0. But $\omega_0 \neq 0$ because it is constant.

Further, if $\omega_{n-1}(t) = 0$ then $\omega_n(t)$ and $\omega_{n-2}(t)$ have opposite signs. From the recurrence relationship we have that

$$\omega_n(t) = -\frac{a_n a_{n-2}}{a_{n-1}^2} \omega_{n-2}(t)$$

where the leading coefficients $a_n, a_{n-1}, \ldots$ of $\omega_n, \omega_{n-1}, \ldots,$ must be positive (see Theorem 5.1.2).

In order to complete the proof we proceed by induction. Let $\omega_1(s) = 0$ so that $\omega_2(s) < 0$ since $\omega_0 > 0$. $\omega_2 = a_2 x^2 + b_2 x + c_2$ has two distinct real zeros t_1, t_2 in $(-1, 1)$. Hence, sign $\omega_2(\pm 1)$ is positive so that ω_2 changes sign in $(-1, s)$ and in $(s, 1)$ giving $t_1 \in (-1, s)$ so that s separates t_1 and t_2.

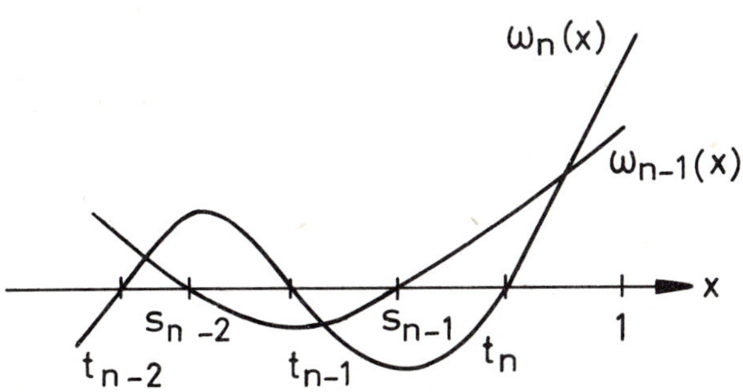

FIG. 5.1.1. Nodes of two consecutive orthogonal polynomials.

Assume Theorem 5.1.5 is true for $n-1>1$. *Let* $s_1, \ldots, s_{n-1}$ be the zeros of $\omega_{n-1}, t_1, \ldots, t_n$ be the zeros of ω_n, and $u_1, \ldots, u_{n+1}$ be the zeros of ω_{n+1}.

Now, $\operatorname{sign}(\omega_{n-1}(1)) = \operatorname{sign}(\omega_{n-1}(\infty)) > 0$. Analogously, $\operatorname{sign}(\omega_n(1)) = \operatorname{sign}(\omega_{n+1}(1))$, is positive. Hence ω_{n-1} is positive in $(s_{n-1}, 1)$ and ω_n is positive in $(t_n, 1)$, while sign $(\omega_{n+1}(t_n)) < 0$ for $\omega_n(t_n) = 0$ and $\omega_{n-1}(t_n) > 0$. Hence ω_{n+1} changes sign in $(t_n, 1]$, and in fact in $(t_n, 1)$ since $x = 1$ cannot be a zero of ω_{n+1} (if that were so we would have $\omega_{n+2}(1) < 0$, which is impossible). Hence, the zero u_{n+1} of ω_{n+1} is situated in the *open* interval $(t_n, 1)$.

We now consider the interval (t_{n-1}, t_n) containing the zero s_{n-1} of ω_{n-1}. At $x = t_{n-1}$ the polynomials ω_{n-1} and ω_{n+1} again have opposite sign. Since $\omega_{n-1}(s_{n-1}) = 0$ both ω_{n-1} and ω_{n+1} change sign in the interval (t_{n-1}, t_n). Hence ω_{n+1} has a zero $u_n \in (t_{n-1}, t_n)$. Proceeding in that manner we find further zeros of ω_{n+1}: $u_{n-1} \in (t_{n-2}, t_{n-1}), \ldots, u_1 \in (-1, t_1)$. Hence we have exactly $n+1$ zeros of ω_{n+1} which separate those of ω_n. Further nodes cannot exist and hence the theorem is proved. ■

5.2 Special systems of orthogonal polynomials

The most important general properties of orthogonal polynomials have been proved in Section 5.1. We now proceed to some special systems of orthogonal polynomials and consider some of their properties.

5.2.1 Legendre polynomials

Let $W(x) \equiv 1$. The corresponding system of orthogonal polynomials with respect to $(\cdot\,,\cdot)$ is called the system of *Legendre polynomials* $L_n(x)$ (or $l_n(x)$ if these are normalised). Because $\omega_n(x) \in \dot{\mathbb{P}}_n$, and $W(x) \equiv 1$, the function ${}^{(n)}\omega_n(x)$ is in $\dot{\mathbb{P}}_{2n}$ and has the n-fold zeros ± 1. Hence, Rodriguez' formula becomes

$$L_n(x) = A_n \frac{\mathrm{d}^n}{\mathrm{d}x^n}((x^2-1)^n). \tag{5.2.1}$$

The constant A_n may be suitably chosen, for example so that the coefficients of $L_n(x)$ have no common divisor or so that the leading coefficient of $L_n(x)$ has a prescribed value, e.g. 1. From formula (5.2.1) we immediately have an explicit representation of $L_n(x)$:

$$\begin{aligned} L_n(x) &= A_n \frac{\mathrm{d}^n}{\mathrm{d}x^n}\left(\sum_{k=0}^{n} \binom{n}{k}(-1)^k\, x^{2n-2k}\right) \\ &= A_n \sum_{k=0}^{[n/2]} \binom{n}{k}(-1)^k \frac{(2n-2k)!}{(n-2k)!}\, x^{n-2k}. \end{aligned} \tag{5.2.2}$$

The Legendre polynomials will later be considered in the interval [0, 1]. In this case

$$L_n(x) = A_n \frac{\mathrm{d}^n}{\mathrm{d}x^n}(x^n(x-1)^n) = A_n \sum_{k=0}^{n} \binom{n}{k}(-1)^k \frac{(2n-k)!}{(n-k)!} x^{n-k}.$$

However, for the moment we consider the interval $[-1, 1]$. Let us choose A_n so that the leading coefficient of $L_n(x)$ is 1. Then $A_n = n!/(2n)!$, and

$$L_n(x) = \frac{n!n!}{(2n)!} \sum_{k=0}^{[n/2]} (-1)^k \binom{n}{k}\binom{2n-2k}{n-2k} x^{n-2k}. \tag{5.2.3}$$

In order to obtain the normalised polynomial $l_n(x)$ from (5.2.1) we calculate A_n as follows. Integration by parts and use of the boundary conditions gives

$$\begin{aligned}(L_n, L_n) &= -('L_n, L_n') = (''L_n, L_n'') = \cdots = (-1)^n({}^{(n)}L_n, L_n^{(n)})\\ &= (-1)^n(A_n(x^2-1)^n, A_n(2n)!) = (-1)^n A_n^2 (2n)!\ ((x^2-1)^n, 1).\end{aligned}$$

The integrals $((x^2-1)^n, 1)$ may be calculated recursively. We have that

$$((x^2-1)^n, 1) = (x^2(x^2-1)^{n-1}, 1) - ((x^2-1)^{n-1}, 1),$$

and replacing $(x^2-1)^{n-1}$ by $\dfrac{1}{2nx}[(x^2-1)^n]'$ and integrating by parts we find that

$$(x^2(x^2-1)^{n-1}, 1) = \frac{1}{2n}\left(x\frac{\mathrm{d}}{\mathrm{d}x}[(x^2-1)^n], 1\right) = -\frac{1}{2n}((x^2-1)^n, 1).$$

Hence

$$((x^2-1)^n, 1) = -\frac{1}{2n}((x^2-1)^n, 1) - ((x^2-1)^{n-1}, 1)$$

which, with rearrangement, becomes

$$\begin{aligned}((x^2-1)^n, 1) &= -\frac{2n}{2n+1}((x^2-1)^{n-1}, 1)\\ &= \frac{2n}{2n+1}\frac{2n-2}{2n-1}((x^2-1)^{n-2}, 1) = \mp \cdots\\ &= (-1)^n \frac{2n}{2n+1}\frac{2n-2}{2n-1}\cdots\frac{2}{3}2 = (-1)^n \frac{2^{2n+1}\, n!n!}{(2n+1)!}\end{aligned}$$

since $(1, 1) = 2$. Thus

$$(L_n, L_n) = A_n^2 \frac{2}{2n+1}(2^n n!)^2 = 1 \quad \text{if} \quad A_n = \frac{1}{2^n n!}\left(\frac{2n+1}{2}\right)^{\frac{1}{2}}.$$

Hence, the normalised Legendre polynomials are

$$l_n(x)=\frac{1}{2^n n!}\left(\frac{2n+1}{2}\right)^{\frac{1}{2}}\frac{\mathrm{d}^n}{\mathrm{d}x^n}((x^2-1)^n)$$

$$=\frac{1}{2^n n!}\left(\frac{2n+1}{2}\right)^{\frac{1}{2}}\sum_{k=0}^{[n/2]}(-1)^k\binom{n}{k}\frac{(2n-2k)!}{(n-2k)!}x^{n-2k}. \quad (5.2.4)$$

From (5.2.2) we see, that L_n and l_n both contain only the same odd or even powers of x. From (5.2.4) we obtain the coefficients for the three-term recurrence relationship (5.1.7). Thus, $\text{coeff}(x^n)=a_n=\frac{(2n)!}{2^n n!n!}\left(\frac{2n+1}{2}\right)^{\frac{1}{2}}$ and coeff $(x^{n-1})=b_n=0$ and hence

$$l_n(x)=\frac{1}{n}\left(\frac{2n+1}{(2n-1)(2n-3)}\right)^{\frac{1}{2}}((2n-1)(2n-3)^{\frac{1}{2}}xl_{n-1}(x)-(n-1)(2n-1)^{\frac{1}{2}}l_{n-2}(x)),$$

and, because

$$l_n(x)=\frac{(2n)!}{2^n n!n!}\left(\frac{2n+1}{2}\right)^{\frac{1}{2}}L_n(x),$$

$$L_n(x)=xL_{n-1}(x)-\frac{(n-1)^2}{(2n-1)(2n-3)}L_{n-2}(x). \quad (5.2.5)$$

These recurrence formulae need the initial values $L_0=1$, $L_1=x$ or $l_0=\sqrt{(1/2)}$, $l_1=\sqrt{(3/2)}x$.

5.2.2 Tschebyscheff polynomials

We consider the specific cases where $W(x)=1/\sqrt{(1-x^2)}$ and $W(x)=\sqrt{(1-x^2)}$. The corresponding orthogonal polynomials with respect to $(\cdot\,,\cdot)$ are now, respectively, the *Tschebyscheff polynomials* of the *first kind* and of the *second kind*. These polynomials can be derived in a manner similar to that for the Legendre polynomials. However, there is a simpler way, as we shall see in this section.

The Tschebyscheff polynomials of the first kind are denoted by $T_n(x)$ (the normalised polynomials by $t_n(x)$), and those of the second kind by $U_n(x)$ (the normalised polynomials by $u_n(x)$). These satisfy

$$\int_{-1}^{1}\frac{1}{(1-x^2)^{\frac{1}{2}}}T_n(x)T_m(x)\,\mathrm{d}x=0,$$

$$\int_{-1}^{1}(1-x^2)^{\frac{1}{2}}U_n(x)U_m(x)\,\mathrm{d}x=0 \quad \text{for} \quad n\neq m. \quad (5.2.6)$$

Under the substitution $x = \cos\varphi$ (5.2.6) becomes

$$\int_0^\pi T_n(\cos\varphi)T_m(\cos\varphi)\,d\varphi = 0,$$

$$\int_0^\pi \sin^2\varphi U_n(\cos\varphi)U_m(\cos\varphi)\,d\varphi = 0, \qquad n \neq m$$

which is valid if

$$T_n(\cos\varphi) = \cos(n\varphi), \qquad U_n(\cos\varphi) = \sin((n+1)\varphi)/\sin\varphi, \qquad n = 0(1)\ldots.$$

This is immediatly verified by calculation. $U_n(\cos\varphi)$ is indeed a polynomial and

$$\begin{aligned} U_n(\cos\varphi) &= \sin((n+1)\varphi)/\sin\varphi = \sin(n\varphi)\cos\varphi/\sin\varphi + \cos(n\varphi) \\ &= \sin((n-1)\varphi)\cos^2\varphi/\sin\varphi + \cos((n-1)\varphi)\cos\varphi + \cos(n\varphi) \\ &= \ldots \\ &= \sin\varphi\cos^n\varphi/\sin\varphi + \cos^{n-1}\varphi\cos\varphi \\ &\quad + \cdots + \cos((n-1)\varphi)\cos\varphi + \cos(n\varphi) \end{aligned}$$

or

$$\begin{aligned} U_n(x) &= x^n + x^{n-1}T_1(x) + x^{n-2}T_2(x) + \cdots + xT_{n-1}(x) + T_n(x) \\ &= \sum_{k=0}^{n} x^{n-k}T_k(x) \in \dot{\mathbb{P}}_n, \end{aligned}$$

because $T_n(x) = \cos(n \arccos x) \in \mathbb{P}_n$ and $T_1 = x$. Analogously, $U_n(x) = \sin((n+1)\arccos x)/\surd(1-x^2)$. From the properties of trigonometric functions we have that

$$T_0 = 1, \qquad t_0 = 1/\sqrt{\pi} \qquad U_0 = 1, \qquad u_0 = 1/\sqrt{\pi}$$

$$T_1 = x, \qquad t_1 = \sqrt{2/\pi}\,x \qquad U_1 = 2x, \qquad u_1 = \sqrt{2/\pi}\,x,$$

and, since $\cos((n\pm 1)\varphi) = \cos n\varphi \cos\varphi \mp \sin n\varphi \sin\varphi$, that

$$\cos((n+1)\varphi) + \cos((n-1)\varphi) = 2\cos n\varphi\cos\varphi$$

or

$$T_{n+1}(x) = 2xT_n(x) - T_{n-1}(x).$$

Similarly, since $\sin((n\pm 1)\varphi) = \sin n\varphi\cos\varphi \pm \cos n\varphi \sin\varphi$, or

$$\sin((n+1)\varphi))\sin\varphi + \sin((n-1)\varphi)/\sin\varphi = 2\sin(n\varphi)\cos\varphi/\sin\varphi$$

$$U_{n+1}(x) = 2xU_n(x) - U_{n-1}(x).$$

From these recurrence relationships we see that the leading coefficient of $T_n(x)$ is 2^{n-1} and that of $U_n(x)$ is 2^n. Further, from $\int_0^\pi \cos^2(n\varphi)\,d\varphi = \int_0^\pi \sin^2(n\varphi)\,d\varphi = \pi/2$ we have that $t_n(x) = \sqrt{2/\pi}T_n(x)$ and $u_n(x) = \sqrt{2/\pi}U_n(x)$ for $n > 0$.

The Tschebyscheff polynomials have the useful property that their zeros can be easily determined; $\cos n\varphi$ has the n zeros $\varphi_i = \frac{2i-1}{n}\frac{\pi}{2}$, $i = 1(1)n$ in $(0, \pi)$ and hence $T_n(x)$ has the n zeros $x_i = \cos\left(\frac{2i-1}{n}\frac{\pi}{2}\right)$ in $(-1, 1)$, $\sin((n+1)\varphi)/\sin\varphi$ has the n zeros $\varphi_i = i\pi/(n+1)$, $i = 1(1)n$ in $(0, \pi)$ and hence $U_n(x)$ has the n zeros $x_i = \sin(i\pi/(n+1))$ in $(-1, 1)$.

As proved in Section 3.4.1, $U_n/2^n$ minimises $\int_{-1}^1 |f(x)|\mathrm{d}x$ for the polynomials $f \in \dot{\mathbb{P}}_n$ of the form $f = x^n + a_{n-1}x^{n-1} + \cdots + a_0$ with leading coefficient 1.

5.2.3 Jacobi polynomials

The Legendre polynomials and the Tschebyscheff polynomials are special cases of the more general *Jacobi polynomials*, which are orthogonal polynomials with respect to the weight function $W(x) = (1-x)^\alpha(1+x)^\beta$, $\alpha, \beta > -1$ and to the inner product $(\cdot\,,\cdot)$. We denote them by $P_n^{\alpha,\beta}(x)$, and by $p_n^{\alpha,\beta}(x)$ if they are orthonormal.

The Rodriguez formula for the Jacobi polynomials is given by

$$p_n^{\alpha,\beta}(x) = \frac{A_n}{W(x)}\frac{\mathrm{d}^n}{\mathrm{d}x^n}(W(x)(1-x^2)^n) = \frac{A_n}{W(x)}\frac{\mathrm{d}^n}{\mathrm{d}x^n}((1-x)^{\alpha+n}(1+x)^{\beta+n}). \tag{5.2.7}$$

For $\alpha = \beta = 0$ we obtain the Legendre polynomials, and for $\alpha = \beta = \pm\frac{1}{2}$ the Tschebyscheff polynomials. Generally, for $\alpha = \beta$ the Jacobi polynomials are called *ultraspherical polynomials*. The Rodriguez formula gives

$$P_0^{\alpha,\beta}(x) = A_0$$
$$P_1^{\alpha,\beta}(x) = A_1\left[-2x + \frac{W'(x)}{W(x)}(1-x^2)\right] = A_1[-2x + (1-x)\beta - (1+x)\alpha]$$

because of $W'(x) = W(x)[(1-x)\beta - (1+x)\alpha]/(1-x^2)$. Hence $P_1^{\alpha,\beta}(x) = A_1[\beta - \alpha - (\alpha+\beta+2)x]$.

Using the Leibniz differentiation rule in (5.2.7) we have

$$P_n^{\alpha,\beta} = A_n\sum_{j=0}^n (-1)^j\binom{n}{j}\frac{\Gamma(\alpha+n+1)\Gamma(\beta+n+1)}{\Gamma(\alpha+n-j+1)\Gamma(\beta+j+1)}(1-x)^{n-j}(1+x)^j. \tag{5.2.8}$$

Requiring that $P_n^{\alpha,\beta}(1)=1$ we obtain $A_n=\dfrac{(-1)^n\Gamma(\alpha+1)}{2^n\Gamma(\alpha+n+1)}$.

The *orthonormal Jacobi polynomials* $p_n(x)$ are given by

$$p_n^{\alpha,\beta}=\frac{\Gamma(\alpha+n+1)}{n!\,\Gamma(\alpha+1)}\left(\frac{(2n+\alpha+\beta+1)n!\,\Gamma(n+\alpha+\beta+1)}{2^{\alpha+\beta+1}\Gamma(n+\alpha+1)\Gamma(n+\beta+1)}\right)^{\frac{1}{2}}P_n^{\alpha,\beta}.$$

The polynomials $P_n(x)$ satisfy the recurrence relationship

$$P_{n+1}^{\alpha,\beta}=\frac{\gamma[\alpha^2-\beta^2+(\gamma^2-1)x]P_n^{\alpha,\beta}-2(n+\alpha)(n+\beta)(\gamma+1)P_{n-1}^{\alpha,\beta}}{2(n+1)(\gamma-n)(\gamma-1)}$$

with $\gamma=2n+\alpha+\beta+1$ and the initial values are $P_0^{\alpha,\beta}=1$, $P_1^{\alpha,\beta}=\dfrac{1+\alpha}{2}(x+1)+\dfrac{1+\beta}{2}(x-1)$.

The proof of the last two statements is left to the reader (cf. Problem P.5.6). The three-term recurrence relationships are very useful for computer evaluations with non-numerical compilers. We present here a FORMAC program which lists the Jacobi polynomials.

```
JACPOL: PROC OPTIONS (MAIN);
FORMAC-OPTIONS:
PRINT-OUT (P(0)=1; P(1)=(1+A)/2*(X+1)+(1+B)/2*(X-1));
DO N=1 TO 8;
LET (N="N"; G=2*N+A+B+1);
PRINT-OUT (P(N+1)=(G*(A**2-B**2+(G**2-1)*X)*P(N)
            -2*(N+A)*(N+B)*(G+1)*P(N-1))/
            (2*(N+1)*(G-N)*(G-1)));
END;
END JACPOL;
```

5.2.4 Hermite polynomials and Laguerre polynomials

The Hermite polynomials (sometimes denoted as Hermite–Tschebyscheff polynomials) are a system of orthogonal polynomials for $W(x)=\exp(-x^2)$ with respect to the inner product $(f,g)=\int_{-\infty}^{\infty}e^{-x^2}f(x)g(x)\,dx$. We denote them by $H_n(x)$ or by $h_n(x)$ if they are normalised.

Rodriguez' formula now gives

$$H_n(x)=e^{x^2}\frac{d^n}{dx^n}(e^{-x^2}).$$

Hence, we have that $H_0=1$, $H_1=-2x$, $H_2=4x^2-2$, $H_3=-8x^3+12x$, etc. The leading coefficient is $(-2)^n$ as is easily proved by induction. Further,

$$(H_n, H_n)=(-2)^n \int_{-\infty}^{\infty} (-1)^n n!\, e^{-x^2}\, dx = n!2^n\sqrt{\pi}$$

so that

$$h_n(x)=H_n(x)/(n!2^n\sqrt{\pi})^{\frac{1}{2}}$$

and finally,

$$h_n=-\left(\frac{2}{n}\right)^{\frac{1}{2}} x h_{n-1}-\left(\frac{n-1}{n}\right)^{\frac{1}{2}} h_{n-2} \quad \text{and} \quad H_n=-2xH_{n-1}-2(n-1)H_{n-2}.$$

The polynomials which are orthogonal with respect to the weight function $W(x)=x^\alpha\, e^{-x}$, $\alpha>-1$, and with respect to the inner product $(f, g)=\int_0^\infty x^\alpha\, e^{-x} f(x)g(x)\, dx$, are called *Laguerre polynomials* (sometimes Laguerre–Tschebyscheff polynomials). We denote them by $\Lambda_n^\alpha(x)$ and by $\lambda_n^\alpha(x)$ if they are orthonormal.

Rodriguez' formula gives

$$\Lambda_n^\alpha(x)=x^{-\alpha} e^x \frac{d^n}{dx^n}(x^{n+\alpha} e^{-x})$$

Hence

$$\Lambda_0^\alpha=1, \qquad \Lambda_1^\alpha=-x+(\alpha+1), \qquad \Lambda_2^\alpha=x^2-2(\alpha+2)x+(\alpha+1)(\alpha+2).$$

$$(\Lambda_n^\alpha, \Lambda_n^\alpha)=(-1)^n \int_0^\infty x^n \frac{d^n}{dx^n}(x^{n+\alpha} e^{-x})\, dx = n!\,\Gamma(n+\alpha+1)$$

so that

$$\lambda_n^\alpha=(n!\,\Gamma(n+\alpha+1))^{-\frac{1}{2}}\Lambda_n^\alpha$$

and finally,

$$\lambda_n^\alpha(x)=-\left(\frac{1}{n(n+\alpha)}\right)^{\frac{1}{2}}(x-2n-\alpha+1)\lambda_{n-1}^\alpha(x)-\left(\frac{n-1}{n}\,\frac{n+\alpha-1}{n+\alpha}\right)^{\frac{1}{2}}\lambda_{n-2}^\alpha(x),$$

$$\Lambda_n^\alpha(x)=-(x-2n-\alpha+1)\Lambda_{n-1}^\alpha-(n-1)(n+\alpha-1)\Lambda_{n-2}^\alpha.$$

5.3 Implicitly-defined orthogonal polynomials

The orthogonal polynomials in Sections 5.1 and 5.2 are such that $W(x)$ does not depend on the orthogonal polynomials themselves. However, there exist important classes of orthogonal polynomials whose weight function does depend on ω_n so that $W(x)=W(x,\omega_n(x))=W(x,\omega_n)$. We say that W depends on ω_n if it depends on at least one zero of ω_n. Usually, however, $W(x,\omega_n)$ depend on the total of n zeros of ω_n.

5.3.1 Existence

We first prove the existence of such polynomials.

Theorem 5.3.1 *Let $W(x,\cdot)\geqslant 0$ (or $W(x,\cdot)\leqslant 0$) be a continuous weight function. Then there exists a polynomial $\omega_n \in \dot{\mathbb{P}}_n$ such that $\omega_n \perp \mathbb{P}_{n-1}$ with respect to $W(x,\omega_n(x))$.*

Proof Consider the space $\hat{\mathbb{P}}_n$ of polynomials with n distinct real zeros in $(-1,1)$. Without loss of generality we can assume that the leading coefficients of all polynomials which we are interested in is 1 (since, as before, we are only interested in the zeros of these polynomials).

For every $g_n \in \mathbb{P}_n$ there exists a system of orthogonal polynomials h_k, $k=1(1)n$ such that

$$(h_n, x^k)_g \equiv \int_0^1 W(x,g_n)h_n x^k \,\mathrm{d}x = 0 \quad \text{for} \quad k=0(1)n-1$$

and

$$(h_n, h_n)_g \neq 0.$$

This is a consequence of Theorem 5.1.1.

Hence, we have a mapping T_n: $\hat{\mathbb{P}}_n \to \hat{\mathbb{P}}_n$ which is continuous and unique. In order to see this we assume first that it is not unique. Then there exists a polynomial g_n for which at least two polynomials h_n and p_n satisfy $(h_n, x^k)_g = 0$ and $(p_n, x^k)_g = 0$, $k=0(1)n-1$, for the same g_n. Hence $(h_n - p_n, x^k)_g = 0$ for $k=0(1)n-1$. However, $h_n - p_n \in \mathbb{P}_{n-1}$. Because $p_n \perp \mathbb{P}_{n-1}$ and $h_n \perp \mathbb{P}_{n-1}$ we also have that $h_n - p_n \perp \mathbb{P}_{n-1}$. Hence, $h_n - p_n = 0$ identically and the mapping T_n is unique.

The mapping T_n is also continuous. Assume that this is wrong. We then choose a convergent sequence of polynomials $\omega_{n,m} \to \omega_n$ as $m \to \infty$ where $\omega_{n,m}$, $\omega_n \in \hat{\mathbb{P}}_n$. Then the mapping generates a sequence $h_{n,m} \in \hat{\mathbb{P}}_n$. Let h_n be the picture of ω_n. Let us denote $\|h_n\|^2 \equiv \int_0^1 W(x,\omega_n)h_n^2(x)\,\mathrm{d}x$. Then

$$\|h_{n,m} - h_n\|^2 = \int_0^1 W(x,\omega_n)(h_{n,m} - h_n)(h_{n,m} - h_n)\,\mathrm{d}x$$

where $g(x) \equiv h_{n,m} - h_n \in \mathbb{P}_{n-1}$. Hence

$$\|h_{n,m} - h_n\|^2 = \int_0^1 W(x,\omega_n)h_{n,m} g\,\mathrm{d}x - \int_0^1 W(x,\omega_n)h_n g\,\mathrm{d}x.$$

The last integral vanishes since $h_n \perp \mathbb{P}_{n-1}$. For the same reason

$$\int_0^1 W(x, \omega_{n,m})h_{n,m}g \, dx = 0,$$

and hence

$$\|h_{n,m} - h_n\|^2 = \int_0^1 W(x, \omega_n)h_{n,m}g \, dx - \int_0^1 W(x, \omega_{n,m})h_{n,m}g \, dx$$
$$= \int_0^1 (W(x, \omega_n) - W(x, \omega_{n,m}))h_{n,m}g \, dx \to 0 \quad \text{for} \quad m \to \infty$$

since $W(x, \omega)$ is continuous, and $h_{n,m}g$ is bounded. Hence T_n is a unique, continuous mapping.

In order to prove the existence of a fixed point $\omega_n(x) \in \hat{\mathbb{P}}_n$ of T_n such that

$$\int_0^1 W(x, \omega_n)\omega_n(x)x^k \, dx = 0, \qquad k = 0(1)n-1,$$

we notice the one-to-one correspondence between the space $\hat{\mathbb{P}}_n$ of polynomials with the zeros $0 < x_1 < x_2 < \cdots < x_n < 1$, and the space $S_n \subset \mathbb{R}^n$ of n-dimensional vectors whose components $x_1, \ldots, x_n$ satisfy $0 < x_1 < x_2 < \cdots < x_n < 1$. Clearly, S_n is an n-dimensional open simplex such as plotted in Fig. 5.3.1 for $n = 3$.

Hence we have that T_n also maps S_n into S_n. The mapping T_n cannot have a fixed point on the boundary ∂S_n of S_n. This can be seen as follows:

Assume that the polynomial $q_n \in \mathbb{P}_n \backslash \hat{\mathbb{P}}_n$ corresponds to a point of the

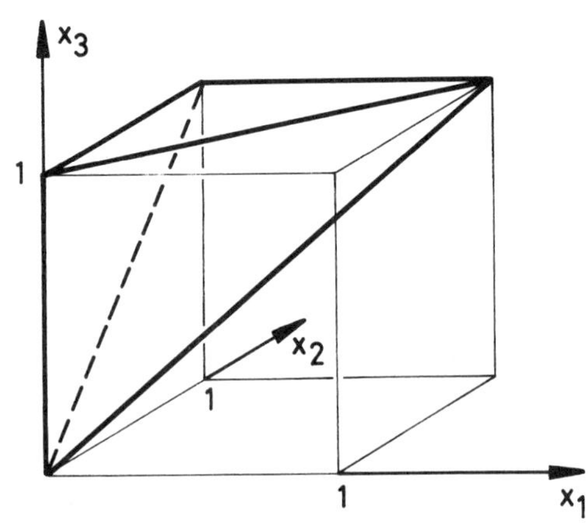

FIG. 5.3.1. Simplex S_3 with $0 < x_1 < x_2 < x_3 < 1$.

boundary ∂S_n of S_n. Nevertheless, $W(x, q_n)$ is an admissible weight function and hence T_n maps q_n to the polynomial $T_n q_n \in \hat{\mathbb{P}}_n$. Thus, q_n cannot be a fixed point of T_n and a fixed point can only lie in S_n. However, there must exist at least one fixed point of T_n in S_n as a consequence of the Brouwer fixed point theorem:

Theorem 5.3.2 (Brouwer's fixed point theorem) *Every continuous unique mapping T of any closed, bounded and convex subset S of the n-dimensional Euclidian space $\mathbb{R}^n$ possesses at least one fixed point in S.*

For a proof see [13] p. 11.

As we have already seen, S_n is bounded. It is also convex since for every two points $p, q \in S_n$ all intermediate points $ap+(1-a)q$, $0<a<1$, belong to S_n. Further, T_n is continuous and unique. However S_n is not closed. This is no longer necessary since we showed that there cannot exist a fixed point on ∂S_n. Thus Theorem 5.3.1 is proved. ■

Note that we did not prove the uniqueness of a fixed point. Possibly, T_n has several fixed points in S_n. We present:

Example 5.3.1 For the weight function $W(x, \omega_2)=(1+d)^2-x^2$ with $d=1-x_2=x_1+1$ $(x_2=-x_1)$ which is related to the interval $[-1, 1]$ the symmetric polynomial $\omega_2(x)=x^2-x_1^2$ must satisfy

$$\int_{-1}^{1} W(x, \omega_2)\omega_2(x)x^k \, dx=0 \quad \text{for} \quad k=0, 1.$$

From this equation we obtain (cf. Section 7.4.3) two different values of d and hence two different values of $x_1=-x_2$. Thus, the corresponding mapping T_2 has two fixed points in S_2 (and $\hat{\mathbb{P}}_2$). ■

In general it is impossible to construct the implicitly-defined orthogonal polynomials explicitly as we have done for the orthogonal polynomials in Section 5.2. The only way to determine them is by computing the zeros x_i, $i=1(1)n$ of $\omega_n \equiv \prod_{i=1}^{n}(x-x_i)$. To this end we proceed as in Section 5.1.3: we integrate $(\omega_n, 1)$ n times by parts and obtain the system of nonlinear equations ${}^{(k)}\omega_n(1)=0$, $k=1(1)n$, for the zeros x_i, $i=1(1)n$.

For quadratures the following theorem is very important because it guarantees certain minimum properties of the error coefficients of interpolatory quadratures (see Chapter 7).

Theorem 5.3.3 *Let $\omega_n \in \hat{\mathbb{P}}_n$ and $g_n \in \mathbb{P}_n$ where ω_n is a fixed point of T_n. Then*

$$\min_{g_n \in \mathbb{P}_n} \int_0^1 W(x, \omega_n) g_n^2(x)\, dx = \int_0^1 W(x, \omega_n)\omega_n^2\, dx$$

and the minimiser ω_n is unique.

Proof We have that

$$\begin{aligned}\int_0^1 W(x, \omega_n)(g_n^2 - \omega_n^2)\, dx &= \int_0^1 W(x, \omega_n)(g_n^2 - 2g_n\omega_n + \omega_n^2)\, dx \\ &\quad + 2\int_0^1 W(x, \omega_n)\omega_n(g_n - \omega_n)\, dx \\ &= \int_0^1 W(x, \omega_n)(g_n - \omega_n)^2\, dx \geqslant 0\end{aligned}$$

since $\omega_n \perp (g_n - \omega_n) \in \mathbb{P}_{n-1}$. Hence $\int_0^1 W(x, \omega_n) g_n^2\, dx \geqslant \int_0^1 W(x, \omega_n)\omega_n^2\, dx$, and equality arises for $g_n \equiv \omega_n$.

Assume that the minimiser ω_n is not unique. Then another $\tilde{\omega}_n \in \hat{\mathbb{P}}_n$ exists such that $\omega_n \not\equiv \tilde{\omega}_n$ and that

$$0 < \int_0^1 W(x, \omega_n)\omega_n^2\, dx = \int_0^1 W(x, \omega_n)\tilde{\omega}_n^2\, dx.$$

Hence

$$\int_0^1 W(x, \omega_n)(\omega_n^2 - \tilde{\omega}_n^2)\, dx = \int_0^1 W(x, \omega_n)(\omega_n - \tilde{\omega}_2)^2\, dx = 0$$

as seen before. However, since $W(x, \omega_n) \geqslant 0$ this is only possible if $\omega_n \equiv \tilde{\omega}_n$, and hence the minimiser is unique. Note that this holds for *every* fixed point of T_n. ■

5.3.2 Special cases

We choose $W(x, \omega_n) = W(x)\omega_n^{2s}(x)$ where $s > 0$ is an integer. We then obtain the *s-orthogonal polynomials* in the sense of Ghizzetti and Ossicini (book, p. 74 ff.). In this case existence can also be proved without using Theorem 5.3.2. Such polynomials first appeared in a paper by Turàn [14] for $s = 1$ in connection with quadrature formulae containing first and second derivatives of the integrand. The extension to $s > 1$ seems to be due to Ghizzetti and Ossicini.

Table 5.3.1 contains the zeros of the first s-orthogonal polynomials for $s = 1, 2$ and 3. Here $W(x, \omega_n) = \omega_n^{2s}(x)$.

TABLE 5.3.1.

s	n	Zeros
1	1	0
	2	±0.62921 11283 49909
	3	0;±0.81443 91855 77763
	4	±0.30550 35800 42618; ±0.87090 65246 47102
	5	0;±0.56086 74191 61637; ±0.92711 78696 09891
2	1	0
	2	±0.65002 77534 75829
	3	0;±0.82923 27380 69265
	4	±0.36592 43547 58919; ±0.89982 92126 75958
	5	0;±0.56903 24387 58243; ±0.93436 89752 99190
3	1	0
	2	±0.66151 46369 54320
	3	0;±0.83709 88523 58573
	4	±0.36966 62875 51275; ±0.90512 38320 65297
	5	0;±0.57330 37859 07091; ±0.93810 61928 43488

A further special case is connected with the quadratures that have optimal error constants in the sense of Section 3.5.3 for functions which are analytic in the circle $|z|<R=1/\varepsilon$. We shall see later that these quadratures are interpolatory quadratures with respect to the rational weight function

$$W(x, \omega_n)=\frac{1}{(\varepsilon^2x)^{2n}\omega_n^2(1/\varepsilon^2x)}=\frac{1}{(1-\varepsilon^2xx_1)^2\cdots(1-\varepsilon^2xx_n)^2}, \quad 0\leqslant\varepsilon<1,$$

where the zeros x_i of ω_n are inside $[0, 1]$. $W(x, \omega_n)$ is not singular for $x\in[0, 1]$. $W(x, \omega_n)$ has n distinct real poles of order 1 outside $[0, R]$.

In order to determine the zeros of $\omega_n(x)$ we proceed by induction. We have to evaluate the integrals

$$\int_0^1 \prod_{j=1}^{n} \frac{x-x_j}{(1-\varepsilon^2xx_j)^2}\, x^k\, dx=0, \qquad k=0(1)n-1.$$

We define

$$g(x)\equiv\prod_{j=1}^{n}\frac{x-x_j}{(1-\varepsilon^2xx_j)^2}=\frac{\omega_n(x)}{(\varepsilon^2x)^{2n}\omega_n^2(1/\varepsilon^2x)}.$$

In order to integrate $g(x)x^k$, we first expand $g(x)$ into partial fractions:

$$g(x)=\sum_{i=1}^{n}\left(\frac{A_i}{1-\varepsilon^2xx_i}+\frac{B_i}{(1-\varepsilon^2xx_i)^2}\right)=\frac{\omega_n(x)}{(\varepsilon^2x)^{2n}\omega_n^2(1/\varepsilon^2x)}.$$

so that

$$g(x)\prod_{j=1}^{n}(1-\varepsilon^2xx_j)^2=\sum_{i=1}^{n}\left[(A_i(1-\varepsilon^2xx_i)+B_i)\prod_{\substack{j=1\\ \neq i}}^{n}(1-\varepsilon^2xx_j)^2\right]=\omega_n(x). \tag{5.3.1}$$

Substituting $x=1/\varepsilon^2x_k$ we have that

$$\omega_n(1/\varepsilon^2x_k)=\sum_{i=1}^{n}\left[(A_i(1-x_i/x_k)+B_i)\prod_{\substack{j=1\\ \neq i}}^{n}(1-x_j/x_k)^2\right]$$

$$=B_k\prod_{\substack{j=1\\ \neq k}}^{n}(1-x_j/x_k)^2=B_k[\omega_n'(x_k)]^2/x_k^{2n-2}.$$

Hence

$$B_k=[x_k^{n-1}/\omega_n'(x_k)]^2\omega_n(1/\varepsilon^2x_k)=\frac{1-\varepsilon^2x_k^2}{\varepsilon^2x_k}g_k(1/\varepsilon^2x_k),\qquad k=1(1)n, \tag{5.3.2}$$

where

$$g_k(x)\equiv\frac{(1-\varepsilon^2xx_k)^2}{x-x_k}g(x).$$

In order to calculate A_k we differentiate (5.3.1) and again substitute $x=1/\varepsilon^2x_k$. Thus,

$$\left[g(x)\prod_{j=1}^{n}(1-\varepsilon^2xx_j)^2\right]'_{x=1/\varepsilon^2x_k}=\omega_n'(1/\varepsilon^2x_k)=0$$

and

$$\left(\sum_{i=1}^{n}(A_i(1-\varepsilon^2xx_i)+B_i)\prod_{\substack{j=1\\ \neq i}}^{n}(1-\varepsilon^2xx_j)^2\right)'_{x=1/\varepsilon^2x_k}=0$$

$$=\sum_{i=1}^{n}\left\{-\varepsilon^2x_iA_i\prod_{\substack{j=1\\ \neq i}}^{n}(1-x_j/x_k)^2+(A_i(1-x_i/x_k)+B_i)\left(\prod_{\substack{j=1\\ \neq i}}^{n}(1-\varepsilon^2xx_j)^2\right)'_{x=1/\varepsilon^2x_k}\right\}$$

since

$$\left(\prod_{\substack{j=1\\ \neq i}}^{n}(1-\varepsilon^2xx_j)^2\right)'_{x=1/\varepsilon^2x_k}=\frac{2\varepsilon^2}{x_k^{2n-3}}\omega'(x_k)$$

$$\times\{(n-1)\omega_n'(x_k)-x_k\omega_n''(x_k)\}=0\quad\text{for}\quad i=k,$$

and 0 for $i \neq k$, we finally have that

$$\omega_n'(1/\varepsilon^2 x_k) = \sum_{i=1}^{n} \varepsilon^2 \Big\{ -x_i A_i \prod_{\substack{j=1\\ \neq i}}^{n} (1 - x_j/x_k)^2$$

$$+ (A_i(1 - x_i/x_k) + B_i) \frac{2}{x_k^{2n-3}} \omega_n'(x_k)$$

$$\times [(n-1)\omega_n'(x_k) - x_k \omega_n''(x_k)] \Big\}$$

$$= \frac{-\varepsilon^2 A_k}{x_k^{2n-3}} [\omega_n'(x_k)]^2 + \frac{2\varepsilon^2 B_k}{x_k^{2n-3}} \omega_n'(x_k) [(n-1)\omega_n'(x_k) - x_k \omega_n''(x_k)].$$

Hence

$$A_k = 2B_k \frac{(n-1)\omega_n'(x_k) - x_k \omega_n''(x_k)}{\omega_n'(x_k)} - \frac{\omega_n'(1/\varepsilon^2 x_k)}{\varepsilon^2 x_k} \left[\frac{x_k^{n-1}}{\omega_n'(x_k)} \right]^2$$

or

$$A_k = \frac{B_k}{x_k} \left(2x_k(n-1) - 2x_k^2 \frac{\omega_n''(x_k)}{\omega_n'(x_k)} - \frac{\omega_n'(1/\varepsilon^2 x_k)}{\varepsilon^2 \omega_n(1/\varepsilon^2 x_k)} \right) \tag{5.3.3}$$

by substituting (5.3.2). We calculate the various terms of (5.3.3)

$$\omega_n'(x) = \sum_{k=1}^{n} \prod_{\substack{j=1\\ \neq k}}^{n} (x - x_j) \quad \text{for} \quad \omega_n'(x_k) = \prod_{\substack{j=1\\ \neq k}}^{n} (x_k - x_j)$$

$$\omega_n''(x) = \sum_{i=1}^{n} \sum_{\substack{k=1\\ \neq i}}^{n} \prod_{\substack{j=1\\ \neq k,i}}^{n} (x - x_j),$$

hence

$$\omega_n''(x_k) = \sum_{\substack{i=1\\ \neq k}}^{n} \prod_{\substack{j=1\\ \neq k,i}}^{n} (x_k - x_j) = \omega_n'(x_k) \sum_{\substack{i=1\\ \neq k}}^{n} \frac{1}{x_k - x_i}$$

$$\omega_n(1/\varepsilon^2 x_k) = \frac{1}{(\varepsilon^2 x_k)^n} \prod_{j=1}^{n} (1 - \varepsilon^2 x_k x_j),$$

$$\omega_n'(1/\varepsilon^2 x_k) = \sum_{l=1}^{n} \frac{1}{(\varepsilon^2 x_k)^{n-1}} \prod_{\substack{j=1\\ \neq l}}^{n} (1 - \varepsilon^2 x_k x_j)$$

and obtain

$$\frac{\omega_n''(x_k)}{\omega_n'(x_k)} = \sum_{\substack{i=1\\ \neq k}}^{n} \frac{1}{x_k - x_i} \quad \text{and} \quad \frac{\omega_n'(1/\varepsilon^2 x_k)}{\omega_n(1/\varepsilon^2 x_k)} = \varepsilon^2 x_k \sum_{i=1}^{n} \frac{1}{1 - \varepsilon^2 x_k x_i}.$$

Replacing $n-1$ by $\sum_{\substack{i=1\\ \neq k}}^{n} 1$ we have, finally, that

$$A_k = -B_k\left(2\sum_{\substack{i=1\\ \neq k}}^{n} \frac{x_i}{x_k - x_i} + \sum_{i=1}^{n} \frac{1}{1-\varepsilon^2 x_k x_i}\right)$$

$$= -B_k\left[\frac{1}{1-\varepsilon^2 x_k^2} + \sum_{\substack{i=1\\ \neq k}}^{n} \left(\frac{1}{1-\varepsilon^2 x_k x_i} + \frac{2x_i}{x_k - x_i}\right)\right]. \quad (5.3.4)$$

After having calculated the expansion of $g(x)$ into partial fractions, we are able to establish the system of equations for the nodes x_i, $i = 1(1)n$.

The orthogonality conditions are ${}^{(k)}\omega_n(1) = 0$ for $k = 1(1)n$. For the sake of simplicity we restrict our attention to the special interval $[0, 1]$ and to $\varepsilon = 1$. More general cases are treated similarly.

In the first step we obtain

$$'g(x) = \sum_{i=1}^{n} \frac{1}{x_i}\left\{-A_i \ln(1-xx_i) + B_i\left[\frac{1}{1-xx_i} - 1\right]\right\}$$

and hence

$$'g(1) = 0 = \sum_{i=1}^{n} \frac{1}{x_i}\left\{-A_i \ln(1-x_i) + B_i\left[\frac{1}{1-x_i} - 1\right]\right\}.$$

We integrate $'g(x)$ and obtain

$$''g(x) = \sum_{i=1}^{n} \frac{1}{x_i^2}\{A_i[(1-xx_i)\ln(1-xx_i) + 1 - (1-xx_i)] - B_i[xx_i + \ln(1-xx_i)]\}.$$

By subtraction of $x \cdot {}'g(x)$ the most inconvenient term with respect to further integration cancels, and we have that

$$''g(x) - x\cdot{}'g(x) = \sum_{i=1}^{n} \frac{1}{x_i^2}\left\{A_i[\ln(1-xx_i) + xx_i] - B_i\left[\frac{xx_i}{1-xx_i} + \ln(1-xx_i)\right]\right\} \quad (5.3.5)$$

and

$$''g(1) = 0 = \sum_{i=1}^{n} \frac{1}{x_i^2}\left\{A_i[\ln(1-x_i) + x_i] - B_i\left[\frac{x_i}{1-x_i} + \ln(1-x_i)\right]\right\}.$$

In the next step we obtain

$$\int_0^x (''g(t) - t\cdot{}'g(t))\,dt = 2\cdot{}''g(x) - x\cdot{}''g(x)$$

$$= \sum_{i=1}^{n} \frac{1}{x_i^3}\{-A_i([1-xx_i)\ln(1-xx_i) - [(1-xx_i) - 1] - \tfrac{1}{2}x^2x_i^2]$$

$$+ B_i[\ln(1-xx_i) + (1-xx_i)\ln(1-xx_i) - [(1-xx_i) - 1] + xx_i]\}. \quad (5.3.6)$$

Multiplying (5.3.5) by x and subtracting from (5.3.6) we have that

$$2\cdot{}''g(x)-2x\cdot{}''g(x)+x^2\cdot{}'g(x)=\sum_{i=1}^{n}\frac{1}{x_i^3}\left\{-A_i\left[\ln(1-xx_i)+xx_i+\frac{1}{2}(xx_i)^2\right]\right.$$

$$\left.+B_i\left[2\ln(1-xx_i)+\frac{1}{1-xx_i}-1+xx_i\right]\right\}.$$

Note that

$$'g(x)=\int_0^x g(t)\,\mathrm{d}t,\qquad ''g(x)-x\cdot{}'g(x)=\int_0^x tg(t)\,\mathrm{d}t,$$

$$2\cdot{}''g(x)-2x\cdot{}''(x)+x^2\cdot{}'g(x)=\int_0^x t^2g(t)\,\mathrm{d}t.$$

Proceeding in this way we find that

$$\int_0^x t^m g(t)\,\mathrm{d}t=\sum_{k=0}^{m}{}^{(k+1)}g(x)(-1)^k k!\binom{m}{k}x^{m-k}=$$

$$(-1)^m[m!\,{}^{(m+1)}g(x)-m!\,{}^{(m)}g(x)x+\tfrac{1}{2}m!\,{}^{(m-1)}g(x)x^2+\cdots]$$

Since ${}^{(j)}g(1)=0$ for $j=1(1)m$ we have that ${}^{(m+1)}g(1)=0$ is equivalent to $\int_0^1 t^m g(t)\,\mathrm{d}t=0$ and, by induction,

$$\int_0^x t^{m-1}g(t)\,\mathrm{d}t=\sum_{i=1}^{n}\frac{(-1)^m}{x_i^m}\left\{A_i[\ln(1-xx_i)+p_{m-1}(xx_i)]\right.$$

$$\left.-B_i\left[(m-1)\ln(1-xx_i)+\frac{1}{1-xx_i}+q_{m-2}(xx_i)\right]\right\}.$$

Here $p_{m-1}(x)\in\mathbb{P}_{m-1}$ is the polynomial part of $\ln(1-x)$ and $q_{m-2}(x)\in\mathbb{P}_{m-2}$ is the polynomial part of $(m-1)\cdot\ln(1-x)+1/(1-x)$. Hence $p_0=0$, $\ln(1-xx_i)+p_{m-1}(xx_i)=\sum_{k=m}^{\infty}a_k(xx_i)^k$ with $a_m\neq0$, for $m>1$, and $q_{-1}=-1$; $q_0=0$; $m\ln(1-xx_i)+1/(1-xx_i)+q_{m-1}(xx_i)=$

$$\sum_{k=m}^{\infty}b_k(xx_i)^k\quad\text{with}\quad b_m\neq0,\quad\text{for}\quad m\geqslant2.$$

Introducing the series expansions of $\ln(1-xx_i)$ and $1/(1-xx_i)$ we obtain

$$\ln(1-xx_i)+p_{m-1}(xx_i)=-\sum_{k=1}^{\infty}\frac{(xx_i)^k}{k}+\sum_{k=0}^{m-1}a_k(xx_i)^k=O(x^m)$$

and from this equation the coefficients $a_0=0$, $a_k=1/k$, $k>0$ of $p_{m-1}(x)$ are obtained. By a similar calculation with

$$(m-1)\ln(1-xx_i)+1/(1-xx_i)+q_{m-2}(xx_i)=-(m-1)\sum_{k=1}^{\infty}\frac{(xx_i)^k}{k}$$

$$+\sum_{k=0}^{\infty}(xx_i)^k+\sum_{k=0}^{m-2}b_k(xx_i)^k=O(x^{m-1})$$

we obtain the coefficients $b_0=-1$ and $b_j=(m-j-1)/j$ for $j=1(1)m-2$ by comparing the powers of x. Hence

$$q_{m-2}(xx_i)=-1+(m-2)xx_i+\frac{m-3}{2}(xx_i)^2$$

$$+\frac{m-4}{3}(xx_i)^3+\cdots+\frac{1}{m-2}(xx_i)^{m-2}.$$

We collect these results:

Theorem 5.3.4 *The zeros of the orthogonal polynomials ω_n with respect to the weight function $1/(x^n\omega_n(1/x))^2$ and to the interval $[0,1]$ are the solutions of the system of equations*

$$\sum_{i=1}^{n}\frac{1}{x_i^m}\Big\{A_i[\ln(1-x_i)+p_{m-1}(x_i)]$$

$$-B_i\Big[(m-1)\ln(1-x_i)+\frac{1}{1-x_i}+q_{m-2}(x_i)\Big]\Big\}=0 \quad \text{for} \quad m=1(1)n \quad (5.3.7)$$

where A_i and B_i are the coefficients of the fractional part representation of $\omega_n(x)/[x^n\omega_n(1/x)]^2$, which are given by (5.3.2) and (5.3.4) for $\varepsilon=1$.

The derivation of (5.3.7) for $\varepsilon\neq 1$ for $[0,1]$ or for $[-1,1]$ can be performed in the same way. However, the equations are more complicated.

For practical calculations the coefficients of A_i and B_i can be expressed in another way. Let us define

$$s_m(x_i)\equiv x_i^{-m}[\ln(1-x_i)+p_{m-1}(x_i)]=-\sum_{k=m}^{\infty}\frac{1}{k}x_i^{k-m}$$

and

$$t_m(x_i)\equiv x_i^{-m}\Big[(m-1)\ln(1-x_i)+\frac{1}{1-x_i}+q_{m-2}(x_i)\Big]$$

$$=\frac{1}{x_i}\sum_{k=m-1}^{\infty}\Big(-\frac{m-1}{k}x_i^{k-m+1}+x_i^{k-m+1}\Big)=\frac{1}{x_i}\Big((m-1)s_{m-1}(x_i)+\frac{1}{1-x_i}\Big).$$

Hence we have the system

$$\sum_{i=1}^{n}\Big\{A_i s_m(x_i)+B_i\Big[-(m-1)s_m(x_i)+\frac{1}{1-x_i}\Big]\Big\}=0, \qquad m=1(1)n,$$

TABLE 5.3.2. Zeros of the implicitly-defined orthogonal polynomials $\omega_n(x)$ for

$$W(x, \omega_n) = \left[\frac{1}{(1-xx_1)(1-xx_2)\cdots(1-xx_n)}\right]^2.$$

n	zeros
1	6.83802623752020681154195788794 (−1)
2	4.847005558887724664064183034610 (−1)
	9.557919082340107418998201515215 (−1)
3	3.718781864605394925256213768208 (−1)
	8.848695349416572124830483404107 (−1)
	9.922658962997714786841630705824 (−1)
4	3.006594816073145588894403459037 (−1)
	8.130009109679515232610786600264 (−1)
	9.740785334417990073746577464926 (−1)
	9.983331592549651202923011092337 (−1)
5	2.519193914799296554123216815317 (−1)
	7.472011462748665950036650982440 (−1)
	9.494099767422589670329268755786 (−1)
	9.934257850989132638806072395657 (−1)
	9.995811030712120731876180264749 (−1)
6	2.165672464183651782206950725286 (−1)
	6.888830064445490820678065398640 (−1)
	9.212465638828346680931947986873 (−1)
	9.854569462661215806912573039316 (−1)
	9.981402071064274019646751578544 (−1)
	9.998817592877494702930913955957 (−1)
7	1.897957017907582060466422641476 (−1)
	6.377011475539656025611154521939 (−1)
	8.915504258538161228205771813910 (−1)
	9.749595041263052033714152075343 (−1)
	9.954800752647562307683108133430 (−1)
	9.994245203975243970731425924843 (−1)
	9.999634345380904851952241943689 (−1)
8	1.688398607581053925741326781971 (−1)
	5.928231416504065978260404817013 (−1)
	8.615471567488862536858657185855 (−1)
	9.625091050429875452931844477089 (−1)
	9.915906449434648417242370470479 (−1)
	9.984925066642247046945380694918 (−1)
	9.998883176199769577109427722500 (−1)
	9.999878228610345579105211179482 (−1)

or, by substituting (5.3.4) for $\varepsilon = 1$,

$$\sum_{i=1}^{n} B_i\left\{-s_m(x_i)\left[m-\frac{x_i^2}{1-x_i^2}\right.\right.$$

$$\left.\left.+\sum_{\substack{j=1\\ \neq i}}^{n}\left(\frac{1}{1-x_ix_j}+\frac{2x_j}{x_i-x_j}\right)\right]+\frac{1}{1-x_i}\right\}=0, \qquad m=1(1)n. \quad (5.3.8)$$

From practical calculations there results the interesting heuristic approximation

$$x_i^{(n)} \cong 1-\exp\left(\frac{-i(i+\alpha_i)^i\sqrt{i}}{n+1}\right), \qquad i=1(1)n$$

with $\alpha_1 = 0.65$, $\alpha_2 = 0.85$, $\alpha_3 = 1.1$, $\alpha_4 \cdots \alpha_{n-2} = 1.2$, $\alpha_{n-1} = 1.3$, $\alpha_n = 1.8$ which especially give the right number of nines (cf. Table 5.3.2). So we obtain good starting values for Newton's iteration procedure applied to (5.3.8). The numerical condition of (5.3.8) is extremely bad. Hence we give only the zeros for $n = 1(1)8$.

5.4 Numerical calculation of the zeros of orthogonal polynomials

As indicated in Section 5.3, we shall use the zeros of suitable orthogonal polynomials as nodes for quadrature formulae; we therefore need reliable methods for computing these zeros. The most effective tools for this purpose are embedding methods, asymptotic expansions, symmetry properties, and last but not least the separation property of two consecutive orthogonal polynomials together with iterative methods. These last usually need very good starting values in order to achieve convergence.

5.4.1 Extrapolation techniques

In order to use the iterative methods we proceed as follows. Assume we have computed the zeros of some consecutive orthogonal polynomials. Plotting them, we obtain a sketch as in Fig. 5.4.1. If we combine, for example, the smallest zeros of these consecutive polynomials we obtain a monotonic function as shown in Fig. 5.4.2, and extrapolation to the next value of n is easily possible. The same is true for further zeros following the smallest one, and we can repeat this procedure by starting with the largest zero and going on with the following smaller zeros.

Let us assume that we shall perform such an extrapolation process. We then have to decide what kind of interpolation we should take. Clearly,

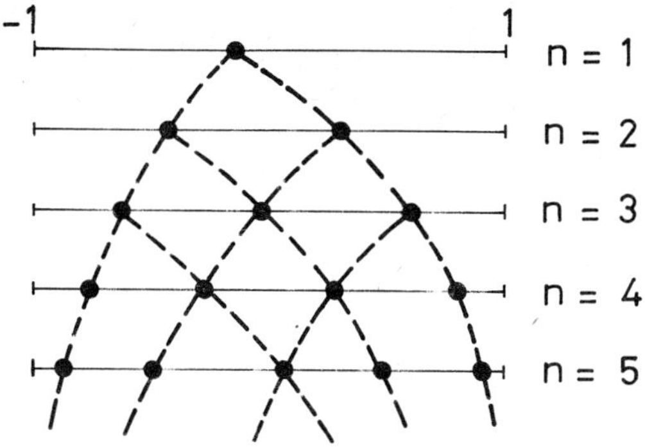

FIG. 5.4.1. Zeros of consecutive orthogonal polynomials.

for example, $x_1^{(n)} \to 0$ monotonically for $n \to \infty$. Hence we should choose an interpolation which, if not strictly then at least approximately, satisfies this property. Further, we do not necessarily need interpolation since approximation may be at least as good.

Example 5.4.1 Calculate approximate values for the largest zeros $x_n^{(n)}$ of the orthogonal polynomials with respect to the weight function $W(x)=1-x$ for $x \in [-1, 1]$.

Rational extrapolation by $f(n)=(a+nb)/(c+n) \approx x_n^{(n)}$ where the coefficients a, b, c are determined (by requiring that $f_{n-i}=f(n-i)=x_{n-i}^{(n-i)}$ for $i=1, 2, 3$) to be $c=-(n-1)-2(f_{n-2}-f_{n-3})/(f_{n-1}-2f_{n-2}+f_{n-3})$, $b=f_{n-2}+(c+n-1)(f_{n-1}-f_{n-2})$, and $a=(n-1+c)f_{n-1}-(n-1)b$ gives the

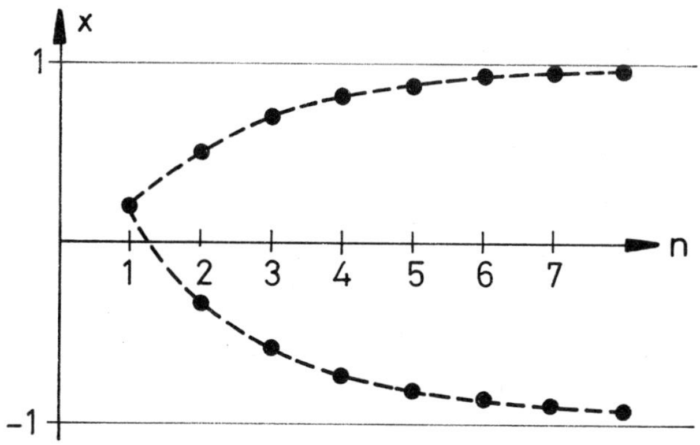

FIG. 5.4.2. Extrapolation of the largest and smallest zero of consecutive orthogonal polynomials.

following values:

n	$x_n^{(n)}$	f_n
3	0.822 824 081	
4	0.885 791 608	
5	0.920 380 286	
6	0.941 367 146	0.942 2470
7	0.955 041 227	0.955 4571

Obviously the extrapolated approximate values $f_n \approx x_n^{(n)}$ are rather good. Moreover, the following simple rule is often acceptable: add $x_0^{(n)} = -1$ and $x_{n+1}^{(n)} = 1$ to the zeros of ω_n. Then the averages $(x_i^{(n)} + x_{i+1}^{(n)})/2$, $i = 0(1)n$ approximate the zeros $x_i^{(n+1)}$ of ω_{n+1}. ■

5.4.2 Embedding methods

Assume that we know the zeros of the orthogonal polynomials with respect to $W(x)$ and we want to compute those for $\tilde{W}(x) \neq W(x)$. We then consider the orthogonal polynomials with respect to the weight functions $W_\varepsilon \equiv (1-\varepsilon)W(x) + \varepsilon \tilde{W}(x)$, $0 \leqslant \varepsilon \leqslant 1$. For $\varepsilon = 0$ we have the known zeros corresponding to $W(x)$ and for $\varepsilon = 1$ we obtain the unknown zeros. Hence, for $\varepsilon_1 \ll 1$ we can expect that the nodes differ only slightly from the known zeros. Then the zeros for $\varepsilon = 0$ are good starting values for the zeros of the polynomials belonging to $\varepsilon = \varepsilon_1$. Having calculated these zeros we increase ε_1 and do the same for ε_1 instead of ε. In this way we usually reach $\varepsilon = 1$ in a small number of steps. This embedding method is not only simple but also powerful. Its disadvantage is the large amount of work necessary if the steps happen to become very small.

Example 5.4.2 The zeros of the Legendre polynomial $L_2(x) \perp \mathbb{P}_1$ with respect to $W(x) \equiv 1$ are $\pm 1/\sqrt{3}$. We wish to find the zeros of the Jacobi polynomial $P_2^{0,1}(x) \perp \mathbb{P}_1$ with respect to $W(x) = 1 + x$. Hence, we consider the weight function $W_\varepsilon(x) = 1 - \varepsilon + \varepsilon(1+x) = 1 + \varepsilon x$. Then the system

$$\int_{-1}^{1} (1+\varepsilon x)(a + bx + x^2)x^\nu \, dx = 0, \qquad \nu = 0, 1$$

has to be solved for some increasing values of ε. Although this can be performed explicitly in this simple case, this is generally not so, for example for the implicitly-defined orthogonal polynomials of the preceding section. Here we have the results:

ε	x_1	x_2
0	−0.57735	0.57735
0.2	−0.54783	0.60188
0.4	−0.51066	0.62333
0.6	−0.46157	0.64339
0.8	−0.39299	0.66412
1.0	−0.28990	0.68990

5.4.3 Asymptotic expansions of the zeros of Legendre polynomials

The methods described previously are very useful for establishing a table for the zeros of (nearly) all consecutive polynomials of a system of orthogonal polynomials. If we are interested in the zeros of only one fixed $\omega_n(x)$, then we need different methods. Principally we can use all special algorithms for the calculation of zeros of polynomials. They should be assumed to work well because the nodes are all real and distinct. However, notice that for large n we have only one big cluster of zeros in $(-1, 1)$. Hence, the amount of work can be large and the loss of accuracy due to round-off errors may be dramatic. For the Legendre polynomials we now calculate asymptotic expansions of the zeros, which give very good approximations of the zeros of the L_n without any knowledge about $L_0, \ldots, L_{n-1}$. For the sake of simplicity we choose first the interval $[0, 1]$. Then, from the Rodriguez formula (cf. Section 5.2.1), we derive explicitly

$$\begin{aligned}
L_n(x) &= \frac{(-1)^n}{n!}\frac{d^n}{dx^n}(x^n(1-x)^n) = \sum_{k=0}^{n}\binom{n}{k}\binom{n+k}{k}(-1)^{n+k}x^k \\
&= 1-(n(n+1)x)+\frac{1}{(2!)^2}(n(n+1)x)^2\left(1-\frac{1\cdot 2}{n(n+1)}\right) \\
&\quad -\frac{1}{(3)!^2}(n(n+1)x)^3\left(1-\frac{1\cdot 2}{n(n+1)}\right)\left(1-\frac{2\cdot 3}{n(n+1)}\right)\pm\ldots \\
&= 0 \text{ for every zero } x = x_i,\ i = 1(1)n.
\end{aligned}$$

From this representation it can be concluded that for every zero there should hold an asymptotic expansion of the form

$$x = x_i = \frac{a_i}{n(n+1)}+\frac{b_i}{n^2(n+1)^2}+\frac{c_i}{n^3(n+1)^3}+\ldots, \qquad i = 1(1)n.$$

Introducing $x = a/(n(n+1))$ and rearranging in powers of $1/(n(n+1))$ we obtain by induction

$$L_n\left(\frac{a}{n(n+1)}\right) = 1 - a + \frac{a^2}{2!2!} - \frac{a^3}{3!3!} + \frac{a^4}{4!4!} - + \ldots + \frac{(-1)^n}{n!n!} a^n$$

$$- \frac{1}{n(n+1)}\left[\frac{1 \cdot 2}{2!2!} a^2 - \frac{1 \cdot 2 + 2 \cdot 3}{3!3!} a^3 + \frac{1 \cdot 2 + 2 \cdot 3 + 3 \cdot 4}{4!4!} a^4 - + \ldots\right]$$

$$+ \frac{1}{n^2(n+1)^2}\left[\frac{1 \cdot 2 \cdot 2 \cdot 3}{3!3!} a^3\right.$$

$$\left. - \frac{1 \cdot 2 \cdot 2 \cdot 3 + 1 \cdot 2 \cdot 3 \cdot 4 + 2 \cdot 3 \cdot 3 \cdot 4}{4!4!} a^4 + \ldots\right] + O(1/n^6).$$

We denote the first sum by $s_n(a) \equiv \sum_{k=0}^{n} \frac{(-a)^k}{k!k!}$, and look for positive zeros a of $s_n(a)$. We find the following result:

n	Smallest positive zero
1	1
2	2 (double zero)
3	1.429 993 47
4	1.446 776 58
5	1.445 756 87
6	1.445 797 67
7	1.445 796 46
8	1.445 796 49

For small values of n the nodes of the Legendre polynomials can be easily obtained from tables. Hence we are interested in large n. The result shows that for large n we can drop the dependence of n and replace $s_n(a)$ by the series

$$s(a) = \sum_{k=0}^{\infty} (-1)^k \frac{a^k}{k!k!} = J_0(2\sqrt{a}) \tag{5.4.1}$$

where $J_0(x)$ is the Bessel function with index zero. $J_0(x)$ has infinitely many real positive zeros. Following our considerations, the first n zeros of $J_0(2\sqrt{a})$ may approximate the n zeros of $L_n(x)$. But because of the symmetry property $x_i^{(n)} = 1 - x_{n-i+1}^{(n)}$, $i = 1(1)[n/2]$. If n is odd, then 1/2 is also a zero. Hence, it suffices to consider the first $[n/2]$ zeros of $J_0(2\sqrt{a})$ in order to obtain approximations of $x_i^{(n)}$ for $i=1(1)[n/2]$. Clearly these approximations are not very good, particularly as i increases, because

$$L_n\left(\frac{a}{n(n+1)}\right)=J_0(2\sqrt{a})+O\left(\frac{1}{n(n+1)}\right).$$

We now improve our first approximation of the zeros and we replace $a/(n(n+1))$ by

$$r=\frac{a}{n(n+1)}+\frac{b}{n^2(n+1)^2}+\frac{c}{n^3(n+1)^3}+\dots.$$

Then, in the same way, we find that

$$\begin{aligned}L_n(r)=s(a)&+\frac{1}{n(n+1)}\left[-b+\frac{2ab}{2!2!}-\frac{1\cdot 2}{2!2!}a^2-\frac{3}{3!3!}a^2b+\frac{1\cdot 2+2\cdot 3}{3!3!}a^3\right.\\ &\left.+\frac{4}{4!4!}a^4b-\frac{1\cdot 2+2\cdot 3+3\cdot 4}{4!4!}a^4\pm\dots\right]\\ &+\frac{1}{n^2(b+1)^2}\left[-c+\frac{b^2}{2!2!}-\frac{3ab^2}{3!3!}-\frac{3a^2c}{3!3!}+3\frac{1\cdot 2+2\cdot 3}{3!3!}a^2b\right.\\ &+\frac{2ac}{2!2!}-2\frac{1\cdot 2}{2!2!}ab-\frac{1\cdot 2\cdot 2\cdot 3}{3!3!}a^3+\frac{6a^2b^2}{4!4!}+\frac{4a^3c}{4!4!}\\ &\left.-4\frac{1\cdot 2+2\cdot 3+3\cdot 4}{4!4!}a^3b+\frac{1\cdot 2\cdot 2\cdot 3+1\cdot 2\cdot 3\cdot 4+2\cdot 3\cdot 3\cdot 4}{4!4!}a^4+\dots\right]\\ &+O(1/n^6)\end{aligned}$$

We now study the coefficients:

$$\begin{aligned}\text{coeff}\,(1/n(n+1))=&\left[-\frac{1\cdot 2}{2!2!}a^2+\frac{1\cdot 2+2\cdot 3}{3!3!}a^3-\frac{1\cdot 2+2\cdot 3+3\cdot 4}{4!4!}a^4\pm\dots\right]\\ &+b\left[-1+\frac{1}{1!2!}a-\frac{1}{2!3!}a^2+\frac{1}{3!4!}a^3-+\dots\right].\end{aligned}$$

The first sum is denoted by $t(a)$, the second one is immediately seen to be $bs'(a)$ when compared with (5.4.1). Hence coeff $(1/n(n+1))=t(a)+bs'(a)$. Here, we again extend the sums into finite series as (5.4.1) because we are mainly interested in large values of n. In particular, we have that

$$t(a)=-\sum_{k=2}^{\infty}\frac{(-a)^k}{k!k!}\left(\sum_{j=1}^{k-1}j(j+1)\right).$$

The interior sum can be evaluated in closed form to give

$$\sum_{j=1}^{k-1} j(j+1)=\sum_{j=1}^{k-1} j^2+\sum_{j=1}^{k-1} j=\frac{(k-1)k(2k-1)}{6}+\frac{(k-1)k}{2}=\frac{1}{3}(k-1)k(k+1).$$

Hence

$$t(a)=-\sum_{k=2}^{\infty}\frac{(-a)^k}{(k-2)!k!}\frac{k+1}{3}=\frac{a^2}{3}(s''(a)-s'(a)).$$

The coefficient of $1/n^2(n+1)^2$ is somewhat more complicated. Separating a, b, c we have that

$$\begin{aligned}
\text{coeff }(1/n^2(n+1)^2)=&\left[-\frac{1\cdot 2\cdot 2\cdot 3}{3!3!}a^3\right.\\
&\left.+\frac{1\cdot 2\cdot 2\cdot 3+1\cdot 2\cdot 3\cdot 4+2\cdot 3\cdot 3\cdot 4}{4!4!}a^4-+\dots\right]\\
&+b\left[-2\frac{1\cdot 2}{2!2!}a+3\frac{1\cdot 2+2\cdot 3}{3!3!}a^2-4\frac{1\cdot 2+2\cdot 3+3\cdot 4}{4!4!}a^3+-\dots\right]\\
&+b^2\left[\frac{1}{2!2!}-\frac{3}{3!3!}a+\frac{6}{4!4!}a^2-+\dots\right]\\
&+c\left[-1+\frac{2}{2!2!}a-\frac{3}{3!3!}a^2+\frac{4}{4!4!}a^3-+\dots\right].
\end{aligned}$$

We denote the first sum by $u(a)$, the second sum equals $t'(a)$, the third sum equals $\frac{1}{2}s''(a)$, and the fourth sum equals $s'(a)$ if we again replace the finite sums by infinite series as in (5.4.1). First, we consider

$$\begin{aligned}
u(a)&=-\frac{1\cdot 2\cdot 2\cdot 3}{3!3!}a^3+\frac{1\cdot 2\cdot 2\cdot 3+1\cdot 2\cdot 3\cdot 4+2\cdot 3\cdot 3\cdot 4}{4!4!}a^4-+\dots\\
&=\sum_{k=3}^{\infty}\frac{(-a)^k}{k!k!}\left\{\sum_{i=3}^{k}4\binom{k-i+3}{2}\sum_{j=0}^{k-i}\binom{j+2}{2}\right\}\\
&=\sum_{k=3}^{\infty}\frac{(-a)^k}{k!k!}\sum_{i=3}^{k}4\binom{k-i+3}{2}\binom{k-i+3}{3}\\
&=\sum_{k=3}^{\infty}\frac{(-a)^k}{k!k!}\frac{1}{3}\left\{\sum_{i=1} i^3(i-1)^2-2\sum_{i=1}^{k} i^2(i-1)^2\right\}.
\end{aligned}$$

The two interior sums are again summable analogously to $\sum i(i+1)$. We obtain

$$u(a)=\sum_{k=3}^{\infty}\frac{(-a)^k}{k!k!}\frac{1}{90}(k+1)k(k-1)(k-2)(5k^2+k-3)$$
$$=\frac{1}{90}\sum_{k=3}^{\infty}\frac{(-a)^k}{(k-3)!(k-3)!}\left\{5+\frac{21}{k-2}+\frac{9}{(k-1)(k-2)}-\frac{3}{k(k-1)(k-2)}\right\}$$
$$=-\frac{a^3}{90}\{5s(a)-21s'(a)+9s''(a)+3s'''(a)\}$$

Note that $s''(a)=-(s(a)+s'(a))/a$ and, therefore, $s'''(a)=-s'(a)/a+2(s(a)+s'(a))/a^2$. We replace s'' and s'''. Then $u(a)=\frac{a}{90}(-5a^2+9a-6)s(a)+\frac{a}{30}(7a^2+4a-2)s'(a)$. Finally, we have that

$$L_n(r)\doteq s(a)+\frac{1}{n(n-1)}[t(a)+bs'(a)]+\frac{1}{n^2(n+1)^2}$$
$$\times\left[u(a)+bt'(a)+\frac{b^2}{2}s''(a)+cs'(a)\right]+O\left(\frac{1}{n^6}\right).$$

We are now able to determine the zeros of the right-hand side up to $O(1/n^6)$ by requiring

$$s(a)=0$$
$$t(a)bs'(a)=0$$
$$u(a)+bt'(a)+\frac{b^2}{2}s''(a)+cs'(a)=0.$$

From the first equation we determine a, from the second one b, and from the third one c. This calculation is facilitated by introducing the terms $s(a)$ and $s'(a)$. It is easy to verify that

$$t(a)=\frac{a}{3}(a+1)s'(a)+\frac{a}{3}s(a),\qquad t'(a)=\frac{2}{3}as'(a)-\frac{a}{3}s(a).$$

Hence, for every zero $\bar{a}$ of $s(a)$ we have that $\bar{a}$ is a solution of $s(a)=0$, from which it follows that

$$b=-t(\bar{a})/s'(\bar{a})=-\frac{\bar{a}}{3}(\bar{a}+1)\quad\text{and}\quad c=-\left[u(\bar{a})+bt'(\bar{a})+\frac{b^2}{2}s''(\bar{a})\right]/s'(\bar{a}).$$

Because

$$t(\bar{a})=\frac{\bar{a}}{3}(\bar{a}+1)s'(\bar{a}), \qquad t'(\bar{a})=\frac{2}{3}\bar{a}s'(\bar{a}),$$

$$s''(\bar{a})=-\frac{1}{\bar{a}}s'(\bar{a}), \qquad u(\bar{a})=\frac{\bar{a}}{30}(7\bar{a}^2+4\bar{a}-2)s'(\bar{a})$$

we obtain: $\bar{a}$ such that $s(\bar{a})=J_0(2\sqrt{\bar{a}})=0$,

$$b=-\frac{\bar{a}}{3}(\bar{a}+1) \quad \text{and} \quad c=\frac{1}{90}\bar{a}(4\bar{a}^2+18\bar{a}+11).$$

Hence, we have an approximation of order $O(1/n^6)$ for the zeros of the n-th degree Legendre polynomial which only needs the zeros of $J_0(2\sqrt{a})$.

If we improve these expansions we obtain further coefficients of r:

$$d=\frac{-\bar{a}}{11340}(36\bar{a}^3+432\bar{a}^2+964\bar{a}+610)$$

$$e=\frac{\bar{a}}{340200}(48\bar{a}^4+1200\bar{a}^3+8696\bar{a}^2+6654\bar{a}+11551)$$

$$f=\frac{-\bar{a}}{11226600}(48\bar{a}^5+2160\bar{a}^4+7512\bar{a}^3+308556\bar{a}^2-331790\bar{a}+430489)$$

$$g=\frac{\bar{a}}{91945854000}(8640\bar{a}^6+635040\bar{a}^5+96211248\bar{a}^4-1068289752\bar{a}^2$$
$$+6605805220\bar{a}^2-11250181134\bar{a}+7416279161).$$

However, it is not necessary to calculate the zeros of $J_0(x)$ as we can use MacMahon's asymptotic expansion of the zeros a_i of $J_0(x)$, i.e.

$$a_i\approx\frac{1}{4}\left[\frac{1}{4}(4i-1)\pi+\frac{1}{2(4i-1)\pi}-\frac{31}{6(4i-1)^3\pi^3}\right.$$
$$\left.+\frac{3779}{15(4i-1)^5\pi^5}-\frac{62\,77237}{210(4i-1)^7\pi^7}+\dots\right]^2$$

i	Exact zeros	MacMahon's approximation with 5 terms
1	1.44579649073669613	1.44232546
2	7.61781558591552160	7.61780623
3	18.72175169767379586	18.7217514
4	34.76007110661496225	34.7600711
5	55.73307590440853924	55.7330759
6	81.6408382330821140	81.6408383

In exactly the same way we set up an asymptotic expansion for the zeros in the middle of the interval. First we consider the case of an even number of nodes. In $[-1, 1]$ we obtain, from the Rodriguez formula, the normalised Legendre polynomials $L_{2n}(x)$ with smallest coefficient 1 and $2n = N$

$$L_{2n}(x) = 1-(N(N+1)x^2)/2! + (N(N+1)x^2)^2\left(1-\frac{2\cdot 3}{N(N+1)}\right)\Big/4!$$
$$-(N(N+1)x^2)^3\left(1-\frac{2\cdot 3}{N(N+1)}\right)\left(1-\frac{4\cdot 5}{N(N+1)}\right)\Big/6!$$
$$+(N(N+1)x^2)^4\left(1-\frac{2\cdot 3}{N(N+1)}\right)\left(1-\frac{4\cdot 5}{N(N+1)}\right)\left(1-\frac{6\cdot 7}{N(N+1)}\right)\Big/8!\ldots.$$

From this expansion we see that the zeros of $L_{2n}(x)$ have the expansion

$$x_i^2 = \frac{a}{N(N+1)} + \frac{b}{N^2(N+1)^2} + \frac{c}{N^3(N+1)^3} + \ldots.$$

and we obtain the corresponding series expansions

$$s(a) = 1 - \frac{a}{2!} + \frac{a^2}{4!} - \frac{a^3}{6!} + - \ldots = \cos\sqrt{a}$$

$$t(a) = -\frac{2\cdot 3}{4!}a^2 + \frac{2\cdot 3 + 4\cdot 5}{6!}a^3 - \frac{2\cdot 3 + 4\cdot 5 + 6\cdot 7}{8!}a^4 + - \ldots$$

$$u(a) = -\frac{2\cdot 3\cdot 4\cdot 5}{6!}a^3 + \frac{2\cdot 3\cdot 4\cdot 5 + 2\cdot 3\cdot 6\cdot 7 + 4\cdot 5\cdot 6\cdot 7}{8!}a^4 - + \ldots$$

....

These are to be handled in the same way as before and thus we have

$\bar{a}$ such that $s(\bar{a}) = \cos\sqrt{\bar{a}} = 0$, i.e. $\bar{a} = \left(\frac{2i-1}{2}\pi\right)^2$

$$b = \frac{-\bar{a}}{6}(2\bar{a}+3)$$

$$c = \frac{\bar{a}}{360}(16\bar{a}^2 + 90\bar{a} + 135)$$

$$d = \frac{-\bar{a}}{5040}(16\bar{a}^3 + 224\bar{a}^2 + 210\bar{a} + 2835)$$

$$e = \bar{a}\left(\frac{2}{14175}\bar{a}^4 + \frac{1}{252}\bar{a}^3 + \frac{61}{720}\bar{a}^2 - \frac{47}{64}\bar{a} + \frac{243}{128}\right)$$

$$f=-\bar{a}\left(\frac{2}{467775}\bar{a}^5+\frac{1}{4725}\bar{a}^4-\frac{43}{1680}\bar{a}^3+\frac{61}{72}\bar{a}^2-\frac{867}{128}\bar{a}+\frac{3159}{256}\right)$$

$$g=\bar{a}\left(\frac{4}{4256725}\bar{a}^6+\frac{1}{133650}\bar{a}^5+\frac{817}{56700}\bar{a}^4-\frac{150007}{221760}\bar{a}^3\right.$$
$$\left.+\frac{949}{80}\bar{a}^2-\frac{39843}{512}\bar{a}+\frac{132759}{1024}\right)$$

so that in $[-1, 1]$ for $i=1(1)\ldots$

$$x_i=\left(\frac{\bar{a}}{N(N+1)}+\frac{b(\bar{a})}{N^2(N+1)^2}+\frac{c(\bar{a})}{N^3(N+1)^3}+\ldots+O\left(\frac{1}{N^8(N+1)^8}\right)\right)^{\frac{1}{2}}$$

and in $[0, 1]$ the approximate zeros $\bar{x}_i=(1+x_i)/2$.

If we compare these approximations with values of an extended table of zeros of Legendre polynomials we see that the errors of the two approximations are equal in the region of $i=n/3$. Hence, for $0<i\leqslant n/3$ we choose the first approximation, for $n/3<i\leqslant n/2$ the second, and for $i>n/2$ we use the symmetry $x_i=-x_{n-i+1}$. For example, for $n=256$ all nodes are obtained in this way to at least 13 correct decimal places.

The same expansions do not hold for an odd number of zeros, say $2n+1=N$. In this case we obtain, in the same way,

$$L_{2n+1}(x)=x-(N-1)(N+2)x^3/3!$$
$$+((N-1)(N+2))^2x^5/5!\left(1-\frac{2\cdot 5}{(N-1)(N+2)}\right)$$
$$-((N-1)(N+2))^3x^7/7!\left(1-\frac{2\cdot 5}{(N-1)(N+2)}\right)\left(1-\frac{4\cdot 7}{(N-1)(N+2)}\right)$$
$$+((N-1)(N+2))^4x^9/9!\left(1-\frac{2\cdot 5}{(N-1)(N+2)}\right)$$
$$\times\left(1-\frac{4\cdot 7}{(N-1)(N+2)}\right)\left(1-\frac{6\cdot 9}{(N-1)(N+2)}\right)$$
$$-+\ldots.$$

and thus we now have an asymptotic expansion of the zeros in powers of $1/[(N-1)(N+2)]$ which requires far-reaching changes. Hence this case is omitted.

Appendix

At the end of Section 7.3.1 we obtain the result $A_i = \dfrac{x_i(1-x_i)}{n^2 L_{n-1}^2(x_i)}$, $i = 1(1)n$ (with normalisation $L_n(1) = 1$) for the n weights of the Gaussian quadrature formula $Q_n f$ with respect to $[0, 1]$. The n zeros of $L_n(x)$ are the nodes of $Q_n f$. Since we expanded x_i in powers of $n(n+1)$ we can also expand $L_{n-1}(x_i)$ and, hence, A_i in the same way. We outline the result only for the first weight A_1 and for the first power of $n(n+1)$ in order to obtain a result which will be needed in Section 6.6.1.

The polynomials $L_n(x)$ of this section satisfy $L_n(1) = 1$ and are appropriate to $[0, 1]$. (If they are set up for $[-1, 1]$ then the normalisation is the same.) For $x_1 = \dfrac{a}{n(n+1)} + O(n^{-4})$ with $a = 1.44579649$ and, by rearrangement of the terms, we obtain

$$L_{n-1}\left(\frac{a}{n(n+1)} + O\left(\frac{1}{n^4}\right)\right)$$

$$\doteq 1 - \frac{n-1}{n+1}\left(\sum_{k=0}^{\infty} \frac{(-1)^k a^k}{k!\,k!} + \frac{1}{n}\sum_{k=2}^{\infty} \frac{(-1)^k a^k}{k!\,k!}\sum_{i=2}^{k} i - \frac{1}{n+1}\sum_{k=3}^{\infty} \frac{(-1)^k a^k}{k!\,k!}\sum_{i=1}^{k-2} i\right)$$

$$\doteq 1 - \left(1 - \frac{2}{n+1}\right)\left(1 - s(a) + \frac{1}{2n}[2 + as'(a) + (a+2)s(a)]\right.$$

$$\left. + \frac{1}{2(n+1)}[2 + 3as'(a) + (a-2)s(a)]\right).$$

Since $s(a) = 0$ we have that

$$L_{n-1}\left(\frac{a}{n(n+1)} + O\left(\frac{1}{n^4}\right)\right) \doteq -\frac{as'(a)}{n+1} + O\left(\frac{1}{n^2}\right) \equiv \frac{C_n}{n(n+1)} \quad \text{for} \quad [0, 1].$$

Hence, because $s'(a) = -0.43175481$, we obtain for $[-1, 1]$

$$A_1 \doteq \frac{2a}{n(n+1)(as'(a))^2} = \frac{7.4207613}{n(n+1)}$$

as required in Section 6.6.1.

5.5 Orthogonal polynomials in the complex plane

In Section 3.5 we derived error bounds for analytic functions. Our considerations there were strongly influenced by orthogonal polynomials in the complex plane. We list some facts which are useful in this field. As we have seen the monomials $(z/|z|)^k$, $k=0(1)\infty$ form an orthonormal polynomial system with respect to the inner product $(f, g)=(1/2\pi)\int_{|z|=R\geq 1} f(z)\overline{g(z)}\,d\varphi$ with $z=Re^{i\varphi}$. Turning to the inner product $(f, g)=\int_{|z|=R} f(z)\overline{g(z)}\,ds$, where s is the arc length of $|z|=R$, $s=R\varphi$, we have the orthogonal polynomial system $z^k/(2\pi R^{2k+1})^{\frac{1}{2}}$, $k=0(1)\infty$. This can be verified by straightforward calculation.

We can use not only contour integrals but also area integrals in order to define inner products. Let us define the inner product

$$(f, g)=\iint_{|z|\leq R} f(z)\overline{g(z)}\,dx\,dy=\int_0^{2\pi}\int_0^R f(re^{i\varphi})\overline{g(re^{i\varphi})}r\,dr\,d\varphi$$

$$\text{for}\quad z=x+iy=re^{i\varphi}.$$

Here again the monomials are orthogonal. If we normalise them, we obtain $((k+1)/\pi)^{\frac{1}{2}}(z^k/R^{k+1})$, $k=0(1)\infty$. For the ellipses E_ρ having foci at ± 1 and semi-axis $a=\frac{1}{2}(\sqrt{\rho}+1/\sqrt{\rho})>b=\frac{1}{2}(\sqrt{\rho}-1/\sqrt{\rho})$, $\rho=(a+b)^2>1$ the Tschebyscheff polynomials of the first kind $T_n(z)$ are orthogonal with respect to the inner product

$$(f, g)=\int_{E_\rho} |1-z^2|^{-\frac{1}{2}} f(z)\overline{g(z)}\,ds$$

where s is the arc length of E_ρ.

The same is true for the Tschebyscheff polynomials of the second kind with respect to the inner product

$$(f, g)=\iint_{E_\rho}(1-z^2)^{\frac{1}{2}} f(z)\overline{g(z)}\,dx\,dy.$$

(For a proof of these properties see Davis [1] pp. 240, 241.)

Moreover, let C be a curve with finite arc length in the complex plane with $z=x+iy$. Then the monomials z^k, $k=0(1)\ldots$ are linearly-independent functions. (From $a_k z^k+a_{k-1}z^{k-1}+\ldots+a_0\equiv 0$, where a_k are complex numbers, we conclude $a_k=\ldots=a_o=0$.) Hence

$$(f, g)=\int_C f(z)g(z)\,ds,\quad s \text{ arc length of } C$$

is an inner product, $(f, f)=0$ if and only if $f=0$. The system of monomials z^k, $k=0(1)\ldots$ may be orthogonalised by the Gram–Schmidt orthogonalisation procedure, and thus we have that for every curve with finite arc length in the complex plane there exists an orthogonal system of polynomials. Clearly, there can exist further systems of orthogonal functions as we have seen for the rectangle.

Davis [1] p. 242 gives the first eight orthogonal polynomials for a rectangle. However, there is no general representation for these polynomials.

5.6 N-Dimensional orthogonal polynomials

The orthogonal polynomials in two or more dimensions are of comparable importance for cubature formulae to the one-dimensional orthogonal polynomials for quadrature formulae. But here the situation is more complicated. For $N=2$ dimensions we already have $(n+1)(n+2)/2$ linearly-independent polynomials of $\mathbb{P}_n$ and $n+1$ linearly-independent polynomials of $\dot{\mathbb{P}}_n$. Hence orthogonality of such a system of polynomials means at least the orthogonality of $n+1$ linearly-independent polynomials of $\dot{\mathbb{P}}_n$ to $\mathbb{P}_{n-1}$ and with respect to the same weight function $W(\mathbf{x})$. For N dimensions we have $\binom{N+n-1}{n}$ linearly-independent polynomials of $\dot{\mathbb{P}}_n$ orthogonal to $\mathbb{P}_{n-1}$ in N dimensions. We shall now give some details.

5.6.1 Existence of orthogonal polynomials

Theorem 5.6.1 *Let $\Omega \subseteq \mathbb{R}^N$ be a given domain, and $W(x_1, \ldots, x_N)$ a non-negative weight function ($W \geq 0$ for $\mathbf{x} \in \Omega$). There exists a system of $\binom{N+n-1}{n}$ linearly-independent polynomials $p_n \in \dot{\mathbb{P}}_n$, with $p_n \perp \mathbb{P}_{n-1}$ such that*

$$(p_n, q) \equiv \int_\Omega W(\mathbf{x}) p_n(\mathbf{x}) q(\mathbf{x})\, d\Omega = 0 \quad \textit{for every} \quad q \in \mathbb{P}_{n-1}.$$

Proof First, we look for the number of orthogonal polynomials. For degree 0 we have one polynomial: the constant polynomial $p_0 = 1$. For degree 1 we have the N additional monomials $x_1, \ldots, x_N$. For degree 2 the monomials $x_1^2, x_1x_2, \ldots, x_1x_N, x_2^2, x_2x_3, \ldots, x_2x_N, \ldots, x_N^2$ appear. These are the combinations of two from N monomials $x_1, \ldots, x_N$ considering commutivity. Hence, from combinatorics, these are $N(N+1)/2 = \binom{N+1}{2}$ monomials of degree 2. Proceeding by induction we see that $\binom{N+n-1}{n}$ is all right.

We now prove the existence. In order to keep the formulae readable we choose $N=2$ and $n=2$. The proof is similar to that for $N=1$. Let $x_1=x$ and $x_2=y$. The monomials 1, x, y, x^2, xy, y^2 are linearly independent. If they were not, then the system

$$\begin{aligned}(P_2(x, y), 1) &= (a_0+a_1x+a_2y+a_3x^2+a_4xy+a_5y^2, 1)=0\\ (P_2(x, y), x) &= (P_2(x, y), y)=(P_2(x, y), x^2)=(P_2(x, y), xy)\\ &= (P_2(x, y), y^2)=0 \quad (5.6.1)\end{aligned}$$

would have non-trivial solutions. This results by taking the inner products respectively of the first, second, . . . equation with 1, x, y, ... The system (5.6.1) is thus equivalent with the system

$$\begin{aligned}&a_0(1, 1)+a_1(x, 1)+a_2(y, 1)+a_3(x^2, 1)+a_4(xy, 1)+a_5(y^2, 1)=0\\ &a_0(1, x)+a_1(x, x)+a_2(y, x)+a_3(x^2, x)+a_4(xy, x)+a_5(y^2, x)=0\\ &a_0(1, y)+a_1(x, y)+a_2(y, y)+a_3(x^2, y)+a_4(xy, y)+a_5(y^2, y)=0\\ &a_0(1, x^2)+a_1(x, x^2)+a_2(y, x^2)+a_3(x^2, x^2)+a_4(xy, x^2)+a_5(y^2, x^2)=0\\ &a_0(1, xy)+a_1(x, xy)+a_2(y, xy)+a_3(x^2, xy)+a_4(xy, xy)+a_5(y^2, xy)=0\\ &a_0(1, y^2)+a_1(x, y^2)+a_2(y, y^2)+a_3(x^2, y^2)+a_4(xy, y^2)+a_5(y^2, y^2)=0\end{aligned}$$

In this system we multiply the i-th equation with a_{i-1}, $i=1(1)n$, and add all resulting equations. We then have that

$$(P_2(x, y), P_2(x, y))=0. \tag{5.6.2}$$

However, for $W(\mathbf{x})\geqslant 0$ we have $(P_2, P_2)=\iint_\Omega W(\mathbf{x})P_2^2(\mathbf{x})\,\mathrm{d}\Omega>0$, which contradicts (5.6.2).

The same considerations hold for arbitrary degree.

Note that Theorem 5.6.1 is not valid for the discrete inner product $\langle P_2, P_2\rangle=\sum_{k=0}^{3} W_kP_2^2(\mathbf{x}_k)>0$ since the polynomial P_2 has, in general, at least one zero line. If we choose the nodes $\mathbf{x}_k$ on this line, we have $\langle P_2, P_2\rangle=0$.

The proof of existence is now clear by applying the Gram–Schmidt orthogonalisation procedure to the monomials 1, x, y, x^2, xy, y^2, x^3, x^2y, xy^2 and y^3. This leads to four polynomials of degree 3 orthogonal to $\mathbb{P}_2$. Because these are linearly independent, we can normalise them as

$$\begin{aligned}p^{3,0}(x, y)&=x^3+a_5x^2+a_4xy+a_3y^2+a_2x+a_1y+a_0\\ p^{2,1}(x, y)&=x^2y+b_5x^2+b_4xy+b_3y^2+b_2x+b_1y+b_0\\ p^{1,2}(x, y)&=xy^2+c_5x^2+c_4xy+c_3y^2+c_2x+c_1y+c_0\\ p^{0,3}(x, y)&=y^3+d_5x^2+d_4xy+d_3y^2+d_2x+d_1y+d_0. \quad \blacksquare\end{aligned}$$

For the determination of the coefficients we only need the linear independence of $1, x, y, x^2, xy$ and y^2. However, certain combinations of these polynomials can also be orthogonal to $\mathbb{P}_2$.

Example 5.6.1 Let us consider the square Ω: $|x|, |y| \leq 1$. Then the conditions that $p^{2,0}, p^{1,1}, p^{0,2} \perp 1, x, y$ have to be valid. Hence

$$\iint_{\Omega} p^{2,0}\, d\Omega = 4/3 + 4a_0 = 0, \quad \iint_{\Omega} p^{2,0} x\, d\Omega = (4/3)a_2 = 0,$$

$$\iint_{\Omega} p^{2,0} y\, d\Omega = (4/3)a_1 = 0$$

and $a_1 = a_2 = 0$ and $a_0 = -1/3$, and $p^{2,0} = x^2 - 1/3$. Similarly, we have that $b_1 = b_2 = b_0 = 0$ and $c_1 = c_2 = 0$, $c_0 = -1/3$ and therefore $p^{2,0} = x^2 - 1/3$, $P^{1,1} = xy$, $P^{0,2} = y^2 - 1/3$. However, also $x^2 + y^2 - 2/3$, $x^2 - y^2$, $(y \pm x)^2 - 2/3 \perp \mathbb{P}_1$. All these polynomials have zero lines, which are sketched in Fig. 5.6.1 for $p^{2,0}$, $x^2 + y^2 - 2/3$, $p^{0,2}$ as well as for $(y - x)^2 - 2/3$, xy and $x^2 + y^2 - 2/3$. We find that these orthogonal polynomials have some zero lines intersecting all others at the four points marked. These are to be used as nodes in a cubature formula of degree 2. ■

5.6.2 Common zeros of orthogonal polynomials

It may happen that the orthogonal polynomials of degree n have a set of common zeros. We shall see later that under certain restrictions these can be used to construct cubature formulae of high degree if compared with the number of nodes. We do not have such good results as for $N = 1$ dimension. We only have the following weak similar result.

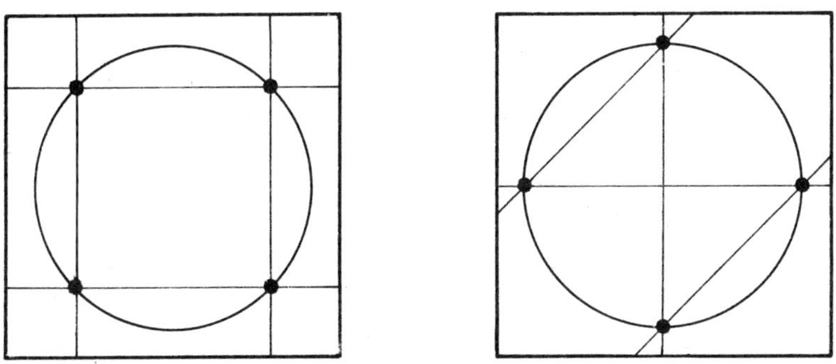

FIG. 5.6.1. Zero lines of orthogonal polynomials of degree 2.

Theorem 5.6.2 *Assume we are given the factorisation* $p_n(\mathbf{x}) = p_{n,1}(\mathbf{x})p_{n,2}(\mathbf{x}) \ldots p_{n,k}(\mathbf{x}) \perp \mathbb{P}_{n-1}$, *of* $p_n(\mathbf{x})$, *such that all factors are real and irreducible in the real field. Then all* $p_{n,i}(\mathbf{x})$ *are distinct and the zero line* $p_{n,i}(\mathbf{x}) = 0$ *is not completely outside of* Ω *for* $i = 1(1)k$.

Proof Assume, that $p_{n,1}/p_{n,2} = \text{constant} \neq 0$. Note that $p_n(\mathbf{x}) \perp p_{n,3}(\mathbf{x}) \ldots p_{n,k}(\mathbf{x})$. Hence with appropriate $c > 0$

$$0 = c(p_n, p_{n,3} \ldots p_{n,k}) = \iint_\Omega (p_{n,2}(\mathbf{x}) \ldots p_{n,k}(\mathbf{x}))^2 \, d\Omega > 0$$

which is a contradiction. If more than two factors coincide, the arguments are similar.

Assume that the zero line $p_{n,1}(\mathbf{x}) = 0$ does not intersect Ω. Then the polynomial $p_{n,1} \cdot p_{n,2}^2 \ldots p_{n,k}^2$ has constant sign in Ω. Hence

$$\iint_\Omega p_{n,1} p_{n,2}^2 \ldots p_{n,k}^2 \, d\Omega \neq 0.$$

On the other hand the orthogonality condition

$$0 = (p_n, p_{n,2} \ldots, p_{n,k}) = \iint_\Omega p_{n,1} p_{n,2}^2 \ldots p_{n,k}^2 \, d\Omega$$

holds, which again gives a contradiction. Hence the proof is finished. ■

5.6.3 Special systems of orthogonal polynomials in N dimensions

We shall now present some special systems of orthogonal polynomials for some standard regions. We give only the *essential* polynomials. The remaining polynomials are obtained from symmetry by changing the variables $x_1, \ldots, x_N$ of $\mathbf{x}$ cyclically.

Sometimes there exists a Rodriguez formula which facilitates the calculation of the polynomials.

The N-sphere: Here we have the Rodriguez formula (with $i = n_1 + n_2 + \ldots + n_N$)

$$P_{n_1,\ldots,n_N}(\mathbf{x}) = K_i \frac{\partial^i}{\partial x_1^{n_1} \ldots \partial x_N^{n_N}} (x_1^2 + x_2^2 + \ldots + x_N^2 - 1)^i.$$

The corresponding system of polynomials is (cf. Stroud, book, p. 71)

$$
\begin{aligned}
p_{0,\ldots,0} &= 1\\
p_{1,0,\ldots,0} &= x_1\\
p_{2,0,\ldots,0} &= x_1^2 - 1/(N+1)\\
p_{1,1,0,\ldots,0} &= x_1x_2\\
p_{3,0,\ldots,0} &= x_1^3 - 3x_1/(N+4)\\
p_{2,1,0,\ldots,0} &= x_1^2x_2 - x_2/(N+4)\\
p_{1,1,1,0,\ldots,0} &= x_1x_2x_3\\
p_{4,0,\ldots,0} &= x_1^4 - 6x_1^2/(N+6) + 3/[(N+4)(N+6)]\\
p_{3,1,0,\ldots,0} &= x_1^3x_2 - 3x_1x_2/(N+6)\\
p_{2,2,0,\ldots,0} &= x_1^2x_2^2 - (x_1^2+x_2^2)/(N+6) + 1/[(N+4)(N+6)]\\
p_{2,1,1,0,\ldots,0} &= x_1^2x_2x_3 - x_2x_3/(N+6)\\
p_{1,1,1,1,0,\ldots,0} &= x_1x_2x_3x_4.
\end{aligned}
$$

Let us consider functions of the form

$$p_{n_1,\ldots,n_N} = p_{n_1}(r)p_{n_2}(\sin\varphi_1)p_{n_3}(\sin\varphi_2)\ldots p_{n_N}(\sin\varphi_{N-1}).$$

For $N=2$, for example, we have the orthogonality conditions

$$\int_0^{2\pi}\int_0^1 p_{n_1}(r)p_{n_2}(\sin\varphi)Q_{m_1}(r)Q_{m_2}(\sin\varphi)r\,\mathrm{d}r\,\mathrm{d}\varphi = 0 \quad \text{for} \quad m_1<n_1,\ m_2<n_2.$$

Hence we need only take the Jacobi polynomials $p_{n_1}^{1,0}$ for $p_{n_1}(r)$, and $p_{n_2}(\sin\varphi)\equiv\sin(n_2\varphi)$ in order to have the desired orthogonality property.

For $N=3$ we have that

$$\int_0^{2\pi}\int_0^{2\pi}\int_0^1 p_{n_1}(r)p_{n_2}(\sin\varphi_1)p_{n_3}(\sin\varphi_2)Q_{m_1}(r)Q_{m_2}(\sin\varphi_1)$$
$$\times Q_{m_3}(\sin\varphi_1)r^2\sin\varphi_1\,\mathrm{d}r\,\mathrm{d}\varphi_2\,\mathrm{d}\varphi_2 = 0 \quad \text{for} \quad m_i<n_i,\ i=1(1)3.$$

In order to have orthogonality we choose the Jacobi polynomials $p_{n_1}^{2,0}$ for $p_{n_1}(r)$, further $p_{n_2}(\sin\varphi_1)\equiv\sin((n_2+1)\varphi_1)$ and $p_{n_3}(\sin\varphi_2)\equiv\sin(n_3\varphi_2)$, and in general the choice

$$
\begin{aligned}
&\text{Jacobi polynomial } p_{n_1}^{N-1,0} \quad \text{for} \quad p_{n_1}(r)\\
&p_{n_2}(\sin\varphi_1) = \sin((n_2+N-2)\varphi_1)\\
&p_{n_3}(\sin\varphi_2) = \sin((n_3+N-3)\varphi_2)\\
&\ldots\\
&p_{n_{N-1}}(\sin\varphi_{N-2}) = \sin((n_{N-1}+1)\varphi_{N-2})\\
&p_{n_N}(\sin\varphi_{N-1}) = \sin(\varphi_{N-1})
\end{aligned}
$$

establishes an orthogonal function system for the N-sphere. The zero lines of this system of functions are easily obtained and their points of intersection can be taken as nodes of cubature formulae. This system is not a system of orthogonal *polynomials* in the classical sense but, thinking of trigonometric polynomials as polynomials in a more general sense, we have another system of orthogonal "polynomials" for the N-sphere.

The N-simplex: For the N-simplex there exists the Rodriguez formula

$$S_{n_1,n_2,\ldots,n_N}(\mathbf{x}) = K_i \frac{\partial^i}{\partial x_1^{n_1} \partial x_2^{n_2} \ldots \partial x_N^{n_N}} \{x_1^{n_1} x_2^{n_2} \ldots x_N^{n_N} (1 - x_1 - x_2 - \ldots - x_N)^i\}$$

with $i = n_1 + n_2 + \ldots + n_N$. Let us define $\binom{N}{k}^{-1} \equiv \begin{bmatrix} N \\ k \end{bmatrix}$. Then (cf. Stroud, book, p. 73)

$$S_{0,\ldots,0} = 1$$

$$S_{1,0,\ldots,0} = x_1 - \begin{bmatrix} N+1 \\ 1 \end{bmatrix}$$

$$S_{2,0,\ldots,0} = x_1^2 - 4x_1 \begin{bmatrix} N+3 \\ 1 \end{bmatrix} + \begin{bmatrix} N+3 \\ 2 \end{bmatrix}$$

$$S_{1,1,0,\ldots,0} = x_1 x_2 - (x_1 + x_2) \begin{bmatrix} N+3 \\ 1 \end{bmatrix} + \frac{1}{2} \begin{bmatrix} N+3 \\ 2 \end{bmatrix}$$

$$S_{3,0,\ldots,0} = x_1^3 - 9x_1^2 \begin{bmatrix} N+5 \\ 1 \end{bmatrix} + 9x_1 \begin{bmatrix} N+5 \\ 2 \end{bmatrix} - \begin{bmatrix} N+5 \\ 3 \end{bmatrix}$$

$$S_{2,1,0,\ldots,0} = x_2^2 x_2 - x_1^2 \begin{bmatrix} N+5 \\ 1 \end{bmatrix} - 4x_1 x_2 \begin{bmatrix} N+5 \\ 1 \end{bmatrix} + 2x_1 \begin{bmatrix} N+5 \\ 2 \end{bmatrix}$$

$$S_{1,1,1,0,\ldots,0} = x_1 x_2 x_3 - (x_1 x_2 + x_1 x_3 + x_2 x_3) \begin{bmatrix} N+5 \\ 1 \end{bmatrix} + \frac{1}{2!}(x_1 + x_2 + x_3) \begin{bmatrix} N+5 \\ 2 \end{bmatrix} + \frac{1}{3!} \begin{bmatrix} N+5 \\ 3 \end{bmatrix}$$

The N-cube: The N-cube is a product space. Hence we easily obtain a system of orthogonal polynomials by multiplying the Legendre polynomials.

$$L_{n_1,\ldots,n_N}(\mathbf{x}) = L_{n_1}(x_1) \ldots L_{n_N}(x_N).$$

Having a weight function which can by factorised in the same way, the corresponding system of orthogonal polynomials is the direct product of the orthogonal polynomials with respect to each coordinate weight factor.

Example 5.6.2 The system of orthogonal polynomials with respect to the inner product

$$\int_{-\infty}^{\infty}\int_{-1}^{1}\int_{0}^{\infty} e^{-x_1^2-x_3}\sqrt{1-x_2^2}f(\mathbf{x})g(\mathbf{x})\,dx_3\,dx_2\,dx_1$$

is

$$p_{n_1,n_2,n_3}(\mathbf{x}) = L_{n_1}(x_1)U_{n_2}(x_2)H_{n_3}(x_3)$$

where $L_{n_1}(x_1)$ are Laguerre polynomials, $U_{n_2}(x_2)$ are Tschebyscheff polynomials of the second kind, and $H_{n_3}(x_3)$ are Hermite polynomials.

Note, that this product construction is impossible, if the weight function cannot be factorised. ■

Entire N-space: The N-dimensional Euclidian space is again a product space and hence the Hermite polynomials are a system of orthogonal polynomials for this space. The product structure again appears in the product-type Rodriguez formula

$$H_{n_1\dots n_N}(\mathbf{x}) = K_i\frac{\partial^i}{\partial x_1^{n_1}\dots\partial x_N^{n_N}}\{\exp(-x_1^2-\dots-x_N^2)\}.$$

For the weight function $W(\mathbf{x}) = (\mathbf{x},\mathbf{x})^{\frac{1}{2}}$ we do not have a product structure. The corresponding system of orthogonal polynomials is due to Stroud and Don Secrest (cf. Stroud, book, p. 74)

$$\begin{aligned}
L_{0,\dots,0} &= 1\\
L_{1,0,\dots,0} &= x_1\\
L_{2,0,\dots,0} &= x_1^2-(N+1)\\
L_{1,1,0,\dots,0} &= x_1x_2\\
L_{3,0,\dots,0} &= x_1^3-3(N+3)x_1\\
L_{2,1,0,\dots,0} &= x_1^2x_2-(N+3)x_2\\
L_{1,1,1,0,\dots,0} &= x_1x_2x_3\\
L_{4,0,\dots,0} &= x_1^4-6(N+5)x_1^2+12(x_1^2+\dots+x_N^2)/(2N+3)\\
&\quad +3(N+1)(N+3)(2N+7)/(2N+3)\\
L_{3,1,0,\dots,0} &= x_1^3x_2-3(N+5)x_1x_2\\
L_{2,2,0,\dots,0} &= x_1^2x_2^2-(N+5)(x_1^2+x_2^2)+4(x_1^2+\dots+x_N^2)/(2N+3)\\
&\quad +(N+1)(N+3)(2N+7)/(2N+3)\\
L_{2,1,1,0,\dots,0} &= x_1^2x_2x_3-(N+5)x_2x_3\\
L_{1,1,1,1,0,\dots,0} &= x_1x_2x_3x_4.\\
&\dots
\end{aligned}$$

In Section 5.7 we shall see how to use such systems of orthogonal polynomials in order to obtain cubature formulae.

Finally, we mention here that the orthogonal polynomials for complex contour integrals or complex domain integrals as inner products for product spaces are again obtained by products of orthogonal polynomials in one complex variable.

5.7 Applications to cubature problems

In Section 2.3.4 we stated some problems which appear in cubature but not in quadrature; for example, the question of finding cubature formulae of prescribed degree with minimal number of nodes.

We now consider how to determine the minimal number of nodes in a cubature formula for positive weights and in this case how to obtain such formulae.

5.7.1 The determination of the minimal number of nodes of cubature formulae of odd degree in two dimensions

As we shall see in Section 7.3, the Gaussian qudrature formulae with n nodes achieve degree $2n-1$. Hence, it seems natural to seek similar cubature formulae with odd degree and the minimal number of nodes. Clearly, this number must be larger for $N>1$ dimensions than for $N=1$. The problem of finding the minimal number of nodes for any given domain and for every given degree is still unsolved. For the two-dimensional case Mysovskih [7]–[11] (see also Radon [12]) determined cubature formulae with the minimal number of nodes for degrees 1, 3 and 5. Möller [5], [6], who generalised Mysovskih's and Radon's method, obtained general results for odd degree cubature formulae for arbitrary dimension N and region Ω. These ideas are now followed (cf. [6]) but for the sake of simplicity we first restrict our attention to $N=2$ dimensions although later the case $N>2$ is discussed. The main result of this section is to bound the minimal number of nodes. Since the methods are constructive we are able either to construct cubatures with minimal number of nodes or to decide whether or not the lower bound is sharp. Special results are obtained for symmetric regions. Thus, we define:

Definition 5.7.1 The domain Ω is said to possess *central symmetry* if $I_\Omega f=0$ for every monomial $f=x^i y^k$, $0<i, k$ where $i+k$ is odd. ■

This holds, in particular, if for every given $\mathbf{x}\in\Omega$ we also have $-\mathbf{x}\in\Omega$ (and if the weight function $W(\mathbf{x})$ is even).

We work with triplets of orthogonal polynomials. Let $\mathbb{K}_r$ be the space of orthogonal polynomials $f \in \dot{\mathbb{P}}_r$ such that $f \perp \mathbb{P}_{r-1}$ (cf. Section 5.6). These polynomials can be uniquely determined by $P^{r-i,i}(\mathbf{x}) = x^{r-i}y^i + g(\mathbf{x})$ with $g \in \mathbb{P}_{r-1}$. If this is done, then $\mathbb{K}_r$ is a linear space with basis $\{P^{r,0}, P^{r-1,1}, \ldots, P^{0,r}\}$ as is shown by Jackson [4] (cf. Problem P.5.12). For symmetric regions we have:

Lemma 5.7.1 If Ω possesses central symmetry, then if r is even (odd) $\mathbb{K}_r$ contains only even (odd) polynomials $P(\mathbf{x})$. ■

(See the special systems of N-dimensional orthogonal polynomials in Section 5.6.3.)

In order to establish this lemma, let $P_+(\mathbf{x}) \equiv (P(\mathbf{x}) + P(-\mathbf{x}))/2$, and $P_-(\mathbf{x}) \equiv (P(\mathbf{x}) - P(-\mathbf{x}))/2$, $\mathbf{x} \in \Omega$. In this case $P(\mathbf{x}) = P_+(\mathbf{x}) + P_-(\mathbf{x})$, and $P_+(\mathbf{x})$ is even, while $P_-(\mathbf{x})$ is odd. Assume that the degree m is even, then $P_-(\mathbf{x}) \in \mathbb{P}_{m-1}$ and $P_+(\mathbf{x}) \in \mathbb{P}_m$. Hence, for $P(\mathbf{x}) \in \mathbb{K}_m$ we have that $0 = I_\Omega(PP_-) = I_\Omega(P_+P_-) + I_\Omega(P_-P_-) = I_\Omega(P_-P_-)$ since P_+P_- is odd. Hence we have $I_\Omega(P_-P_-) = 0$, which is only possible if $P_-(\mathbf{x}) = 0$, and which implies that $P(\mathbf{x}) = P_+(\mathbf{x})$ is even. The alternative case (n odd) may be treated in the same way.

Let $\mathbb{L}$ be the space of triples (P_0, P_1, P_2), where $P_0 = xP_1 + yP_2$ so that $\mathbb{L} = \mathbb{K}_s^3 = \mathbb{K}_s \times \mathbb{K}_s \times \mathbb{K}_s$. The condition $P_0 \in \mathbb{K}_s$ implies that P_1 and P_2 respectively do not contain x^s and y^s. There is also the connection between P_1 and P_2 in that $\mathbb{L}$ is isomorphic to a subspace of $\mathbb{R}_s$.

Lemma 5.7.2 There exists a one-to-one correspondence between the spaces $\mathbb{L}$ and $\bar{\mathbb{R}}^s \subset \mathbb{R}^s$ of s-dimensional vectors $\mathbf{a}^T = (a_1, a_2, \ldots, a_s)$ whose components satisfy

$$\mathbf{Aa} \equiv \sum_{i=1}^{s} a_i I_\Omega\{P^{s+1-j,j-1}P^{s-i,i} - P^{s-j,j}P^{s+1-i,i-1}\} = \mathbf{0}, \qquad j = 1(1)s. \quad ■ \tag{5.7.1}$$

Proof Consider the triplets (P_0, P_1, P_2), where P_1 and P_2 have the unique representation $P_1 = \sum_{i=1}^{s} a_i P^{s-i,i}$ and $P_2 = \sum_{i=1}^{s} b_i P^{s+1-i,i-1}$. In this case

$$P_0 = xP_1 + yP_2 = \sum_{i=1}^{s} (a_i + b_i)x^{s+1-i}y^i + \text{lower degree polynomials.}$$

Hence $P_0 \in \mathbb{P}_s$ if and only if $b_i = -a_i$ thus giving a connection between P_1 and P_2. It follows from the representation $P_0 = xP_1 + yP_2$ that $P_0 \perp \mathbb{P}_{s-1}$ which is equivalent with $I_\Omega(P_0 x^{s-j}y^{j-1}) = 0$ for $j = 1(1)s$. Thus, replacing

P_0 by its defining equation, we have that

$$I_\Omega[(xP_1+yP_2)x^{s-j}y^{j-1}]=I_\Omega\left[\left(x\sum_{i=1}^{n}a_iP^{s-i,i}-y\sum_{i=1}^{n}a_iP^{s+1-i,i-1}\right)x^{s-j}y^{j-1}\right]$$

and, from the identities,

$$I_\Omega(P^{s-i,i}x^{s+1-j}y^{j-1})=I_\Omega(P^{s-i,i}P^{s+1-j,j-1}),$$
$$I_\Omega(P^{s+1-i,i-1}x^{s-j}y^{j})=I_\Omega(P^{s+1-i,i-1}P^{s-j,j})$$

the result (5.7.1) follows. ■

The matrix $\mathbf{A}$ is antisymmetric. For this reason rank($\mathbf{A}$) is even and can be denoted by $2\alpha(s,\Omega)$ and since rank$(\mathbf{A})\leqslant s$ we have that $\alpha(s,\Omega)\in\{0,1,\ldots,[s/2]\}$. Thus, the space $\bar{\mathbb{R}}_s$ of vectors $\mathbf{a}$ satisfying (5.7.1) has dimension $s-2\alpha(s,\Omega)$ and, because of the one-to-one correspondence (Lemma 5.7.2),

$$\text{dimension }(\mathbb{L})=s-2\alpha(s,\Omega). \tag{5.7.2}$$

For certain important special cases $\alpha(s,\Omega)$ can be determined explicitly.

Lemma 5.7.3 In order that $\alpha(s,\Omega)=[s/2]$, either of the following conditions is sufficient:

(i) Ω possesses central symmetry, $s\in\mathbb{N}$,

(ii) Ω is a product-type region so that there exist one-dimensional integrals I_1 and I_2 with $I_\Omega(x^iy^k)=(I_1x^i)(I_2y^k)$, $i\cdot k\in\mathbb{N}$. ■

Proof We have to prove that $\alpha(s,\Omega)=[s/2]$ or, equivalently, that dim $(\mathbb{L})=s-2\alpha(s,\Omega)=s-2[s/2]\leqslant 1$, which means that either dim $(\mathbb{L})=1$ or dim $(\mathbb{L})=0$.

Assume that Ω possesses central symmetry. For every $(P_0,P_1,P_2)\in\mathbb{L}$ we have: if s is even (odd) then P_0, P_1 and P_2 are even (odd) and hence $P_0=xP_1+yP_2$ is odd (even) which is impossible unless $P_0\equiv 0$. Because of the connection between P_1 and P_2 there must further exist a polynomial $Q\in\mathbb{P}_{s-1}$ with $P_1=yQ$ and $P_2=-xQ$. Every polynomial $p(\mathbf{x})=a_{0,0}+a_{1,0}x+a_{0,1}y+\ldots+a_{0,s}y^s$ has the property that $a_{0,0}=p(\mathbf{0})=p(0,0)$. Further, $p(\mathbf{x})-p(\mathbf{0})$ can be split into $xp_1(\mathbf{x})+yp_2(\mathbf{x})$ where $p_1,p_2\in\mathbb{P}_{s-1}$ are suitably chosen. Hence $p(\mathbf{x})=p(\mathbf{0})+xp_1(\mathbf{x})+yp_2(\mathbf{x})$. On integration we obtain

$$\begin{aligned}I_\Omega(Qp)&=p(\mathbf{0})I_\Omega Q+I_\Omega(xQp_1)+I_\Omega(yQp_2)\\&=p(\mathbf{0})I_\Omega Q-I_\Omega(P_2p_1)+I_\Omega(P_1p_2)=p(\mathbf{0})I_\Omega Q\end{aligned}$$

since $P_1,P_2\perp\mathbb{P}_{s-1}$. There always exists a uniquely-determined representer $r\in\mathbb{P}_s$ such that $I_\Omega(rp)=p(\mathbf{0})$. This r may be constructed explicitly (see Problem P.5.13) and its existence may be concluded from the Riesz representation Theorem 3.5.1 since $I_\Omega(fg)=(f,g)$ is an inner product.

If s *is even* then P_1 and P_2 are also even and hence Q is odd, so that $I_\Omega Q=0$, from which it follows that $\dim(\mathbb{L})=0$.

Assume the contrary that $r \equiv Q/I_\Omega Q$ or, equivalently, $Q=r\cdot I_\Omega Q$. Then $\dim(\mathbb{L}) \leqslant \dim(0, yrI_\Omega Q, xrI_\Omega Q)=1$ and since $s-\dim(\mathbb{L})$ is even it now follows that $\dim(\mathbb{L})=1$.

In order to prove (ii) we assume that Ω is a product region. Then there exist systems of orthogonal polynomials relative to each variable. Assume that $R_k(x)$ and $S_k(y)$, $k=1(1)\ldots$ are such systems. Then the products $R_i(x)S_j(y)$ are orthogonal polynomials in Ω (cf. Section 5.6). Hence we have $P^{s-i,i}=R_{s-i}(x)S_i(y)$ if we normalise such that the leading coefficients are unity and the matrix **A** of (5.7.1) has the elements

$$a_{j,i}=I_\Omega(R_{s+1-j}S_{j-1}R_{s-i}S_i-R_{s-j}S_jR_{s+1-i}S_{i-1}).$$

Because of the orthogonality properties of $R_i(x)$ and $S_j(y)$ we have that $a_{j,i}\neq 0$ only if $i-j=\pm 1$ and hence the matrix has the form

$$\mathbf{A}=\begin{bmatrix} 0 & a_{1,2} & & \cdots & 0 \\ -a_{1,2} & 0 & a_{2,3} & \cdots & 0 \\ \cdots & & & & \cdots \\ \cdots & & & & \cdots \\ 0 & 0 & 0 & \cdots \quad -a_{s-1,s} & 0 \end{bmatrix}, \quad a_{i,i+1}\neq 0, \quad i=1(1)s-1.$$

This matrix has rank s if s is even and rank $s-1$ if s is odd, so that it always has rank $2[s/2]$. ■

Of course, we immediately ask whether $\alpha(s,\Omega)$ can be smaller than $[s/2]$. A sufficient condition for $\alpha(s,\Omega)=0$ is that

$$I_\Omega(P^{s+1-j,j-1}P^{s-i,i})=I_\Omega(P^{s-i,j}P^{s+1-i,i-1}) \quad \text{for} \quad i,j=1(1)s.$$

If this holds, then the matrix **A** of (5.7.1) contains only zeros and hence its rank is trivially zero. Clearly this is a constraint on the domain Ω. Such domains are given in [8] and [11] and in the following example.

Example 5.7.1 Let $\Omega=\{(x,y)\mid 0\leqslant|x|\leqslant A, \quad 0\leqslant y\leqslant B\}\cup\{(x,y)\mid 0\leqslant |x|\leqslant a, 0\geqslant y\geqslant -b\}$ which is symmetric with respect to the y axis, but which does not possess central symmetry. By choosing the three nodes $(\pm 1, 1)$ and $(0,\beta)$ and the corresponding weights A_1, A_1, and A_2, there are seven unknowns A, a, B, b, β, A_1, and A_2 which can be used to produce a formula of degree 3. On applying the method of undetermined coefficients we obtain a nonlinear system of equations for these unknowns which has infinitely many solutions if $0<\beta<4.195$; for example, $A=1.805\,275$, $B=2.068\,749$, $a=2.540\,185$, $b=6.214\,685$, $A_1=4.091\,065$, $A_2=2.444\,484$. ■

We now turn to cubature formulae $C_nf = \sum_{k=1}^{n} A_k f(\mathbf{x}_k)$ for arbitrary Ω and denote by $\mathbb{B}_s \equiv \{p \in \mathbb{P}_s \mid p(\mathbf{x}_k) = 0,\ k = 1(1)n\}$ ($\mathbb{B}_{s+1}$ analogously) the set of nodal polynomials for which the nodes of the cubature formula belong to the set of zero points. $\mathbb{B}_s$ is a linear space and since $\dim(\mathbb{P}_s) = \binom{s+2}{2}$, we have that

$$\dim(\mathbb{B}_s) \geq \binom{s+2}{2} - n. \tag{5.7.3}$$

Further, $\mathbb{B}_s \subseteq \mathbb{K}_s$, as an immediate consequence of:

Lemma 5.7.4 Let degree $(C_nf) = d$. If a polynomial $p(\mathbf{x}) \in \mathbb{P}_m$ has a zero at every node of C_nf, then $p(\mathbf{x}) \perp \mathbb{P}_{d-m}$. ■

Proof Assume that $p(\mathbf{x}) \in \dot{\mathbb{P}}_m$. Then, for every $q(\mathbf{x}) \in \mathbb{P}_{d-m}$ we have that $I_\Omega pq = C_n pq = 0$ since $qp \in \mathbb{P}_d$ and degree $(C_nf) = d$. Hence $p(\mathbf{x}) \perp \mathbb{P}_d$. If, in particular, $d = 2s-1$ and $m = s$, we have that $\mathbb{B}_s \subseteq \mathbb{K}_s$.

Let $\{P_1, \ldots, P_m\}$ be a basis of $\mathbb{B}_s$, and denote by $\mathbb{D}$ the space which is spanned by the polynomials P_i, xP_i, yP_i, $i = 1(1)m$. We show that $\dim(\mathbb{D})$ is bounded:

$$\dim(\mathbb{D}) \leq \dim(\mathbb{B}_{s+1}) \leq m + s + 2. \tag{5.7.4}$$

The inequality (5.7.4) is true since $\mathbb{B}_s = \mathbb{B}_{s+1} \cap \mathbb{P}_s \subseteq \mathbb{B}_{s+1} \subseteq \mathbb{P}_{s+1}$ so that every $p \in \mathbb{B}_{s+1}$ is the sum of a polynomial of $\mathbb{P}_{s+1}$ and an element of $\mathbb{B}_s$, and $\dim(\mathbb{P}_{s+1}) - \dim(\mathbb{P}_s) = \binom{s+3}{2} - \binom{s+2}{2} = s+2$. The inequality $\dim(\mathbb{D}) \leq \dim(\mathbb{B}_{s+1})$ follows from $\mathbb{D} \subseteq \mathbb{B}_{s+1}$. In order to obtain a lower bound of $\dim(\mathbb{D})$ we consider that the $(3m - \dim(\mathbb{D}))$-dimensional space of all vectors $\mathbf{b} \in \mathbb{R}^{3m}$ satisfying

$$\sum_{i=1}^{m} b_i P_i = x \sum_{i=1}^{m} b_{m+i} P_i + y \sum_{i=1}^{m} b_{2m+i} P_i$$

is in a one-to-one correspondence with the space $\mathbb{L} \cap (\mathbb{B}_s \times \mathbb{B}_s \times \mathbb{B}_s)$ because of the mapping $\mathbf{b} \to (\Sigma b_i P_i, \Sigma b_{m+i} P_i, \Sigma b_{2m+i} P_i)$. Hence, using (5.7.2), we see that

$$3m - \dim(\mathbb{D}) = \dim\{(\mathbb{L}) \cap (\mathbb{B}_s \times \mathbb{B}_s \times \mathbb{B}_s)\} \leq \dim(\mathbb{L}) = s - 2\alpha(s, \Omega) \tag{5.7.5}$$

so that $\dim(\mathbb{D}) \geq 3m - s + 2\alpha(s, \Omega)$ is obtained. From this, and from (5.7.4), it follows that $\dim(\mathbb{B}_s) = m \leq s + 1 - \alpha(s, \Omega)$. In addition, use of

(5.7.3) gives that $n \geq \binom{s+1}{2} + \alpha(s, \Omega)$. In the special case where $n = \binom{s+1}{2} + \alpha(s, \Omega)$ we have that $m = s+1-\alpha(s, \Omega)$, $\mathbb{D} = \mathbb{B}_{s+1}$, $\mathbb{L} \subseteq (\mathbb{B}_s \times \mathbb{B}_s \times \mathbb{B}_s)$. These results are amalgamated into a single theorem:

Theorem 5.7.1 (Möller) *If m is the largest number of linearly-independent polynomials of $\mathbb{P}_s$ which have zeros at all nodes of $C_n f$ with degree $(C_n f) = 2s-1$, then the inequalities*

$$n \geq \binom{s+1}{2} + \alpha(s, \Omega), \qquad m \leq s+1-\alpha(s, \Omega)$$

hold, where $2\alpha(s, \Omega)$ is the rank of A in (5.7.1).

If $n = \binom{s+1}{2} + \alpha(s, \Omega)$, then $m = s+1-\alpha(s, \Omega)$ and the three polynomials P_0, P_1, $P_2 \in \mathbb{K}_s$ (with $P_0 = xP_1 + yP_2$) have the common zeros $\mathbf{x}_k$ (which are the nodes of $C_n f$).

For regions Ω which possess central symmetry or for product regions we thus have the following minimal number of nodes for a cubature formula:

Degree of $C_n f$	s	Minimal n
1	1	1
3	2	4
5	3	7
7	4	12
9	5	17
11	6	24
13	7	31
15	8	40
17	9	49

We now turn to the problem of constructing cubature formulae of degree $2s-1$ with minimal number of nodes $\binom{s+1}{2} + \alpha(s, \Omega)$.

5.7.2 The construction of cubature formulae with minimal number of nodes

The basic result for constructing cubature formulae with minimal numbers of nodes is:

Theorem 5.7.2 (Möller) *Let degree* $(C_n f) = 2s-1$, *and* $n = \binom{s+1}{2} + \alpha(s, \Omega)$. *There then exist* $m = s+1-\alpha(s, \Omega)$ *linearly-independent polynomials*

$$P_i = \sum_{j=0}^{s} a_{i,j} P^{s-j,j}, \qquad i = 1(1)m, \tag{5.7.6}$$

having common zeros at the nodes of $C_n f$ *with the following properties:*

(i) *from* $xQ_1 + yQ_2 = Q_0$, $Q_i \in \mathbb{K}_s$, $i = 0, 1, 2$, *it follows that* Q_i *depend linearly on* $P_1, \ldots P_m$,

(ii)

$$\operatorname{rank} \begin{bmatrix} a_{1,0} & a_{1,1} & \cdots & & a_{1,s} & 0 \\ \cdots & \cdots & \cdots & \cdots & \cdots & \cdots \\ a_{m,0} & a_{m,1} & \cdots & & a_{m,s} & 0 \\ \cdots & \cdots & \cdots & \cdots & \cdots & \cdots \\ 0 & a_{1,0} & a_{1,1} & \cdots & a_{1,s-1} & a_{1,s} \\ 0 & a_{m,0} & a_{m,1} & \cdots & a_{m,s-1} & a_{m,s} \end{bmatrix} = s+2,$$

(iii) *there exist exactly* $2s - 3\alpha(s, \Omega)$ *linearly-independent vectors* $\mathbf{b} \in \mathbb{R}^{3m}$ *such that the polynomial*

$$F_{\mathbf{b}} = x^2 \sum_{i=1}^{m} b_i P_i + xy \sum_{i=1}^{m} b_{m+1} P_i + y^2 \sum_{i=1}^{m} b_{2m+i} P_i \in \mathbb{P}_{s+1},$$

which is rewritten as

$$F_{\mathbf{b}} = \sum_{i=1}^{m} L_i(\mathbf{x}) P_i, \qquad L_i(\mathbf{x}) \in \mathbb{P}_1 \quad \textit{for} \quad i = 1(1)m.$$

Proof It follows immediately from Theorem 5.7.1 that there exist m linearly-independent polynomials $P_1, \ldots, P_m \in \mathbb{P}_s$ having common zeros at the nodes of $C_n f$, and satisfying (i). Since $C_n f$ has degree $2s-1$ these are such that $P_i \in \mathbb{K}_s$ and hence they have the representation (5.7.6).

From the proof of Theorem 5.7.1 we see that dim $(\mathbb{D}) = m + s + 2$, and that dim $(\mathbb{B}_s) = $ dim $(\mathbb{D} \cap \mathbb{P}_s) = m$. Hence, dim $(\mathbb{D}/\mathbb{P}_s) = s+2$, where $\mathbb{D}/\mathbb{P}_s$ is generated by the polynomials xP_i and yP_i whose components in $\mathbb{P}_s$ are neglected. In addition, the vectors of the coefficients of the P_i are the rows of the matrix in (ii) so that (ii) holds.

The space $\mathbb{D}/\mathbb{P}_s$ is a subspace of $\mathbb{P}_{s+1}/\mathbb{P}_s$. But, since both spaces, have dimension $s+2$, they are equal: $\mathbb{D}/\mathbb{P}_s = \mathbb{P}_{s+1}/\mathbb{P}_s$. Let

$\bar{\mathbb{D}} \equiv \left\{ \sum_{i=1}^{m} Q_i P_i / Q_i \in \mathbb{P}_2 \right\}$. Then $x\mathbb{D}$, $y\mathbb{D} \subseteq \mathbb{D}$ and because of $x\mathbb{P}_{s+1}/\mathbb{P}_{s+1} + y\mathbb{P}_{s+1}/\mathbb{P}_{s+1} = \mathbb{P}_{s+2}/\mathbb{P}_{s+1}$ we find that $\bar{\mathbb{D}}/\mathbb{P}_{s+1} = \mathbb{P}_{s+2}/\mathbb{P}_{s+1}$; in particular, $\dim(\bar{\mathbb{D}}/\mathbb{P}_{s+1}) = s+3$. There thus exist exactly $3m-(s+3) = 2s - 3\alpha(s, \Omega)$ linearly-independent vectors $\mathbf{b} \in \mathbb{R}^{3m}$ with

$$F_{\mathbf{b}} = x^2 \sum_{i=1}^{m} b_i P_i + xy \sum_{i=1}^{m} b_{m+i} P_i + y^2 \sum_{i=1}^{m} b_{2m+i} P_i \in \mathbb{P}_{s+1}.$$

For each of these vectors $\mathbf{b}$ we have that $F_{\mathbf{b}}(\mathbf{x}_k) = 0$, which means that $F_{\mathbf{b}} \in \mathbb{B}_{s+1} = \mathbb{D}$ so that $F_{\mathbf{b}}$ is a linear combination of P_i, xP_i, and yP_i, $i = 1(1)m$. The statement of (iii) follows immediately. ■

Theorem 5.7.3 (Möller) *If the linearly-independent polynomials* $P_i = \sum_{j=0}^{s} a_{i,j} P^{s-j,j}$, $i = 1(1)m = s+1-\alpha(s, \Omega)$, *satisfy* (i), (ii) *and* (iii) *of Theorem* 5.7.2, *then they have* $n = \binom{s+1}{2} + \alpha(s, \Omega)$ *common zeros. If these are real and distinct they can be taken as nodes of a cubature formula* $C_n f$ *of degree* $2s-1$.

Although this theorem is very important for our purposes we cannot prove it in full here, since the proof needs much auxiliary material mainly from algebraic geometry, which cannot be treated here. For this reason the reader is referred to [5] for details. An outline of those parts of the proof which are subsequently needed is given.

Arguments similar to those of the proof of Theorem 5.7.2 allow us to conclude from (ii) that $\mathbb{D}/\mathbb{P}_s = \mathbb{P}_{s+1}/\mathbb{P}_s$. Let $\mathbb{D}_r \equiv \left\{ \sum_{i=1}^{m} Q_i P_i \mid Q_i \in \mathbb{P}_{r-s} \right\}$ for $r \geq s$, then $\mathbb{D} = \mathbb{D}_{s+1}$ and $x\mathbb{D}_r$, $y\mathbb{D}_r \subseteq \mathbb{D}_{r+1}$, and, by induction, it follows that $\mathbb{D}_{r+1}/\mathbb{P}_r = \mathbb{P}_{r+1}/\mathbb{P}_r$ for $r \geq s$. The space $\mathbb{D}_s$ is spanned by $P_1, \ldots, P_m$, and hence $\dim(\mathbb{D}_s) = m$. From this, $\dim(\mathbb{D}_r) = m + \sum_{i=s}^{r-1} \dim(\mathbb{P}_{i+1}/\mathbb{P}_i) = \binom{r+2}{2} - n$, $r \geq s$ is obtained.

The set of nodal polynomials having common zeros at the nodes of $C_n f$ is a linear subspace $\mathbb{B}$ (an ideal) of the ring of polynomials with the (ideal) basis $P_1, P_2, \ldots, P_m$. The equation $\dim(\mathbb{D}_r) = \binom{r+2}{2} - n$ for $r \geq s$ (together with (i) and (iii) of Theorem 5.7.2) lead (by elementary considerations) to the following:

Let $q \in \mathbb{B} \cap \mathbb{P}_r$, then there exist polynomials $Q_i \in \mathbb{P}_{r-s}$ ($\mathbb{P}_k = \{0\}$ for $-k \in \mathbb{N}$) with $q = \sum_{i=1}^{m} Q_i P_i$. As an immediate consequence $\mathbb{D}_r = \mathbb{B} \cap \mathbb{P}_r$ for $r \geqslant s$ and hence the polynomials of $\mathbb{B}$ form a space of codimension n in $\mathbb{P}_r$. It can be shown by considerations which can be taken from the theory of ideals that in this case $P_1, \ldots, P_m$ have exactly n common zeros which can happen to be not all distinct or real. If, in fact, these n common zeros are real and distinct, then the corresponding functionals $L_i : f(\mathbf{x}) \to f(\mathbf{x}_i)$ are linearly independent in $\mathbb{P}_{2s-1}$. Considering, in addition, the equality $\mathbb{D}_{2s-1} = \mathbb{B} \cap \mathbb{P}_{2s-1}$ and the properties that $L_i(q) = 0$, $i = 1(1)n$ for $q \in \mathbb{P}_{2s-1}$ and that $I_\Omega q = I_\Omega\left(\sum_{i=1}^{n} Q_i P_i\right) = 0$ (since the P_i are orthogonal) then I_Ω depends linearly on $L_1, \ldots, L_n$ (in $\mathbb{P}_{2s-1}$) and hence there exists a cubature formula of degree $2s-1$.

If one basis of an n-dimensional subspace of $\mathbb{K}_s$ satisfies any requirement of Theorem 5.7.3, then every further basis satisfies this. We say that $\mathbb{E}$ satisfies the requirements of Theorem 5.7.3 if $\mathbb{E}$ is an m-dimensional subspace of $\mathbb{K}_s$ and if every requirement of Theorem 5.7.3 is satisfied by a basis of $\mathbb{E}$.

In view of Theorems 5.7.2 and 5.7.3 it suffices to construct the m-dimensional subspaces of $\mathbb{K}_s$ satisfying the requirements of Theorem 5.7.3 in order to construct all cubature formulae of degree $2s-1$ with minimal number of nodes $\binom{s+1}{2} + \alpha(s, \Omega)$. This is done as follows.

First, we set up the space $\mathbb{L}$ of all triplets $(Q_0, Q_1, Q_2) \in \mathbb{K}_s^3$ with $Q_0 = xQ_1 + yQ_2$. The components of these generate a linear subspace $\mathbb{E}_0$ of $\mathbb{K}_s$. (The m-dimensional space $\mathbb{E} \subseteq \mathbb{K}_s$ satisfies (i) of Theorem 5.7.2 if and only if $\mathbb{E}_0 \subseteq \mathbb{E}$.) Next, we consider the space $\mathbb{F}$ of all 6-tuplets $(R_1, \ldots, R_6) \in \mathbb{K}_s^6$ with $x^2R_1 + xyR_2 + y^2R_3 = xR_4 + yR_5 + R_6$. If $(Q_0, Q_1, Q_2) \in \mathbb{L}$ then the 6-tuplet $(Q_1, Q_2, 0, Q_0, 0, 0) \in \mathbb{K}_s^6$ belongs to $\mathbb{F}$ since $x(xQ_1 + yQ_2) = xQ_0$ holds. Analogously, $(0, Q_1, Q_2, 0, Q_0, 0) \in \mathbb{K}_s^6$, since it is equivalent to $y(xQ_1 + yQ_2) = yQ_0$.

These two 6-tuplets generate a subspace $\mathbb{F}_0$ of $\mathbb{F}$. (On construction $\mathbb{E}_0$ is that space which is spanned by the components of the elements of $\mathbb{F}_0$.) The subspaces $\mathbb{F}_1$ of $\mathbb{F}$ with $\mathbb{F}_0 \subseteq \mathbb{F}_1$ are now determined such that the following holds: The triplets which contain the first three components of each element of $\mathbb{F}_1$ establish a subspace of $\mathbb{K}_s^3$ with dimension $2s - 3\alpha(s, \Omega)$, *and* the components of the elements of $\mathbb{F}_1$ span a subspace $\mathbb{E}_1$ of $\mathbb{K}_s$ with $\dim(\mathbb{E}_1) \leqslant m$. ($\mathbb{E}_0 \subseteq \mathbb{E}_1$ since $\mathbb{F}_0 \subseteq \mathbb{F}_1$.) Thus, the m-dimensional space $\mathbb{E} \subseteq \mathbb{K}_s$ satisfies (i) and (iii) of Theorem 5.7.2 if and only if there exist spaces $\mathbb{F}_1$ and $\mathbb{E}_1$ with the above-mentioned properties, and such that

$\mathbb{E}_1 \subseteq \mathbb{E}$. It is now easy to test whether the remaining requirements of Theorem 5.7.3 are fulfilled.

In order to become familiar with the previous theoretical results we give an example:

Example 5.7.2 Let $\Omega = \{(x, y) \mid |x|, |y| \leq 1\}$. As already seen in Section 5.6, the corresponding system of orthogonal polynomials is that of products $P^{i,j}(\mathbf{x}) = L_i(x)L_j(y)$ of Legendre polynomials. These are assumed to be normalised so that their leading coefficients are unity. Since Ω possesses central symmetry we have that the minimal number of nodes for the cubature formula is $n = \binom{s+1}{2} + [s/2]$. Taking the special case $n = 17$ for $s = 5$, we determine the nodes and weights of the corresponding ninth-degree cubature formula $C_{17}f$, starting with the calculation of Q_0, Q_1 and Q_2. We have to set up and to solve the system of equations (5.7.1) which has the matrix of coefficients

$$\mathbf{A} = \begin{bmatrix} 0 & -(L_1L_1)(L_4L_4) & 0 & 0 & 0 \\ (L_1L_1)(L_4L_4) & 0 & -(L_2L_2)(L_3L_3) & 0 & 0 \\ 0 & (L_2L_2)(L_3L_3) & 0 & -(L_2L_2)(L_3L_3) & 0 \\ 0 & 0 & (L_2L_2)(L_3L_3) & 0 & -(L_1L_1)(L_4L_4) \\ 0 & 0 & 0 & (L_1L_1)(L_4L_4) & 0 \end{bmatrix}$$

where (L_iL_k) are inner products of Legendre polynomials. Hence (5.7.1) has the solution

$$a_2 = a_4 = 0, \qquad a_1 = a_5 = \frac{c}{(L_1L_1)(L_4L_4)}, \quad \text{and} \quad a_3 = \frac{c}{(L_2L_2)(L_3L_3)}$$

where c is an arbitrary real constant. From Section 5.2.1 we have that

$$(L_nL_n) = (L_n, L_n) = \frac{1}{2n+1} \frac{(n!)^4 2^{2n+1}}{((2n)!)^2} = \frac{2^{2n-1}}{2n+1} \left[\frac{(n-1) \cdot \cdots \cdot 1}{(2n-1) \cdots (n+1)} \right]^2.$$

The choice $c = 64/1575$ produces the smallest integer solution

$$\begin{aligned} Q_1 &= \sum_{i=1}^{5} a_i P^{5-i,i} = 21P^{4,1} + 20P^{2,3} + 21P^{0,5} \\ Q_2 &= -\sum_{i=1}^{5} a_i P^{6-i,i-1} = -21P^{5,0} - 20P^{3,2} - 21P^{1,4} \end{aligned} \tag{5.7.7}$$

and hence $Q_0 = xQ_1 + yQ_2 = 0$. Therefore, the space $\mathbb{L}$ contains only the triplet $(0, Q_1, Q_2)$ and its constant multiples, and thus $\mathbb{E}$ is spanned by Q_1 and Q_2. Note that the polynomials of $\mathbb{K}_5$ are odd. Hence, in $x^2R_1 + xyR_2 + y^2R_3 = xR_4 + yR_5 + R_6$, $R_i \in \mathbb{K}_5$, the polynomials x^2R_1,

xyR_2, y^2R_3, and R_6 are odd and the polynomials xR_4 and yR_5 are even. Thus, $\mathbb{F}$ is the direct sum of the spaces

$$\mathbb{F}' = \{(R_1, R_2, R_3, 0, 0, R_6) \mid x^2R_1 + xyR_2 + y^2R_3 = R_6\}$$

and

$$\mathbb{F}'' = \{(0, 0, 0, R_4, R_5, 0) \mid xR_4yR_5 = 0\}.$$

The components of the elements of the second space belong to $\mathbb{E}_0$. Since $m = s + 1 - \alpha(s, \Omega) = 4$ for $s = 5$, we have $\dim(\mathbb{E}) = m = 4$ causing only the first space $\mathbb{F}'$ to be important for the determination of $\mathbb{E} \subseteq \mathbb{K}_5$. In a similar manner as in the proof of Lemma 5.7.2 it can be seen that there exists a one-to-one correspondence between the space of 6-tuplets $(R_1, R_2, R_3, 0, 0, R_6) \in \mathbb{F}$ and the space of vectors $\mathbf{a} \in \mathbb{R}^{10}$, satisfying the relationships

$$R_1 = \sum_{i=1}^{5} a_i P^{5-i,i}, \qquad R_3 = \sum_{i=0}^{4} a_{i+6} P^{5-i,i}$$

$$R_2 = -\sum_{i=0}^{4} a_{i+1} P^{5-i,i} - \sum_{i=1}^{5} a_{i+5} P^{5-i,i}$$

and

$$\begin{bmatrix} 0 & 20 & 0 & 0 & 0 & 20 & 0 & -21 & 0 & 0 \\ -20 & 0 & 21 & 0 & 0 & 0 & 21 & 0 & -21 & 0 \\ 0 & -21 & 0 & 21 & 0 & 0 & 0 & 21 & 0 & -20 \\ 0 & 0 & -21 & 0 & 20 & 0 & 0 & 0 & 20 & 0 \end{bmatrix} \cdot \begin{bmatrix} a_1 \\ \cdot \\ \cdot \\ a_{10} \end{bmatrix} = 0. \tag{5.7.8}$$

The element $f_1 = (Q_1, -Q_2, 0, 0, 0, 0) \in \mathbb{F}'$ corresponds to $\mathbf{a}^{(1)} = (21, 0, 20, 0, 21, 0, 0, 0, 0, 0)^T$ and $f_2 = (0, Q_1, -Q_2, 0, 0, 0) \in \mathbb{F}'$ to $\mathbf{a}^{(2)} = (0, 0, 0, 0, 0, 21, 0, 20, 0, 21)^T$. It is easy to see that $\mathbb{F}_1$ satisfies the above conditions if and only if $\mathbb{F}_1 \cap \mathbb{F}'$ possesses a basis, which contains f_1, f_2 and two further elements, and if their components generate a space having at most dimension two, ignoring multiples of $\mathbb{E}_0$.

Hence, we have to find all linearly-independent vectors $\mathbf{a}^{(3)}$ and $\mathbf{a}^{(4)} \in \mathbb{R}^{10}$ omitting all linear combinations of $\mathbf{a}^{(1)}$ and $\mathbf{a}^{(2)}$, such that $\mathbf{a}^{(3)}$ and $\mathbf{a}^{(4)}$ are solutions of (5.7.8) and satisfy

$$\operatorname{rank} \begin{bmatrix} 21 & 0 & 20 & 0 & 21 & 0 \\ 0 & 21 & 0 & 20 & 0 & 21 \\ b_{0,0} & b_{1,0} & b_{2,0} & b_{3,0} & b_{4,0} & b_{5,0} \\ \cdot & \cdot & \cdot & \cdot & \cdot & \cdot \\ b_{0,7} & b_{1,7} & b_{2,7} & b_{3,7} & b_{4,7} & b_{5,7} \end{bmatrix} \leqslant 4 \tag{5.7.9}$$

where the 6-tuplet $(R_1^{(i)}, R_2^{(i)}, R_3^{(i)}, 0, 0, R_0^{(i)}) \in \mathbb{F}'$, $i=3,4$, corresponds to $\mathbf{a}^{(i)}$ with

$$R_k^{(3)} = \sum_{j=0}^{5} b_{j,k} P^{5-j,j}, \qquad R_k^{(4)} = \sum_{i=0}^{5} b_{j,k+4} P^{5-j,j}, \qquad k=0(1)3.$$

After some tedious calculation we see that for every $\mathbb{F}_1$ having the above properties the space $\mathbb{E}_1$ (which is generated by the components of the elements of $\mathbb{F}_1$) is spanned by Q_1, Q_2, Q_3, and Q_4 with Q_1, Q_2 as in (5.7.7), $Q_3 = 3P^{5,0} - \sigma P^{4,1} - 3P^{3,2}$, $Q_4 = -3P^{2,3} + \sigma P^{1,4} + 3P^{0,5}$, if $\sigma^2 = 43/18$. If $\mathbb{E} \subseteq \mathbb{K}_s$ for any four-dimensional space satisfying (i) and (iii) of Theorem 5.7.2, then either $\mathbb{E} = \mathbb{E}_1^+$ or $\mathbb{E} = \mathbb{E}_1^-$, where the only difference is the sign of σ. We see that in both cases (ii) of Theorem 5.7.2 is satisfied by $\mathbb{E}$, and the common zeros of the polynomials of $\mathbb{E}$ are real and distinct.

For $\mathbb{E} = \mathbb{E}_1^-$ the nodes and weights of the resulting cubature formula of degree 9 with minimal number of nodes 17 are

Nodes	Weights	Values of $x_1, x_2, \ldots, y_4$
$(0, 0)$	0.52674 89711 93415 63786	
$\pm(x_1, y_1)$, $\pm(-y_1, x_1)$	0.08887 93781 70198 70697	$x_1 = 0.96884\ 99663\ 61977\ 72072$ $y_1 = 0.63068\ 01197\ 31668\ 85417$
$\pm(x_2, y_2)$, $\pm(-y_2, x_2)$	0.11209 96021 29596 48528	$x_2 = 0.75027\ 70999\ 78900\ 53354$ $y_2 = 0.92796\ 16459\ 59569\ 66740$
$\pm(x_3, y_3)$, $\pm(-y_3, x_3)$	0.39828 24392 62070 09528	$x_3 = 0.52373\ 58202\ 14429\ 33604$ $y_3 = 0.45333\ 98211\ 35647\ 19076$
$\pm(x_4, y_4)$, $\pm(-y_4, x_4)$	0.26905 13376 39780 80301	$x_4 = 0.07620\ 83281\ 92617\ 17318$ $y_4 = 0.85261\ 57293\ 33662\ 30775$

Correspondingly, for $\mathbb{E} = \mathbb{E}_1^+$, the coordinates of each node are interchanged. From the method of construction it can be concluded that there is no other real cubature formula of degree 9 with the minimal number of nodes 17. ■

5.7.3 Lower bounds for the minimal number of nodes for N-dimensional cubature formulae

While Section 5.7.2 was concerned with constructive aspects for $N = 2$ dimensions, we now consider the existence of cubature formulae with minimal number of nodes in the general N-dimensional situation.

Denote the monomials by $\mathbf{x}^{\mathbf{i}}=x_1^{i_1}\dots x_N^{i_N}$ with $x_i\in\mathbb{R}$, $\mathbf{x}\in\mathbb{R}_N$, $\mathbf{i}\in\mathbb{N}_0\times\mathbb{N}_0\times\dots\times\mathbb{N}_0=\mathbb{N}_0^N$. For the multi-index $\mathbf{i}$ we define $|\mathbf{i}|=\sum_{j=1}^{N} i_j$. Further, let $\mathbf{e}_r^T=(\delta_{r,1},\delta_{r,2},\dots,\delta_{r,N})$, $r\in\{1,\dots,N\}$ be the N-dimensional generalisation of the Kronecker symbol $\delta_{r,k}$. It is not necessary to prove Theorem 5.7.1 again for $N>2$ dimensions since this extension is easily performed as follows:

Assume that we are given the domain Ω for the N-dimensional integral $I_\Omega f$, and the degree $2s-1$ of a cubature formula $C_n f$ which approximates $I_\Omega f$. The space $\mathbb{K}$ of N-dimensional orthogonal polynomials possesses a basis $\{P^{\mathbf{i}}\mid \mathbf{i}\in\mathbb{N}_0^N, |\mathbf{i}|=s\}$ such that each polynomial $P^{\mathbf{i}}$ is uniquely determined by $P^{\mathbf{i}}(\mathbf{x})=\mathbf{x}^{\mathbf{i}}+Q_{\mathbf{i}}(\mathbf{x})$ with $Q_{\mathbf{i}}\in\mathbb{P}_{s-1}$. For arbitrary $l,t\in\{1,\dots,N\}$ with $l\neq t$ we obtain from

$$x_l P_1+y_t P_2=P_0,\qquad P_0,P_1,P_2\in\mathbb{K}_s \tag{5.7.10}$$

the equations

$$P_1=\sum_{|\mathbf{r}|=s-1}\mathbf{a}_{\mathbf{r}}P^{\mathbf{r}+\mathbf{e}_t},\qquad P_2=\sum_{|\mathbf{r}|=s-1}-\mathbf{a}_{\mathbf{r}}P^{\mathbf{r}+\mathbf{e}_l},$$

and the equations

$$0=\sum_{|\mathbf{r}|=s-1}\mathbf{a}_{\mathbf{r}}I_\Omega(P^{\mathbf{r}+\mathbf{e}_t}P^{\mathbf{q}+\mathbf{e}_l}-P^{\mathbf{r}+\mathbf{e}_l}P^{\mathbf{q}+\mathbf{e}_t},\qquad |\mathbf{q}|=s-1, \tag{5.7.11}$$

which generalise the system of equations (5.7.1). The matrix of the system (5.7.11) is also antisymmetric and hence has even rank $2\alpha_{l,t}$ with

$$\alpha_{l,t}=\alpha_{l,t}(s,\Omega)\in\left\{0,1,\dots,\left[\frac{1}{2}\binom{s+N-2}{N-1}\right]\right\}. \tag{5.7.12}$$

The upper limit $\left[\frac{1}{2}\binom{s+N-2}{N-1}\right]$ corresponds to the $[s/2]$ used for $N=2$. In (5.7.11) the number of equations is obtained by taking into account all combinations of N integers b_i with $0\leqslant b_i\leqslant s-1$ such that $b_1+\dots+b_N=s-1$. From combinatorics we have that there exist $\binom{s+N-2}{N-1}$ such combinations.

Recall the linear space $\mathbb{B}_s=\{P\in\mathbb{P}_s\mid P(\mathbf{x}_k)=0,\ k=1(1)m\}$. For every basis $\{P_1,\dots,P_m\}$ of $\mathbb{B}_s$ we consider the totality of solutions of (5.7.10) and conclude that from the set of polynomials $\mathbb{D}=\{P_i,\ x_lP_i,\ x_tP_i,\ i=1(1)m\}$ at least $3m-\binom{s+N-2}{N-1}+2\alpha(s,\Omega)$ are linearly independent. However, $\mathbb{B}$ is a subspace of the space of those polynomials which is spanned by certain monomials of $\mathbb{P}_{s+1}$. These belong either to $\mathbb{P}_s$ or possess a factor x_l or x_t and, hence, the dimension of this larger space is $\binom{s+N}{N}+\binom{s+N-1}{N-1}+\binom{s+N-2}{N-2}$. Thus, at most $m+\binom{s+N-1}{N-1}+\binom{s+N-2}{N-2}$ polynomials of $\mathbb{D}$ can be linearly independent so that $m \leqslant \binom{s+N-1}{N-1}-\alpha_{l,t}$ and:

Theorem 5.7.4 (Möller) *For a cubature formula $C_n f$ with degree $2s-1$ the number of nodes is bounded by $n \geqslant \binom{s+N-1}{N}+\alpha(s,\Omega)$ for arbitrary $l, t \in \{1,\ldots,N\}$, $l \neq t$, where $2\alpha_{l,t}$ is the rank of the matrix of the system* (5.7.11).

If $n=\binom{s+N-1}{N}+\alpha_{l,t}$, then any three polynomials P_0, P_1, P_2 which satisfy (5.7.10) *have common zeros at the nodes $\mathbf{x}_k$, $k=1(1)n$ of $C_n f$.*

Example 5.7.3 Assume that Ω is the N-dimensional simplex $S_N=\left\{\mathbf{x}\in\mathbb{R}^N \,\middle|\, \sum_{i=1}^{N} x_i \leqslant 1, x_i \geqslant 0\right\}$, and let $s=2$. We construct an N-dimensional third-degree cubature formula for S_N. The space $\mathbb{K}_s$ has the basis $\{P^{\mathbf{i}} \mid |\mathbf{i}|=2\}$ with

$$P^{\mathbf{e}_m+\mathbf{e}_n}=x_m x_n-\frac{1}{N+3}(x_m+x_n)+\frac{1}{(N+2)(N+3)}\Bigg\} \quad m\neq n;\quad m,n=1(1)N.$$

$$P^{2\mathbf{e}_m}=x_m^2-\frac{4}{N+3}x_m+\frac{2}{(N+2)(N+3)}.$$

After some calculation we see that $\alpha_{m,n}=1$, and that the space of triplets (P_0, P_1, P_2) which satisfy (5.7.10) has a basis $\{(Q_l^{(m,n)}, R_l^{(m,n)}, S_l^{(m,n)}) \mid l \in \{1,\ldots,N\},\ n\neq l\neq m\}$ for $N\geqslant 3$ where

$$R_l^{(m,n)}=P^{2\mathbf{e}_n}+P^{\mathbf{e}_n+\mathbf{e}_m}+(N-1)P^{\mathbf{e}_n+\mathbf{e}_l}.$$

$$-S_l^{(m,n)}=P^{2\mathbf{e}_m}+P^{\mathbf{e}_m+\mathbf{e}_n}+(N-1)P^{\mathbf{e}_n+\mathbf{e}_l}$$

$$Q_l^{(m,n)}=(R_l^{(m,n)}+S_l^{(m,n)})/(N+3).$$

Hence, from Theorem 5.7.4 the third-degree cubature formulae have at least $N+2$ nodes, and for $N\geqslant 3$ the nodes of any third-degree formula

$C_{N+2}f$ are the common zeros of the polynomials

$$\left.\begin{aligned} R_l^{(m,n)} &= \left(x_n - \frac{1}{N+3}\right)(x_m + x_n + (N-1)x_l - 1) \\ -S_l^{(m,n)} &= \left(x_m - \frac{1}{N+3}\right)(x_m + x_n + (N-1)x_l - 1) \end{aligned}\right\} \begin{array}{l} m, n, l \in \{1, \ldots, N\}; \\ m \neq n \neq l; \quad m \neq l, \end{array}$$

These polynomials have exactly $N+2$ common zeros so that all third-degree cubature formulae with only $N+2$ nodes have the same set of nodes. The weights are to be obtained by the method of undetermined coefficients so that the following third-degree cubature formula for the N simplex, $N>2$, is obtained. ■

Nodes	Weights
$\mathbf{x}_1^T = \left(\frac{1}{N+1}, \ldots, \frac{1}{N+1}\right)$	$-\frac{1}{4}(N+1)^3/(N+2)!$
$\mathbf{x}_2^T = \left(\frac{1}{N+3}, \ldots, \frac{1}{N+3}\right)$	
$\mathbf{x}_3^T = \left(\frac{3}{N+3}, \frac{1}{N+3}, \frac{1}{N+3}, \ldots, \frac{1}{N+3}\right)$	$\frac{1}{4}(N+3)^3/(N+3)!$
. . . .	
$\mathbf{x}_{N+2}^T = \left(\frac{1}{N+3}, \ldots, \frac{1}{N+3}, \frac{3}{N+3}\right)$	

Note that there is a negative weight, which cannot arise in quadrature formulae with minimal number of nodes (cf. Chapter 7).

I thank Dr. Möller for private communication concerning Section 5.7. He suggested several improvements.

5.8 Remarks, problems, references

5.8.1 Remarks

R.5.1: Möller [6] also shows, that, in addition to Lemma 5.7.3,

$$\alpha(s, \Omega) = [s/2] \quad \text{if} \quad I_\Omega p = \int_0^1 \int_0^{1-y} p(x, y)\, dx\, dy, \qquad p(\mathbf{x}) \in \mathbb{P}_{2s},$$

is valid for $s = 1(1)6$. This can be shown by straightforward calculation (cf. [6] and Section 5.7.1).

R.5.2: For the quadrature problem the nodes of minimal node formulae are the uniquely-determined zeros of explicitly-defined orthogonal polynomials. For n nodes we obtain degree $2n-1$. The corresponding situation in cubature is different. We have seen that the minimal number of nodes of a cubature formula of degree 7 is 12. There are, however, at least three essentially different formulae of this type, as given in Table 5.8.1.

TABLE 5.8.1

Nodes x_i	Nodes y_i	Weights
$6/7$	0	$98/405$
$\pm((114\mp 3\sqrt{583})/287)^{\frac{1}{2}}$	$\pm(114\mp 3\sqrt{583})/287)^{\frac{1}{2}}$	$\dfrac{178981\pm 2769\sqrt{583}}{472230}$
$\pm\sqrt{3/5}$	$\pm\sqrt{3/5}$	$25/81$
0	$\pm\frac{1}{2}(3(1\mp\sqrt{11/35}))^{\frac{1}{2}}$	$\frac{4}{81}(7\pm 3\sqrt{25/11})$
$\pm\frac{1}{2}(3(1\mp\sqrt{11/35}))^{\frac{1}{2}}$	0	$\frac{4}{81}(7\pm 3\sqrt{25/11})$
±0.917 117 822 313	±0.547 931 206 828	0.213 057 211 621
±0.611 268 766 465	±0.938 843 256 659	0.174 009 488 947
±0.529 422 802 043	0	0.635 853 883 443
0	±0.627 041 373 780	0.590 012 715 421

The first formula is due to Tylor (1953) [15], the second one to Engels (1966) [2] and the third to Haegemans and Piessens (1975) [3]. There is not only a lack of uniqueness (which is impossible for Gaussian quadrature formulae), but also the further disadvantage that the second formula has four nodes outside Ω. This is also a surprising result, which is not to be expected from the behaviour of quadrature formulae (cf. Section 5.7.).

5.8.2 Problems

P.5.1: Orthonormalise the system of functions $\{1, \sin\frac{1}{2}\pi x, \cos\frac{1}{2}\pi x\}$ with respect to the inner products (cf. Section 5.2)

$$(f, g)=\int_0^1 fg\,dx \quad \text{and} \quad \langle f, g\rangle=\tfrac{1}{4}[f(0)g(0)+2f(\tfrac{1}{2})g(\tfrac{1}{2})+f(1)g(1).$$

P.5.2: Use a recurrence relationship in order to calculate the Legendre polynomials $L_n(x)$ and $l_n(x)$ for $n=0(1)6$ (cf. Section 5.2.1).

P.5.3: Expand x^5 into a sum of Legendre polynomials $L_n(x)$ (cf. Formula (5.1.5)).

P.5.4: Calculate explicitly the Tschebyscheff polynomials $T_n(x)$ by solving the second-order difference equation

$$T_{n+1}(x) = 2xT_n(x) - T_{n-1}(x)$$

with initial conditions $T_0 = 1$, $T_1 = x$. (Hint: try $T_n(x) = At^n$ with constant A (cf. Section 5.2.2).)

P.5.5: Show that the functions $f_k(x) = \cos k\pi x$, $k = 0(1)n$ are orthogonal with respect to the inner product $\langle f, g\rangle = \sum_{r=0}^{n}{}'' f\left(\frac{r}{n}\right) g\left(\frac{r}{n}\right)$.

The index $''$ indicates that the first and last terms are to be halved (cf. Section 5.1).

P.5.6: Show (cf. Section 5.2.3):

(i) If the Jacobi polynomials are given by

$$P_n^{\alpha,\beta} = \frac{(-1)^n}{2^n n!} \sum_{j=0}^{n} \binom{n}{j} \frac{\Gamma(n+\alpha+1)\Gamma(n+\beta+1)}{\Gamma(n+\alpha-j+1)\Gamma(\beta+j+1)} (1-x)^{n-j}(1+x)^j(-1)^j$$

then the normalised Jacobi polynomials are

$$p_n^{\alpha,\beta} = \left(\frac{(2n+\alpha+\beta+1)n!\,\Gamma(n+\alpha+\beta+1)}{2^{\alpha+\beta+1}\Gamma(n+\alpha+1)\Gamma(n+\beta+1)}\right)^{\frac{1}{2}} P_n^{\alpha,\beta}.$$

(ii) The Jacobi polynomials satisfy

$$P_{n+1}^{\alpha,\beta} = \frac{\gamma[\alpha^2-\beta^2+(\gamma^2-1)x]P_n^{\alpha,\beta} - 2(n+\alpha)(n+\beta)(\gamma+1)P_{n-1}^{\alpha,\beta}}{2(n+1)(\gamma-n)(\gamma-1)}$$

with $\gamma = 2n+\alpha+\beta+1$ and

$$P_0^{\alpha,\beta} = 1, \qquad P_1^{\alpha,\beta} = \tfrac{1}{2}(1+\alpha)(x+1) + \tfrac{1}{2}(1+\beta)(x-1)$$

(cf. Davis [1]).

P.5.7: Show: the Tschebyscheff polynomials of the first kind satisfy (cf. Section 5.2.2)

$$\int_{-1}^{1} T_k(x)\,dx = \begin{cases} 0 & \text{for} \quad k \text{ odd} \\ -2/(k^2-1) & \text{for} \quad k \text{ even.} \end{cases}$$

P.5.8: Prove: the Jacobi polynomials satisfy the second-order differential equation (cf. Section 5.2.3)

$$W(x)P_n^{\alpha,\beta}(x) + \frac{1}{n(n+\alpha+\beta+1)}\frac{d}{dx}\left[(1-x^2)W(x)\frac{d}{dx}P_n^{\alpha,\beta}(x)\right],$$

where $W(x)$ is the Jacobi weight function.

P.5.9: Calculate the weights of the quadrature formulae based upon the zeros of Tables 5.3.1 and 5.3.2 as nodes with respect to the corresponding weight functions (cf. Section 5.3.2).

P.5.10: Orthonormalise the monomials x^k, $k=0(1)\ldots$ for the inner products (cf. Sections 5.1 and 5.5)

$$(f, g) = \int_{|z|=R} f(z)\overline{g(z)}\, ds, \qquad z = x + iy = |z|e^{is}, \quad s \text{ arc length of } |z| = R,$$

and

$$(f, g) = \int_{|z|\leq R} f(z)\overline{g(z)}\, dx\, dy.$$

P.5.11: Decide whether there exist orthogonal polynomials of degree 3 having common zeros (a, b), $(a, -b)$, $(-a, b)$, and $(-a, -b)$ with $a^2+b^2=2/3$ for the square $|x|, |y|\leq 1$ (cf. Fig. 5.6.1).

P.5.12: Show that the space $\mathbb{K}_r$ in Section 5.7.1 is a linear space with basis $\{P^{r,0}, P^{r-1,1}, \ldots, P^{0,r}\}$. Here $P^{r-i,i} = P^{r-i,i}(x, y) = x^{r-i}y^i + g(x, y)$, $g \in \mathbb{P}_{r-1}$, $i = 0(1)r$ (Jackson [4]).

P.5.13: Let $p(\mathbf{x}) = 1 + 2x + xy + 3y^2$. Determine the representer $r \in \mathbb{P}_2$ with the property $(r, p) = I_\Omega(r, p) = p(0, 0)$ by calculating directly its coefficients. Ω is the square $0 \leq |x|, |y| \leq 1$. How do you proceed if $p \in \mathbb{P}_s$ is a given polynomial and Ω is not a square but possesses central symmetry? (Section 5.7.)

5.8.3 References

1. Davis, P. (1963). "Interpolation and approximation." Blaisdell, New York.
2. Engels, H. (1966). "Neue Ergebnisse zur numerischen Kubatur." Thesis, Aachen.
3. Haegemans, A. and Piessens, R. (1975). Cubature formulas of degree nine for symmetric planar regions. *Math. Comput.*, **29,** 810–815.
4. Jackson, D. (1936). Formal properties of orthogonal polynomials in two variables. *Duke Math. J.*, **2,** 423–434.
5. Möller, H. M. (1973). "Polynomideale und Kubaturformeln." Thesis, Dortmund.
6. Möller, H. M. (1976). Kubaturformeln mit minimaler Knotenzahl. *Num. Math.*, **25,** 185–200.
7. Mysovskih, I. P. (1967). Radon's paper on the cubature formula. *USSR Comput. Math.*, **7** (4), 232–236.
8. Mysovskih, I. P. (1969). Cubature formulae and orthogonal polynomials. *USSR Comput. Math.*, **9** (2), 217–228.
9. Mysovskih, I. P. (1970). Numerical characteristics of orthogonal polynomials in two variables. *Vestnik Leningrad Univ.*, **4** (19), 46–53.
10. Mysovskih, I. P. (1972). Interpolatory cubature formulae and orthogonal polynomials. *Vopr. vicisl. i pricl. mat., Taschkent,* **14,** 90–98.
11. Mysovskih, I. P. and Cernina, V. J. (1971). The answer to a question of Radon., *Soviet Math. Doklady*, **12** (3), 852–854.
12. Radon, J. (1948). Zur mechanischen Kubatur. *Monatshefte Math.* **52,** 286–300.
13. Smart, D. R. (1974). "Fixed Point Theorems." Cambridge University Press, Cambridge.

14. Turàn, P. (1950). On the theory of mechanical quadrature. *Acta Sci. Math.*, **12,** 30–37.
15. Tylor, G. W. (1953). Numerical integration of functions of several variables. *Can. J. Math.*, **5,** 393–412.

6

Interpolatory Quadrature and Cubature Formulae–Preassigned Nodes or Weights

We now treat some quadrature procedures in which either the nodes or the weights are preassigned. In this case the procedures do not usually have the highest possible degree of precision. Further, they are required to be interpolatory; i.e. they are integrals of an interpolant of f.

6.1 Classification and general properties

We consider polynomial interpolatory quadrature formulae with given or preassigned nodes and weight function $W \equiv 1$. The most important class of such formulae is that of the *Newton–Cotes formulae* which are based upon the preassigned *equidistant* nodes $x_i = -1 + 2(i-1)/(n-1)$, $i = 1(1)n$. In this case ± 1 are nodes and we denote the corresponding quadrature formulae as *closed Newton–Cotes formulae.* If the nodes ± 1 are omitted, the formulae become *open Newton–Cotes formulae.* For $x_i = -1 + 2i/(n+1)$, $i = 1(1)n$, the corresponding formulae are denoted as *Maclaurin open formulae.* These formulae can be easily extended to certain N-dimensional regions. This is always possible for product-type regions where tensor product interpolants of the integrand can be used. Extension to simplexes, spheres or spherical shells can also be done; the last by using mixed interpolation consisting of polynomial interpolation in the radial direction and trigonometric interpolation in the angular direction, thus again producing a product-type interpolation.

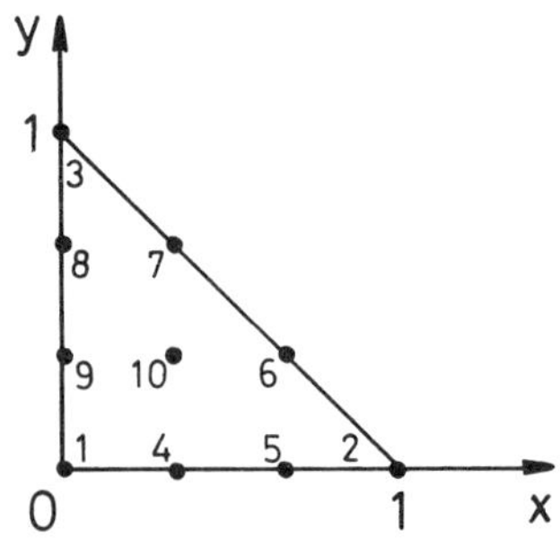

FIG. 6.1.1. 2-Simplex with ten equidistant nodes.

For the 2-simplex (the extension to $N>2$ is trivial) we have the following third-degree interpolation formula $N_3(f, x, y)$ of Newton type for ten nodes (Fig. 6.1.1).

The ten coefficients of

$$N_3(f, x, y) = a_{0,0} + a_{1,0}x + a_{0,1}y + a_{2,0}x^2 + a_{1,1}xy + a_{0,2}y^2 + a_{3,0}x^3 + a_{2,1}x^2y + a_{1,2}xy^2 + a_{0,3}y^3$$

are uniquely determined by the interpolation requirement that $N_3(f, i/3, k/3) = f(i/3, k/3)$, $0 \leqslant i+k \leqslant 3$. On integration, a cubature formula $C_{10}f$ results.

If the weights, rather than the nodes, are preassigned then another kind of quadrature procedure arises. For $W(x) \equiv 1$ and for *equal weights* we obtain the *Tschebyscheff equally-weighted quadrature formulae* which have already been listed in Section 2.4.3. The construction of corresponding equally-weighted cubature formulae is not straightforward as there is again the difficulty that the degree of such formulae does not only depend on the number of nodes but also on the shape of Ω. It is more convenient and in some sense more natural to derive equally-weighted cubature formulae by requiring only that they have the highest possible degree of precision. The resulting cubature formulae may or may not be interpolatory quadrature formulae.

We mention that quadrature formulae with given nodes may, in addition, contain derivatives of the integrand if they are derived from a suitable Hermitian interpolating operator. Some properties of interpolatory quadrature formulae are now stated. As already mentioned, these formulae are obtained by integrating an interpolating operator, specifically by integration of an interpolating polynomial. From the error representation of polynomial interpolation we have

Theorem 6.1.1 *The degree of a polynomial interpolatory quadrature formula using n nodes is at least $n-1$.*

The proof is performed by remembering that the quadrature error satisfies

$$E_n f = \int_{-1}^{1} e_n(f, x)\,\mathrm{d}x = \int_{-1}^{1} \frac{f^{(n)}(t)}{n!}\,\omega_n(x)\,\mathrm{d}x, \qquad t \in [-1, 1],$$

where $e_n(f, x)$ is the error of polynomial interpolation and $\omega_n(x)$ is the nodal polynomial. ■

If we are given some quadrature formula with n nodes we can decide whether it is a polynomial interpolatory formula if we know its degree. This follows on account of the inverse theorem with respect to Theorem 6.1.1:

Theorem 6.1.2 *If degree* $(Q_n f) \geq n-1$, *then* $Q_n f$ *is a polynomial interpolatory quadrature formula.*

Proof: Assume that degree $(Q_n f) \geq n-1$. Then

$$Q_n \lambda_i(x) = \int_{-1}^{1} \lambda_i(x)\,\mathrm{d}x, \qquad i = 1(1)n,$$

where $\lambda_i(x)$ are the Lagrange cardinal polynomials belonging to $\mathbb{P}_{n-1}$. From the property $\lambda_i(x_k) = \delta_{ik}$ it follows that $Q_n \lambda_i(x) = A_i$ for $i = 1(1)n$ and, hence, that

$$A_i = \int_{-1}^{1} \lambda_i(x)\,\mathrm{d}x,$$

which completes the proof. ■

6.2 The Newton–Cotes quadrature formulae

6.2.1 Closed Newton–Cotes formulae

From the definition in Section 6.1, the Newton–Cotes formulae have weights

$$A_i = \int_{-1}^{1} \lambda_i(x)\,\mathrm{d}x = \int_{-1}^{1} \left(\prod_{\substack{j=1 \\ \neq i}}^{n} \frac{x - x_j}{x_i - x_j} \right) \mathrm{d}x \qquad i = 1(1)n,$$

$$\text{with } x_i = -1 + \frac{2(i-1)}{n-1}$$

which, when evaluated, are as in Table 6.2.1 (cf. [11]).

TABLE 6.2.1. Newton–Cotes formulae.

n	Numerator of A_i, $(A_{n-i+1}=A_i)$	Denominator	Degree
2	1; 1	1	1
3	1; 4; 1	3	3
4	1; 3; 3; 1	4	3
5	7; 32; 12; 32; 7	45	5
6	19; 75; 50; 50; 75; 19	144	5
7	41; 216; 27; 272; 27; 216; 41	420	7
8	751; 3577; 1323; 2989; ...	8640	7
9	989; 5888; −928; 10496; −4540; ...	14175	9
10	2857; 15741; 1080; 19344; 5778; ...	44800	9

Note the three negative weights for $n=9$. If the table is continued we see that for $n>10$ some of the weights are always negative. This property is not good for the convergence of the Newton–Cotes formulae. The crucial term for convergence is $\|Q_n\|=\sum_{i=1}^{n}|A_i|$, where some values of this norm are listed for different n.

n	9	10	11	21	31	41	51
$\|Q_n\|$	2.90	2	6.13	1.09(+3)	4.24(+5)	2.20(+8)	1.34(+11)

It turns out that the growth of $\|Q_n\|$ is very rapid, and hence convergence cannot be expected for every function $f\in C^0[-1,1]$.

6.2.2 The error of the Newton–Cotes formulae

For an *even number of nodes* we have, from Section 2.5.3, the error representation

$$E_{2n}f=\int_{-1}^{1}\frac{f^{(2n)}(t)}{(2n)!}\,\omega_{2n}(x)\,\mathrm{d}x,\qquad t\in(-1,1),$$

where the nodes are $-1,\ -1+\dfrac{2}{2n-1},\ -1+\dfrac{4}{2n-1},\ \ldots,\ -\dfrac{1}{2n-1},$ $\dfrac{1}{2n-1},\ldots,1$. Using symmetry we have that

$$\omega_{2n}(x)=\prod_{i=1}^{n}\left[x^2-\left(1-\frac{2i-2}{2n-1}\right)^2\right].$$

It is seen from Table 6.2.1 of Newton–Cotes formulae that degree $(Q_{2n}f)=$ degree $(Q_{2n-1}f)$ and hence, for an *odd* number $2n-1$ of nodes, we should obtain the same degree $2n$. Hence the interpolation formula presumably contains one derivative, which does not appear in the final quadrature formula. Because of the symmetry, we choose Hermite interpolation at $x=0$ and Lagrange interpolation in the points -1, $-1+\dfrac{1}{n-1},\ldots,-\dfrac{1}{n-1},\dfrac{1}{n-1},\ldots,1$ so that the mixed Lagrange Hermite interpolating operator is

$$LH_{2n-2,1}(f,x)=L_{2n-2}(f,x)+\bar{\omega}_{2n-2}(x)H_1\left(\frac{f(x)-L_{2n-2}(f,x)}{\bar{\omega}_{2n-2}(x)},x\right).$$

Here $H_1(g,x)=g(0)+xg'(0)$, and

$$\bar{\omega}_{2n-2}(x)=\prod_{i=1}^{n-1}\left[x^2-\left(1-\frac{i-1}{n-1}\right)^2\right].$$

We thus have that

$$\begin{aligned}\int_{-1}^{1}LH_{2n-2,1}(f,x)\,dx&=\int_{-1}^{1}L_{2n-2}(f,x)\,dx\\&+\int_{-1}^{1}\bar{\omega}_{2n-2}(x)\frac{f(0)-L_{2n-2}(f,0)}{\bar{\omega}_{2n-2}(0)}\,dx\\&+\int_{-1}^{1}x\bar{\omega}_{2n-2}(x)\left[\frac{f(x)-L_n(f,x)}{\bar{\omega}_{2n-2}(x)}\right]'_{x=0}dx.\end{aligned}$$

The last integral vanishes because the integrand is odd and we obtain the Newton–Cotes quadrature formula for an odd number of nodes as

$$Q_{2n-1}f=\frac{1}{\bar{\omega}_{2n-2}(0)}\left[\int_{-1}^{1}[L_{2n-2}(f,x)\bar{\omega}_{2n-2}(0)-L_{2n-2}(f,0)\bar{\omega}_{2n-2}(x)]\,dx+f(0)\int_{-1}^{1}\bar{\omega}_{2n-2}(x)\,dx\right].$$

where the error is given by (cf. Section 2.5.6)

$$E_{2n-1}f=\int_{-1}^{1}\frac{f^{(2n)}(t)}{(2n)!}x^2\bar{\omega}_{2n-2}(x)\,dx=\int_{-1}^{1}\frac{f^{(2n)}(t)}{(2n)!}x\omega_{2n-1}(x)\,dx,\qquad t\in(-1,1);$$

here

$$\omega_{2n-1}(x)=x\prod_{i=1}^{n-1}\left[x^2-\left(1-\frac{i-1}{n-1}\right)^2\right]$$

is the nodal polynomial for the $2n-1$ equidistant nodes $x_1, \ldots, x_{2n-1}$. Hence $x_n = 0$. We now show that the quadrature formula $Q_{2n-1}f$ which is obtained on integration of $LH_{2n-2,1}(f, x)$ is indeed the Newton–Cotes formula with $2n-1$ nodes.

Let $\omega_{2n-1}(x) \equiv x\bar{\omega}_{2n-2}(x)$. Then for $i \neq n$

$$\lambda_i(x) = \frac{\omega_{2n-1}(x)}{\omega'_{2n-1}(x_i)(x-x_i)} = \frac{x\bar{\omega}_{2n-2}(x)}{x_i\bar{\omega}'_{2n-2}(x_i)(x-x_i)} \quad \text{since} \quad \bar{\omega}_{2n-2}(x_i) = 0.$$

Clearly, $\omega_{2n-1}(x)$ is the nodal polynomial and $\lambda_i(x)$, $i = 1(1)2n-1$, are the Lagrange cardinal polynomials with respect to the $2n-1$ nodes of $Q_{2n-1}f$. In particular $\lambda_n(x) = (\bar{\omega}_{2n-2}(x))/\bar{\omega}'_{2n-2}(0)$.

We now expand $L_{2n-2}(f, x)$ and thus obtain

$$Q_{2n-1}f = \frac{1}{\bar{\omega}_{2n-2}(0)} \left\{ \sum_{\substack{i=1 \\ \neq n}}^{2n-1} [\bar{\lambda}_i(x)\bar{\omega}_{2n-2}(0) - \bar{\lambda}_i(0)\bar{\omega}_{2n-2}(x)] f(x_i) \right.$$
$$\left. + f(x_n) \int_{-1}^{1} \bar{\omega}_{2n-2}(x)\, \mathrm{d}x \right\}$$

from which it is first seen that coeff $(f(x_n) = f(0)) = \int_{-1}^{1} \lambda_n(x)\, \mathrm{d}x = A_n^{(2n-1)}$ and, secondly, it is found that for $i \neq n$

$$\frac{\bar{\lambda}_i(x)\bar{\omega}_{2n-2}(0) - \bar{\lambda}_i(0)\bar{\omega}_{2n-2}(x)}{\bar{\omega}_{2n-2}(0)} = \frac{\bar{\omega}_{2n-2}(x)\bar{\omega}_{2n-2}(0)}{\bar{\omega}'_{2n-2}(x_i)\bar{\omega}_{2n-2}(0)} \left[\frac{1}{x-x_i} - \frac{1}{-x_i} \right] = \lambda_i(x).$$

Thus, coeff $(f(x_i)) = \int_{-1}^{1} \lambda_i(x)\, \mathrm{d}x = A_i^{(2n-1)}$ for $i = 1(1)n-1$ and also for $n+1(1)2n-1$, which completes the proof. We therefore have:

Theorem 6.2.1 *The n-point Newton–Cotes formula Q_nf has degree* $2[(n+1)/2]-1$.

Although the nodal polynomials $\omega_n(x)$ change sign in $[-1, 1]$ the following theorem is valid.

Theorem 6.2.2 *For every $n > 0$ there exists a point $t \in (-1, 1)$ such that*

$$E_{2n}f = \frac{f^{(2n)}(t)}{(2n)!} \int_{-1}^{1} \omega_{2n}(x)\, \mathrm{d}x \tag{6.2.1}$$

and

$$E_{2n-1}f = \frac{f^{(2n)}(t)}{(2n)!} \int_{-1}^{1} x\omega_{2n-1}(x)\, \mathrm{d}x \tag{6.2.2}$$

Proof Some basic properties of the nodal polynomials $\omega_k(x)$ are first stated. We consider only the case $k=2n-1$. Let $'\omega_k(x) \equiv \int_{-1}^{x} \omega_k(t)\,dt$, $k=2n-1$, so that $'\omega_k(-1)=0$ and, in addition, $'\omega_k(1)=0$ since $'\omega_{2n-1}(x)$ is even. We designate the nodes as $x_{-i}=-i/(n-1)$, $x_0=0$, $x_i=i/(n-1)$, $i=1(1)n-1$. Then, for $0<\theta<1$ we have for the negative nodes that

$$\begin{aligned}
\left|\omega_{2n-1}\left(x_{-i}-\frac{\theta}{n-1}\right)\right| &= \left|x_{-i}-\frac{\theta}{n-1}\right| \left|\prod_{j=1}^{n-1}\left(x_{-i}-\frac{\theta}{n-1}\right)^2-\left(\frac{j}{n-1}\right)^2\right| \\
&= \frac{1}{n-1}|i+\theta| \prod_{j=1}^{n-1}\left(\frac{1}{n-1}\right)^2 |i^2+2i\theta+\theta^2-j^2| \\
&> \frac{1}{n-1}|i-\theta| \prod_{j=1}^{n-1}\left(\frac{1}{n-1}\right)^2 |i^2-2i\theta+\theta^2-j^2| \\
&= \left|\omega_{2n-1}\left(x_{-i}+\frac{\theta}{n-1}\right)\right|, \qquad i=1(1)n-1
\end{aligned}$$

while for the positive nodes x_i the reverse inequality holds. Hence

$$\left|\int_{x_{-i}}^{x_{-(i+1)}} \omega_{2n-1}(x)\,dx\right| > \left|\int_{x_{-(i-1)}}^{x_{-i}} \omega_{2n-1}(x)\,dx\right|, \qquad i=1(1)n-2$$

and (6.2.3)

$$\left|\int_{x_i}^{x_{i+1}} \omega_{2n-1}(x)\,dx\right| > \left|\int_{x_{i-1}}^{x_i} \omega_{2n-1}(x)\,dx\right|, \quad i=1(1)n-2.$$

Clearly, for $x<-1$ we have that $\omega_{2n-1}(x)<0$ and hence that $\omega_{2n-1}(x)>0$ in $(-1,-1+1/(n-1))$. It follows from (6.2.3) that $'\omega_{2n-1}(x)$ has positive sign in $(-1,1)$. Similarly, $'\omega_{2n}(x)\leq 0$ in $[-1,1]$.

The proof of the theorem can now be established. For $k=2n-1$ we obtain from the error of the Lagrange interpolating polynomial the quadrature error

$$E_{2n-1}f=\int_{-1}^{1} \omega_{2n-1}(x)\frac{f^{(2n-1)}(t(x))}{(2n-1)!}\,dx, \qquad t(x)\in(-1,1)$$

which, on integration by parts, becomes

$$\begin{aligned}
E_{2n-1}f &= {}'\omega_{2n-1}(x)\frac{f^{(2n-1)}(t(x))}{(2n-1)!}\bigg|_{-1}^{1} - \int_{-1}^{1} {}'\omega_{2n-1}(x)\left(\frac{d}{dx}\frac{f^{(2n-1)}(t(x))}{(2n-1)!}\right)dx \\
&= -\int_{-1}^{1} {}'\omega_{2n-1}(x)\left(\frac{d}{dx}\frac{f^{(2n-1)}(t(x))}{(2n-1)!}\right)dx \quad \text{since} \quad {}'\omega_{2n-1}(\pm 1)=0.
\end{aligned}$$

In order to calculate the derivative we remember that the point $t(x)$ was obtained by applying Rolle's theorem to the error $e_{k+1}^{(k)}(f, x)$ of interpolation with the $k+1$ nodes $x_1, x_2, \ldots, x_k, x$. Since Rolle's theorem is a mean-value theorem we can assume that $t(x)$ is the intermediate point in this mean-value theorem. For $k=1$ we then have that

$$f(x)=f(\bar{x})-(x-\bar{x})f'(t(x)) \quad \text{or} \quad f'(t(x))=\frac{f(x)-f(\bar{x})}{x-\bar{x}},$$

and that

$$\frac{d}{dx}f'(t(x))=\frac{d}{dx}\frac{f(x)-f(\bar{x})}{x-\bar{x}}$$

$$=\begin{cases}\dfrac{(x-\bar{x})f'(x)-(f(x)-f(\bar{x}))}{(x-\bar{x})^2} & \text{for} \quad x\neq\bar{x}\\[2mm] \tfrac{1}{2}f''(x) \quad \text{for} \quad x=\bar{x} & \text{(by l'Hospital's rule).}\end{cases}$$

For $x\neq\bar{x}$ there exists an intermediate point $s(x)$ such that

$$\frac{d}{dx}f'(t(x))=\tfrac{1}{2}f''(s(x)).$$

Proceeding by induction we have that there exists a point $s(x)\in(-1, 1)$ with $\dfrac{d}{dx}f^{(k)}(t(x))=\dfrac{1}{k+1}f^{(k+1)}(s(x))$ and hence that

$$E_{2n-1}f=-\int_{-1}^{1}{}'\omega_{2n-1}(x)\frac{f^{(2n)}(s(x))}{(2n)!}\,dx.$$

Since $'\omega_{2n-1}$ does not change sign in $[-1, 1]$ for suitable $t\in(-1, 1)$

$$E_{2n-1}f=-\frac{f^{(2n)}(t)}{(2n)!}\int_{-1}^{1}{}'\omega_{2n-1}(x)\,dx=\frac{f^{(2n)}(t)}{(2n)!}\int_{-1}^{1}x\omega_{2n-1}(x)\,dx.$$

The last equality is the result of another partial integration:

$$\int_{-1}^{1}{}'\omega_{2n-1}(x)\,dx=x\,{}''\omega_{2n-1}(x)\Big|_{-1}^{1}-\int_{-1}^{1}x\omega_{2n-1}(x)\,dx=-\int_{-1}^{1}x\omega_{2n-1}(x)\,dx. \tag{6.2.4}$$

For $k=2n$ we only outline the proof. In this case the interval may be split into $[-1, 1-2/(2n-1)]$ and $[1-2/(2n-1), 1]$ where in the second interval $\omega_{2n}(x)\leq 0$. For the first interval the situation as for $k=2n-1$ is reconstructed and similar arguments are used. Finally, the error is split into two parts relevant to each interval $E_{2n}f=c_1f^{(2n)}(t_1)+c_2f^{(2n)}(t_2)$ with $c_1c_2>0$. Then there exists a point $s\in(-1, 1)$ such that $E_{2n}f=c_3f^{(2n)}(s)$ is valid. (For a detailed proof see for example [15].) ∎

Remarks:

The corresponding error representations for the open Newton–Cotes formulae are obtained analogously so that

$$E_{2n-1}f = \frac{f^{(2n)}(t)}{(2n)!}\int_{-1}^{1} x\omega_{2n-1}(x)\,dx, \qquad \text{open Newton–Cotes formulae}$$

$$E_{2n-1}f = \frac{f^{(2n)}(t)}{(2n)!}\left(\frac{2n-1}{2n-2}\right)^{2n+1}\int_{-1}^{1} x\omega_{2n-1}(x)\,dx, \quad \text{Maclaurin formulae.}$$

6.2.3 Divergence and convergence of the Newton–Cotes quadrature procedure

In Section 6.2.1 some values of $\|Q_n\|$ for the closed Newton–Cotes formulae have been listed. These indicate an unfortunate error growth as $n \to \infty$ so that the following divergence result holds:

Theorem 6.2.3 *The Newton–Cotes quadrature procedure does not converge for every $f \in C^0[-1, 1]$.*

Proof The first condition of the Banach–Steinhaus Theorem 4.3.2 is not fulfilled. It is therefore possible to construct a continuous function f for which $Q_n f$ cannot converge. Assume the contrary, that $Q_n f$ converges. It thus converges for every polynomial, while $\|Q_n\| \not\leqslant C$ for every n.

We choose continuous functions $f_n(x)$ such that

$$Q_n f_n = \|Q_n\| \quad \text{and} \quad |f_n(x)| \leqslant 1$$

(e.g. f_n is linear between two subsequent nodes and $f_n(x_i) = \text{sign}\, A_i$). Since $\|Q_n\|$ is not bounded there exists a sequence $n_k \in \mathbb{N}$ such that

$$\|Q_{n_k}\| \leqslant k!\, k \to \infty \quad \text{as} \quad k \to \infty,$$

which contradicts the convergence, as has already been pointed out in the proof of the Banach–Steinhaus Theorem 4.3.2. Hence, the proof is completed if we succeed in showing that $\|Q_n\|$ is unbounded. This is easily done by using the asymptotic expansions of the weights A_i given by Ouspenski [18]:

$$A_1 = A_n = \frac{2}{n[\ln n]^2}(1 + s_n) \quad \text{with} \quad s_n \to 0 \quad \text{for} \quad n \to \infty$$

and

$$A_k = -\frac{2}{n[\ln n]^2}\binom{n-1}{k-1}\left[\frac{(-1)^{k-1}}{k-1} + \frac{(-1)^{n-k}}{n-k}\right](1 + t_n) \text{ with } t_n \to 0 \text{ as } n \to \infty$$

for $k = 2(1)n-1$.

Summing the absolute values we find that

$$\|Q_n\| = \sum_{k=1}^{n} |A_i| \geqslant \frac{2}{n(\ln n)^2}\left[2 + \sum_{k=2}^{n-1}\binom{n-1}{k-1}\left|\frac{(-1)^{k-1}}{k-1} + \frac{(-1)^{n-k}}{n-k}\right|\right] b_n$$

with $b_n \equiv \min\{|1-|s_n||, |1-|t_n||\} \leqslant 1$ and $b_n \to 1$ as $n \to \infty$ where we assume that n is large enough to ensure that in addition $b_n > 0$. Let

$$a_k^{(n)} \equiv \left|\frac{(-1)^{k-1}}{k-1} + \frac{(-1)^{n-k}}{n-k}\right| = \frac{1}{(k-1)(n-k)}|(n-k)(-1)^{k-1} + (k-1)(-1)^{n-k}|.$$

We observe first that the $a_k^{(n)}$ are monotonically decreasing for $k \leqslant [n/2]$ (increasing for $k \geqslant [n/2]$). For $n = 2N+1 > 2$ we have that

$$a_k^{(n)} = \frac{1}{(k-1)(2N+1-k)}|(2N+1-k)(-1)^{k-1} + (k-1)(-1)^{k-1}|$$

$$= \frac{2N}{(k-1)(2N+1-k)} = \frac{n-1}{(k-1)(n-k)} \geqslant \frac{1}{(n-2)^2}, \qquad k = 2(1)n-1,$$

while for $n = 2N$ we have that

$$a_k^{(n)} = \frac{1}{(k-1)(2N-k)}|(2N-k)(-1)^{k-1} + (k-1)(-1)^k| = \frac{|2N-2k+1|}{(k-1)(2N-k)}$$

$$= \frac{|n-2k+1|}{(k-1)(n-k)} \geqslant \frac{1}{(n-2)^2}, \qquad k = 2(1)n-1.$$

Hence we are able to determine the inequality

$$\sum_{k=1}^{n} |A_i| \geqslant \frac{2}{n(\ln n)^2}\left\{2 + \sum_{k=2}^{n-1}\binom{n-1}{k-1}\frac{1}{(n-2)^2}\right\} b_n$$

$$= \frac{2}{n(\ln n)^2}\left\{2 + \frac{2^{n-1}-2}{(n-2)^2}\right\} b_n$$

$$\geqslant \frac{2^n - 4}{n(n-2)^2(\ln n)^2}\, b_n = \frac{2^n}{n(n-2)^2(\ln n)^2}\, \bar{b}_n$$

with $\bar{b}_n \equiv b_n(1-2^{2-n}) \leqslant 1$ and $\bar{b}_n \to 1$ as $n \to \infty$.

Thus $\|Q_n\| = \sum |A_i|$ is unbounded for the Newton–Cotes formulae and convergence for every $f \in C^0[-1, 1]$ cannot be obtained. ■

For the following theorem we need an upper bound for $\|Q_n\|$. With $c_n \equiv \max\{1+|s_n|, 1+|t_n|\} \geq 1$ and $c_n \to 1$ as $n \to \infty$ we have that

$$\sum_{k=1}^{n} |A_k| \leq \frac{2}{n(\ln n)^2}\left\{2+\sum_{k=2}^{n-1}\binom{n-1}{k-1}\frac{n-1}{(k-1)(n-k)}\right\} c_n$$

$$\leq \frac{2}{n(\ln n)^2}\left\{2+\frac{n-1}{n-2}(2^{n-1}-2)\right\} c_n$$

since $(k-1)(n-k) \geq (n-2)$ for $2 \leq k \leq n-1$ and $n \geq 3$. With $\bar{c}_n \equiv \left[2^{2-n}+\dfrac{n-1}{n-2}(1-2^{2-n})\right] c_n \geq 1$ and $\bar{c}_n \to 1$ as $n \to \infty$ we have that

$$\|Q_n\| = \sum_{i=1}^{n} |A_i| \leq \frac{2^n}{n(\ln n)^2}\,\bar{c}_n.$$

Using this bound we are able to prove:

Theorem 6.2.4 *The Newton–Cotes quadrature procedure converges for functions $f(z)$ which are analytic for $|z| \leq 2$.*

Proof We use the error theory of Section 3.5 for functions which are analytic for $|z| < R$.

Hence

$$\|E_n\| = \sum_{k=n(2)}^{\infty} \varepsilon^{2k}\left(\int_{-1}^{1} x^k\,dx - \sum_{j=1}^{n} A_i x_i^k\right)^2, \qquad \varepsilon = 1/R < 1,\ n \text{ even.}$$

If n is odd, then $k = n+1(2)\infty$. We now demand a bound for $\|E_n\|$ so that

$$\|E_n\| = \sum_{k=n(2)}^{\infty}\left[\left(\frac{2\varepsilon^k}{k+1}\right)^2 - \frac{4}{k+1}\sum_{i=1}^{n} A_i(\varepsilon^2 x_i)^k + \sum_{i=1}^{n}\sum_{j=1}^{n} A_i A_j \varepsilon^{2k} x_i^k x_j^k\right]$$

$$< \sum_{k=n(2)}^{\infty}\left(\frac{2\varepsilon^k}{(k+1)}\right)^2 + \frac{4\varepsilon^{2n} a_n}{n}\sum_{i=1}^{n}\frac{1}{1-\varepsilon^2 x_i} + \varepsilon^{2n} a_n^2 \sum_{i=1}^{n}\sum_{j=1}^{n}\frac{1}{1-\varepsilon^2 x_i x_j}$$

$$< 4\varepsilon^{2n}\sum_{k=n(2)}^{\infty}\left(\frac{1}{k+1}\right)^2 + \frac{4\varepsilon^{2n} a_n}{1-\varepsilon^2} + \frac{n^2\varepsilon^{2n} a_n^2}{1-\varepsilon^2} \quad \left(\text{where } \ a_n \equiv \sum_{i=1}^{n} |A_i|\right)$$

since $1/(k+1) < 1/n$, $x_i \leq 1$ and

$$\sum_{i=1}^{n}\frac{1}{1-\varepsilon^2 x_i} < \frac{n}{1-\varepsilon^2}; \qquad \sum_{i=1}^{n}\sum_{j=1}^{n}\frac{1}{1-\varepsilon^2 x_i x_j} < \frac{n^2}{1-\varepsilon^2}.$$

Hence $\|E_n\| < \dfrac{\varepsilon^{2n}}{1-\varepsilon^2}\left[4\sum_{k=n(2)}^{\infty}\left(\dfrac{1}{k+1}\right)^2 + 4a_n + n^2 a_n^2\right]$ where the sum within the brackets tends to 0 as $n \to \infty$. The crucial term for convergence is

$n^2\varepsilon^{2n}a_n^2$ and, using the asymptotic expansion for a_n, we have that $n\varepsilon^n a_n \sim \varepsilon^n 2^n/(\ln n)^2 \to 0$ as $n\to\infty$ if $\varepsilon \leqslant \frac{1}{2}$, which means that we have convergence if $R \geqslant 2$. ■

6.2.4 Historical remarks about Newton–Cotes formulae

The Newton–Cotes quadrature procedure has an interesting history which is worth mentioning briefly.

It is the oldest quadrature procedure and since its conception numerous efforts have been made to obtain formulae with more than ten nodes. The question of convergence seems first to have been raised by Stieltjes in his correspondence with Hermite. (Stieltjes was the first person to prove the convergence of the Gauss quadrature procedure for Riemann integrable functions.) Ouspenski [18] in 1925 derived asymptotic expressions for the weights and concluded from the growth of $\|Q_n\|$ that the Newton–Cotes quadrature is not a good quadrature procedure for practical applications. Some years later Kusmin [17] proved the divergence for $f \in C^0[-1, 1]$ and then, in 1933, Polya [19] showed that the Newton–Cotes procedure is not even convergent for every analytic function. His counterexample involved the function

$$g(z) = -\sum_{k=4}^{\infty} a^{k!}\frac{\sin k![(z+1)/2]}{\cos \pi[(z+1)/2]}, \qquad \tfrac{1}{2} < a < 1,$$

where $g(z)$ is analytic in the strip $|\mathrm{Im}\,(z)| < \dfrac{2}{\pi}\log 2 = 0.4412712$.

Subsequently, Davis succeeded in proving that the Newton–Cotes procedure converges at least for every $f \in \mathbb{E}_\rho$ for $\rho > 4$; for $\rho = 4$ E_ρ is the ellipse with foci ± 1 and semiaxes 1.25 and 0.75 so that $g(z) \notin \mathbb{E}_\rho$. He thus revived the fortunes of the Newton–Cotes procedure. Nevertheless, the rapid growth of $\|Q_n\|$ does limit its applicability and the presence of round-off errors in $f(x_i)$ may cause dramatic loss of accuracy even if n is quite small. (For Gauss quadratures the use of $n \sim 100$ is common.)

6.3 Further quadratures with equidistant nodes

Previously, we have integrated a Lagrangian interpolating polynomial with equidistant nodes, to produce our quadrature formulae. We now integrate different interpolating functions to produce quadrature formulae with equidistant nodes. These no longer have polynomial accuracy and their degree behaves differently from that of Newton–Cotes formulae. This is true if, for example, we integrate an interpolating cubic spline with equidistant nodes. If the second derivatives $S''(x)$ of this spline satisfy $S''_n(f,0)=S''_n\left(f,\dfrac{1}{n-1}\right)$ and $S''_n\left(f,\dfrac{n-2}{n-1}\right)=S''_n(f,1)$, then the following Table 6.3.1 of quadrature formulae is obtained (cf. Problem P.6.2). For further details the reader is referred to the book [1] of Section 2.8.3.

TABLE 6.3.1 Spline-based quadratures.

Number of nodes	Weights: Numerators	Weights: Denominator	Degree
2	1, 1	2	1
3	1, 4, 1	6	3
4	3, 9, 9, 3	27	3
5	13, 44, 30, 44, 13	144	3
6	35, 115, 90, . . . ($\hat{=}$ symmetric)	480	3
7	16, 53, 40, 46, . . .	264	3
8	131, 433, 330, 366, . . .	2520	3
9	179, 592, 450, 504, 486, . . .	3936	3
10	163, 539, 410, 458, 446, . . .	4032	3

For another alternative we have seen in Chapter 3 that the quadrature formula Q_nf for analytic functions integrates exactly the n functions $1/(1-xx_i)$ for $i=1(1)n$. Using the nodes $x_i=-1+2i/(n+1)$, $i=1(1)n$, the resulting formulae do not have polynomial accuracy. For such formulae there exist error constants σ_n such that $|E_nf|\leqslant\sigma_n\,\|f\|$. We list some of these in Table 6.3.2 for the interval $[-1,1]$ and for $\varepsilon-1$. They are necessarily open formulae, because the end-points cannot be used when $\varepsilon=1$ (cf. [9]).

It can be seen that the norm $\|Q_n\|$ grows very rapidly and hence, because of the unfortunate accumulation of round-off errors, the formulae are not useful in practice.

6.4 Newton–Cotes cubature formulae

For simple domains the Newton–Cotes quadrature formulae can be extended to more than one dimension by integrating multidimensional interpolating polynomials. Some simple cases are illustrated.

6.4.1 Newton–Cotes cubature formulae for simplices

We begin by treating the standard two-dimensional simplex $S_2=\Omega$, a triangle. For S_2 we can easily find cardinal polynomials (as is often done

TABLE 6.3.2. Quadratures with minimal error norm.

n	Nodes	Weights	σ_n	$\|Q_n\|$
1	0	2	0.967	2
2	±1/2	1.029 949 021	0.639	2.06
3	±2/3 0	0.841 256 139 0.317 487 721	0.488	2.00
4	±3/4 ±1/4	0.617 364 390 0.380 300 425	0.421	2.00
5	±4/5 ±2/5 0	0.587 422 921 −0.300 391 130 1.425 936 778	0.359	3.20
6	±5/6 ±3/6 ±1/6	0.482 251 981 −0.109 522 941 0.627 483 341	0.327	2.44
7	±6/7 ±4/7 ±2/7 0	0.476 788 710 −0.832 495 216 3.131 080 798 −3.550 748 548	0.293	12.43

in finite element techniques) for a grid consisting of all points of intersection of the lines $y = i/n$, $x = i/n$, $y + x = i/n$, $i = 0(1)n$. For $n = 1$ we obtain the three vertices of S_2. The corresponding cardinal functions are

$$\lambda_{1,1}(x, y) = -(x + y - 1), \qquad \lambda_{1,2}(x, y) = x, \qquad \lambda_{1,3}(x, y) = y.$$

hence

$$L_1(f, x, y) \equiv (1 - x - y)f(0, 0) + xf(1, 0) + yf(0, 1).$$

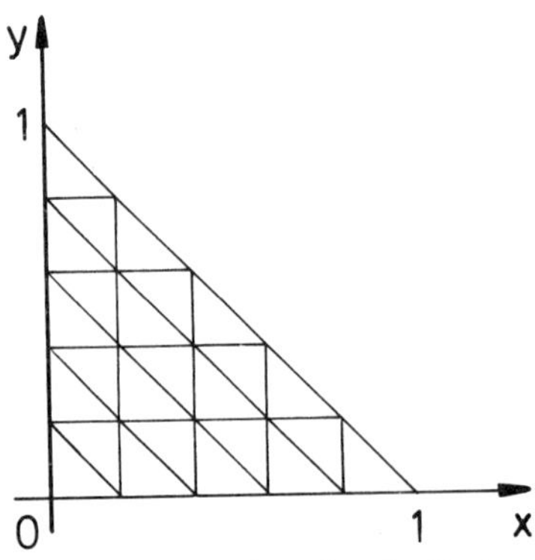

FIG. 6.4.1. Grid lines for the 2-simplex.

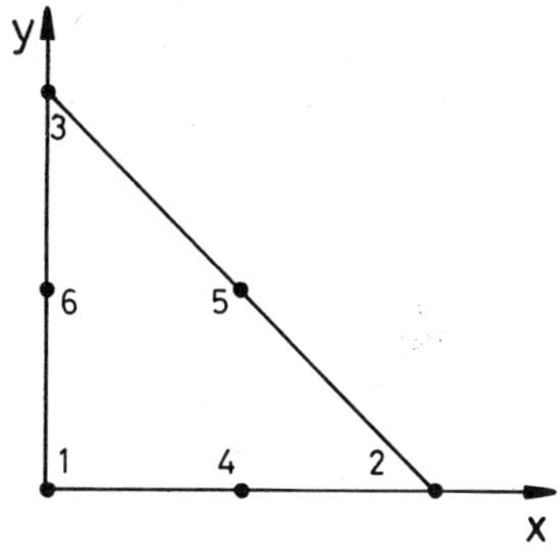

FIG. 6.4.2. Nodes of $C_6 f$ for $\Omega = S_2$.

Note that every grid line is a zero line of one of the cardinal functions. Integration leads to the first-degree formula

$$C_3 f = \int_0^1 \int_0^{1-x} L_1(f, x, y)\, dy\, dx = \frac{1}{6}[f(0, 0) + f(1, 0) + f(0, 1)].$$

For $n = 2$ we obtain six nodes (cf. Fig. 6.4.2). Denoting these nodes by $1, 2, \ldots, 6$ we have the corresponding cardinal functions

$$\lambda_{2,1} = 2(x + y - 1)(x + y - \tfrac{1}{2}), \qquad \lambda_{2,2} = 2x(x - \tfrac{1}{2}),$$

$$\lambda_{2,3} = 2y(y - \tfrac{1}{2}), \qquad \lambda_{2,4} = -4x(x + y - 1),$$

$$\lambda_{2,5} = 4xy, \qquad \lambda_{2,6} = -4y(x + y - 1).$$

It is immediately seen that $\lambda_{2,k}(P_i) = \delta_{i,k}$, if P_i is the i-th node. Again, we have used the grid lines as zero lines of $\lambda_{2,k}$. The interpolating function is $L_2(f, x, y) \equiv \sum_{i=1}^{6} \lambda_{2,i} f(P_i)$ and integration leads to the second-degree formula

$$C_3 f = \tfrac{1}{6}[f(\tfrac{1}{2}, 0) + f(\tfrac{1}{2}, \tfrac{1}{2}) + f(0, \tfrac{1}{2})].$$

Further, for $n = 3$ we have (cf. Fig. 6.1.1) ten nodes and the corresponding cardinal functions are

$$\lambda_{3,1} = -\tfrac{9}{2}(x + y - 1)(x + y - \tfrac{2}{3})(x + y - \tfrac{1}{3}), \qquad \lambda_{3,2} = \tfrac{9}{2}x(x - \tfrac{1}{3})(x - \tfrac{2}{3}),$$

$$\lambda_{3,3} = \tfrac{9}{2}y(y - \tfrac{1}{3})(y - \tfrac{2}{3}), \qquad \lambda_{3,4} = \tfrac{27}{2}x(x + y - 1)(x + y - \tfrac{2}{3}),$$

$$\lambda_{3,5} = -\tfrac{27}{2}x(x - \tfrac{1}{3})(x + y - 1), \qquad \lambda_{3,6} = \tfrac{27}{2}xy(x - \tfrac{1}{3}),$$

$$\lambda_{3,7} = \tfrac{27}{2}xy(y - \tfrac{1}{3}), \qquad \lambda_{3,8} = -\tfrac{27}{2}y(y - \tfrac{1}{3})(x + y - 1),$$

$$\lambda_{3,9} = \tfrac{27}{2}y(x + y - 1)(x + y - \tfrac{2}{3}), \qquad \lambda_{3,10} = -27xy(x + y - 1).$$

Hence $L_{10}(f, x, y) \equiv \sum_{i=1}^{10} \lambda_{3,i} f(P_i)$, where P_i are the nodes as marked in Fig. 6.1.1. The corresponding third-degree cubature formula is

$$C_{10}f = \tfrac{1}{240}[4(f(P_1)+f(P_2)+f(P_3))+9(f(P_4)+\ldots+f(P_9))+54f(P_{10})].$$

The manner of proceeding to higher order interpolating polynomials with $15, 21, \ldots, \frac{1}{2}(n+1)(n+2)$ nodes is now clear. The amount of work is reduced enormously by the fact that the formulae $C_d f$, where $d = \frac{1}{2}(n+1)(n+2)$, which are obtained in the general case, have only n different weights. This is due to the fact that all three directions x, y, $x+y$ have equal prominence. Hence $C_{10}f$ can be obtained by integration only of $\lambda_{3,2}$, $\lambda_{3,6}$ and $\lambda_{3,10}$. In the next step for $n=4$ we find that (cf. Fig. 6.4.3) the essential cardinal polynomials are

$$\lambda_{4,2} = \tfrac{32}{3}x(x-\tfrac{1}{4})(x-\tfrac{2}{4})(x-\tfrac{3}{4}) \qquad \lambda_{4,5} = 64xy(x-\tfrac{1}{4})(y-\tfrac{1}{4})$$

$$\lambda_{4,8} = -\tfrac{128}{3}x(x-\tfrac{1}{4})(x-\tfrac{2}{4})(x+y-1) \quad \lambda_{1,14} = -128xy(x-\tfrac{1}{4})(x+y-1)$$

and the weights are obtained by integrating these over S_2 so that the fourth-degree formula is

$$C_{12}f = \tfrac{1}{90}[4(f(P_7)+\ldots+f(P_{12}))-(f(P_4)+\ldots+f(P_6))+8(f(P_{13})+\ldots+f(P_{15}))].$$

Here again the weights for P_1, P_2, P_3 vanish, and for the first time negative weights occur. Continuing in the same way, we find that

$$C_{21}f = \tfrac{1}{2016}[11(f(P_1)+\ldots+f(P_3))+25(f(P_4)+\ldots+f(P_9)+f(P_{10})+\ldots+f(P_{15})) \\ +200(f(P_{16})+\ldots+f(P_{18}))+25(f(P_{19})+\ldots+f(P_{21}))].$$

Here, the interior nodes $P_{16}, \ldots, P_{21}$ are numbered in the same manner as those of Fig. 6.4.3 and the process is extended to higher values of n so that by integration of $L_n(f, x, y)$ we obtain in each case a formula C_d (d as

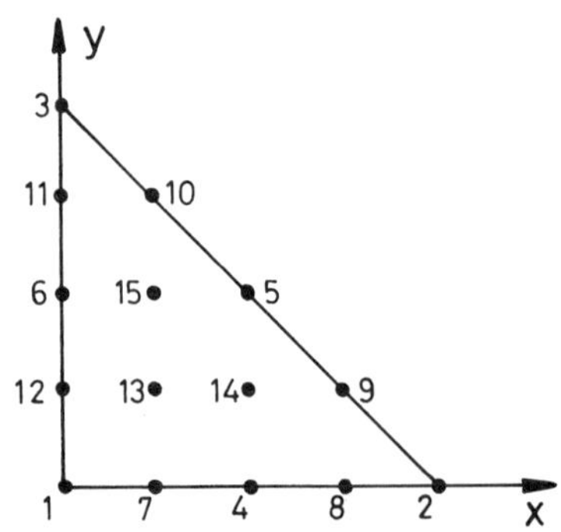

FIG. 6.4.3. Nodes for $n=4$ in S_2.

above) with n different weights. We see by induction that we have $[n/3]$ interior triangles in S_2 containing regularly-spaced nodes on their boundary. Here, an empty triangle consisting of one point only is allowed. Thus we have a sequence of $[n/3]+1$ concentric triangles, each with the left lower vertex $(i/n, i/n)$, $i=0(1)[n/3]$ (cf. Figs 6.1.1, 6.4.1, 6.4.2, 6.4.3 for $n=3, 5, 2, 4$ respectively).

On the boundary of each of these triangles we have $[n/2]+1$, $[(n-3)/2]+1$, $[(n-6)/2]+1, \ldots$ nodes with *different* weights ($[(n-m)/2]\equiv -1$ here if $m>n$). In Table 6.4.1 we list only the essential nodes and corresponding weights starting with the left lower vertex of the largest triangle proceeding to the smallest interior triangle. The formulae for $n=1(1)5$ of Table 6.4.1 have been given explicitly in the preceding text.

It turns out that the same procedure can be used for simplices in $N>2$ dimensions. For example, when $N=3$ we have the standard simplex $S_3=\Omega$ and again we construct a grid from the intersections of the planes $x=i/n$, $y=i/n$, $z=i/n$, $x+y+z=i/n$, $i=0(1)n$. For $n=1$ the vertices of S_3 are as in Fig. 6.4.4. The cardinal functions are now

$$\lambda_{1,1}(x, y, z)=1-x-y-z, \qquad \lambda_{1,2}(x, y, z)=x$$

$$\lambda_{1,3}(x, y, z)=y, \qquad \lambda_{1,4}(x, y, z)=z$$

TABLE 6.4.1. Newton–Cotes formulae for the 2-simplex.

Degree $=n$	Essential nodes	Weights of $C_{\binom{n+2}{2}}f$: Numerators	Denomiator	Total number of nodes
1	$(0, 0)$	1	6	3
2	$(0, 0), (\frac{1}{2}, 0)$	0, 1	6	6 (actually 3)
3	$(0, 0), (\frac{1}{3}, 0)$ $(\frac{1}{3}, \frac{1}{3})$	4, 9 54	240	10
4	$(0, 0), (\frac{1}{4}, 0), (\frac{1}{2}, 0)$ $(\frac{1}{4}, \frac{1}{4})$	0, 4, −1 8	90	15 (actually 12)
5	$(0, 0), (\frac{1}{5}, 0), (\frac{2}{5}, 0)$ $(\frac{1}{5}, \frac{1}{5}), (\frac{2}{5}, \frac{1}{5})$	11, 25, 25 200, 25	2016	21
6	$(0, 0), (\frac{1}{6}, 0)(\frac{2}{6}, 0)(\frac{3}{6}, 0)$ $(\frac{1}{6}, \frac{1}{6}), (\frac{2}{6}, \frac{1}{6})$ $(\frac{2}{6}, \frac{2}{6})$	0, 36, −27, 64 72, 72 −54	1680	28 (actually 25)
7	$(0, 0)(\frac{1}{7}, 0), (\frac{2}{7}, 0), (\frac{3}{7}, 0)$ $(\frac{1}{7}, \frac{1}{7}), (\frac{2}{7}, \frac{1}{7}), (\frac{3}{7}, \frac{1}{7})$ $(\frac{2}{7}, \frac{2}{7})$	1336, 2989, 3577, 2695 32242, −6860, 44590 3430	518400	36
8	$(0, 0)(\frac{1}{8}, 0)(\frac{2}{8}, 0)(\frac{3}{8}, 0)(\frac{4}{8}, 0)$ $(\frac{1}{8}, \frac{1}{8})(\frac{2}{8}, \frac{1}{8})(\frac{3}{8}, \frac{1}{8})$ $(\frac{2}{8}, \frac{2}{8})(\frac{3}{8}, \frac{2}{8})$	0, 368, −468, 1136, −1083 704, 832, 672 −1448, 1472	28350	45 (actually 42)

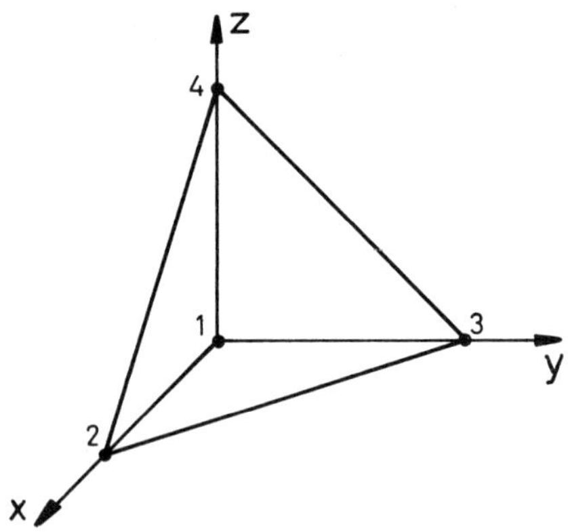

FIG. 6.4.4. Nodes for $n=1$ in S_3.

and, on integration of the interpolating polynomial

$$L_1(f, x, y, z) = \sum_{i=1}^{4} \lambda_{1,i}(x, y, z) f(P_i)$$

where P_i denotes the i-th node, we obtain the first-degree cubature formula

$$C_4 f = \tfrac{1}{24}\{f(P_1) + f(P_2) + f(P_3) + f(P_4)\}.$$

When $n=2$, then the nodes are as in Fig. 6.4.5 and the cardinal

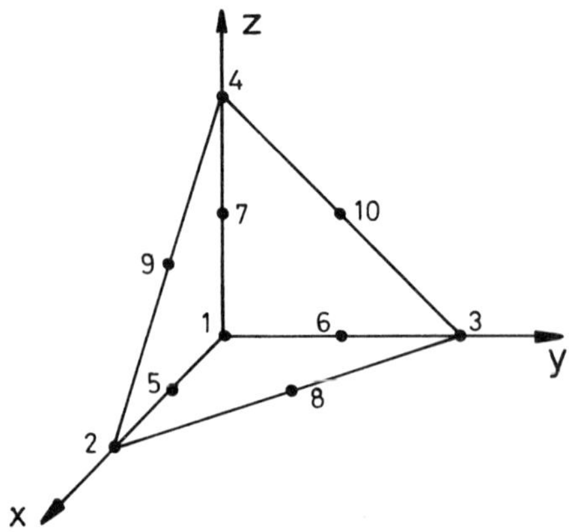

FIG. 6.4.5. Nodes for $n=2$ in S_3.

polynomials are

$$\lambda_{2,1}=2(x+y+z-1)(x+y+z-\tfrac{1}{2}),\qquad \lambda_{2,2}=2x(x-\tfrac{1}{2}),$$

$$\lambda_{2,3}=2y(y-\tfrac{1}{2}),\qquad \lambda_{2,4}=2z(z-\tfrac{1}{2}),$$

$$\lambda_{2,5}=-4x(x+y+z-1),\qquad \lambda_{2,6}=-4y(x+y+z-1),$$

$$\lambda_{2,7}=-4z(x+y+z-1),\qquad \lambda_{2,8}=4xy,$$

$$\lambda_{2,9}=4xz,\qquad \lambda_{2,10}=4yz.$$

Hence, because of $\lambda_{2,i}(P_k)=\delta_{i,k}$,

$$L_2(f, x, y, z)=\sum_{i=1}^{10}\lambda_{2,i}(x, y, z)f(P_i)$$

is the Lagrange interpolating polynomial of degree 2. On integration we obtain

$$C_{10}f=\int_0^1\int_0^{1-x}\int_0^{1-x-y}L_2(f, x, y, z)\,dz\,dy\,dx$$

and, performing this integration, we see that we have only two different weights in the second-degree cubature formula

$$C_{10}f=\frac{1}{120}\{-[f(P_1)+\ldots+f(P_4)]+4[f(P_5)+\ldots+f(P_{10})]\}.$$

As for the two-dimensional case, the essential nodes and weights are listed in Table 6.4.2. The formulae for $n=1$ and $n=2$ of Table 6.4.2 have been

TABLE 6.4.2. Newton–Cotes formulae for the 3-simplex.

Degree $=n$	Essential nodes	Weights of $C_d f$ Numerators	Denominator	Total number of nodes
1	$(0, 0, 0)$	1	24	4
2	$(0, 0, 0)(\tfrac{1}{2}, 0, 0)$	$-1, 4$	120	10
3	$(0, 0, 0), (\tfrac{1}{3}, 0, 0)$ $(\tfrac{1}{3}, \tfrac{1}{3}, 0)$	$1, 0$ 9	240	20 actually 8
4	$(0, 0, 0), (\tfrac{1}{4}, 0, 0), (\tfrac{2}{4}, 0, 0)$ $(\tfrac{1}{4}, \tfrac{1}{4}, 0)$ $(\tfrac{1}{4}, \tfrac{1}{4}, \tfrac{1}{4})$	$-5, 16, -12$ 16 128	2520	35
5	$(0, 0, 0), (\tfrac{1}{5}, 0, 0), (\tfrac{2}{5}, 0, 0)$ $(\tfrac{1}{5}, \tfrac{1}{5}, 0), (\tfrac{2}{5}, \tfrac{1}{5}, 0)$ $(\tfrac{1}{5}, \tfrac{1}{5}, \tfrac{1}{5})$	$33, -35, 35$ $275, -75$ 375	24192	56
6	$(0, 0, 0), (\tfrac{1}{6}, 0, 0,), (\tfrac{2}{6}, 0, 0), (\tfrac{3}{6}, 0, 0)$ $(\tfrac{1}{6}, \tfrac{1}{6}, 0), (\tfrac{2}{6}, \tfrac{1}{6}, 0), (\tfrac{3}{6}, \tfrac{1}{6}, 0)$ $(\tfrac{1}{6}, \tfrac{1}{6}, \tfrac{1}{6}), (\tfrac{2}{6}, \tfrac{1}{6}, \tfrac{1}{6})$	$-7, 24, -30, 40$ $0, 30, -45$ $180, 0$	8400	84 actually 66

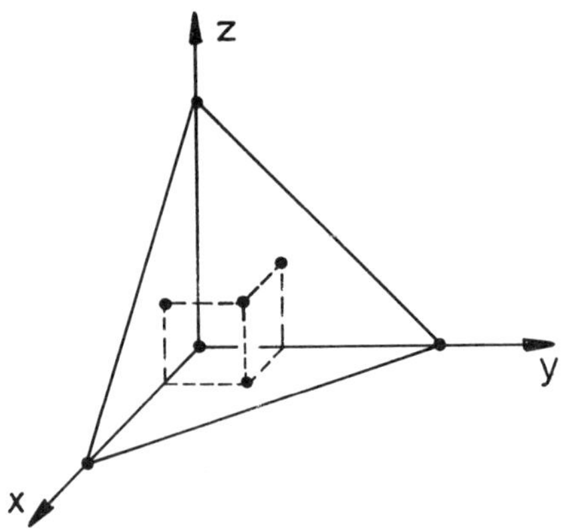

FIG. 6.4.6. Nodes for $C_8 f$ for $n=3$.

given explicitly in the preceding text. We list some formulae $C_d f$ where now $d=(n/6)(n^2+5n+6)$.

It is remarkable that the case $n=3$ produces a very simple formula with only 8 nodes (see Fig. 6.4.6), whereas for $n=2$ the corresponding formula has 10 nodes; this is due to the zero weights for 12 nodes.

The exact origin of every one of these formulae in Table 6.4.2 is uncertain. However, the first extended table of the formulae seems to have been given by Silvester [21], who uses simplex coordinates and a different order of nodes. Guenther and Roetmann [14] derive some formulae for the N-simplex (with the weights given as functions of N) together with the leading error terms up to degree 6. The strict proof that these formulae exist for every $n \in \mathbb{N}$ follows from the proof of Theorem 6.5.1.

6.4.2 Further Newton–Cotes cubature formulae–Convergence

We now take advantage of n-fold tensor products to obtain formulae for product-type regions (rectangles, cuboids, spheres, . . .). The tensor product interpolation formula with nine nodes for the case $n=2$ (2-cube) arises from the cardinal polynomials which are obtained from the one-dimensional quadratic interpolating polynomial

$$L_2(f, x)=(1-x)(1-2x)f(0)+4x(1-x)f(\tfrac{1}{2})+(2x-1)xf(1).$$

nodes	Cardinal functions
$P_1=(0,0)$	$\lambda_{2,1}=4(1-x)(\frac{1}{2}-x)(1-y)(\frac{1}{2}-y)$
$P_5=(\frac{1}{2},0)$	$\lambda_{2,5}=8x(1-x)(1-y)(\frac{1}{2}-y)$
$P_2=(1,0)$	$\lambda_{2,2}=4x(x-\frac{1}{2})(1-y)(\frac{1}{2}-y)$
$P_8=(0,\frac{1}{2})$	$\lambda_{2,8}=8y(1-y)(1-x)(\frac{1}{2}-x)$
$P_9=(\frac{1}{2},\frac{1}{2})$	$\lambda_{2,9}=16xy(1-x)(1-y)$
$P_6=(1,\frac{1}{2})$	$\lambda_{2,6}=8xy(x-\frac{1}{2})(1-y)$
$P_4=(0,1)$	$\lambda_{2,4}=4y(x-\frac{1}{2})(y-\frac{1}{2})(x-1)$
$P_7=(\frac{1}{2},1)$	$\lambda_{2,7}=-8xy(1-x)(\frac{1}{2}-y)$
$P_3=(1,1)$	$\lambda_{2,3}=4xy(x-\frac{1}{2})(y-\frac{1}{2})$

Thus, $L_2(f,x,y)=\sum_{i=1}^{9}\lambda_{2,i}f(P_i)$ is a biquadratic interpolating polynomial, and hence, on integration, the third-degree cubature formula $C_9f=\frac{1}{36}\{f(P_1)+\ldots+f(P_4)+4[f(P_5)+\ldots+f(P_8)]+16f(P_9)\}$ is produced. The original degree 2 is raised to 3 because $C_9f=I_\Omega f=0$ for every polynomial which is odd with respect to the centre of Ω.

The question of convergence of Newton–Cotes-type cubature procedures is easily answered. Since if C_nf is a product-type Newton–Cotes formula, C_nf cannot converge for every $f\in C^0(\bar{\Omega})$. This is due to the fact that a one-dimensional Newton–Cotes procedure does not converge for every continuous function and this property is simply extended into the tensor product procedure.

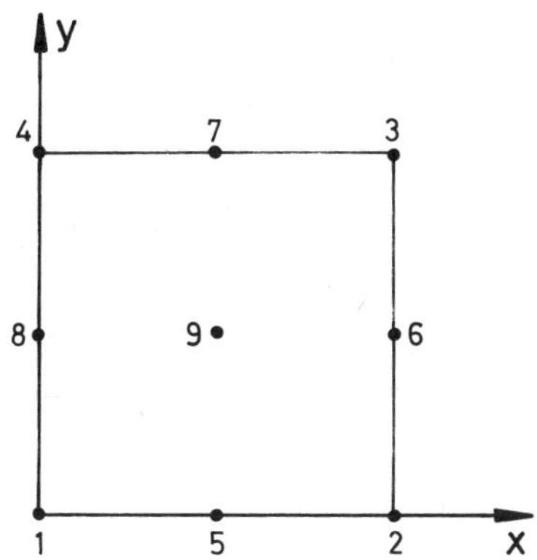

FIG. 6.4.7. Nodes of C_9f.

6.5 The existence of positive cubature formulae with given nodes

The Newton–Cotes cubature formulae of the previous section were derived by integrating interpolation formulae with regularly-spaced nodes on a uniform grid. A result of this method of construction is the existence of C_nf. Existence is less certain for an arbitrary domain Ω and, particularly, for *positive* cubature formulae. We now set out to prove the existence of positive formulae for arbitrary Ω. The proof is done constructively and produces a set of nodes appropriate to the positive formula. Thus, the nodes are not known prior to the construction; if they were preassigned the following result would contradict the properties of the Newton–Cotes formulae in one dimension.

Assume we are given a closed connected region Ω such that the momentum integrals exist as far as we need them. We then state:

Theorem 6.5.1 (Davis) *It is always possible to prescribe* $\binom{N+k}{k}$ *nodes* $\mathbf{x}_i \in \Omega$ *such that the corresponding cubature formula of degree* k *exists and has positive weights.*

We follow the proof of Davis [4] and show first the existence of a positive cubature formula with *more* than $\binom{N+k}{k}$ nodes. Having done this, we finally demonstrate that there exists a positive cubature formula of degree k having *exactly* $\binom{N+k}{k}$ nodes. The number $\binom{N+k}{k}$ arises from the fact that we have exactly $\binom{N+k}{k}$ monomials of degree k. The method of undetermined coefficients then gives a k-th degree formula provided that the linear system of equations for the weights has a solution.

Proof Within the non-empty domain Ω we can choose $\binom{2+k}{k}$ distinct points $\mathbf{x}_i \in \Omega$ which do not lie on a "curve" $C_k(\mathbf{x}) = 0$. (*For the sake of simplicity we consider only the case* $N = 2$.) We show that this is possible: We take three points $\mathbf{x}_1, \mathbf{x}_2, \mathbf{x}_3$ which do not lie on a straight line and a line $C_1(\mathbf{x}) = 0$ with $C_1(\mathbf{x}_i) \neq 0$, $i = 1, 2, 3$, together with three different points $\mathbf{x}_4, \mathbf{x}_5, \mathbf{x}_6$ being on it. These $\mathbf{x}_1, \ldots, \mathbf{x}_6$ cannot lie on a curve $C_2(\mathbf{x}) = 0$ of second degree. This is a consequence of the well-known result of algebraic geometry (cf. [24]) that two curves $P_k(\mathbf{x}) = 0$ and $P_m(\mathbf{x}) = 0$ with more than km common zeros necessarily have a non-constant common factor. In this case $C_2(\mathbf{x}_i) = 0$, $i = 1(1)6$, would imply that Q_1 is a factor of $C_2(\mathbf{x})$, so that $C_2(\mathbf{x})$ would be the product of two linear factors. This is impossible since $\mathbf{x}_1, \mathbf{x}_2, \mathbf{x}_3$ do not lie on a straight line.

The next points $\mathbf{x}_7, \ldots, \mathbf{x}_{10}$ are again chosen on a first-order curve $Q_1(\mathbf{x})=0$ with $Q_1(\mathbf{x}_i)\neq 0$, $i=1(1)6$, and the $\mathbf{x}_1, \ldots, \mathbf{x}_{10}$ do not lie on a third-degree curve $C_3(\mathbf{x})=0$. If they did, $C_3(\mathbf{x})$ would necessarily have the factorisation $C_3(\mathbf{x})=Q_1(\mathbf{x})C_2(\mathbf{x})$ up to a constant multiplier. Hence, $\mathbf{x}_1, \ldots, \mathbf{x}_6$ must lie on $C_2(\mathbf{x})$, which has already been shown to be impossible.

Proceeding by induction we see that we can choose $\binom{2+k}{k}$ distinct nodes $\mathbf{x}_i$ which do not lie on a curve $C_k(\mathbf{x})=0$ of order k. The extension to N dimensions is obvious.

We now present such a system of points by choosing a regular grid G of equidistant nodes with integer components for the N-simplex. We denote these by $G(k, N, \mathbf{x})\equiv\{\mathbf{x}\mid 0\leq x_1+\ldots+x_N\leq k\}$. For example, we have

$$G(2, 2, \mathbf{x})=\{\mathbf{x}\mid 0\leq x_1+x_2\leq 2\}=\{(0,0); (1,0); (0,1); (2,0); (1,1); (0,2)\}.$$

Further, let $\mathbf{a}^T=(a_1, a_2, \ldots, a_N)\in\mathbb{R}^N$. Then

$$G(k, N, \mathbf{x}^{\mathbf{a}})\equiv\{x_1^{a_1}, x_2^{a_2}, \ldots, x_N^{a_N}\mid \mathbf{a}\in G(k, N, \mathbf{x})\}$$

is the set of monomials of degree $\leq k$, for example $G(2, 2, \mathbf{x}^{\mathbf{a}})=\{1; x_1; x_2; x_1^2; x_1x_2; x_2^2\}$ if $\mathbf{a}=G(2, 2, \mathbf{x})$, and finally if $\mathbf{b}^T=(b_0, b_1, \ldots, b_N)\in\mathbb{R}^N$ with distinct components b_i then $G(k, N, \mathbf{b}_{\mathbf{x}})\equiv\{(b_{x_1}, b_{x_2}, \ldots, b_{x_N})\mid\mathbf{x}\in G(k, N, \mathbf{x})\}$ generalises $G(k, N, \mathbf{x})$. For example, let $b_0=0.2$, $b_1=0.3$ and $b_2=0.6$ so that $G(2, 2, \mathbf{b}_{\mathbf{x}})=\{(0.2, 0.2); (0.2, 0.3); (0.2, 0.6); (0.3, 0.2); (0.3, 0.3); (0.6, 0.2)\}=\{(b_{x_1}, b_{x_2})\mid(x_1, x_2)\in G(2, 2, \mathbf{x})\}$.

There is a one-to-one correspondence between the three grids and, in particular, for every polynomial there is a corresponding grid-point. The converse also holds. Taking such a grid to be the set of nodes of the cubature formula we have to show that the resulting matrix of the system of equations for the weights in non-singular. The formula C_nf has degree k if it is exact for every element of $G(k, N, \mathbf{x}^{\mathbf{a}})$ where the nodes are the grid-points $G(k, N, \mathbf{x})$ (or, more generally, $G(k, N, \mathbf{b}_{\mathbf{x}})$. The weights $\mathbf{A}^T=(A_1, \ldots, A_{\binom{N+k}{k}})$ are the solutions of the matrix equation $\mathbf{M}(k, N)\mathbf{A}=\frac{1}{|\Omega|}\mathbf{B}$ where the components B_i of $\mathbf{B}$ are momentum integrals of the elements of $G(k, N, \mathbf{x})$. We then have that

$$\det M(k, N)=\prod_{i=0}^{k-1}[(k-i)!]^{N\binom{N-1}{i}}\neq 0.$$

The proof is again performed by induction. (For the more general grid $G(k, N, \mathbf{b}_{\mathbf{x}})$ the proof of non-singularity is due to Stroud, book). As an example we apply $C_6 f$ with the nodes of $G(2, 2, \mathbf{x})$ to the elements of $G(2, 2, \mathbf{x}^{\mathbf{a}})$ and obtain the matrix

$$\mathbf{M}(2,2)=\begin{bmatrix} 1 & 1 & 1 & 1 & 1 & 1 \\ 0 & 1 & 0 & 2 & 1 & 0 \\ 0 & 0 & 1 & 0 & 1 & 2 \\ 0 & 1 & 0 & 4 & 1 & 0 \\ 0 & 0 & 0 & 0 & 1 & 0 \\ 0 & 0 & 0 & 1 & 1 & 4 \end{bmatrix}, \quad \det \mathbf{M}(2,2)=4 \neq 0.$$

By an affine transformation the grid $G(k, N, \mathbf{x})$ may be introduced into every N-dimensional simplex. Hence, we choose a simplex in the interior of Ω which has a regular grid $G(k, N, \mathbf{x})$. Let us denote the grid-points by $\mathbf{x}_i$. Again we restrict our attention to $N=2$ dimensions in order to simplify the proof. Such a grid is illustrated in Fig. 6.5.1 (points 1 to 10). A rectangle with centre $\mathbf{x}_i$ is constructed around each such point so that no two rectangles overlap and the whole rectangle is contained in Ω (cf. Fig. 6.5.1). Within each rectangle we choose a positive-product cubature formula $C_r f$ with degree $(C_r f) \geq k$. This can always be done, for example, using tensor products of positive quadrature formulae. We also require that the centre $\mathbf{x}_i$ of the i-th rectangle r is a node of $C_r f$ with weight D_i.

The union of the set of rectangles so far constructed is a subset of Ω. We therefore now choose a further set of non-overlapping rectangles in Ω so that the difference (Ω) between the area of Ω and the union of the total number of rectangles is such that $|(\Omega)| \leq d$, (an example is given in Fig. 6.5.1, rectangles 11 to 17), where d is specified below. For each of the new rectangles we construct a positive-product formula $C_s f$ with degree $(C_s f) \geq k$. This introduces extra nodes and now all nodes which are not centres of rectangles are denoted by $\mathbf{y}_i$, $i=1(1)\bar{N} \geq 0$ and corresponding weights B_i. This construction ensures that all weights are positive and that all nodes are inside Ω. The addition of all the cubature formulae $C_r f$ and $C_s f$ produces a cubature formula with positive weights which is valid only for the part of the domain Ω covered by rectangles. Using, again, the grid-points $\mathbf{x}_i$, we construct a (not necessarily positive) cubature formula for the remaining part (Ω) of Ω having weights A_i, $i=1(1)\binom{2+k}{k}$. These satisfy

$$(A_1, \ldots, A_{\binom{2+k}{k}})^T = M(k,2)^{-1}(I_{0,0}, I_{1,0}, \ldots, I_{0,k})^T \big|_{(\Omega)}$$

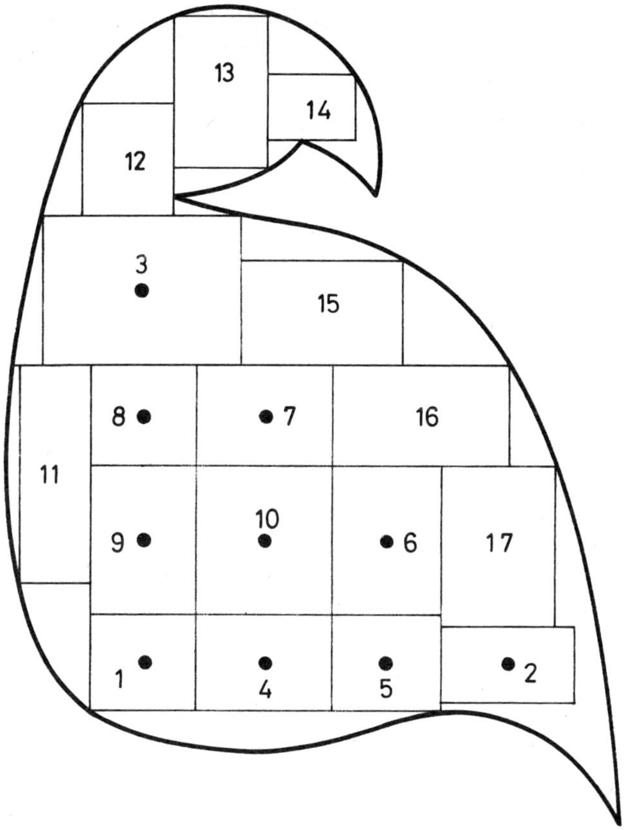

FIG. 6.5.1. Construction of a positive cubature formula.

where $I_{i,k}$ are the momentum integrals with respect to (Ω). This formula is positive if

$$\max_i |A_i| \leq \|M(k, 2)^{-1}\| \max_{0\leq n,m\leq k} |I_{n.m}|$$

$$\leq \|M(k, 2)^{-1}\| \, |(\Omega)| \max_{0\leq p+q\leq k} \sup_{\mathbf{x}\in(\Omega)} |x_1^p x_2^q| < \min_i (D_i)$$

holds. Obviously this is true (since then $D_i + A_i \geq D_i - |A_i| > 0$), if $|(\Omega)|$ is so small that

$$0 \leq |(\Omega)| < \frac{\min_i (D_i)}{\|M(k, 2)^{-1}\| \max_{p+q\leq k} \sup_{\mathbf{x}\in(\Omega)} |x_1^p x_2^q|} \equiv d.$$

Should this not be true, then further rectangles can be chosen as above until the inequality is valid. If, in addition, (Ω) is contained in the unit square, then $0 \leqslant |(\Omega)| < \min_i (D_i)/\|M(k, 2)^{-1}\| \equiv d$ since then $\max_{p+q\leqslant k} \sup_{\mathbf{x}\in(\Omega)} |x_1^p x_2^q| = 1$, for example if $p = q = 0$. This result can be immediately extended to $N > 2$ dimensions.

The last step of the proof can now be performed because if we have any positive cubature formula with *more* than $\binom{N+k}{k}$ nodes we can find a further positive cubature formula having *exactly* $\binom{N+k}{k}$ nodes. In order to see this we assume that the cubature formulae are written as a vector equation

$$\mathbf{I} = \sum_{i=1}^{N} E_i \mathbf{z}_i, \qquad n > \binom{N+k}{k} = \tilde{N},$$

where E_i are the positive weights and for $N = 2$

$$\mathbf{I}^T = (I_{0,0}, I_{1,0}, \ldots, I_{0,k}) \in \mathbb{R}^{\tilde{N}}$$

$$\mathbf{z}_i^T = (1, x_{1_i}, x_{2_i}, x_{1_i}^2, \ldots, x_{2_i}^k) \in \mathbb{R}^{\tilde{N}},$$

since the formula is exact for every polynomial $f \in \mathbb{P}_k$. In general, the $\tilde{N}$-dimensional vectors $\mathbf{z}_i$ and $\mathbf{I}$ are linearly dependent since, at most, $\tilde{N}$ vectors in $\mathbb{R}^{\tilde{N}}$ are linearly independent. Hence there exist real numbers $s_1, \ldots, s_n$, not all zero such that $s_1\mathbf{z}_1 + s_2\mathbf{z}_2 + \ldots + s_n\mathbf{z}_n = \mathbf{0}$, $n > \tilde{N}$, and for which at least one of the coefficients s_i is positive. Hence, if $m \equiv \max_i (s_i/E_i) > 0$ then

$$\mathbf{I} = \sum_{i=1}^{n} \left[E_i \mathbf{z}_i - \frac{s_i}{m} \mathbf{z}_i \right] = \sum_{i=1}^{n} \left[\frac{(mE_i - s_i)}{m} \right] \mathbf{z}_i$$

is another linear combination of the $\mathbf{z}_i$ with non-negative coefficients and at least one of which vanishes. If now exactly $\binom{N+k}{k}$ coefficients are different from zero then the proof is complete. If this is not so, this process is repeated at most a finite number of times. ■

6.6 Equally-weighted quadrature formulae – Tschebyscheff's problem

We now set up quadrature and cubature formulae using the reverse approach of specifying the weights A_i and calculating the nodes. The most important choice of weight specification is to make them equal so that $A_i = (1/n)\int_{-1}^{1} W(x)\,dx$, $i = 1(1)n$. This is known as *Tschebyscheff's problem* of equally-weighted quadrature formulae which was treated in Section 2.4 where some formulae have been listed. We noted then that such formulae with real nodes do not exist for arbitrary n and $W(x)$. Tschebyscheff constructed the formulae only for $n = 1(1)7$ and $W(x) = 1$. Hence, he did not notice that they do not exist for $n = 8$ and $n > 9$. Some years later this was proved by Bernstein. In order to prove the non-existence of real equally-weighted quadrature formulae for the weight function $W(x) \equiv 1$ we need some preliminary results.

We compare the equally-weighted formula $Q_n f = (2/n) \sum_{i=1}^{n} f(x_i)$ and the Gaussian formula $Q_m f = \sum_{i=1}^{m} A_i f(t_i)$ of degree $2m-1$ (cf. Section 7.3). $t_1 < \ldots < t_m$ are the m zeros of the m-th degree Legendre polynomial.

Lemma 6.6.1 If $E_n f = 0$ for $f \in \mathbb{P}_{2m-1}$, where $n > m$, then $x_1 < t_1$. ■

Proof For the polynomial $p_{2m-1} = L_m^2/(x - t_1)$ we have $Ip_{2m-1} = 0$ since L_m is orthogonal to $L_m/(x - t_1) \in \mathbb{P}_{m-1}$. Further, $Q_n p_{2m-1} = 0$ since we assumed that $E_n p_{2m-1} = 0$. The equation $Q_n p_{2m-1} = 0$ is equivalent to

$$\sum_{k=1}^{n} p_{2m-1}(x_k) = 0.$$

Since p_{2m-1} has $m < n$ zeros only, not all terms of this sum vanish and hence negative values $p_{2m-1}(x_k)$ exist. But, since $p_{2m-1} < 0$ only for $x < t_1$, at least the smallest node x_1 satisfies $x_1 < t_1$. ■

Lemma 6.6.2 If $Q_n f$ is equally-weighted, if degree $(Q_n f) = 2m - 1$ with $m < n$ and if degree $\left(Q_m f = \sum_{i=1}^{m} A_i f(t_i)\right) = 2m - 1$, then $2/n < A_1$. ■

Proof Let $p_{2m-2}(x) \equiv \left[\dfrac{L_m(x)}{L_m'(t_1)(x - t_1)}\right]^2$. Then

$$Ip_{2m-2}(x) = Q_n p_{2m-2}(x) = \frac{2}{n} \sum_{i=1}^{n} p_{2m-2}(x_i) > 0$$

and

$$Ip_{2m-2}(x) = Q_m p_{2m-2}(x) = A_1,$$

since $L_m(t_i)=0$ for $i>1$, $L_m(t_1)\neq 0$ and $p_{2m-2}(t_1)=1$. Hence, since $p_{2m-2}(x)\geq 0$, we have that $A_1=\frac{2}{n}\sum_{i=1}^{n} p_{2m-2}(x_i)\geq\frac{2}{n}p_{2m-2}(x_1)$.

As the polynomial $p_{2m-2}(x)$ has the $m-1$ double roots $t_2,\ldots,t_m$, by Rolle's theorem $p'_{2m-2}(x)$ has the simple roots $t_2,\ldots,t_m$ together with $m-2$ simple roots in (t_2, t_m). Since $p'_{2m-2}(x)$ has an odd degree, and no root smaller than t_2, we have that $p'_{2m-2}<0$ for $x<t_2$ (note that $p'_{2m-2}(-\infty)=-\infty$). Hence, $p_{2m-2}(x)>p_{2m-2}(t_1)=1$ for $x<t_1$. From Lemma 6.6.1 we have that $x_1<t_1$ and, therefore, that $p_{2m-2}(x_1)>1$ so that $A_1>2/n$. ∎

From Section 5.4.3 we see that for the interval $[0, 1]$ we have $(t_1=t_1^{(m)})$ that $0<t_1<a/m(m+1)$ with $a=1.445\,7965$ and hence, for the interval $[-1, 1]$, we have the inclusion $-1<t_1<-1+2a/m(m+1)$. A bound is also needed for the first weight A_1 of $Q_m f$. In the appendix of Section 5.4.3 we saw that $A_1=A_1^{(m)}=\frac{1}{m(m+1)}C_m$. Comparison with a table of Gaussian quadrature formulae (e.g. Stroud and Secrest, book) shows that C_m depends only very weakly on m, that C_m grows monotonically, and that Richardson extrapolation gives the correct result for C_∞ in Table 6.6.1. Hence

$$A_1\leq\frac{C_\infty}{m(m+1)}. \tag{6.4.1}$$

TABLE 6.6.1.

m	$1/m$	C_m					
16	$\frac{1}{16}$	7.385 468 960					
			7.420 379 907				
32	$\frac{1}{32}$	7.411 652 170		7.420 735 220			
			7.420 713 013		7.420 758 470		
64	$\frac{1}{64}$	7.418 447 803		7.420 758 107		7.420 761 016	
			7.420 755 290		7.420 761 005		7.420 761 326 $=C_\infty$
128	$\frac{1}{128}$	7.420 178 417		7.420 760 960		7.420 761 325	
			7.420 760 607		7.420 761 325		
256	$\frac{1}{256}$	7.420 615 060		7.420 761 320			
			7.420 761 273				
512	$\frac{1}{512}$	7.420 724 719					

We are now able to prove Bernstein's theorem:

Theorem 6.6.1 (Bernstein) *No equally-weighted quadrature formulae Q_nf with $W(x)\equiv 1$ with real nodes* exist *for $n=8$ and for $n>9$.*

Proof The nodes of the Tschebyscheff formula Q_nf satisfy (due to the method of undetermined coefficients)

$$x_1^k+x_2^k+\ldots+x_n^k=\frac{n}{2}\frac{1+(-1)^k}{k+1}=\begin{cases}0 & \text{for odd } k\\ n/(k+1) & \text{for even } k\end{cases}\quad k=0(1)n$$

from which it is clear that the solution is symmetric, i.e. $x_1=-x_n$, $x_2=-x_{n-1}$, etc. Hence, if n is even then the additional equation $x_1^{n+1}+x_2^{n+1}+\ldots+x_n^{n+1}=0$ is also satisfied, and for n odd, degree $(Q_nf)=n$. Hence, for $n=2m-2$ and for $n=2m-1$ we have that degree $(Q_nf)=2m-1$. These two cases are treated separately.

Let $n=2m-1$ be an odd integer.

Combining the statement of Lemma 6.6.2 with (6.4.1) we have that

$$\frac{2}{n}<\frac{C_\infty}{m(m+1)}=\frac{4C_\infty}{(n+1)(n+3)}.$$

Hence $(n+1)(n+3)<2nC_\infty$, which means that $n^2+2n(2-C_\infty)+3<0$ and that n is situated between the two roots of this second-order polynomial. One root is 0.284 161 93 while the second is 10.557 360 72. Hence, $n\leqslant 9$ for *odd n*.

Analogously for $n=2m-2$ we obtain

$$\frac{2}{n}<\frac{C_\infty}{m(m+1)}=\frac{4C_\infty}{(n+2)(n+4)}$$

so that $(n+2)(n+4)<2nC_\infty$ or, equivalently, $n^2+2n(3-C_\infty)+8<0$. The positive roots are 7.818 279 641 and 1.023 243 01 and hence $2\leqslant n\leqslant 6$ for *n even.* ■

Note that the result is sharp and determines the existence of real nodes of Q_nf for $n=1(1)7$ and $n=9$, while Bernstein's proof required the detailed discussion of several special cases arising from the fact that Bernstein's bounds for A_1 and t_1 were less sharp.

The number of real equally-weighted formulae which exist when $W(x)\not\equiv 1$ depends on the form of $W(x)$ and may be either greater or less than the number for $W(x)\equiv 1$. This has already been seen in Section 2.4 and the corresponding tables of Section 2.8.2. Classes of weight functions for which equally-weighted Q_nf with real nodes exist have been treated by Bourgoyne [3], Gatteschi [12], Salzer [20], Ullmann [23] and Wilf [25]. These papers are concerned with both finite and infinite intervals of integration.

6.7 Equally-weighted cubature formulae

The extension of equally-weighted quadrature to equally-weighted cubature formulae is not straightforward since the interpolating problem in two or more dimensions is very delicate as the nodes are not known *a priori*. We therefore use the method of undetermined coefficients in order to obtain two-dimensional equally-weighted cubature formulae for some standard domains. Expanding $f(x, y)$ in Taylor series about $\mathbf{0}$, we have a triangular scheme of monomials

$$\begin{array}{ccccc} & & 1 & & \\ & x & & y & \\ x^2 & & xy & & y^2 \\ & & \cdots\cdots & & \end{array}$$

whose elements must be integrated exactly by C_nf. For a k-th degree cubature formula $C_nf = \frac{|\Omega|}{n} \sum_{i=1}^{n} f(x_i, y_i)$ with $W(\mathbf{x}) \equiv 1$ we therefore have $\binom{k+2}{2} - 1$ monomials for which $E_nf = 0$ has to be valid; note that $C_n 1 = |\Omega|$. This gives $\binom{k+2}{2} - 1$ equations for the $2n$ unknown values x_i, y_i, $i = 1(1)n$. We now define:

Definition 6.7.1 The real equally-weighted cubature formula C_nf is called a *strict T-formula* if $\binom{k+2}{2} - 1 = 2n$, an *optimal T-formula* if $\binom{k+2}{2} - 1 > 2n$ and a *T-formula* if $\binom{k+2}{2} - 1 < 2n < \binom{k+3}{3} - 1$ (cf. Engels [5]). ■

We will treat some special cases with $W(\mathbf{x}) \equiv 1$.

6.7.1 Equally-weighted cubature formulae for triangles

We refer to the standard triangle given by Fig. 2.3.1. The moments are

$$\left.\begin{aligned} I_{k,0}&=\frac{2h^{k+2}}{\sqrt{3}(k+1)(k+2)}[2^{k+2}+(-1)^k(3k+5)],\quad I_{k,1}=0,\\ I_{k,l}&=\frac{1}{3(k+1)(k+2)}\Big[(-1)^k(1+(-1)^l)h^{k+l+2}(\sqrt{3})^{l+1}\\ &\qquad\times\left(\frac{3(k+2)}{l+1}-1\right)+l(l-1)I_{k+2,l-2}\Big] \end{aligned}\right\}\begin{aligned} k&=0(1)\ldots,\\ l&=2(1)\ldots \end{aligned}$$

and, as a result, many of the momentum integrals vanish and we seek node distributions such that the same terms vanish within the expansion of C_nf. The following formulae have this property:

(i) degree (C_nf) *odd*

$$I_\Omega f=\frac{3\sqrt{3}h^2}{(k^2-1)/4+k}\left\{\sum_{i=1}^{k}f(x_i,0)+\sum_{i=k+1}^{(k^2-1)/8+k}(f(x_i,y_i)+f(x_i,-y_i))\right\}$$

(ii) degree (C_nf) is a *multiple of* 4

$$I_\Omega f=\frac{3\sqrt{3}h^2}{k^2/4+k}\left(\sum_{i=1}^{k}f(x_i,0)+\sum_{i=k+1}^{k^2/8+k}(f(x_i,y_i)+f(x_i,-y_i))\right\}$$

(iii) degree (C_nf) *even but not a multiple of* 4

$$I_\Omega f=\frac{3\sqrt{3}h^2}{k+k^2/4+1}\left\{\sum_{i=1}^{k}f(x_i,0)+\sum_{i=k+1}^{k+\frac{1}{2}(k^2/4+1)}(f(x_i\,y_i)+f(x_i,-y_i)\right\}$$

The first k nodes on the x-axis are used in all three cases to obtain $E_nx^m=0$ for $m=1(1)k$. In case (i) we have $(k-1)+(k-3)+\ldots+2=(k^2-1)/4$ remaining conditions for the determination of the $(k^2-1)/4$ unknowns x_i, y_i. Hence we can expect strict T-formulae. For case (ii) we have $(k-1)+(k-3)+\ldots+1=k^2/4$ conditions and these are needed for determining the $k^2/4$ unknowns. We again expect strict T-formulae. In case (iii), $k^2/4$ is an odd integer so that $k^2/4+1$ is even and we have $k^2/4+1$ unknowns but only $k^2/4$ conditions. Hence, we expect T-formulae because one additional parameter is not sufficient to produce the next higher degree of precision. If we solve the systems for the nodes (x_i, y_i) then we obtain the following formulae (Table 6.7.1) (cf. Engels [6]).

TABLE 6.7.1 Equally-weighted cubature formulae for the triangle.

Degree	Quality	Nodes x_i	Nodes y_i	Case
1	strict	0	0	1
2	–	$\pm h$	0	3
		0	$\pm h$	
	optimal	$h\cos(\varphi+2\pi i/3)$	$h\sin(\varphi+2\pi i/3)$	–
		$i=1, 2, 3,\quad \varphi$ arbitrary		
3	strict	$1.158\,662\,134\,133\,79\, h$	0	
		$0.442\,443\,677\,709\,54\, h$	0	1
		$-0.801\,095\,811\,843\,34\, h$	0	
		$-0.4\, h$	$\frac{1}{2}\sqrt{5}\, h$	
4	strict	$1.396\,576\,168\,236\,918\, h$	0	2
		$0.218\,068\,700\,163\,556\, h$	0	
		$-0.172\,103\,828\,307\,169\, h$	0	
		$-0.822\,336\,937\,206\,577\, h$	0	
		$\frac{1}{10}(\sqrt{6}-4\pm\sqrt{30})h$	$(1\pm 1/\sqrt{5})^{\frac{1}{2}}h$	

These seem to be the only available formulae of this type (cf. Engels [5]).

6.7.2 Equally-weighted cubature formulae for some further planar regions

If Ω is not a triangle, there is no general method for choosing nodes and weights *a priori* in such a way that the terms of the expansion of $C_n f$ vanish if and only if the corresponding momentum integrals vanish.

We list some results.

Square (Fig. 2.3.1: cf. Engels [7])

TABLE 6.7.2. Equally-weighted cubature formulae for the square $\max(|x|, |y|) \leq h$.

Degree	Quality	Nodes x_i	Nodes y_i	
1	strict	0	0	
2		$\sqrt{\frac{2}{3}}\,h\cos\left(\varphi+\frac{2\pi i}{3}\right)$	$\sqrt{\frac{2}{3}}\,h\sin\left(\varphi+\frac{2\pi i}{3}\right)$	$i=0, 1, 2$
		φ arbitrary		
3	optimal	$\sqrt{\frac{2}{3}}\,h\cos\left(\varphi+\frac{\pi}{2}i\right)$	$\sqrt{\frac{2}{3}}\,h\sin\left(\varphi+\frac{2\pi i}{2}\right)$,	$i=0, 2, 3$
		φ arbitrary		

Circle (Fig. 2.3.1; cf. Engels [8])

TABLE 6.7.3. Equally-weighted cubature formulae for the circle $x^2+y^2 \leq h^2$.

Degree	Quality	Nodes x_i	y_i	
1	strict	0	0	
2		$\sqrt{\frac{1}{2}}h\cos\left(\varphi+\frac{2\pi i}{3}\right)$	$\sqrt{\frac{1}{2}}h\sin\left(\varphi+\frac{2\pi i}{3}\right)$,	$i=0, 1, 2$
		φ arbitrary		
3	optimal	$\sqrt{\frac{1}{2}}h\cos\left(\varphi+\frac{\pi i}{2}\right)$	$\sqrt{\frac{1}{2}}h\sin\left(\varphi+\frac{\pi i}{2}\right)$,	$i=0, 1, 2, 3$
		φ arbitrary		
$k=5(1)7$	optimal for $k=7$	$r_j h\cos\left(\varphi_j+\frac{2\pi i}{k+1}\right)$	$r_j h\sin\left(\varphi_j+\frac{2\pi i}{k+1}\right)$,	$j=1, 2$ $i=0(1)k$
		$r_{1,2}=(1\pm 1/\sqrt{3})/2)^{\frac{1}{2}}$	$\varphi_{1,2}$ arbitrary	

Hexagon (Fig. 2.3.1; cf. Engels [8])

TABLE 6.7.4. Equally-weighted cubature formulae for the hexagon.

Degree	Quality	Nodes x_i	y_i	
1	strict	0	0	
2		$rh\cos\left(\varphi+\frac{2\pi i}{3}\right)$	$rh\sin\left(\varphi+\frac{2\pi i}{3}\right)$	$i=0, 1, 2$
		φ arbitrary		$r=\frac{1}{2}\sqrt{\frac{5}{3}}$
3		$rh\cos\left(\varphi+\frac{\pi i}{2}\right)$	$rh\sin\left(\varphi+\frac{\pi i}{2}\right)$	$i=0(1)3$
		φ arbitrary		
		0	0	$r=5\sqrt{3}/12$
		$rh\cos\left(\varphi+\frac{\pi i}{2}\right)$	$rh\sin\left(\varphi+\frac{\pi i}{2}\right)$	$i=0(1)3$
		φ arbitrary		
5		$r_j h\cos\left(\varphi_j+\frac{\pi i}{3}\right)$	$r_j h\sin\left(\varphi_j+\frac{\pi i}{3}\right)$	$i=0(1)6$
		$r_{1,2}=((5\pm\sqrt{43/5})/12)^{\frac{1}{2}}$,	$\varphi_{1,2}$ arbitrary	$j=1, 2$

6.8 Remarks, problems, references

6.8.1 Remarks

R.6.1: In 1969 Johnson and Riess [16] improved Davis' result for the convergence of the Newton–Cotes formulae. They used Ouspenski's asymptotic expansions of the weights but replaced the contour inner product for the space E_ρ by the area inner product

$$(f, g) \equiv \int_{E_\rho} f(z)\overline{g(z)}\,dx\,dy, \qquad z = x + iy,$$

to prove that if $f(z)$ is analytic in E_ρ, then convergence of the Newton–Cotes quadrature is assured even for $\rho > 2$ (cf. Section 6.2). E_2 is the ellipse with major semiaxis $a = 1.06066$ and minor semiaxis $b = 0.35355$. This seems to contradict Polya's counterexample because Polya's function $g(z)$ is regular even for $|\text{Im}(z)| < 0.4412712$ (cf. Section 6.3.4). In the same paper Johnson and Riess show that if

$$f \in C^\infty[-1, 1] \quad \text{and} \quad \frac{\|f^{(n)}\|}{n!} \leqslant K|n \log n| \quad \text{then} \quad \lim_{n\to\infty} E_n f = 0,$$

where $\|\cdot\|$ is the sup-norm.

R.6.2: Newton–Cotes formulae may also be derived for weight functions $W(x) \neq 1$. Such formulae were calculated by Engels and Meuer [11] for

$$\begin{aligned} &W(x) = x^A(1-x)^B, && A, B > -1, && x \in [0, 1] \\ &W(x) = x^A \ln(1/x), && A > -1, && x \in [0, 1] \\ &W(x) = x^A e^{-x}, && A > -1, && x \in [0, \infty). \end{aligned}$$

From these the following formulae have fewer nodes than might be expected (cf. Section 6.1).

In $\displaystyle\int_0^1 x^A(1-x)^{A-1} f(x)\,dx \doteq \sum_{i=1}^{2n+1} A_i f\left(\frac{i-1}{2n}\right)$ the weight $A_{2n+1} = 0$,

$$\int_0^1 \left(\frac{1-x}{x}\right)^{\frac{1}{2}} f(x)\,dx \doteq \tfrac{1}{6}[2f(0) + 3f(\tfrac{1}{4}) + f(\tfrac{3}{4})] \quad \text{has degree 4},$$

$$\int_0^1 \frac{f(x)}{(x(1-x))^{\frac{1}{2}}}\,dx \doteq \tfrac{1}{6}[f(0) + 2f(\tfrac{1}{4}) + 2f(\tfrac{3}{4}) + f(1)] \quad \text{has degree 5}.$$

6.8.2 Problems

P.6.1: Apply Davis' theorem to construct a positive three-point cubature formula of degree 1 for the domain (cf. Section 6.5)

$$\Omega = \{(x, y) \mid x \geqslant 0, y \geqslant 0, x \leqslant 1, y \leqslant 1, x + y \leqslant 7/4\}.$$

Note that this is possible since in this case $\binom{N+k}{k} = \binom{3}{1} = 3$.

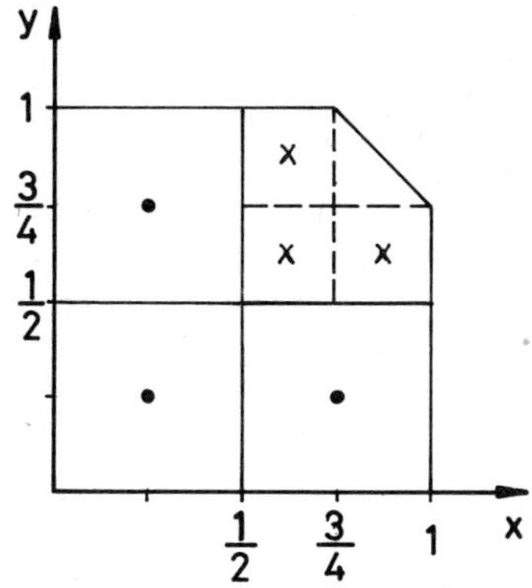

FIG. 6.8.1.

Hint: For simplicity choose the initial grid $\mathbf{x}_1^T = (\frac{3}{4}, \frac{1}{4})$, $\mathbf{x}_2^T = (\frac{1}{4}, \frac{1}{4})$ and $\mathbf{x}_3^T = (\frac{1}{4}, \frac{3}{4})$ as in Fig. 6.8.1. The midpoint formulae are then positive first-degree cubature formulae and hence fit the requirements, and we have $D_1 = D_2 = D_3 = 1/4$.

P.6.2: Determine the quadrature formula for the interval $[0, 1]$ which is obtained by integration of the cubic interpolating polynomial spline $S_n(f, x)$ with the nodes $x_i = i/n$, $i = 0(1)n$ and the boundary conditions $S_n''(f, 0) = S_n''\left(f, \frac{1}{n}\right)$ and $S_n''\left(f, \frac{n-1}{n}\right) = S_n''(f, 1)$. In particular, choose $n = 4$ (cf. Table 6.3.1).

P.6.3: Determine a cubature formula of degree 4 for the two-dimensional simplex S_2 (cf. Section 6.4.1).

(i) The nodes (x_i, y_i) are the points of intersection of the lines $y = i/4$, $x = i/4$ and $x + y = i/4$, $i = 0(1)4$.
(ii) Calculate the cardinal functions $\lambda_{4,k}(x, y)$, $k = 1(1)15$. How many cardinal functions are essential?
(iii) Determine the weights of the cubature formula by integration of the interpolating polynomial $L(x, y) \equiv \sum_{k=1}^{15} \lambda_{4,k}(x, y) f(P_k)$.

P.6.4: In the Newton–Cotes cubature formulae for the two-dimensional simplex S_2 the nodes are situated on concentric triangles. These nodes are the grid-points $(x_i, y_i) = (i/n, k/n)$, $0 \leq i + k \leq n$. Show that (cf. Section 6.4.1)

(i) there exist exactly $[n/3] + 1$ such triangles,
(ii) on the sides of the k-th triangle we have exactly $[(n - 3k + 3)/2] + 1$ essential nodes. (For the boundary of S_2 $k = 1$.)

P.6.5: Consider the Newton–Cotes cubature formulae on the two-dimensional simplex S_2 with vertices P_1, P_2, P_3. Show that the corresponding weights vanish if n is even. (We have a total of $\binom{n+2}{2}$ nodes on S_2 (cf. Section 6.4.1).)

6.8.3 References

1. Bernstein, S. N. (1937). On quadrature formulas of Cotes and Tschebyscheff. [Russian] *Dokl. Akad. Nauk SSSR*, **14,** 323–326.
2. Bernstein, S. N. (1932). On formulas of approximate integration of Tschebyscheff. [Russian] *Izvestia Akad. Nauk SSSR, Math. a. Phys. Sc. Ser.*, 1219–1227.
3. Bourgoyne, F. D. (1963). The nonexistence of certain Laguerre–Tschebyscheff quadrature formulas. *MTAC*, **17,** 196–197.
4. Davis, Ph. (1962). Errors of numerical approximation for analytic functions. *In* "Survey of Numerical Analysis," (J. Todd., Ed.). pp. 468–484. McGraw-Hill, New York.
5. Engels, H. (1971). Gleichgewichtete Kubaturformeln fur ein Dreieckiges ebenes Integrationsgebiet. *ZAMM*, **51,** T49–T51.
6. Engels, H. (1970). Über gleichgewichtete Kubaturformeln für ein Dreiecksgebiet. *Elektronische datenverarbeitung*, 535–539.
7. Engels, H. (1968). Gleichgewichtete Kubaturformeln für quadratisches Integrationsgebiet. *Elektronische datenverarbeitung*, 121–124.
8. Engels, H. (1970). Über gleichgewichtete Kubaturformeln für Kreis- und Sechseck-Gebiet. *Elektronische datenverarbeitung*, 216–223.
9. Engels, H. and Mangold, R. (1971). "Über eine Klasse Wilf'scher Quadraturformeln." Report Jül–789–MA. Nuclear Research Centre Jülich.
10. Engels, H. and Meuer, H. W. (1970). "Die Entwicklung von Interpolationsquadratur- und Differentiationsformeln zu gegebenen Stützstellen mit PL/1-FORMAC." Report Jül–662–MA. Nuclear Research Centre Jülich.
11. Engels, H. and Meuer, H. W. (1970). Quadraturformeln mit äquidistanten Abszissen zur Berechnung von Integralen $\int_0^1 x^A(1-x)^B f(x)\,dx$ für spezielle Werte von A und B." Report Jül–648–MA. Nuclear Research Centre Jülich.
12. Gatteschi, L. (1964). Su di un problema connesso alle formule di quadratura di Tschebyscheff. [Italian] *Rend. Sem. Matem. Univ. Pol. Torino*, **23,** 75–87.
13. Gatteschi, L. (1965). Sulla non existenca di certe formule di quadratura. [Italian]. *Rend. Sem. Matem. Univ. Pol. Torino*, **24,** 157–172.
14. Guenther, R. B. and Roetmann, E. L. (1970). Newton–Cotes Formulas in *n* dimensions. *Num. Math.*, **14,** 330–345.
15. Isaacson, E. and Keller, H. (1966). "Analysis of Numerical Methods." J. Wiley & Sons, New York.
16. Johnson, L. W. and Riess, R. D. (1969). A note on a theorem of Davis. *Math. Zeitschrift*, **110,** 211–212.
17. Kusmin, R. O. (1931). Concerning the theory of mechanical quadratures. [Russian] *Leningrad, Corr. polyt. inst.* (division of techn. nat. sc., and math.) **33,** 5–14.
18. Ouspenski, J. (1925). Sur les valeurs asymptotiques des coefficients de Cotes. *Bull. Am. Math. Soc.*, **31,** 145–156.
19. Polya, G. (1933). Über die Konvergenz von Quadraturverfahren. *Math. Zeitschrift*, **37,** 264–286.
20. Salzer, H. E. (1955). Equally weighted quadrature formulas over semi-infinite and infinite intervals. *J. Math. Phys.*, **34,** 54–63.
21. Silvester, P. (1970). Symmetric quadrature formulae for simplexes. *Math. Comput.*, **24,** 95–100.

22. Tschebyscheff, P. L. (1874). Sur les quadratures. *J. Math. Pures Appl.*, **19,** 19–34.
23. Ullmann, J. L. (1966). A class of weight functions that admit Tschebyscheff quadrature. *Michigan Math. J.*, **13,** 417–423.
24. Walker, R. J. (1962). "Algebraic Curves." Dover, New York.
25. Wilf, H. S. (1961). The possibility of Tschebyscheff quadrature on infinite intervals. *Proc. Nat. Acad. Sci.*, **47,** 209–213.
26. Wilson, M. W. (1969). A general algorithm for non-negative quadrature formulas. *Math. Comput.*, **23,** 253–258.

7

Refined Interpolatory Quadrature

When general Lagrange interpolation is used to produce quadrature formulae (as in Chapter 6) all the nodes of $Q_n f$ are preassigned and the weights have to be determined. The most famous examples of this technique are the Newton–Cotes formulae. However, these do not possess the highest possible degree of precision $2n-1$. This level of precision is obtained with Gauss quadrature formulae, which are now discussed in detail in two forms. In the first, the n nodes are considered to be free and are determined using Hermite interpolation. In the second, certain nodes are preassigned *a priori* and the corresponding quadrature formulae are obtained using mixed interpolation.

7.1 General Gauss quadratures with free nodes

In Section 2.5 we constructed Hermitian interpolating operators $H_n(f, x)$ which interpolate both the integrand f and its first derivative f' at the same nodes x_i, $i=1(1)n$. By integration of the weighted form $W(x)H_n(f, x)$ we obtain the quadrature procedure

$$Q_n f \equiv \sum_{i=1}^{n} (A_i f(x_i) + B_i f'(x_i)) = \int_{-1}^{1} W(x) H_n(f, x)\,dx \qquad (7.1.1)$$

with weights

$$A_i = \int_{-1}^{1} W(x)\left[1 - 2\lambda_i'(x_i)\frac{h_i(x)}{h_i'(x_i)}\right]\lambda_i^2(x)\,dx$$

$$B_i = \int_{-1}^{1} W(x)\frac{h_i(x)}{h_i'(x_i)}\lambda_i^2(x)\,dx$$

and suitable basis functions $h_i(x)$ and cardinal functions $\lambda_i(x)$ as defined in Section 2.5.

The use of derivatives $f'(x_i)$ is inconvenient for automatic calculations and so it is desirable to have $B_i = 0$, $i = 1(1)n$. Such quadratures $Q_n f$ are most important for practical purposes.

Definition 7.1.1 The quadrature procedure (7.1.1) is said to be a *general Gaussian procedure* if $B_i = 0$ for $i = 1(1)n$, and for all $n \in \mathbb{N}$. ■

Hence we have to satisfy

$$B_i = 0 \quad \text{for} \quad i = 1(1)n \tag{7.1.2}$$

which is a nonlinear system of equations involving as unknowns only the nodes x_i, $i = 1(1)n$. Clearly, we cannot say in general whether this system has a real solution, although in particular cases a solution exists. Note that the following theorems concerning general quadratures are nearly all new in this generality. For special quadratures these are well known.

Once (7.1.2) is satisfied we have the most important consequence:

Theorem 7.1.1 *If $B_i = 0$ for $i = 1(1)n$ then the quadrature $Q_n f$ is positive.*

Proof We have that

$$A_i = \int_{-1}^{1} W(x)\left(1 - 2\lambda_i'(x_i)\frac{h_i(x)}{h_i'(x_i)}\right)\lambda_i^2(x)\,dx = \int_{-1}^{1} W(x)\lambda_i^2(x)\,dx - 2\lambda_i'(x_i)B_i$$

$$= \int_{-1}^{1} W(x)\lambda_i^2(x)\,dx > 0$$

since $W(x) \geqslant 0$ and $\lambda_i^2(x) \geqslant 0$. ■

This theorem facilitates the construction of positive quadratures. We remember that positivity of $Q_n f$ is desirable for such reasons as convergence, damping of round-off errors, applicability to functions with low order of differentiability, and so on.

If the nodes are known, we have in general two possibilities for calculating the weights. First, we can integrate $W(x)\lambda_i^2(x)$ and, second, we can obtain the weights from the linear system of equations $Q_n k_i(x) = Ik_i(x)$, $i = 1(1)n$, if the functions $k_i(x)$ are linearly independent and satisfy $H_n(k_i, x) \equiv k_i$, $i = 1(1)n$. For example, the functions $\lambda_i^2(x)$, $i = 1(1)n$ (as well as every n linearly-independent linear combination of $\lambda_i^2(x)$, $i = 1(1)n$) have this property.

There is yet another remarkable property of the weights A_i. Assume that we are given the nodes x_i such that $B_i = 0$, $i = 1(1)n$. We then construct the *Lagrange* interpolating operator $L_n(f, x)$ with the same nodes x_i, $i = 1(1)n$. Integration gives

$$Q_n f = \sum_{i=1}^{n} \bar{A}_i f(x_i) \quad \text{with} \quad \bar{A}_i = \int_{-1}^{1} W(x)\lambda_i(x)\,dx.$$

In most special cases the cardinal functions $\lambda_i(x)$ also satisfy $H_n(\lambda_i(x), x) \equiv \lambda_i(x)$, $i = 1(1)n$. Hence, in this case, $A_i = \bar{A}_i$, $i = 1(1)n$.

Theorem 7.1.2 (Engels) *If (7.1.2) is valid and if $H_n(\lambda_i(x), x) \equiv \lambda_i(x)$ then the weights A_i are determined by*

$$A_i = \int_{-1}^{1} W(x)\lambda_i^2(x)\,dx = \int_{-1}^{1} W(x)\lambda_i(x)\,dx, \qquad i = 1(1)n.$$

Sometimes equation (7.1.2) is connected with a minimality property of the quadrature error

$$E_n f = \int_{-1}^{1} W(x)e_n(f, x)\,dx = \int_{-1}^{1} W(x)\frac{f^{(2n)}(t) - H^{(2n)}(f, t)}{(\omega_n^2(x))^{(2n)}\big|_{x=t}}\,\omega_n^2(x)\,dx.$$

The function $W(x)\omega_n^2(x)$ has constant sign in $[-1, 1]$. Hence there exists another point $s \in (-1, 1)$ such that

$$E_n f = \frac{f^{(2n)}(s) - H^{(2n)}(f, s)}{(\omega_n^2(x))^{(2n)}\big|_{x=s}} \int_{-1}^{1} W(x)\omega_n^2(x)\,dx,$$

and we can ask for the minimal value of the integral $\sigma_n \equiv \int_{-1}^{1} W(x)\omega_n^2(x)\,dx$ with respect to the zeros $x_1, \ldots, x_n$ of ω_n. A necessary condition for minimality is that $\dfrac{\partial}{\partial x_i}\displaystyle\int_{-1}^{1} W(x)\omega_n^2(x)\,dx = 0$, $i = 1(1)n$, since the integral depends only on the nodes x_i, $i = 1(1)n$, if these are not yet fixed. We then have the necessary condition

$$2\int_{-1}^{1} W(x)\omega_n(x)\frac{\partial}{\partial x_i}\omega_n(x)\,dx = 0, \qquad i = 1(1)n,$$

which may be written as (cf. (7.1.2))

$$\int_{-1}^{1} W(x)\frac{h_i(x)}{h_i'(x_i)}\left(\frac{\omega_n(x)h_i'(x_i)}{\omega_n'(x_i)h_i(x)}\right)^2 dx = \frac{h_i'(x_i)}{(\omega_n'(x_i))^2}\int_{-1}^{1} W(x)\omega_n(x)\frac{\omega_n(x)}{h_i(x)}\,dx = 0. \tag{7.1.3}$$

Now (7.1.2) is equivalent with (7.1.3) if

$$\frac{\partial}{\partial x_i}\,\omega_n(x) = C_i\,\frac{\omega_n(x)}{h_i(x)}, \qquad i = 1(1)n.$$

Here $C_i \neq 0$ are unimportant constants. Hence we have:

Theorem 7.1.3 (Engels) *A necessary condition for σ_n to be minimal and for $B_i = 0$, $i = 1(1)n$ to be valid simultaneously is that*

$$\frac{\partial}{\partial x_i}\,\omega_n(x) = C_i\,\frac{\omega_n(x)}{h_i(x)}, \qquad i = 1(1)n$$

where $C_i \neq 0$ are constants.

If, for example, $\omega_n(x) = G(x)\prod_{i=1}^{n} h_i(x)$ with $G(x)$ independent of x_i, then

$$\frac{\partial}{\partial x_i}\,\omega_n(x) = \left(G(x)\prod_{\substack{j=1\\ \neq i}}^{n} h_j(x)\right)\frac{\partial}{\partial x_i}\,h_i(x).$$

Further, if $\frac{\partial}{\partial x_i}\,h_i(x) = C_i = \text{constant}$, then

$$\frac{\partial}{\partial x_i}\,\omega_n(x) = C_i G(x)\prod_{\substack{j=1\\ \neq i}}^{n} h_j(x) = C_i\,\frac{\omega_n(x)}{h_i(x)}$$

and Theorem 7.1.3 applies.

The most important example of this type is that $h_i(x) = x - x_i$. Further examples are $h_i(x) = g(x) - g(x_i)$ for suitable functions $g(x)$.

In general it is impossible to present a sufficient condition for the minimality of σ_n which is consistent with (7.1.2).

7.2 Gauss quadratures with preassigned nodes

In some problems it is desirable to assign one or more nodes so as to take into account at these points some specific property of the integrand; for example, a sharp peak, or (if f is considered as a measured function) some more than ordinary reliable value. These characteristics can be reflected in the quadrature formula by using a special Lagrange–Hermite interpolation as described below (see also Engels [6]). If, in particular, the limits of the range of integration are preassigned then, for interpolating *polynomials*, we obtain the quadratures of Radau and Lobatto. This section contains the derivation of general formulae reflecting this behaviour.

7.2.1 One preassigned node

If one node z_1 is preassigned, then integration of the mixed interpolating operator

$$HL_{n,1}(f, x) = H_n(f, x) - \frac{\omega_n^2(x)}{\omega_n^2(z_1)}(H_n(f, z_1) - f(z_1))$$

leads to the quadrature formula

$$Q_{n+1}f = \sum_{i=1}^{n} (A_i f(x_i) + B_i f'(x_i)) + C_1 f(z_1),$$

where

$$B_i = \frac{1}{h_i'(x_i)\omega_n^2(z_1)} \int_{-1}^{1} W(x)[h_i(x)\lambda_i^2(x)\omega_n^2(z_1) - h_i(z_1)\lambda_i^2(z_1)\omega_n^2(x)]\,dx.$$

The requirement that $B_i = 0$, $i = 1(1)n$, is equivalent to the conditions

$$\int_{-1}^{1} W(x)\frac{\omega_n^2(x)}{h_i(x)}\,dx = \int_{-1}^{1} W(x)\frac{\omega_n^2(x)}{h_i(z_1)}\,dx, \qquad i = 1(1)n, \qquad (7.2.1)$$

which are called *characteristic integral relationships*. The solution of (7.2.1) produces the nodes x_i, $i = 1(1)n$. If it is assumed that this system has a solution, then the weight

$$C_1 = \int_{-1}^{1} W(x)\frac{\omega_n^2(x)}{\omega_n^2(z_1)}\,dx > 0 \quad \text{is always positive and for } i = 1(1)n$$

$$A_i = \frac{1}{\omega_n^2(z_1)} \int_{-1}^{1} W(x)[\lambda_i^2(x)\omega_n^2(z_1) - \lambda_i^2(z_1)\omega_n^2(x)]\,dx - 2\lambda_i'(x_i)B_i.$$

Substituting for $\lambda_i(x)$ the expression $\omega_n(x)h_i'(x_i)/[\omega_n'(x_i)h_i(x)]$, $i = 1(1)n$, and since $B_i = 0$

$$A_i = \left(\frac{h_i'(x_i)}{\omega_n'(x_i)}\right)^2 \int_{-1}^{1} W(x)\omega_n^2(x)\left(\frac{1}{h_i^2(x)} - \frac{1}{h_i^2(z_1)}\right)dx$$

which, using the integral relationship (7.2.1), becomes

$$A_i = \left(\frac{h_i'(x_i)}{\omega_n'(x_i)}\right)^2 \int_{-1}^{1} W(x)\frac{\omega_n^2(x)}{h_i(x)}\left(\frac{1}{h_i(x)} - \frac{1}{h_i(z_1)}\right)dx.$$

Finally, by the definition of λ_i we find that

$$A_i = \int_{-1}^{1} W(x)\lambda_i^2(x)\left(1-\frac{h_i(x)}{h_i(z_1)}\right)dx.$$

This representation shows that subject to a condition on $h_i(x)$ the weights A_i are always positive so that the quadrature formula

$$Q_{n+1}f = \sum_{i=1}^{n} A_i f(x_i) + C_1 f(z_1)$$

is positive. We thus have the following theorem:

Theorem 7.2.1 *A sufficient condition for $Q_{n+1}f$ to be positive is that $h_i(z_1) \geq h_i(x)$, $i=1(1)n$, for $x \in [-1, 1]$.*

Since $h_i(x_i)=0$ and $h_i(x_k) \neq 0$ for $i \neq k$, the basis functions $h_i(x)$ are usually strictly monotonic functions (see for example Fig. 7.2.1 for $z_1 = 1$). Assume that these are strictly monotonically *decreasing*, so that

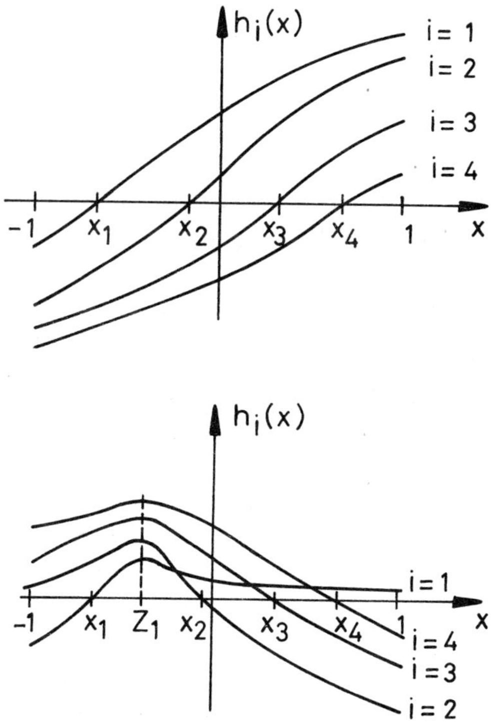

FIG. 7.2.1. (a) Monotonic basis functions. (b) Non-monotonic basis functions.

clearly we must have $z_1 \leqslant -1$. Conversely, if the basis functions strictly monotonically increase, then $z_1 \geqslant 1$.

Remark: Another, somewhat unnatural choice of $h_i(x)$, which is sufficient for the positivity of Q_{n+1}, is given in Figure 7.2.1. If z_1 is a maximum point of *all* $h_i(x)$, $i=1(1)n$, then z_1 may be an interior point of $[-1, 1]$ without destroying the positivity of $Q_{n+1}f$. Hence we have:

Corollary 7.2.1 If the basis functions h_i, $i=1(1)n$ are strictly monotonic functions then $|z_1| \geqslant 1$ is a sufficient condition for $Q_{n+1}f$ to be positive. ■

By the usual methods of deriving error representations for interpolating operators we see that the quadrature error of $Q_{n+1}f$ is

$$E_{n+1}f = \int_{-1}^{1} W(x) \frac{f^{(2n+1)}(t) - HL_{n,1}^{(2n+1)}(f, t)}{[\omega_n^2(x)(x-z_1)]_{x=t}^{(2n+1)}} \omega_n^2(x)(x-z_1)\,dx, \qquad t \in (-1, 1).$$

For $|z_1| \geqslant 1$ the function $W(x)\omega_n^2(x)(x-z_1)$ has constant sign in $[-1, 1]$ and hence we have, for the quadrature $Q_{n+1}f$, that the error is

$$E_{n+1}f = \frac{f^{(2n+1)}(s) - HL_{n,1}^{(2n+1)}(f, s)}{[\omega_n^2(x)(x-z_1)]_{x=s}^{(2n+1)}} \int_{-1}^{1} W(x)\omega_n^2(x)(x-z_1)\,dx, \qquad s \in (-1, 1).$$

7.2.2 Two preassigned nodes

Let us assume now that the two nodes z_1 and z_2 are preassigned and that the corresponding Lagrange operator is

$$L_2(f, x) = \frac{k_2(x)}{k_2(z_1)} f(z_1) + \frac{k_1(x)}{k_1(z_2)} f(z_2) \quad \text{with} \quad L_2(1, x) \equiv 1,$$

where $k_1(x)$ and $k_2(x)$ are basis functions similar to the $h_i(x)$ of Section 2.5.1, for z_1 and z_2 respectively.

The mixed interpolating operator under discussion is

$$HL_{n,2}(f, x) = H_n(f, x) - \omega_n^2(x) L_2\left(\frac{H_n(f, x) - f(x)}{\omega_n^2(x)}, x\right),$$

which, when integrated, gives

$$Q_{n+2}f = \sum_{i=1}^{n} A_i f(x_i) + C_1 f(z_1) + C_2 f(z_2)$$

provided that, for $i=1(1)n$,

$$B_i = \int_{-1}^{1} W(x) \left\{ \frac{h_i(x)\lambda_i^2(x)}{h_i'(x_i)} - \omega_n^2(x) L_2\left(\frac{h_i(x)\lambda_i^2(x)}{h_i'(x_i)\omega_n^2(x)}, x\right) \right\} dx = 0.$$

The conditions $B_i = 0$, $i = 1(1)n$, are equivalent with

$$\int_{-1}^{1} W(x)\frac{\omega_n^2(x)}{h_i(x)}\,dx = \int_{-1}^{1} W(x)\omega_n^2(x)L_2\left(\frac{1}{h_i(x)}, x\right)dx \tag{7.2.2}$$

which we again call the *characteristic integral relationships*. The weights A_i, $i = 1(1)n$, can then be written as

$$A_i = \int_{-1}^{1} W(x)\left[\lambda_i^2(x) - \omega_n^2(x)L_2\left(\frac{\lambda_i^2(x)}{\omega_n^2(x)}, x\right)\right]dx - 2\lambda_i'(x_i)B_i, \qquad i = 1(1)n,$$

which, taking $B_i = 0$ and by substituting for the λ_i, become

$$A_i = \left(\frac{h_i'(x_i)}{\omega_n'(x_i)}\right)^2 \int_{-1}^{1} W(x)\omega_n^2(x)\left[\frac{1}{h_i^2(x)} - L_2\left(\frac{1}{h_i^2(x)}, x\right)\right]dx.$$

In order to simplify this expression we multiply (7.2.2) first by $1/h_i(z_1)$ and then by $1/h_i(z_2)$ and add the resulting equations to produce

$$\int_{-1}^{1} W(x)\frac{\omega_n^2(x)}{h_i(x)}\left(\frac{1}{h_i(z_1)} + \frac{1}{h_i(z_2)}\right)dx = \int_{-1}^{1} W(x)\omega_n^2(x)L_2\left(\frac{1}{h_i(x)}, x\right) \times\left(\frac{1}{h_i(z_1)} + \frac{1}{h_i(z_2)}\right)dx$$

which, on expanding $L_2(f, x)$ in terms of the cardinal functions $\Lambda_i(x)$, $i = 1, 2$, becomes

$$\int_{-1}^{1} W(x)\omega_n^2(x)\left[\frac{\Lambda_1(x)}{h_i^2(z_1)} + \frac{\Lambda_1(x)}{h_i(z_1)h_i(z_2)} + \frac{\Lambda_2(x)}{h_i(z_2)h_i(z_1)} + \frac{\Lambda_2(x)}{h_i^2(z_2)}\right]dx$$
$$= \int_{-1}^{1} W(x)\omega_n^2(x)L_2\left(\frac{1}{h_i^2(x)}, x\right)dx + \int_{-1}^{1} W(x)\omega_n^2(x)\frac{L_2(1, x)}{h_i(z_1)h_i(z_2)}\,dx.$$

Because of the characteristic integral relationships, the first integral can be replaced so that

$$A_i = \left(\frac{h_i'(x_i)}{\omega_n'(x_i)}\right)^2 \int_{-1}^{1} W(x)\omega_n^2(x)\left[\frac{1}{h_i^2(x)} - \left(\frac{1}{h_i(z_1)} + \frac{1}{h_i(z_2)}\right)\frac{1}{h_i(x)} + \frac{L_2(1, x)}{h_i(z_1)h_i(z_2)}\right]dx$$
$$= \left(\frac{h_i'(x_i)}{\omega_n'(x_i)}\right)^2 \int_{-1}^{1} W(x)\frac{\omega_n^2(x)}{h_i^2(x)}\left(1 - \frac{h_i(x)}{h_i(z_1)}\right)\left(1 - \frac{h_i(x)}{h_i(z_2)}\right)dx$$

where the property that $L_2(1, x) \equiv 1$ has been used. Finally, reintroducing the cardinal functions $\lambda_i(x)$ we find that

$$A_i = \int_{-1}^{1} W(x)\lambda_i^2(x)\left(1-\frac{h_i(x)}{h_i(z_1)}\right)\left(1-\frac{h_i(x)}{h_i(z_2)}\right)dx.$$

The weights C_1 and C_2, which belong to the preassigned nodes z_1 and z_2, are easily obtained and have the form

$$C_i = \frac{1}{\omega_n^2(z_i)} \int_{-1}^{1} W(x)\omega_n^2(x)\Lambda_i(x)\,dx, \qquad i=1,2.$$

The following theorem has thus been proved:

Theorem 7.2.2 *Sufficient conditions for the quadrature procedure*

$$Q_{n+2}f = \sum_{i=1}^{n} A_i f(x_i) + C_1 f(z_1) + C_2 f(z_2),$$

with two preassigned nodes z_1 and z_2 to be positive are that

$$\left(1-\frac{h_i(x)}{h_i(z_1)}\right)\left(1-\frac{h_i(x)}{h_i(z_2)}\right) \geqslant 0, \qquad i=1(1)n,$$

and

$$\Lambda_i(x) \geqslant 0, \qquad i=1,2 \quad in \quad [-1,1].$$

It is found that for strictly monotonic $\Lambda_i(x)$ the preassigned nodes z_1 and z_2 cannot be interior points of $[-1, 1]$. Further, if the basis functions $h_i(x)$ are also monotonic, then the product $(1-h_i(x)/h_i(z_1))(1-h_i(x)/h_i(z_2))$ is always non-negative if $|z_1|, |z_2| \geqslant 1$ for $x \in [-1, 1]$, and $z_1 z_2$ must be negative. If $h_i(x)$ is monotonically increasing then for $x \in [-1, x_i]$ we have that $1-h_i(x)/h_i(z_1) > 0$ and that $1-h_i(x)/h_i(z_1) = 1 + h_i(x)/|h_i(z_1)| > 0$ for $x \in [x_i, 1]$. Analogously, $1-h_i(x)/h_i(z_2) \geqslant 0$. It is impossible for z_1 and z_2 both simultaneously to be greater than or equal to unity since in this case we would have that $\Lambda_1(x) = k_2(x)/k_2(z_1) \geqslant 0$ and $\Lambda_2(x) = k_1(x)/k_1(z_2) \geqslant 0$ for $x \in [-1, 1]$. This cannot happen as is seen immediately from Fig. 7.2.2. Hence, either $z_1 \leqslant -1$, $z_2 \geqslant 1$ or $z_1 \geqslant 1$, $z_2 \leqslant -1$. (Analogously, it is seen that $z_1, z_2 \leqslant -1$ is impossible.)

The situation is summarised as follows:

Corollary 7.2.2 Let $\Lambda_i(x)$, $i=1,2$ be monotonic (positive in $(-1, 1)$) cardinal functions of $L_2(f, x)$ with $L_2(1, x) \equiv 1$, interpolating at the preassigned nodes $z_1 \neq z_2$ with $z_1 z_2 < 0$, and $z_1, z_2 \notin (-1, 1)$. If, further, $h_i(x)$ is monotonic, then $Q_{n+2}f$ is positive. ■

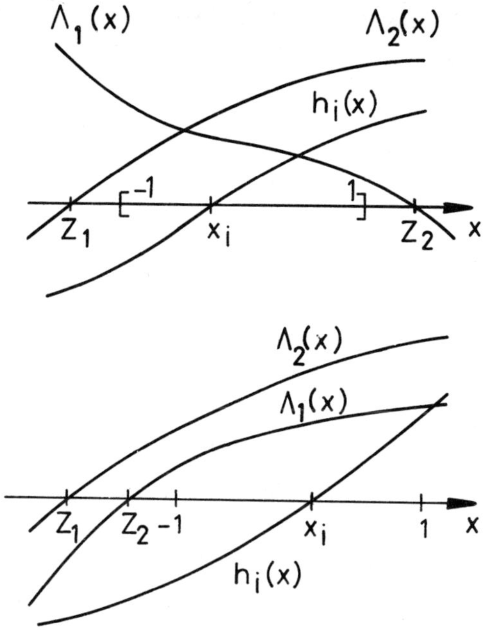

FIG. 7.2.2. Monotonic basis functions for two preassigned nodes.

The error $E_{n+2}f$ of $Q_{n+2}f$ is

$$E_{n+2}f = \int_{-1}^{1} W(x) \frac{f^{(2n+2)}(t) - HL_{n,2}^{(2n+2)}(f, t)}{[\omega_n^2(x)\Omega_2(x)]_{x=t}^{(2n+2)}} \omega_n^2(x)\Omega_2(x)\,dx$$

where $\Omega_2(x) = k_1(x)k_2(x)$. Further, there exists a point $s \in (-1, 1)$ such that

$$E_{n+2}f = \frac{f^{(2n+2)}(s) - HL_{n,2}^{(2n+2)}(f, s)}{[\omega_n^2(x)\Omega_2(x)]_{x=s}^{(2n+2)}} \int_{-1}^{1} W(x)\omega_n^2(x)\Omega_2(x)\,dx$$

if $\Omega_2(x) \geq 0$ in $[-1, 1]$.

7.2.3 Several preassigned nodes

We extend the previous results by induction. Assume that we are given the m preassigned nodes z_j, $j = 1(1)m$, and the corresponding Lagrange interpolating operator $L_m(f, x)$ with basis functions $k_j(x)$, $j = 1(1)m$. Let $\Omega_m(x) \equiv k_1(x) \ldots k_m(x)$ and let $\Lambda_j(x)$ be the Lagrange cardinal functions.

Note that we have no T-formula of degree 4 in Tables 6.7.2, 6.7.3, and 6.7.4.

The integration of the mixed-type interpolating operator

$$HL_{n,m}(f, x) = H_n(f, x) - \omega_n^2(x) L_m\left(\frac{H_n(f, x) - f(x)}{\omega_n^2(x)}, x\right)$$

with respect to the weight function $W(x)$ generates the quadrature formula

$$Q_{n+m}f = \sum_{i=1}^{n} (A_i f(x_i) + B_i f'(x_i)) + \sum_{j=1}^{m} C_j f(z_j)$$

with

$$B_i = \int_{-1}^{1} W(x)\left[\frac{h_i(x)}{h_i'(x_i)}\lambda_i^2(x) - \omega_n^2(x) L_m\left(\frac{h_i(x)}{h_i'(x_i)}\frac{\lambda_i^2(x)}{\omega_n^2(x)}, x\right)\right] dx, \qquad i = 1(1)n.$$

Requiring that $B_i = 0$ and replacing $\lambda_i(x)$, $i = 1(1)n$, as before, we have the *characteristic integral relationships*

$$\int_{-1}^{1} W(x)\frac{\omega_n^2(x)}{h_i(x)}\,dx = \int_{-1}^{1} W(x)\omega_n^2(x) L_m\left(\frac{1}{h_i(x)}, x\right) dx, \qquad i = 1(1)n. \tag{7.2.3}$$

The weights A_i, $i = 1(1)n$, of $Q_{n+m}f$ are

$$A_i = \int_{-1}^{1} W(x)\left[\lambda_i^2(x) - \omega_n^2(x) L_m\left(\frac{\lambda_i^2(x)}{\omega_n^2(x)}, x\right)\right] dx - 2\lambda_i'(x_i) B_i, \qquad i = 1(1)n,$$

which can be written as

$$A_i = \left(\frac{h_i'(x_i)}{\omega_n'(x_i)}\right)^2 \int_{-1}^{1} W(x)\omega_n^2(x)\left[\frac{1}{h_i^2(x)} - L_m\left(\frac{1}{h_i^2(x)}, 1\right)\right] dx, \qquad i = 1(1)n,$$

if we replace the cardinal functions λ_i by ω_n. The weights A_i can again be simplified, using the characteristic integral relationships. To this end we proceed as before, multiplying (7.2.3) respectively by $1/h_i(z_1), \ldots, 1/h_i(z_m)$ and adding the resultant equations. Thus, by induction,

$$\begin{aligned}
&\int_{-1}^{1} W(x)\frac{\omega_n^2(x)}{h_i(x)}\left(\sum_{j=1}^{m}\frac{1}{h_i(z_j)}\right) dx \\
&= \int_{-1}^{1} W(x)\omega_n^2(x) L_m\left(\frac{1}{h_i(x)}, x\right)\left(\sum_{j=1}^{m}\frac{1}{h_i(z_j)}\right) dx \\
&= \int_{-1}^{1} W(x)\omega_n^2(x)\left(\sum_{k=1}^{m}\sum_{j=1}^{m}\frac{\Lambda_k(x)}{h_i(z_k)h_i(z_j)}\right) dx
\end{aligned}$$

$$= \int_{-1}^{1} W(x)\omega_n^2(x)\Bigg[\sum_{k=1}^{m} \frac{\Lambda_k(x)}{h_i^2(z_k)} + \sum_{\substack{j,l=1\\ j<l}}^{m} \frac{1}{h_i(z_j)h_i(z_l)} \sum_{k=1}^{m} \Lambda_k(x)$$

$$- \sum_{\substack{j,l,r=1\\ j<l<r}}^{m} \frac{1}{h_i(z_j)h_i(z_l)h_i(z_r)} \sum_{k=1}^{m} h_i(z_k)\Lambda_k(x)$$

$$+ - \ldots + \frac{(-1)^{m-2}}{h_i(z_1)\ldots h_i(z_m)} \sum_{k=1}^{m} h_i^{m-2}(z_k)\Lambda_k(x)\Bigg]\,dx$$

$$= \int_{-1}^{1} W(x)\omega_n^2(x)\Bigg[L_m\left(\frac{1}{h_i^2(x)}, x\right) + L_m(1, x) \sum_{\substack{j,k=1\\ j<k}}^{m} \frac{1}{h_i(x_j)h_i(x_k)}$$

$$- L_m(h_i(x), x) \sum_{\substack{j,k,r=1\\ j<k<r}}^{m} \frac{1}{h_i(z_j)h_i(z_k)h_i(z_r)}$$

$$+ - \ldots + L_m(h_i^{m-2}(x), x)\frac{(-1)^{m-2}}{h_i(z_1)h_i(z_2)\ldots h_i(z_m)}\Bigg]\,dx.$$

Hence,

$$A_i = \left(\frac{h_i'(x_i)}{\omega_n'(x_i)}\right)^2 \int_{-1}^{1} W(x)\frac{\omega_n^2(x)}{h_i^2(x)}\Bigg[1 - \sum_{j=1}^{m} \frac{h_i(x)}{h_i(z_j)} + L_m(1, x) \sum_{\substack{j,k=1\\ j<k}}^{m} \frac{h_i^2(x)}{h_i(z_j)h_i(z_k)}$$

$$- L_m(h_i(x), x) \sum_{\substack{j,k,r=1\\ j<k<r}}^{m} \frac{h_i^2(x)}{h_i(x_j)h_i(x_k)h_i(x_r)}$$

$$+ - \ldots - L_m(h_i^{m-2}(x), x)\frac{(-1)^{m-2}h_i^2(x)}{h_i(z_1)h_i(z_2)\ldots h_i(z_m)}\Bigg]\,dx. \quad (7.2.4)$$

This representation can be simplified by assuming, further, that

$$L_m(h_i^k(x), x) = h_i^k(x) \quad \text{for} \quad k = 0(1)m-2 \quad (7.2.5)$$

(this is true, for example, if $h_i(x) = x - x_i$, $k_i(x) = x - z_i$). Then

$$A_i = \left(\frac{h_i'(x_i)}{\omega_n'(x_i)}\right)^2 \int_{-1}^{1} W(x)\frac{\omega_n^2(x)}{h_i^2(x)}\Bigg[1 - \sum_{j=1}^{m} \frac{h_i(x)}{h_i(z_j)} + \sum_{\substack{j,k=1\\ j<k}}^{m} \frac{h_i^2(x)}{h_i(z_j)h_i(z_k)}$$

$$- \sum_{\substack{j,k,r=1\\ j<k<r}}^{m} \frac{h_i^3(x)}{h_i(z_j)h_i(z_k)h_i(z_r)} \pm \ldots + \frac{(-1)^m h_i^m(x)}{h_i(z_1)h_i(z_2)\ldots h_i(z_m)}\Bigg]\,dx$$

$$= \left(\frac{h_i'(x_i)}{\omega_n'(x_i)}\right)^2 \int_{-1}^{1} W(x)\frac{\omega_n^2(x)}{h_i^2(x)}\left\{\prod_{j=1}^{m}\left(1 - \frac{h_i(x)}{h_i(z_j)}\right)\right\}\,dx.$$

If we again introduce λ_i then, finally,

$$A_i = \int_{-1}^{1} W(x)\lambda_i^2(x)\left\{\prod_{j=1}^{m}\left(1-\frac{h_i(x)}{h_i(z_j)}\right)\right\} dx.$$

If (7.2.5) is not valid then, from (7.2.4), we have that

$$A_i = \int_{-1}^{1} W(x)\lambda_i^2(x)\left[1-\sum_{j=1}^{m}\frac{h_i(x)}{h_i(z_j)}+\sum_{\substack{j,k=1\\ j<k}}^{m}\frac{L_m(1,x)h_i^2(x)}{h_i(z_j)h_i(z_k)}\right.$$
$$\left. -+\ldots+\frac{(-1)^m L_m(h_i^{m-2}(x),x)h_i^2(x)}{h_i(z_1)h_i(z_2)\ldots h_i(z_m)}\right] dx$$

which cannot easily be shown to satisfy a positivity condition. The weights C_j of the preassigned values $f(z_j)$ are

$$C_j = \frac{1}{\omega_n^2(z_j)}\int_{-1}^{1} W(x)\Lambda_j(x)\omega_n^2(x)\, dx, \qquad j=1(1)m.$$

Hence, we have the following theorem:

Theorem 7.2.3 (Engels) *If* $L_m(h_i^k(x),x)\equiv h_i^k(x)$ for $k=0(1)m-2$, *then a sufficient condition for* $A_i>0$, $i=1(1)n$, *is that*

$$\prod_{j=1}^{n}(1-h_i(x)/h_i(z_j))\geqslant 0.$$

In order to ensure that $Q_{n+m}f$ is positive, we need an additional condition for the weights C_j to be positive. It is, however, impossible to guarantee *a priori* $C_j>0$, $j=1(1)m$ (except in the case when the nodes z_j are all outside $(-1,1)$) so that $\Lambda_j(x)\geqslant 0$ for $|x|\leqslant 1$ is a sufficient positivity condition, without imposing over-restrictive conditions. If we assume, again, that the basis functions $h_i(x)$ are monotonic and, in addition, that $z_j\leqslant -1$ or $z_j\geqslant 1$, then $h_i(x)>h_i(z_j)$ and hence we have that the following corollary holds:

Corollary 7.2.3 If $z_j\geqslant 1$ or $z_j\leqslant -1$, $\Lambda_j(x)\geqslant 0$ for $|x|\leqslant 1$, and if $h_i(x)$ are monotonic functions, then $Q_{n+m}f$ is positive. ■

As in the previous special cases, the error is determined from

$$E_{n+m}f = \frac{f^{(2n+m)}(s)-HL_{n,m}^{(2n+m)}(f,s)}{[\omega_n^2(x)\Omega_m(x)]_{x=s}^{(2n+m)}}\int_{-1}^{1} W(x)\omega_n^2(x)\Omega_m(x)\, dx, \qquad s\in(-1,1)$$

provided that $\Omega_m(x)\geqslant 0$ for $|x|\leqslant 1$.

We now specialise these general results to many well-known quadrature procedures.

7.3 Classical Gauss quadratures

We consider classical Gauss quadratures which are based upon polynomial accuracy. For these Gauss quadratures, general properties are first developed after which the nodes and weights are listed in a table. These quadratures, especially for $W(x) \equiv 1$, are intensively used in automatic integration subroutines. They are reliable for practical calculation and are in common use up to approximately $n = 200$.

7.3.1 General properties of classical Gauss quadratures

If $H_n(f, x)$ is now a *polynomial* of degree at most $2n-1$, then integration gives the classical Gauss quadratures for $W(x) \equiv 1$. Gauss, in fact, derived his quadrature method by the use of continued fractions. However, Markoff noticed that these Gaussian quadratures are interpolatory quadratures. Important generalisations, particularly for $W(x) \not\equiv 1$, are due to Jacobi and Christoffel. The properties of those quadratures are summarised below, referring to Section 7.1.

If $h_i(x) = x - x_i$, $\omega_n(x) = (x - x_1) \ldots (x - x_n)$, $\lambda_i(x) = \omega_n(x)/[(x - x_i)\omega_n'(x_i)]$ then

$$B_i = 0 = \frac{1}{\omega_n'(x_i)} \int_{-1}^{1} W(x)\omega_n(x)\lambda_i(x)\,dx, \qquad i = 1(1)n,$$

and $\omega_n(x)$ must be a polynomial which is orthogonal to $\mathbb{P}_{n-1}$ with respect to $W(x)$. For $W(x) \equiv 1$ we obtain the Legendre polynomials, while for $W(x) = (1+x)^\alpha(1-x)^\beta$, $\alpha, \beta > -1$ we have the Jacobi polynomials, of which the Tschebyscheff polynomials are the special case $\alpha = \beta = \pm\frac{1}{2}$. If the range of integration is extended to become $(0, \infty)$ and $W(x) = x^\alpha e^{-x}$, $\alpha > -1$ we obtain the Laguerre polynomials, while for $(-\infty, \infty)$ and $W(x) = \exp(-x^2)$ we obtain the Hermite polynomials. As we have seen in Chapter 5, the orthogonal polynomials $\omega_n(x)$ always have n real distinct roots. Hence, Q_nf *exists* for every n and Q_nf *is positive* due to Theorem 7.1.1. In addition, Q_nf converges for every $f \in C^0[-1, 1]$. Since $H_n(f, x) \in \mathbb{P}_{2n-1}$ and $\omega_n^2(x) \in \mathbb{P}_{2n}$ with leading coefficient 1, the error can be written as

$$E_nf = \frac{f^{(2n)}(s)}{(2n)!} \int_{-1}^{1} W(x)\omega_n^2(x)\,dx = \sigma_n f^{(2n)}(s), \qquad s \in (-1, 1),$$

where σ_n is the error coefficient. This is obtained explicitly for the case $W(x)\equiv 1$ using Rodriguez' formula for Legendre polynomials

$$L_n(x)=\frac{n!}{(2n)!}\frac{\mathrm{d}^n}{\mathrm{d}x^n}((x^2-1)^n)=x^n+\ldots.$$

Hence, from Section 5.2.1, we have that for $W(x)\equiv 1$

$$\sigma_n\equiv\int_{-1}^{1}\omega_n^2(x)\,\mathrm{d}x=\left(\frac{n!}{2n!}\right)^2\int_{-1}^{1}L_n^2(x)\,\mathrm{d}x=\frac{(n!)^4}{(2n!)^3}\frac{2^{2n+1}}{2n+1}.$$

We now derive some useful expressions for the weights A_i of the Gauss quadrature formulae for $W\equiv 1$. If we apply Q_nf, which has degree $2n-1$, to the polynomial $L_n'\cdot\lambda_i\in\mathbb{P}_{2n-2}$ then the quadrature error is zero and thus

$$\sum_{j=1}^{n}A_jL_n'(x_j)\lambda_i(x_j)=L_n'(x_i)A_i=\int_{-1}^{1}L_n'\lambda_i\,\mathrm{d}x=L_n(1)\lambda_i(1)-L_n(-1)\lambda_i(-1)$$

after integration by parts and taking into account that $L_n\perp\lambda_i'$ where $\lambda_i'\in\mathbb{P}_{n-2}$. Hence

$$A_i=\frac{L_n(1)\lambda_i(1)-L_n(-1)\lambda_i(-1)}{L_n'(x_i)}.$$

Normalising L_n such that $L_n(1)=1$ and denoting the normalised form by $\bar{L}_n$, we can write

$$A_i=\frac{\lambda_i(1)-(-1)^n\lambda_i(-1)}{\bar{L}_n'(x_i)}.$$

From the definition of the cardinal functions λ_i we obtain

$$\lambda_i(1)=\frac{1}{\bar{L}_n'(x_i)(1-x_i)}\quad\text{and}\quad\lambda_i(-1)=\frac{(-1)^n}{\bar{L}_n'(x_i)(-1-x_i)},$$

so that

$$A_i=\frac{2}{(\bar{L}_n'(x_i))^2(1-x_i^2)}.$$

From this formula we are now able to calculate the weights A_i, $i=1(1)n$, once the nodes x_i are known, which removes the necessity of computing integrals or of solving a linear system of equations in order to obtain the weights.

If we change the normalisation such that $L_n=x^n+\ldots$, then

$$L_n(x)=\frac{n!(n-1)!2^{n-1}}{(2n-1)!}\bar{L}_n(x)\equiv a_n\bar{L}_n(x)$$

and

$$A_i = \left[\frac{n!(n-1)!2^{n-1}}{(2n-1)!L_n'(x_i)}\right]^2 \frac{2}{1-x_i^2}.$$

Since $L_n' \in \mathbb{P}_{n-1}$ we can expect that there exists a close connection between $\int_{-1}^{1} L_n'\lambda_i \,dx$ and $\int_{-1}^{1} L_{n-1}\lambda_i \,dx$. In order to find this, we calculate both sides of the equality $\int_{-1}^{1} L_{n-1}\lambda_i \,dx = Q_n(L_{n-1}\lambda_i)$ by replacing λ_i with its unique expansion in terms of Legendre polynomials L_k so that

$$\lambda_i(x) = \frac{L_n(x)}{L_n'(x_i)(x-x_i)} = \frac{(x^{n-1}+\ldots)}{L_n'(x_i)} = \frac{L_{n-1} + \sum_{k=0}^{n-2} c_k L_k}{L_n'(x_i)}.$$

Since $L_{n-1} \perp \mathbb{P}_{n-2}$ we obtain that

$$\int_{-1}^{1} L_{n-1}\lambda_i \,dx = \frac{1}{L_n'(x_i)} \int_{-1}^{1} L_{n-1}^2 \,dx = \frac{(L_{n-1}, L_{n-1})}{L_n'(x_i)}$$
$$= \frac{(1/a_n)(L_{n-1}, L_{n-1})}{\bar{L}_n'(x_i)}$$

For $Q_n(L_{n-1}\lambda_i)$ we find that

$$Q_n(L_{n-1}\lambda_i) = A_i L_{n-1}(x_i) = \frac{2L_{n-1}(x_i)}{(\bar{L}_n'(x_i))^2(1-x_i^2)} = \frac{2a_{n-1}\bar{L}_{n-1}(x_i)}{(\bar{L}_n'(x_i))^2(1-x_i^2)}.$$

Comparing these two expressions we see that $(1-x_i^2)\bar{L}_n'(x_i) = \dfrac{2a_n a_{n-1}\bar{L}_{n-1}(x_i)}{(L_{n-1}, L_{n-1})}$.

The inner product has already been calculated in Section 5.2.1 so that the result $(1-x_i^2)\bar{L}_n'(x_i) = n\bar{L}_{n-1}(x_i)$ is obtained. With this, the weights A_i can be calculated so that

$$A_i = \frac{2(1-x_i^2)}{n^2\bar{L}_{n-1}^2(x_i)} = \frac{2(1-x_i^2)}{L_{n-1}^2(x_i)} \left[\frac{(n-1)!(n-2)!2^{n-2}}{(2n-3)!n}\right]^2.$$

The only disadvantage of formulae of this type is that L_n' or L_{n-1} must be known explicitly. This is particularly serious for use of the asymptotic expansions of the nodes x_i derived in Section 5.4.3, because in this case the expansions differ for odd and even n. Thus, we replace $L_{n-1}(x_i)$ by $L_{n-2}(x_i)$ using the three-term recurrence relationship for Legendre polynomials which gives (cf. Section 5.2.1)

$$L_n(x_k) = 0 = x_k L_{n-1}(x_k) - \frac{(n-1)^2}{(2n-1)(2n-3)} L_{n-2}(x_k)$$

for every zero x_k of $L_n(x)$. Thus

$$L_{n-1}(x_i) = \frac{(n-1)^2}{(2n-1)(2n-3)} L_{n-2}(x_i)/x_i$$

and hence

$$A_i = \frac{2(1-x_i^2)x_i^2}{L_{n-2}^2(x_i)} \left[\frac{2n-1}{n(n-1)} \cdot \frac{(n-2)!(n-2)!}{(2n-4)!} 2n^{n-2}\right]^2 = \frac{2(2n-1)^2(1-x_i^2)x_i^2}{n^2(n-1)^2\bar{L}_{n-2}^2(x_i)}.$$

Note that if the interval $[-1, 1]$ is replaced by $[0, 1]$, the same result holds with the factor $\frac{1}{2}x_i(1-x_i)$ instead of $1-x_i^2$.

The limits of integration can be extended to infinity for suitable weight functions $W(x)$ and in this case Theorem 7.1.3 remains valid. Using the properties of *orthogonal polynomials* we can sharpen Theorem 7.1.3 to become:

Theorem 7.3.1 *The condition $B_i = 0$, $i = 1(1)n$, is necessary and sufficient for σ_n to be minimal for polynomial interpolatory quadratures.*

The proof follows immediately from Theorem 5.3.3 (Chapter 5).

For $W(x) = 1/(1-x^2)^{\frac{1}{2}}$ we have an important special case:

Theorem 7.3.2 *For $W(x) = 1/(1-x^2)^{\frac{1}{2}}$ the quadrature $Q_n f$ is equally-weighted, which means that $A_i = \pi/n$, $i = 1(1)n$. The nodes are $x_i = \cos((2i-1)\pi/2n)$.*

Proof The orthogonal polynomials corresponding to this weight function $W = 1/(1-x^2)^{\frac{1}{2}}$ are the Tschebyscheff polynomials of the first kind $\omega_n(x) \equiv T_n(x)$, which are defined (see Section 5.2.2) by $T_n(x) \equiv \cos(n \arccos x) \equiv \cos(n\theta)$. Hence it follows that

$$A_i = \int_{-1}^{1} \frac{\lambda_i^2(x)}{(1-x^2)^{\frac{1}{2}}}\, dx = \int_{-1}^{1} \frac{1}{(1-x^2)^{\frac{1}{2}}} \lambda_i(x)\, dx = \int_{-1}^{1} \frac{\cos(n \arccos x)\, dx}{T_n'(x_i)(x-x_i)(1-x_i^2)^{\frac{1}{2}}}, \qquad i = 1(1)n,$$

where x_i is the i-th zero of $T_n(x)$. The numbers $\theta_i = (2i-1)\pi/2$, $i = 1(1)n$, are zeros of $\cos\theta$. Hence, $n \arccos x_i = (2i-1)\pi/2$ or

$$x_i = \cos\left(\frac{2i-1}{2n}\pi\right), \qquad i = 1(1)n.$$

Substitution of $x = \cos\theta$ gives

$$A_i = \frac{1}{T_n'(x_i)} \int_0^{\pi} \frac{\cos n\theta}{\cos\theta - \cos\theta_i}\, d\theta, \qquad \theta_i = \frac{2i-1}{2n}\pi,$$

where the integrand is a polynomial of degree $n-1$ in $\cos\theta$ since $T_n(x)/(x-x_i)\in\dot{\mathbb{P}}_{n-1}$ in x which can be written in expanded form to give

$$A_i=\frac{1}{T_n'(x_i)}\int_0^{\pi}\left(\sum_{k=0}^{n-1} a_k\cos k\theta\right)\mathrm{d}\theta=\frac{\pi a_0}{T_n'(x_i)},$$

where a_0 remains unknown.

In order to determine a_0, we calculate $\cos n\theta/(\cos\theta-\cos\theta_i)$ at the n points $\theta_i, \theta_i+2\pi/n, \ldots, \theta_i+2\pi(n-1)/n$ and use l'Hospital's rule to obtain

$$\lim_{\theta\to\theta_i}\left(\frac{\cos n\theta}{\cos\theta-\cos\theta_i}\right)=n\frac{\sin n\theta_i}{\sin\theta_i}.$$

Now

$$\frac{\cos n(\theta_i+2\pi k/n)}{\cos(\theta_i+2\pi k/n)-\cos\theta_i}=\frac{\cos(n\theta_i)}{\cos(\theta_i+2\pi k/n)-\cos\theta_i}=0,\qquad k=1(1)n-1,$$

since the denominator remains different from zero and hence

$$\sum_{k=0}^{n-1} a_k\cos k\theta_i=\frac{n\sin n\theta_i}{\sin\theta_i}$$

and

$$\sum_{k=0}^{n-1} a_k\cos k\left(\theta_i+\frac{2\pi j}{n}\right)=0,\qquad j=1(1)n-1.$$

These last n equations can be added to give

$$\sum_{k=0}^{n-1} a_k\sum_{j=0}^{n-1}\cos k(\theta_i+2\pi j/n)=n\frac{\sin n\theta_i}{\sin\theta_i}.$$

The interior summation is now performed for $0<k<n$. By complex calculation we have that (replacing θ_i with θ in order to avoid confusion between the index i and the imaginary unit $i=\sqrt{-1}$)

$$\sum_{j=0}^{n-1}\left[\cos k\left(\theta+\frac{2\pi j}{n}\right)+i\sin k\left(\theta+\frac{2\pi j}{n}\right)\right]=\sum_{j=0}^{n-1}\mathrm{e}^{ik}\left(\theta+\frac{2\pi j}{n}\right)$$

$$=\mathrm{e}^{ik\theta}\sum_{j=0}^{n-1}(\mathrm{e}^{2\pi ik/n})^j=\mathrm{e}^{ik\theta}\frac{\mathrm{e}^{2\pi ik}-1}{\mathrm{e}^{2\pi ik/n}-1}=0.$$

Hence the interior sum vanishes for $0<k<n$, and therefore $na_0=n\dfrac{\sin n\theta_i}{\sin\theta_i}$. Finally,

$$A_i=\pi\frac{\sin n\theta_i}{T_n'(x_i)\sin\theta_i}=\frac{\pi}{n}$$

since $T'_n(x) = n\dfrac{\sin(n \arccos x)}{(1-x^2)^{\frac{1}{2}}} = \dfrac{n \sin n\theta}{\sin\theta}$, which completes the proof. ■

Again we calculate the error coefficient $\sigma_n = I\omega_n^2(x)/(2n)! = IT_n^2(x)/(2n)!$ explicitly. Using Rodriguez' formula as for the case $W(x) \equiv 1$ we find that $\sigma_n = \pi/((2n)!2^{2n-1})$. This error constant is a special case of that of the error constant for the Jacobi weight function $W(x) = (1-x)^\alpha(1+x)^\beta$, $\alpha, \beta > -1$, which we now calculate.

We again use Rodriguez' formula. From

$$J_n^{\alpha,\beta}(x) = a_n x^n + \dots \quad \text{with} \quad a_n = (-1)^n \frac{\Gamma(\alpha+\beta+2n+1)}{\Gamma(\alpha+\beta+n+1)}$$

it follows that

$$\sigma_n = \frac{1}{a_n^2(2n)!}\int_{-1}^{1}(1-x)^\alpha(1+x)^\beta[J_n^{\alpha,\beta}(x)]^2\,dx.$$

Integrating by parts n times we find that

$$\sigma_n = \frac{(-1)^n}{a_n^2(2n)!}\int_{-1}^{1}(1-x)^{\alpha+n}(1+x)^{\beta+n}[J_n^{\alpha,\beta}(x)]^{(n)}\,dx.$$

But it is well known that $(J_n^{\alpha,\beta}(x))^{(n)} = a_n n!$ and, with the transformation $x = 2t-1$, we have that

$$\int_{-1}^{1}(1-x)^{\alpha+n}(1+x)^{\beta+n}\,dx = 2^{\alpha+\beta+2n+1}\int_0^1(1-t)^{\alpha+n}t^{\beta+n}\,dt$$
$$= 2^{\alpha+\beta+2n+1}\frac{\Gamma(\alpha+n+1)\Gamma(\beta+n+1)}{\Gamma(\alpha+\beta+2n+2)}.$$

Hence

$$\sigma_n = \frac{(-1)^n n!}{a_n(2n)!}2^{\alpha+\beta+2n+1}\frac{\Gamma(\alpha+n+1)\Gamma(\beta+n+1)}{\Gamma(\alpha+\beta+2n+2)}$$
$$= \frac{2^{\alpha+\beta+2n+1}n!\Gamma(\alpha+n+1)\Gamma(\beta+n+1)\Gamma(\alpha+\beta+n+1)}{(2n)!\Gamma(\alpha+\beta+2n+2)\Gamma(\alpha+\beta+2n+1)} \tag{7.3.1}$$

is the error constant of the Gauss–Jacobi quadrature formulae.

7.3.2 Table of classical Gauss quadratures

The weights and nodes of some classical Gauss quadratures are given in Table 7.3.1. Note that for symmetry only the non-negative nodes and the corresponding weights are given.

TABLE 7.3.1. Gaussian quadrature formulae

$$\int_{-1}^{1} f(x)\,dx \doteq \sum_{i=1}^{n} A_i f(x_i), \qquad E_n f = \frac{(n!)^4 2^{2n+1}}{((2n)!)^3 (2n+1)} f^{(2n)}(t)$$

n	$x_i = -x_{n-i+1}$		$A_i = A_{n-i+1}$	
2	0.5773502691 8962576451		1	
4	0.8611363115 9405257522		0.3478548451 3745385737	
	0.3399810435 8485626480		0.6521451548 6254614263	
8	0.9602898564 9753623168		0.1012285362 9037625915	
	0.7966664774 1362673959		0.2223810344 5337447054	
	0.5255324099 1632898582		0.3137066458 7788728734	
	0.1834346424 9564980494		0.3626837833 7836198297	
16	0.9894009349 9164993260		0.2715245941 1754094852	(−1)
	0.9445750230 7323257608		0.6225352393 8647892866	(−1)
	0.8656312023 8783174388		0.9515851168 2492784810	(−1)
	0.7554044083 5500303390		0.1246289712 5553387205	
	0.6178762444 0264374845		0.1495959888 1657673208	
	0.4580167776 5722738634		0.1691565193 9500253819	
	0.2816035507 7925891323		0.1826034150 4492358887	
	0.9501250983 7637440185	(−1)	0.1894506104 5506849629	
32	0.9972638618 4948156354		0.7018610009 4700966004	(−2)
	0.9856115115 4526833540		0.1627439473 0905670605	(−1)
	0.9647622555 8750643077		0.2539206530 9262059456	(−1)
	0.9349060759 3773968917		0.3427386291 3021433103	(−1)
	0.8963211557 6605212397		0.4283589802 2226680657	(−1)
	0.8493676137 3256997013		0.5099805926 2376176196	(−1)
	0.7944837959 6794240696		0.5868409347 8535547145	(−1)
	0.7321821187 4028968039		0.6582222277 6361846833	(−1)
	0.6630442669 3021520098		0.7234579410 8848506225	(−1)
	0.5877157572 4076232904		0.7819389578 7070306472	(−1)
	0.5068999089 3222939002		0.8331192422 6946755222	(−1)
	0.4213512761 3063534536		0.8765209300 4403811143	(−1)
	0.3318686022 8212764978		0.9117387869 5763884713	(−1)
	0.2392873622 5213707454		0.9384439908 0804565639	(−1)
	0.1444719615 8279649349		0.9563872007 9274859419	(−1)
	0.4830766568 7738316235	(−1)	0.9654008851 4727800567	(−1)
64	0.9993050417 3577213946		0.1783280721 6964329473	(−2)
	0.9963401167 7195527935		0.4147033260 5624676353	(−2)
	0.9910133714 7674432074		0.6504457968 9783628561	(−2)
	0.9833362538 8462595693		0.8846759826 3639477230	(−2)
	0.9733268277 8991096374		0.1116813946 0131128819	(−1)
	0.9610087996 5205371892		0.1346304789 6718642598	(−1)
	0.9464113748 5840281606		0.1572603047 6024719322	(−1)

TABLE 7.3.1. (Continued).

n	$\int_{-1}^{1} f(x)\,dx \doteq \sum_{i=1}^{n} A_i f(x_i)$, $x_i = -x_{n-i+1}$	$E_n f = \frac{(n!)^4 2^{2n+1}}{((2n)!)^3(2n+1)} f^{(2n)}(t)$, $A_i = A_{n-i+1}$
64	0.9295691721 3193957582	0.1795171577 5697343085 (−1)
	0.9105221370 7850280576	0.2013482315 3530209372 (−1)
	0.8893154459 9511410585	0.2227017380 8383254159 (−1)
	0.8659993981 5409281976	0.2435270256 8710873338 (−1)
	0.8406292962 5258036275	0.2637746971 5054658672 (−1)
	0.8132653151 2279755974	0.2833967261 4259483228 (−1)
	0.7839723589 4334140761	0.3023465707 2402478868 (−1)
	0.7528199072 6053189661	0.3205792835 4851553585 (−1)
	0.7198818501 7161082685	0.3380516183 7141609392 (−1)
	0.6852363130 5423324256	0.3547221325 6882383811 (−1)
	0.6489654712 5465733985	0.3705512854 0240046040 (−1)
	0.6111553551 7239325025	0.3855015317 8615629129 (−1)
	0.5718956462 0263403428	0.3995374113 2720341387 (−1)
	0.5312794640 1989454566	0.4126256324 2623528610 (−1)
	0.4894031457 0705295748	0.4247351512 3653589007 (−1)
	0.4463660172 5346408798	0.4358372452 9323453377 (−1)
	0.4022701579 6399160370	0.4459055816 3756563060 (−1)
	0.3572201583 3766811595	0.4549162792 7418144480 (−1)
	0.3113228719 9021095616	0.4628479658 1314417295 (−1)
	0.2646871622 0876741637	0.4696818281 6210017325 (−1)
	0.2174236437 4000708415	0.4754016571 4830308662 (−1)
	0.1696444204 2399281804	0.4799938859 6458307728 (−1)
	0.1214628192 9612055447	0.4834476223 4802957170 (−1)
	0.7299312178 7799039450 (−1)	0.4857546744 1503426935 (−1)
	0.2435029266 3424432509 (−1)	0.4869095700 9139720383 (−1)

7.4 Modifications of classical Gauss quadratures having some preassigned nodes

Christoffel seems to have been the first person to introduce preassigned nodes into Gauss quadratures. Although he proved existence, he did not derive formulae explicitly. This was done by Radau who preassigned one of the limits of integration, and by Lobatto who preassigned both limits. Bouzitat generalised the results both of Radau and of Lobatto by admitting the Jacobi weight function $W(x)=(1+x)^{\alpha}(1-x)^{\beta}$, α, $\beta>-1$. Filippi preassigned the nodes $-1-d$ and $1+d$ for $d>0$ which are used twice if $Q_n f$ is applied as a composite rule. In the following sections general weight functions are admitted.

7.4.1 Generalised Radau and Bouzitat (first kind) quadrature formulae

Without loss of generality we assume that the point $x=-1$ is preassigned. Hence (7.2.1) applies with $h_i(x)=x-x_i$. The weights C_1, A_i of the corresponding quadrature formula $Q_{n+1}f=C_1f(-1)+\sum_{i=1}^{n} A_if(x_i)$ are positive, which follows from their representation

$$C_1=\int_{-1}^{1} W(x)\frac{\omega_n^2(x)}{[\omega_n(-1)]^2}\,dx>0, \qquad A_i=\int_{-1}^{1} W(x)\frac{1+x}{1+x_i}\lambda_i^2(x)\,dx>0, \qquad i=1(1)n.$$

From $B_i=0$ we obtain again the system of equations

$$\int_{-1}^{1} W(x)[(x-x_i)\lambda_i^2(x)\omega_n^2(-1)+(1+x_i)\lambda_i^2(-1)\omega_n^2(x)]\,dx=0, \qquad i=1(1)n$$

for the nodes. However, this is not the most convenient form. Replacing $\lambda_i(x)$ and rearranging, we find that

$$\int_{-1}^{1} W(x)\left(\frac{1}{x-x_i}+\frac{1}{1+x_i}\right)\omega_n^2(x)\,dx=0, \qquad i=1(1)n,$$

or, after adding and introducing $\lambda_i(x)$ again, that

$$\frac{1}{(1+x_i)\omega_n'(x_i)}\int_{-1}^{1} W(x)(1+x)\omega_n(x)\lambda_i(x)\,dx=0, \qquad i=1(1)n.$$

Hence, the nodes x_i are the zeros of the orthogonal polynomials with respect to the non-negative weight function $W(x)(1+x)$, and therefore $Q_{n+1}f$ exists. For $W(x)\equiv 1$ the nodes x_i are the zeros of the Jacobi polynomials $P_n^{1,0}(x)$. If we preassign $+1$ instead of -1, then x_i are the zeros of $P_n^{0,1}(x)$.

The quadrature error is

$$E_{n+1}f=\frac{f^{(2n+1)}(t)}{(2n+1)!}\int_{-1}^{1} W(x)(1+x)\omega_n^2(x)\,dx, \qquad t\in(-1,1).$$

The integral in this formula becomes minimal if and only if $\omega_n(x)$ is orthogonal to $\mathbb{P}_{n-1}$ with respect to $W(x)(1+x)$. The proof of this statement is the same as for the Gauss quadrature. From this error representation we conclude convergence of $Q_{n+1}f$ because $Q_{n+1}f$ integrates every given polynomial exactly for large enough n. We collect the facts in:

Theorem 7.4.1 *The generalised Radau quadrature exists for every n, it is positive, and it converges for every $f\in C^0[-1,1]$.*

For $W(x) \equiv (1+x)^\alpha (1-x)^\beta$ $\alpha, \beta > -1$, the error coefficient is ($z_1 = 1$ preassigned)

$$\sigma_n = \frac{2^{\alpha+\beta+2n+2}\Gamma(\alpha+\beta+n+2)\Gamma(\alpha+n+1)\Gamma(n+\beta+2)n!}{(2n)!\,\Gamma(\alpha+\beta+2n+2)\Gamma(\alpha+\beta+2n+3)}$$

and, in particular, for $W(x) \equiv 1$ (see (7.3.1) for β replaced by $\beta+1$)

$$\sigma_n = \frac{4^{n+1} n!\,(n+1)!\,(n+1)!\,n!}{(2n)!\,(2n+1)!\,(2n+2)!}.$$

Some formulae for $W(x) \equiv 1$ are given in Table 7.4.1.

7.4.2 Generalised Lobatto and Bouzitat (second kind) quadrature formulae

Here the limits ± 1 of integration are preassigned as nodes (see Lobatto, book). The weights of the corresponding formula

$$Q_{n+2}f = C_1 f(-1) + \sum_{i=1}^{n} A_i f(x_i) + C_2 f(1)$$

are positive since

$$C_1 = \frac{1}{\omega_n^2(-1)} \int_{-1}^{1} W(x)\omega_n^2(x) \frac{1-x}{2}\,dx > 0,$$

$$C_2 = \frac{1}{\omega_n^2(1)} \int_{-1}^{1} W(x)\omega_n^2(x) \frac{1+x}{2}\,dx > 0,$$

because $\Lambda_1(x) = (1-x)/2$ and $\Lambda_2(x) = (1+x)/2$ are not negative in $[-1, 1]$ and since

$$A_i = \int_{-1}^{1} W(x) \frac{1-x^2}{1-x_i^2} \lambda_i^2(x)\,dx > 0, \qquad i = 1(1)n.$$

The nodes x_i are the solutions of the system of equations (see (7.2.2))

$$\int_{-1}^{1} W(x) \frac{\omega_n^2(x)}{x-x_i} \left[1-(x-x_i)L_2\left(\frac{1}{x-x_i}, x\right)\right] dx = 0, \qquad i = 1(1)n.$$

Since

$$L_2(g, x) = \frac{1-x}{2} g(-1) + \frac{1+x}{2} g(1)$$

we have that

$$1-(x-x_i)L_2\left(\frac{1}{x-x_i}, x\right) = \frac{1-x^2}{1-x_i^2},$$

TABLE 7.4.1. Radau quadrature formulae

$$\int_{-1}^{1} f(x)\,dx \doteq A_1 f(-1) + \sum_{i=2}^{n+1} A_i f(x_i)$$

n	x_i		A_i	
2	−1		0.2222222222 2222222222	
	−0.2898979485 5663561964		0.1024971652 3768432277	(1)
	0.6898979485 5663561964		0.7528061254 0093455010	
4	−1		0.8000000000 0000000000	(−1)
	−0.7204802713 1243889570		0.4462078021 6714148881	
	−0.1671808647 3783364011		0.6236530459 5148250816	
	0.4463139727 2375234464		0.5627120302 9892412038	
	0.8857916077 7096463561		0.2874271215 8245188265	
8	−1		0.2469135802 4691358025	(−1)
	−0.9107320894 2006029853		0.1476540190 4631538582	
	−0.7112674859 1570885703		0.2471893782 0459305236	
	−0.4263504857 1113896210		0.3168437756 7043797834	
	−0.9037336960 6853298065	(−1)	0.3482730027 7296659407	
	0.2561356708 3345539514		0.3376939669 7592958580	
	0.5713830412 0873848328		0.2863866963 5723117115	
	0.8173527842 0041208799		0.2005532980 2455195742	
	0.9644401697 0527309637		0.9071450492 3282917013	(−1)
16	−1		0.6920415224 9134948097	(−2)
	−0.9747060412 5208323606		0.4230158951 7309895354	(−1)
	−0.9160488353 8095090909		0.7467143455 1616165038	(−1)
	−0.8262167530 7428254925		0.1044803216 9818765624	
	−0.7082762955 5042126733		0.1307287373 5310479452	
	−0.5662451079 2973616770		0.1525247060 7715060338	
	−0.4049602758 5008208891		0.1691264101 2270482336	
	−0.2299143037 5297903498		0.1799686281 5730987453	
	−0.4706823205 0336641035	(−1)	0.1846821930 7625388553	
	0.1373513128 0874924983		0.1831066171 2863878578	
	0.3170641449 0435386275		0.1752955764 0021977768	
	0.4859503843 0965828015		0.1615150963 8994752684	
	0.6382588806 8423266750		0.1422345171 3693860050	
	0.7688031248 4036811964		0.1181105882 7402897355	
	0.8731381088 3232650901		0.8996545627 6346551632	(−1)
	0.9477129726 3721189604		0.5876112752 1580889255	(−1)
	0.9900054006 7245615791		0.2560658509 3747701971	(−1)

and hence that

$$\frac{1}{1-x_i^2}\int_{-1}^{1} W(x)(1-x^2)\omega_n(x)\lambda_i(x)\,dx=0, \qquad i=1(1)n.$$

Thus, for classical polynomial interpolation, we see that the nodes are the zeros of the orthogonal polynomial $\omega_n(x)\perp\mathbb{P}_{n-1}$ with respect to the weight function $W(x)(1-x^2)\geqslant 0$ and hence $Q_{n+2}f$ *exists* for every $n\in\mathbb{N}$. (For $W(x)\equiv 1$ the nodes x_i are the zeros of the Jacobi-polynomials $P_n^{1,1}(x.)$

The quadrature error of $Q_{n+2}f$ is

$$E_{n+2}f=\frac{f^{(2n+2)}(t)}{(2n+2)!}\int_{-1}^{1} W(x)(x^2-1)\omega_n^2(x)\,dx, \qquad t\in(-1,1).$$

Again, the integral is minimal if and only if $\omega_n(x)$ is orthogonal to $\mathbb{P}_{n-1}$ with respect to the weight function $W(x)(1-x^2)$. The proof is the same as for the Gauss quadrature.

If we compare with the error term of the Gaussian $n+1$ point formula we have that

$$E_{n+1}^{G}f=\frac{f^{(2n+2)}(\bar{t})}{(2n+2)!}\int_{-1}^{1} W(x)\omega_{n+1}^2(x)\,dx, \qquad \bar{t}\in(-1,1)$$

with opposite sign if $f^{(2n+2)}(x)$ does not change sign in $[-1,1]$. Further, for large enough n, $E_{n+2}f=0$ for every given polynomial. Hence $Q_{n+2}f$ converges for every polynomial and, further, for every $f\in C^0[-1,1]$ because of its positivity and the Banach–Steinhaus Theorem 4.3.2. Thus we have:

Theorem 7.4.2 *The generalised Lobatto quadrature exists for every $n\in\mathbb{N}$, it is positive, and it converges for every $f\in C^0[-1,1]$.*

Note, in addition, that if $W(x)\equiv W(-x)$, then we have $C_1=C_2$. For $W(x)=(1-x)^\alpha(1+x)^\beta$, $\alpha,\beta>-1$, we have the error constant

$$\sigma_n=\frac{2^{\alpha+\beta+2n+3}}{(2n)!}\frac{\Gamma(\alpha+\beta+n+3)\Gamma(\alpha+n+2)\Gamma(\beta+n+2)n!}{\Gamma(\alpha+\beta+2n+3)\Gamma(\alpha+\beta+2n+4)}.$$

In the case $W(x)\equiv 1$ this becomes

$$\sigma_n=\frac{2^{2n+3}(n+1)!\,(n+1)!\,(n+2)!\,n!}{(2n)!\,(2n+2)!\,(2n+3)!}$$

from (7.3.1) with the replacement $\alpha+1$, $\beta+1$ respectively for α,β.

For $W(x)\equiv 1$ these formulae were given by Lobatto and for the Jacobi weight function $W(x)=(1+x)^\alpha(1-x)^\beta$ they date back to Bouzitat.

TABLE 7.4.2. Lobatto quadrature formulae

$$\int_{-1}^{1} f(x)\,dx \doteq A_1 f(-1) + \sum_{i=2}^{n+1} A_i f(x_i) + A_{n+2} f(1)$$

n	$x_1 = 1$, x_i	A_i
2	1	0.1666666666 6666666667
	0.4472135954 9995793928	0.8333333333 3333333333
4	1	0.6666666666 6666666667 (−1)
	0.7650553239 2946469285	0.3784749562 9784698031
	0.2852315164 8064509631	0.5548583770 3548635301
8	1	0.2222222222 2222222222 (−1)
	0.9195339081 6645881383	0.1333059908 5107011113
	0.7387738651 0550507500	0.2248893420 6312645212
	0.4779249498 1044449566	0.2920426836 7968375788
	0.1652789576 6638702463	0.3275397611 8389745666
16	1	0.6535947712 4183006536 (−2)
	0.9761055574 1219854286	0.3997062881 0914066138 (−1)
	0.9206491853 4753387384	0.7063716688 5633664999 (−1)
	0.8355935352 1809021371	0.9901627171 7502802394 (−1)
	0.7236793292 8324268131	0.1242105331 3296710026
	0.5885048343 1866176117	0.1454119615 7380226798
	0.4344150369 1212397534	0.1619395172 3760248926
	0.2663626528 7828098417	0.1732621094 8945622601
	0.8974909348 4652111023 (−1)	0.1790158634 3970308229
32	1	0.1782531194 2959001783 (−2)
	0.9934643625 8751620225	0.1096474081 8306551401 (−1)
	0.9781466888 7993794895	0.1964627816 0713431288 (−1)
	0.9542306489 1280752841	0.2815053686 2243503401 (−1)
	0.9219282956 7744892274	0.3640677095 0782552351 (−1)
	0.8815238365 1227326440	0.4434290206 0244342300 (−1)
	0.8333724428 0314261491	0.5188929183 0927120776 (−1)
	0.7778973064 3186741088	0.5897965402 6745226695 (−1)
	0.7155859607 2021900600	0.6555169010 8265120430 (−1)
	0.6469860079 3601462438	0.7154764987 7072822036 (−1)
	0.5727003112 7492284479	0.7691484306 3590982105 (−1)
	0.4933816986 7327095744	0.8160610385 0422383162 (−1)
	0.4097272265 5959274486	0.8558020597 1977586590 (−1)
	0.3224720545 1148899358	0.8880222528 3584032118 (−1)
	0.2323829848 6911684306	0.9124384680 6914487022 (−1)
	0.1402517241 7619385837	0.9288361362 9610646144 (−1)
	0.4688792571 2970680087 (−1)	0.9370711550 4303312002 (−1)

Note the symmetry of the formulae.

7.4.3 Quadrature formulae of Filippi and extensions

Fillippi proposed that the nodes $-1-d$ and $1+d$, $d>0$ should be preassigned in such a way that they are each used twice in composite rules. If

$$Q_{n+2}f = C_1 f(-1-d) + \sum_{i=1}^{n} A_i f(x_i) + C_2 f(1+d)$$

is applied as a composite rule to m equal subintervals, then for $d = 1-x_n$ and symmetric $Q_n f$ we have that

$$Q_{mn+2}f = \frac{1}{m}\left[C_1 f\left(-1-\frac{d}{m}\right) + \sum_{k=0}^{m-1}\sum_{i=1}^{n} \bar{A}_i f\left(-1+\frac{2k+1}{m}+\frac{1}{m}x_i\right) + C_2 f\left(1+\frac{d}{m}\right)\right]$$

(for Gauss quadrature this composite rule includes $m(n+2)$ nodes, there is thus a saving of $2m-2$ function evaluations). Here $\bar{A}_i = A_i$ for $i = 2(1)n-1$ and $\bar{A}_1 = A_1 + C_2$, $\bar{A}_n = A_n + C_1$. From polynomial interpolation we obtain the positive weights

$$A_i = \frac{1}{(1+d)^2 - x_i^2}\int_{-1}^{1} W(x)[(1+d)^2 - x^2]\lambda_i^2(x)\,dx > 0, \qquad i = 1(1)n,$$

$$C_1 = \frac{1}{2(1+d)\omega_n^2(-1-d)}\int_{-1}^{1} W(x)(1+d-x)\omega_n^2(x)\,dx > 0,$$

$$C_2 = \frac{1}{2(1+d)\omega_n^2(1+d)}\int_{-1}^{1} W(x)(1+d+x)\omega_n^2(x)\,dx > 0.$$

The nodes can be obtained from the system of equations

$$\int_{-1}^{1} W(x)\frac{\omega_n^2(x)}{x-x_i}\left[1-(x-x_i)L_2\left(\frac{1}{x-x_i}, x\right)\right]dx = 0, \qquad i = 1(1)n.$$

Since $L_2(g, x) = \dfrac{1}{2(1+d)}[(1+d-x)g(-1-d) + (x+1+d)g(1+d)]$ we have that

$$1-(x-x_i)L_2\left(\frac{1}{x-x_i}, x\right) = \frac{(1+d)^2 - x^2}{(1+d)^2 - x_i^2}$$

and hence

$$\frac{1}{(1+d)^2 - x_i^2}\int_{-1}^{1} W(x)[(1+d)^2 - x^2]\omega_n(x)\lambda_i(x)\,dx = 0, \qquad i = 1(1)n$$

must be satisfied by the nodes. Clearly, the nodes are the zeros of the orthogonal polynomials $\omega_n(x) \perp \mathbb{P}_{n-1}$ with respect to the weight function $W(x)[(1+d)^2-x^2] \geq 0$. If $W(x)=W(-x)$ and $Q_n f$ is symmetrical then, for $d=1-x_n=1+x_1$, the node x_n (or x_1) and hence $\omega_n(x)$ is implicitly defined. In both cases $Q_{n+2}f$ exists, its nodes x_i are real, distinct and situated in $(-1, 1)$. The quadrature error is

$$E_{n+2}f=\frac{f^{(2n+2)}(t)}{(2n+2)!}\int_{-1}^{1} W(x)[x^2-(1+d)^2]\omega_n^2(x)\,dx, \qquad t\in(-1, 1).$$

The integral has its minimum if $\omega_n(x) \perp \mathbb{P}_{n-1}$ with respect to the weight function $W(x)[(1+d)^2-x^2]$.

For every polynomial f we have $E_n f=0$ if n is large enough. Hence, from Banach–Steinhaus Theorem 4.3.2 we obtain convergence of $Q_{n+2}f$ for every $f\in C^0[-1-d, 1+d]$ since $Q_{n+2}f$ is positive. Thus we have:

Theorem 7.4.3 *The Filippi quadrature formulae exist for every $n\in\mathbb{N}$, they are positive, and converge for $f\in C^0[-1-d, 1+d]$.*

We list the three simplest formulae of Filippi for $W(x)\equiv 1$ in Table 7.4.3.

TABLE 7.4.3. Filippi quadrature formulae.

n	d	x_i	A_i
1	1	$-2, 2$	$1/12$
		0	$11/6$
2	$\left(\frac{4}{3}-\frac{2}{3}\sqrt{\frac{14}{5}}\right)^{\frac{1}{2}}$	$-1-d, 1+d$	$\frac{1-3(1-d)^2}{12d}=0.026\,195\,36402$
	$=0.466\,676\,1522$	$-1+d, 1-d$	$-\frac{1-3(1+d)^2}{12d}=0.973\,804\,6358$
3	$\left(\frac{8}{5}-\frac{2}{5}\sqrt{\frac{102}{7}}\right)^{\frac{1}{2}}$	$-1-d, 1+d$	$\frac{3-5(1-d)^2}{60d(1+d)^2}=0.0129\,176\,4552$
	$=0.270\,367\,1134$	$-1+d, 1-d$	$-\frac{3-5(1+d)^2}{60d(1-d)^2}=0.586\,978\,6988$
		0	$2\frac{3-10(1+d^2)+15(1-d^2)^2}{15(1-d^2)^2}=0.800\,207\,3107$

In the first case it is clear by symmetry that 0 is the only node so that $d=1$. In the second case we have $\omega_2(x)=(x-(1-d))(x-(-1+d))=x^2-(1-d)^2$ and $\Omega_2(x)=(x-(1+d))(x-(-1-d))=x^2-(1+d)^2$. Hence d

is a zero of

$$\frac{1}{2}\int_{-1}^{1} \Omega_2(x)\omega_2(x)\,dx = \tfrac{1}{5} - \tfrac{2}{3}(1+d^2) + (1-d^2)^2 = 0.$$

We obtain the two positive zeros

$$d = \left(\frac{4}{3} \pm \frac{2}{3}\sqrt{\frac{14}{5}}\right)^{\frac{1}{2}} = \begin{cases} 0.466\,676\,1522, \\ 1.564\,889\,784, \end{cases}$$

the second of which is inconvenient because it exceeds one.

The third case is obtained with $\omega_3(x) = x(x^2 - (1-d)^2)$ and d is a zero of

$$\frac{1}{2}\int_{-1}^{1} \Omega_2(x)\omega_3(x)x\,dx = \tfrac{1}{7} - \tfrac{2}{5}(1+d^2) + \tfrac{1}{3}(1-d)^2(1+d)^2 = 0.$$

From the four real zeros, two are positive:

$$d = \left(\frac{8}{5} \pm \frac{2}{5}\sqrt{\frac{102}{7}}\right)^{\frac{1}{2}} = \begin{cases} 0.270\,367\,1134 \\ 1.768\,304\,732 \end{cases}$$

of which again the second is inconvenient.

Extending Filippi's method, further exterior nodes can be preassigned as follows. Assume that we have the four interior nodes $x_{4,1} = \pm(1-d_1)$ and $x_{3,2} = \pm(1-d_2)$, and that we preassign the exterior nodes $z_{4,1} = \pm(1+d_1)$ and $z_{3,2} = \pm(1+d_2)$. Then

$$\omega_4(x) = (x^2 - (1-d_1)^2)(x^2 - (1-d_2)^2)$$

and

$$\Omega_4(x) = (x^2 - (1+d_1)^2)(1 - (1+d_2)^2)$$

so that the equations $\int_{-1}^{1} \Omega_4(x)\omega_4(x)x^{2k}\,dx = 0$, $k = 0, 1$, have to be solved.

More generally, the n distances $d_i = 1 - x_{n+1-i}$, $i = 1(1)n$, have to be calculated and the system of equations

$$\int_{-1}^{1} \Omega_{2n}(x)\omega_{2n}(x)x^{2k}\,dx = 0, \qquad k = 0(1)n-1,$$

has to be solved. This gives a completely implicit system for the nodes x_i and the corresponding quadrature with these nodes is

$$\int_{-1}^{1} f(x)\,dx \doteq Q_{4n}f = \sum_{i=1}^{n} \{A_i[f(1-d_i) + f(-1+d_i)] + C_i[f(-1-d_i) + f(1+d_i)]\}.$$

The corresponding composite rule is

$$Q_{2mn+2n}f = \frac{1}{m}\left\{\sum_{i=1}^{n}\left[C_i\left(f\left(-1-\frac{d_i}{m}\right)+f\left(1+\frac{d_i}{m}\right)\right)\right]\right.$$
$$+A_i\left[f\left(-1+\frac{1}{m}-\frac{d_i}{m}\right)+f\left(1-\frac{1}{m}+\frac{d_i}{m}\right)\right]$$
$$\left.+\sum_{k=1}^{m-1}\sum_{i=1}^{n}(A_i+C_i)\left[f\left(-1+\frac{2k}{m}-\frac{d_i}{m}\right)+f\left(-1+\frac{2k}{m}+\frac{d_i}{m}\right)\right]\right\}.$$

In general, a composite rule has $4mn$ nodes so that there is a saving of $2mn-2n$ function evaluations. Tables of such formulae are in fact not yet published.

7.4.4 Formulae of Kronrod–Piessens–Szegö

In [19] Szegö proves the following theorem which was motivated by a similar conjecture of Stieltjes.

Theorem 7.4.4 *Let $P_n^{k,k} \in \mathbb{P}_n$ be the n-th degree polynomial of Jacobi. Then, for $0 \leqslant k \leqslant 2$ there exists a polynomial $\omega_{n+1} \in \mathbb{P}_{n+1}$ which is orthogonal to $\mathbb{P}_n$ in $[-1, 1]$ with respect to the weight function $W(x) = P_n^{k,k}$ although $P_n^{k,k}$ changes sign in $[-1, 1]$. The zeros of ω_{n+1} are all real, distinct and situated in $[-1, 1]$, and every zero of $P_n^{k,k}$ lies between two consecutive zeros of ω_{n+1}.*

Szegö's proof of this theorem is beyond the scope of this book, because many necessary preliminaries have not been discussed here. Theorem 7.4.4 enables us to take as preassigned for the quadrature formula $Q_{2n+1}f$ the n zeros of the Legendre polynomial $L_n(x) \equiv \Omega_n(x) = x^n + \ldots$ (which produces the n-point Gaussian formula Q_nf) and the $n+1$ nodes of $\omega_{n+1}(x)$ which are not yet determined. For these latter, the weight function $W(x) \equiv \Omega_n(x)$ is used. The formula $Q_{2n+1}f$ adds $n+1$ Gaussian nodes to the primary Gaussian n-point formula Q_nf and therefore is very suitable for improving the formula Q_nf without changing its nodes. Such formulae are computed by Kronrod and by Piessens *et al.* (cf. [12], [14]). The formulae are now constructed. Let $Q_nf = \sum_{i=1}^{n} A_i f(z_i)$. With nodes z_i we construct the Lagrange interpolating polynomial $L_n(f, x) = \sum_{i=1}^{n} \Lambda_i(x) f(z_i)$. The Hermite interpolating polynomial with nodes x_i, $i = 1(1)n+1$, leads to the mixed interpolation formula

$$LH_{n,n+1}(f, x) = L_n(f, x) - \Omega_n(x) H_{n+1}\left(\frac{L_n(f, x) - f(x)}{\Omega_n(x)}, x\right)$$

as was pointed out in Chapter 2. Integrating this operator we have the quadrature formula

$$Q_{2n+1} = \sum_{i=1}^{n} \bar{A}_i f(z_i) + \sum_{j=1}^{n+1} (B_j f(x_j) + C_j f'(x_j))$$

with

$$C_j = \int_{-1}^{1} \Omega_n(x)(x-x_j)\lambda_j^2(x)/\Omega_n(x_j)\,dx, \qquad j=1(1)n+1,$$

where $\lambda_j(x) = \omega_{n+1}(x)/[\omega'_{n+1}(x_j)(x-x_j)]$ and $\omega_{n+1}(x) = (x-x_1)\ldots(x-x_{n+1})$. We now require $C_j = 0$, $j=1(1)n+1$. This is equivalent with

$$\int_{-1}^{1} \Omega_n(x)\lambda_j(x)\omega_{n+1}(x)\,dx = 0, \qquad j=1(1)n+1$$

and hence can be satisfied by the orthogonal polynomials $\omega_{n+1}(x)$. We have the error term

$$E_{2n+1}f = \int_{-1}^{1} \frac{f^{(3n+2)}(\xi)}{(3n+2)!}\Omega_n(x)\omega_{n+1}^2(x)\,dx, \qquad \xi \in (-1, 1),$$

and hence $E_{2n+1}f = 0$ for $f \in \mathbb{P}_{3n+1}$. The weights A_i are calculated as follows. We apply the quadrature formula $Q_{2n+1}f$ to the polynomials $\Omega_n \lambda_i \in \mathbb{P}_{2n}$, thus making the quadrature error zero. This produces the weights (Ω_n is essentially the n-th polynomial)

$$B_j = \frac{1}{\Omega_n(x_j)}\int_{-1}^{1} \Omega_n(x)\lambda_j(x)\,dx = \frac{1}{L_n(x_j)}\int_{-1}^{1} L_n(x)\lambda_j(x)\,dx, \qquad j=1(1)n+1.$$

A unique expansion of λ_i exists in terms of Legendre polynomials so that

$$\lambda_j = \frac{1}{\omega'_{n+1}(x_j)}(x^n + \ldots) = \sum_{i=0}^{n} \alpha_{ij} L_i(x) = \alpha_{nj}x^n + \ldots.$$

Comparing the coefficients of x^n we see that $\alpha_{nj} = 1/\omega'_{n+1}(x_j)$. This expansion is introduced into the integral where $L_n \perp \mathbb{P}_{n-1}$. Then

$$B_j = \frac{\alpha_{nj}}{L_n(x_j)}\int_{-1}^{1} L_n^2(x)\,dx = \frac{1}{\omega'_{n+1}(x_j)\Omega_n(x_j)}(L_n, L_n).$$

Since the zeros of ω_{n+1} and Ω_n separate each other and sign $\omega_{n+1}(1) =$ sign $\Omega_n(1)$ it is easy to see that sign $\omega'_{n+1}(x_j) =$ sign $\Omega_n(x_j)$ so that $B_j > 0$ for $j=1(1)n+1$.

On evaluating (L_n, L_n) (cf. Section 7.3) we obtain for $j=1(1)n+1$ that

$$B_j = \frac{2^{2n+1}(n!)^2}{(2n+1)!\,\omega'_{n+1}(x_j)\Omega_n(x_j)} = \frac{D_n}{\omega'_{n+1}(x_j)\Omega_n(x_j)} \quad \text{with} \quad D_n = \frac{2^{2n+1}(n!)^2}{(2n+1)!}.$$

In order to obtain $\bar{A}_i$ we apply $Q_{2n+1}f$ to $f=\Lambda_j\in\mathbb{P}_{n-1}$ which gives

$$\bar{A}_j=\int_{-1}^{1}\Lambda_j(x)\,\mathrm{d}x-\sum_{k=1}^{n+1}B_k\Lambda_j(x_k),\qquad j=1(1)n.$$

Note that the integral equals the weight A_j of the primary n-point Gauss formula with the nodes $z_1,\ldots,z_n$. If we substitute for the integral, $\Lambda_j(x_k)$ and B_k we then have that

$$\bar{A}_j=A_j-D_n\sum_{k=1}^{n+1}\frac{1}{\omega'_{n+1}(x_k)\Omega'_n(z_j)(x_k-z_j)}\cdot\frac{\omega_{n+1}(z_j)}{\omega_{n+1}(z_j)}$$

$$=A_j+\frac{D_n}{\Omega'_n(z_j)\omega_{n+1}(z_j)}\sum_{k=1}^{n+1}\lambda_k(z_j).$$

The polynomials $\lambda_k(x)$ occur in the Lagrange interpolating polynomial when applied to $f\equiv 1$

$$L_{n+1}(f,x)|_{f\equiv 1}=\sum_{k=1}^{n+1}\lambda_k(x)f(x_k)|_{f\equiv 1}=\sum_{k=1}^{n+1}\lambda_k(x).$$

On the other hand, $L_{n+1}(1,x)\equiv 1$ for every x and hence for $x=z_j$ so that

$$\bar{A}_j=A_j+\frac{D_n}{\Omega'_n(z_j)\omega_{n+1}(z_j)},\qquad j=1(1)n.$$

Unfortunately, positivity of the $\bar{A}_j$ cannot be concluded from this result, since $\Omega'_n(z_j)\omega_{n+1}(z_j)<0$ for $j=1(1)n$. It is only possible to say that $\bar{A}_j<A_j$, while in fact $\bar{A}_j>0$. Thus, finally, we have the $3n+1$ degree formula

$$Q_{2n+1}f=Q_nf+D_n\left[\sum_{i=1}^{n}\frac{f(z_i)}{\Omega'_n(z_i)\omega_{n+1}(z_i)}+\sum_{j=1}^{n+1}\frac{f(x_j)}{\omega'_{n+1}(x_j)\Omega_n(x_j)}\right]$$

where Q_nf is a $2n-1$ degree Gauss formula, and the function values $f(z_i)$ are available from Q_nf. For this reason the nodes x_i are referred to as *optimally-added* nodes of Gauss formulae.

Extended tables of formulae with optimally-added nodes have been calculated by Kronrod and Piessens (cf. [12], [14]). For the case $k=1$, which corresponds to the optimal addition of nodes to Lobatto formulae,

Piessens (cf. [14]) gives the weights as

$$\bar{A}_k = \frac{D_n}{2\Omega_n'(z_k)\omega_{n+1}(z_k)} + \frac{2}{(n+1)(n+2)(\Omega_{n+1}(z_k))^2}$$

$$\bar{A}_0 = \bar{A}_{n+1} = \frac{2}{(n+1)(n+2)} - \frac{D_n}{2(n+1)\omega_{n+1}(1)}$$

$$B_k = \frac{n+2}{2(n+1)} \frac{D_n}{(\Omega_n(x_k) - x_k\Omega_{n+1}(x_k))\omega_{n+1}'(x_k)}$$

with

$$D_n = \frac{2^{2n+3}((n+1)!)^2}{(2_n+3)!}$$

where $Q_{2n+3}f = \bar{A}_0 f(-1) + \sum_{k=1}^{n} \bar{A}_k f(z_k) + \bar{A}_{n+1} f(1) + \sum_{k=1}^{n+1} B_k f(x_k)$. The function values $f(-1)$, $f(1)$ and $f(z_k)$, $k = 1(1)n$ are available from the Lobatto formula

$$Q_{n+2}f = A_0 f(-1) + \sum_{k=1}^{n} A_k f(z_k) + A_{n+1} f(1)$$

Example 7.4.1 If $Q_2 f = f(-\sqrt{1/3}) + f(\sqrt{1/3}) = If$ then $z_1 = -1/\sqrt{3}$, $z_2 = 1/\sqrt{3}$ and hence $\Omega_2(x) = x^2 - 1/3$. Further, $\Lambda_1 = (1/2)\sqrt{3}(1/\sqrt{3} - x)$ and $\Lambda_2 = (1/2)\sqrt{3}(1/\sqrt{3} + x)$. We need the polynomial $\omega_3(x)$ which satisfies

$$\int_{-1}^{1} (x^2 - \tfrac{1}{3})\omega_3(x) x^k \, dx = 0, \qquad k = 0, 1, 2.$$

Solution of this system gives $\omega_3(x) = x(x^2 - 6/7)$ with zeros $x_{1,3} = \mp\sqrt{6/7}$, $x_2 = 0$. The cardinal polynomials are

$$\lambda_1 = \frac{7}{12} x \left(x - \sqrt{\frac{6}{7}}\right), \qquad \lambda_2 = \tfrac{7}{6}(\tfrac{6}{7} - x^2), \quad \text{and} \quad \lambda_3 = \frac{7}{12} x \left(x + \sqrt{\frac{6}{7}}\right).$$

The weights

$$\bar{A}_1 = \bar{A}_2 = \tfrac{27}{55}, \qquad B_1 = B_3 = \tfrac{98}{495}, \quad \text{and} \quad B_2 = \tfrac{28}{45}$$

follow immediately. Application of the formulae to $f(x) = 1/(1+x^2)$ gives that $Q_2(1/(1+x^2)) = 1.5$ with error $-7.0796(-2)$.

The computation of the improved value $Q_5(1/(1+x^2))$ is performed with three additional function evaluations, since $f(\pm 1/\sqrt{3})$ is used once more, and gives $Q_5(1/(1+x^2)) = 1.571\,794\,872$ with error $9.9855(-4)$. The exact value of the integral is $2 \cdot \operatorname{arc\,tg} 1 = 1.570\,796\,327$. (See also Example 3.7.3 in Section 3.7.2.) ■

Some formulae with optimally-added nodes are listed in Table 7.4.4.

TABLE 7.4.4. Quadrature formulae of Kronrod–Piessens–Szegö.

n	Nodes	$Q_n f$	Weights of $Q_{2n+1}f$
1	$\pm\sqrt{3/5}$	0	5/9
	0	2	8/9
2	$\pm\sqrt{6/7}$	0	98/495
	$\pm\sqrt{1/3}$	1	27/55
	0	0	28/45
3	$\pm\frac{1}{3}\left(5+2\sqrt{\frac{30}{11}}\right)^{\frac{1}{2}}$	0	$8019/[350(138+49\sqrt{30/11}]$
	$\pm\sqrt{3/5}$	5/9	12500/46557
	$\pm\frac{1}{3}\left(5-2\sqrt{\frac{30}{11}}\right)^{\frac{1}{2}}$	0	$8019/[350(138-49\sqrt{30/11}]$
	0	8/9	22016/48825
4	$\pm\left(\frac{15+\sqrt{\frac{465}{13}}}{22}\right)^{\frac{1}{2}}$	0	$\frac{24\,362\,624}{14175(219263/13+1743\sqrt{465/13})}$
	$\pm\left(\frac{3+2\sqrt{\frac{6}{5}}}{7}\right)^{\frac{1}{2}}$	$\frac{1}{6}\left(3-\sqrt{\frac{5}{6}}\right)$	$\frac{6125(395-9\sqrt{6/5})}{162(66224+17715\sqrt{6/5})}$
	$\pm\left(\frac{15-\sqrt{\frac{465}{13}}}{22}\right)^{\frac{1}{2}}$	0	$\frac{24\,362\,624}{14175(219263/13-1743\sqrt{465/13})}$
	$\pm\left(\frac{3-2\sqrt{\frac{6}{5}}}{7}\right)^{\frac{1}{2}}$	$\frac{1}{6}\left(3+\sqrt{\frac{5}{6}}\right)$	$\frac{6125(395+9\sqrt{6/5})}{162(66224-17715\sqrt{6/5})}$
	0	0	$\frac{201344}{581175}$

7.5 Quadratures with minimal error norm

In [20] Wilf proposes a special case of the quadratures in Section 3.5.4 which minimise

$$\sum_{k=0}^{\infty} [E_n(x^k)]^2 \quad \text{for} \quad E_n f = \overset{*}{I} f - Q_n f.$$

Although this minimality requirement obviously has no connection with interpolation techniques, the corresponding quadrature formulae can be generated by integrating a suitable Hermitian interpolating operator. This approach is followed here and the formulae are denoted briefly as *minimal quadratures.*

7.5.1 Construction and properties

We choose the basis functions $h_i = x - x_i$ and the nodal function

$$\omega_n(x) = \prod_{i=1}^{n} \frac{x - x_i}{1 - \varepsilon^2 x x_i}, \qquad 0 \leqslant \varepsilon \leqslant 1,$$

so that the cardinal functions are

$$\lambda_i(x) = \frac{\omega_n(x)}{\omega_n'(x_i)(x - x_i)}, \qquad i = 1(1)n.$$

The Hermite operator is thus

$$H_n(f, x) = \sum_{i=1}^{n} [(1 - 2\lambda_i'(x_i)(x - x_i))f(x_i) + (x - x_i)f'(x_i)]\lambda_i^2(x) \quad (7.5.1)$$

which, on integration, produces the quadrature formula

$$Q_n f \equiv \int_{-1}^{1} H_n(f, x)\,dx = \sum_{i=1}^{n} A_i f(x_i)$$

provided that

$$\int_{-1}^{1} \omega_n(x)\lambda_i(x)\,dx = 0 \quad \text{or} \quad \int_{-1}^{1} \prod_{j=1}^{n} \left(\frac{x - x_j}{1 - \varepsilon^2 x x_j}\right)^2 \frac{dx}{x - x_i} = 0, \qquad i = 1(1)n.$$

These equivalent conditions mean that $\Omega_n(x) \equiv (x - x_1) \ldots (x - x_n) \perp \mathbb{P}_{n-1}$ with respect to the weight function $W(x) = \prod_{j=1}^{n} 1/(1 - \varepsilon^2 x x_j)^2$. The existence of the quadrature formulae follows from the corresponding theorem for implicitly-defined orthogonal polynomials given in Section 5.3.1. The nodes x_i are the n zeros of $\Omega_n(x)$. It further follows from orthogonal polynomial theory that the nodes are real, distinct and in $(-1, 1)$. The weights are positive, and since the Hermite operator (7.5.1) reproduces the Lagrange cardinal functions $\lambda_i(x)$ identically, they are obtained from

$$A_i = \int_{-1}^{1} \lambda_i^2(x)\,dx = \int_{-1}^{1} \lambda_i(x)\,dx, \qquad i = 1(1)n.$$

Although the minimal quadratures obviously depend on ε so that in fact $Q_n f = (Q_n f)_\varepsilon$ we omit the index ε for simplicity. Hence we have that the following theorem is valid:

Theorem 7.5.1 *For every given $n \in \mathbb{N}$ and for $0 \leqslant \varepsilon \leqslant 1$ there exists a positive quadrature formula $Q_n f$ with minimal error norm.*

Integrating the error of interpolation the quadrature error is found to be

$$E_n f = \frac{f^{(2n)}(t) - H^{(2n)}(f, t)}{(\omega_n^2(x))^{(2n)}(x=t)} \int_{-1}^{1} \omega_n^2(x)\, dx, \qquad t \in (-1, 1).$$

Clearly, this is not a convenient error representation since it involves the $(2n)$-th derivative of the rational function $H_n(f, x)$. Fortunately there is another error representation for functions which are analytic in $|z| \leqslant R = 1/\varepsilon$ which does not need derivatives of $f(x)$ (see Section 3.5.3), and which also applies here as far as the nodes are known. In order to see that the above statement for $H_n(\lambda_i, x) \equiv \lambda_i$ holds for $i = 1(1)n$, the following theorem is first proved:

Theorem 7.5.2 *The operator $H_n(f, x)$ has the fixed elements $1/(1-\varepsilon^2 x x_i)$ and $x/(1-\varepsilon^2 x x_i)^2$, $i = 1(1)n$.*

Proof We have to establish that for $i = 1(1)n$

$$H_n\left(\frac{1}{1-\varepsilon^2 x x_i}, x\right) = \frac{1}{1-\varepsilon^2 x x_i} \quad \text{and} \quad H_n\left(\frac{x}{(1-\varepsilon^2 x x_i)^2}, x\right) = \frac{x}{(1-\varepsilon^2 x x_i)^2}.$$

Let $P_n(x) \equiv (\varepsilon^2 x)^n \Omega_n(1/\varepsilon^2 x) = (1-\varepsilon^2 x x_1) \ldots (1-\varepsilon^2 x x_n)$ and

$$\Lambda_i(x) \equiv \frac{\Omega_n(x)}{\Omega_n'(x_i)(x-x_i)}, \qquad i = 1(1)n.$$

Then since $\omega_n(x) = \Omega_n(x)/P_n(x)$ and $\lambda_i(x_i) = 1$ we have that

$$\lambda_i'(x_i) = \Lambda_i'(x_i) - \frac{P_n'(x_i)}{P_n(x_i)}, \qquad i = 1(1)n.$$

Further,

$$P_n'(x) = \sum_{j=1}^{n} \left(\prod_{\substack{k=1 \\ \neq j}}^{n} (1-\varepsilon^2 x x_k) \right) (-\varepsilon^2 x_j) = -\varepsilon^2 P_n(x) \sum_{j=1}^{n} \frac{x_j}{1-\varepsilon^2 x x_j}$$

and hence

$$\lambda_i'(x_i)=\Lambda_i'(x_i)+\varepsilon^2\sum_{j=1}^{n}\frac{x_j}{1-\varepsilon^2x_ix_j}, \qquad i=1(1)n.$$

Substituting this into the interpolating operator $H_n(f,x)$ of (7.5.1) we have that

$$H_n(f,x)=\frac{1}{P_n^2(x)}\sum_{i=1}^{n}\Bigg\{\Bigg[1-2\Lambda_i'(x_i)(x-x_i)-2\varepsilon^2(x-x_i)$$

$$\times\sum_{j=1}^{n}x_j/(1-\varepsilon^2x_ix_j)\Bigg]f(x_i)+(x-x_i)f'(x_i)\Bigg\}\Lambda_i^2(x)P_n^2(x_i).$$

For the first identity

$$H_n\left(\frac{x}{(1-\varepsilon^2xx_k)^2},x\right)\equiv\frac{x}{(1-\varepsilon^2xx_k)^2}, \qquad k=1(1)n$$

we have that

$$H_n\left(\frac{x}{(1-\varepsilon^2xx_k)^2},x\right)=\frac{1}{P_n^2(x)}\sum_{i=1}^{n}\Bigg\{\Bigg[1-2\Lambda_i'(x_i)(x-x_i)$$

$$-2\varepsilon^2(x-x_i)\sum_{j=1}^{n}\frac{x_ix_j}{(1-\varepsilon^2x_ix_j)(1-\varepsilon^2x_ix_k)^2}$$

$$+(x-x_i)\frac{1+\varepsilon^2x_ix_k}{(1-\varepsilon^2x_ix_k)^3}\Bigg\}\Lambda_i^2(x)P_n^2(x_i)$$

$$=\frac{1}{P_n^2(x)}\sum_{i=1}^{n}\Bigg\{[1-2\Lambda_i'(x_i)(x-x_i)]\frac{x_iP_n^2(x_i)}{(1-\varepsilon^2x_ix_k)^2}$$

$$+(x-x_i)\left[\frac{1+\varepsilon^2x_ix_k}{1-\varepsilon^2x_ix_k}-\sum_{j=1}^{n}\frac{2\varepsilon^2x_ix_j}{1-\varepsilon^2x_ix_j}\right]$$

$$\times\frac{P_n^2(x_i)}{(1-\varepsilon^2x_ix_k)^2}\Bigg\}\Lambda_i^2(x).$$

Consider the terms of the summation. In order to simplify them let

$$Q(x)\equiv\frac{xP_n^2(x)}{(1-\varepsilon^2xx_k)^2}=x\prod_{\substack{j=1\\ \neq k}}^{n}(1-\varepsilon^2xx_j)^2\in\mathbb{P}_{2n-1}.$$

Then

$$Q'(x)=\left(\prod_{\substack{j=1\\ \neq k}}^{n}(1-\varepsilon^2xx_j)^2\right)\left[1+x\sum_{\substack{j=1\\ \neq k}}^{n}\frac{-2\varepsilon^2x_j}{1-\varepsilon^2xx_j}\right]$$

$$=\frac{P_n^2(x)}{(1-\varepsilon^2xx_k)^2}\left[1-\sum_{j=1}^{n}\frac{2\varepsilon^2xx_j}{1-\varepsilon^2xx_j}+\frac{2\varepsilon^2xx_k}{1-\varepsilon^2xx_k}\right]$$

$$=\left[\frac{1+\varepsilon^2xx_k}{1-\varepsilon^2xx_k}-\sum_{j=1}^{n}\frac{2\varepsilon^2xx_j}{1-\varepsilon^2xx_j}\right]\frac{P_n^2(x)}{(1-\varepsilon^2xx_k)^2}$$

As a result

$$H_n\left(\frac{x}{(1-\varepsilon^2xx_k)^2},x\right)=\frac{1}{P_n^2(x)}\sum_{i=1}^{n}\{(1-2\Lambda_i'(x_i)(x-x_i))Q(x_i)$$
$$+(x-x_i)Q'(x_i)\}\Lambda_i^2(x)=\frac{1}{P_n^2(x)}\tilde{H}_n(Q(x),x)$$

where $\tilde{H}_n(f,x)$ is the ordinary polynomial interpolating operator whose fixed elements are all polynomials of degree at most $2n-1$. Since $Q(x)\in\mathbb{P}_{2n-1}$ we have $\tilde{H}_n(Q(x),x)\equiv Q(x)$ and hence that

$$H_n\left(\frac{x}{(1-\varepsilon^2xx_k)^2},x\right)=\frac{1}{P_n^2(x)}Q(x)=\frac{x}{(1-\varepsilon^2xx_k)^2}\quad\text{for}\quad k=1(1)n.$$

The proof of the second identity

$$H_n\left(\frac{1}{1-\varepsilon^2xx_k},x\right)\equiv\frac{1}{1-\varepsilon^2xx_k}\quad\text{for}\quad k=1(1)n,$$

is left to the reader (see Problem P.7.3). ■

As a result of Theorem 7.5.2 we can write

$$E_n\left(\frac{1}{1-\varepsilon^2xx_k}\right)=\int_{-1}^{1}\frac{dx}{1-\varepsilon^2xx_k}-Q_n\left(\frac{1}{1-\varepsilon^2xx_k}\right)$$
$$=\int_{-1}^{1}\left[\frac{1}{1-\varepsilon^2xx_k}-H_n\left(\frac{1}{1-\varepsilon^2xx_k},x\right)\right]dx=0,$$
$$k=1(1)n$$

and

$$E_n\left(\frac{x}{(1-\varepsilon^2xx_k)^2}\right)=0\quad\text{for}\quad k=1(1)n.$$

These are exactly the conditions which are satisfied by the system of equations (3.5.5a), (3.5.5b) which are the necessary conditions for the nodes and weights to minimise $\sigma_n^2=\sum_{k=0}^{\infty}\varepsilon^{2k}(E_n(x^k))^2$ and which, for $\varepsilon=1$, is identical with Wilf's minimising condition.

From Theorem 7.5.1 we obtain immediately:

Corollary 7.5.1 The rational Lagrange cardinal functions $\lambda_i(x)$, $i=1(1)n$ are fixed elements of $H_n(f, x)$ (see Problem P.7.3). ■

In order to prove the convergence of minimal quadrature procedures we apply the Banach–Steinhaus Theorem 4.3.2. Since Q_nf is positive, convergence need only be proved for every polynomial. If $f=x^k$, $k\in\mathbb{N}_0$ then (cf. p. 129)

$$\|f\|=\|x^k\|=\left(\frac{R^{2k+1}}{2k+1}\right)^{\frac{1}{2}}=R^k\left(\frac{R}{2k+1}\right)^{\frac{1}{2}}=\frac{1}{\varepsilon^k(\varepsilon(2k+1))^{\frac{1}{2}}}.$$

and hence

$$|E_nf|=|E_nx^k|\leqslant\sigma_n\,\|x^k\|=\frac{\sigma_n}{\varepsilon^k(\varepsilon(2k+1))^{\frac{1}{2}}}.$$

From Theorem 4.2.8 we know that there exist constants $C_1, C_2\in\mathbb{R}_+$ such that

$$\sigma_n\leqslant\begin{cases}C_1\left(\dfrac{\ln n}{n}\right)^{\frac{1}{2}} & \text{for}\quad \varepsilon=1\\[2ex] C_2\left(\dfrac{1}{n}\right)^{\frac{1}{2}} & \text{for}\quad 0\leqslant\varepsilon<1.\end{cases}$$

Hence $\lim_{n\to\infty}|E_nx^k|\leqslant\dfrac{1}{\varepsilon^k(\varepsilon(2k+1))^{\frac{1}{2}}}\lim_{n\to\infty}\sigma_n=0$, which completes the proof. ■

Remark: The proof may also be established in another way. Since $E_nf=0$ for $f\in\{1/(1-\varepsilon xx_i),\ i=1(1)n\}$, convergence results if we can show that this system of functions is dense in $C^0[-1,1]$ for $n\to\infty$ (see, for example, Engels [5]). We finally have:

Theorem 7.5.3 *The procedure of minimal quadrature formulae is positive, its nodes are real, distinct, and situated in* $(-1,1)$, *and it converges for every* $f\in C^0[-1,1]$.

Minimal quadratures with preassigned nodes or with weight functions $W(x)\not\equiv 1$ have not yet been studied.

7.5.2 Table of Wilf-type quadrature formulae for $\varepsilon = 1$

The formulae given in Table 7.5.1 are called Wilf formulae because the case $\varepsilon = 1$ is that considered by Wilf.

TABLE 7.5.1. Quadrature formulae with minimal error norm.

n	Nodes $x_i = -x_{n-i+1}$	Weights $A_i = A_{n-i+1}$	Error constant σ_n
2	0.665 742 695 953 213	0.969 303 253 001 722	0.258
4	0.519 692 429 593 520	0.867 796 177 894 194	0.327(−1)
	0.961 396 989 463 654	0.129 148 967 191 694	
8	0.386 956 389 897 216	0.702 358 602 319 239	0.137(−2)
	0.856 457 381 612 934	0.248 814 275 604 362	
	0.980 573 199 445 521	0.444 868 610 923 195(−1)	
	0.998 754 713 951 697	0.424 352 840 681 217(−2)	
16	0.279 894 546 102 950	0.532 300 387 974 500	0.121(−4)
	0.704 666 212 158 882	0.304 176 895 049 497	
	0.905 572 909 757 985	0.117 011 443 753 612	
	0.974 901 827 489 048	0.353 727 524 455 307(−1)	
	0.994 398 434 652 100	0.894 854 871 123 805(−2)	
	0.998 997 008 391 720	0.186 692 420 566 952(−2)	
	0.999 872 494 677 604	0.294 800 267 454 647(−3)	
	0.999 991 900 124 766	0.276 187 911 564 492(−4)	
32	0.199 396 291 854 521	0.398 707 264 674 172	0.119 (−7)
	0.543 961 248 203 279	0.289 762 755 178 020	
	0.772 648 139 902 848	0.170 618 738 401 392	
	0.897 196 048 564 481	0.855 509 095 210 455(−1)	
	0.956 657 003 740 345	0.387 119 615 146 697(−1)	
	0.982 722 713 337 943	0.163 396 248 440 913(−1)	
	0.993 461 053 777 161	0.652 549 365 301 590(−2)	
	0.997 655 988 976 113	0.247 305 073 738 048(−2)	
	0.999 209 679 788 747	0.886 061 380 817 004(−3)	
	0.999 752 128 953 849	0.297 584 943 139 505(−3)	
	0.999 928 841 675 622	0.924 569 476 266 997(−4)	
	0.999 981 740 689 427	0.260 584 840 535 418(−4)	
	0.999 995 964 378 747	0.646 500 005 170 292(−5)	
	0.999 999 279 056 313	0.134 260 505 056 767(−5)	
	0.999 999 908 390 099	0.211 821 303 690 764(−6)	
	0.999 999 994 180 757	0.198 424 188 898 392(−7)	

In Table 7.5.1, $\int_{-1}^{1} f(x)\,dx = \sum_{i=1}^{n} A_i f(x_i)$ and the error constants are valid for functions $f \in \mathbb{H}_2$ with $R = 1$ or $\varepsilon = 1$.

7.6 Rectangle and trapezoidal rules

We now show that rectangle and trapezoidal rules considered in Chapter 2 are general Gauss-type formulae for periodic basis or nodal functions (see Engels [4]). Two different types of integral are treated. For the interval $[0, 2\pi]$ we use periodic basis and nodal functions, and for $(-\infty, \infty)$ we choose a periodic nodal function.

7.6.1 The interval $[0, 2\pi]$

Assume that $h_i(x) = \sin(x - x_i)$, $\omega_n(x) = h_1(x)\dots h_n(x)$ and the weight function is $W(x) \equiv 1$. The requirement that the weights B_i of the derivatives of $f(x)$ should vanish leads to the system of equations

$$\int_0^{2\pi} \frac{\omega_n^2(x)\,dx}{\sin(x - x_i)} = 0, \qquad i = 1(1)n.$$

This has the solution $x_i = 2\pi i/n$, $i = 0(1)n-1$, since

$$\omega_n^2(x) = \sin^2 x \sin^2\left(x - \frac{2\pi}{n}\right)\dots\sin^2\left(x - \frac{2\pi}{n}(n-1)\right) = \omega_n^2\left(x - \frac{2\pi}{n}j\right),$$
$$j = \pm 1, \pm 2, \dots$$

from which it follows that

$$\int_0^{2\pi} \frac{\omega_n^2(x)\,dx}{\sin(x - 2\pi i/n)} = \int_0^{2\pi} \frac{\omega_n^2(x - 2\pi i/n)\,dx}{\sin(x - 2\pi i/n)} = \int_{2\pi i/n - \pi}^{2\pi i/n + \pi} \frac{\omega_n^2(x - 2\pi i/n)}{\sin(x - 2\pi i/n)}\,dx = 0,$$
$$i = 0(1)n-1,$$

because the last integrand is odd with respect to the interval $[2\pi i/n - \pi, 2\pi i/n + \pi]$. For these equidistant nodes the corresponding Hermitian operator $H_n(f, x)$ is no longer an interpolating operator but is instead, an averaging operator (see Remark R.2.11). Although neither $\omega_n'(x_i) = 0$ nor $\lambda_i(x_i) = 0$ fit the general theory, we can derive the weights, which prove to be equal, using $Q_n(\cos k\pi x) = 0$ for $k = 0(1)n-1$. Hence, $Q_n f$ is identical with the rectangle rule $R_n f$ which becomes the trapezoidal rule $T_n f$, if the integrand is a 2π-periodic function, since in that case $f(0) = f(2\pi)$. In Chapter 3 error bounds without derivatives for periodic functions were derived for $R_n f$.

7.6.2 The interval $(-\infty, \infty)$

Let the nodal function be $\omega_h(x) = \sin \pi x/h$, and the basis functions be $h_i(x) = x - x_i$, $i = -\infty(1)\infty$. The interval $(-\infty, \infty)$ contains infinitely many nodes so that with index h instead of n

$$R_h f = h \sum_{j=-\infty}^{\infty} f(jh)$$

is the rectangle rule for the approximation of $\underset{\infty}{I}f$.

Theorem 7.6.1 *The rectangle rule $R_h f$ converges for every $f \in C^0(-\infty, \infty)$, which means for $h \to 0$. Further, $R_h f$ is a Gauss-type quadrature for periodic functions.*

Proof It was shown in Section 2.5.4, formula (2.5.20), that

$$H_h(f, x) = \left(\frac{h}{\pi}\right)^2 \sum_{j=-\infty}^{\infty} [f(jh) + (x - jh)f'(jh)] \left(\frac{\sin \pi x/h}{x - jh}\right)^2$$

is a Hermitian operator. If we integrate $H_h(f, x)$ with respect to the weight function $W(x) \equiv 1$, we see that the weights

$$B_j = \left(\frac{h}{\pi}\right)^2 \int_{-\infty}^{\infty} \frac{1}{x - jh} \sin^2 \frac{\pi x}{h} \, dx = \left(\frac{h}{\pi}\right)^2 \int_{-\infty}^{\infty} \frac{1}{x} \sin^2 \frac{\pi x}{h} \, dx = 0$$

of the derivatives vanish since the integrand is odd. Further, if $\pi x/h$ is replaced by t, the weights

$$A_j = \left(\frac{h}{\pi}\right)^2 \int_{-\infty}^{\infty} \frac{1}{(x - jh)^2} \sin^2 \frac{\pi x}{h} \, dx = \left(\frac{h}{\pi}\right)^2 \int_{-\infty}^{\infty} \frac{1}{x^2} \sin^2 \frac{\pi x}{h} \, dx$$

$$= \frac{h}{\pi} \int_{-\infty}^{\infty} \left(\frac{\sin t}{t}\right)^2 dt = \frac{h}{\pi} \int_{-\infty}^{\infty} \frac{\sin t}{t} \, dt = h > 0$$

are positive. Convergence of $R_h f$ can be shown by considering only the convergence of a dense system of functions. In this case the Banach–Steinhaus Theorem 4.3.2 applies. The functions $\exp(-x^2) \cdot x^k$, $k = 0(1) \ldots$ are dense in the space $C^0(-\infty, \infty) \cap \{f \text{ with } |\underset{\infty}{I}f| < \infty\}$ since the system of Laguerre polynomials is a complete orthogonal function system on $(-\infty, \infty)$ with respect to $W(x) = \exp(-x^2)$. Further, these functions are holomorphic in the whole complex plane. Hence, in the error estimate

$$|E_n f| \leq \left(\coth \frac{\pi \sigma}{h} - 1\right) \int_{-\infty + i\sigma}^{+\infty + i\sigma} |f(z)| \, dz$$

the integral is always bounded for $f = x^k e^{-x^2}$, $k = 0(1) \ldots$ and $\sigma > 0$ can be chosen arbitrarily. Then

$$\sigma_h \equiv \coth \frac{\pi\sigma}{h} - 1 = \frac{2}{e^{2\pi\sigma/h} - 1} \to 0 \quad \text{for} \quad h \to 0$$

and we have convergence of $R_h f$ if $h \to 0$. ■

7.7 Approximate computation of indefinite integrals

We now turn to integrals in which the upper limit of integration is *variable.* Thus let

$$F(x) \equiv F(-1) + \int_{-1}^{x} W(t)f(t)\,dt, \qquad x \in [-1, 1]$$

where $f(t)$ is a given continuous function in $[-1, 1]$. We wish to approximate $F(x)$ and use mainly the Fourier expansion of f or F with respect to a given system of orthogonal polynomials. For the special case $x = 1$ the integral reverts to the standard form.

7.7.1 General theory

If $f \in C[-1, 1]$ then $f(t)$ possesses a convergent series expansion of the type

$$f(x) = \sum_{k=0}^{\infty}{}' a_k \omega_k(x). \tag{7.7.1}$$

The polynomials $\omega_k(x)$ are assumed to be orthogonal with respect to a certain weight function $\tilde{W}(x) \not\equiv W(x)$ which will be specified below. Then the polynomial $\omega_n(x) \in \dot{\mathbb{P}}_n$ has n distinct real zeros $x_i \in (-1, 1)$ and

$$Q_n f = \sum_{i=1}^{n} A_i f(x_i) \quad \text{with} \quad A_i = \int_{-1}^{1} \tilde{W}(x)[\omega_n(x)/((x - x_i)\omega_n'(x_i))]^2\,dx > 0$$

is a Gauss quadrature formula of degree $2n-1$. The discrete inner product based upon the zeros of $\omega_n(x)$ is defined as

$$\langle f, g \rangle \equiv \sum_{i=1}^{n} A_i f(x_i) g(x_i).$$

If the product $fg \in \mathbb{P}_{2n-1}$ then

$$(f, g) = \int_{-1}^{1} \tilde{W}(x) f(x) g(x)\,dx = \sum_{r=1}^{n} A_r f(x_r) g(x_r) = \langle f, g \rangle$$

and, in particular,

$$\langle \omega_i(x), \omega_k(x)\rangle = (\omega_i(x), \omega_k(x)) = \begin{cases} 0 & \text{for} \quad i \neq k \\ \|\omega_k\|^2 & \text{for} \quad i = k \end{cases}$$

so that for $0 \leq i+k < 2n$, we have orthogonality in the discrete sense, and the coefficients of the series expansion (7.7.1) are

$$a_k = (f(x), \omega_k(x))/(\omega_k(x), \omega_k(x)), \qquad k = 1(1)\infty.$$

If the series (7.7.1) is truncated after n terms we find that

$$a_k = \langle f(x), \omega_k(x)\rangle/\langle \omega_k(x), \omega_k(x)\rangle, \qquad k = 1(1)n-1$$

and $\frac{1}{2}a_0 = \langle f(x), \omega_0\rangle/\langle \omega_0, \omega_0\rangle$ so that the numerical values of these coefficients in the truncated series expansion

$$f_{n-1}(x) \equiv \sum_{k=0}^{n-1}{}' a_k\omega_k(x) \tag{7.7.2}$$

can be obtained using only elementary computations, i.e. *without calculating the definite integrals* $(f(x), \omega_k(x))$. Thus, we have that

$$F(x) = F(-1) + \sum_{k=0}^{\infty}{}' \left\{ \frac{(f, \omega_k)}{(\omega_k, \omega_k)} \int_{-1}^{x} W(t)\omega_k(t)\, dt \right\}$$

for the infinite series, and

$$F_{n-1}(x) \equiv F(-1) + \sum_{k=0}^{n-1}{}' \left\{ \frac{\langle f, \omega_k\rangle}{\langle \omega_k, \omega_k\rangle} \int_{-1}^{x} W(t)\omega_k(t)\, dt \right\}$$

for the truncated series. Note that for $k > 0$ the polynomials $\omega_k(x)$ are generally not orthogonal to $f \equiv 1$ with respect to $W(x)$, while they are with respect to $\tilde{W}(x)$.

If we denote the convergent series expansion of $F(x)$ in terms of $\omega_k(x)$, $k = 0(1)\infty$ by $F(x) = \sum_{k=0}^{\infty}{}' b_k\omega_k(x)$, then the coefficients b_k are determined by

$$b_k = (F(x), \omega_k(x))/(\omega_k(x), \omega_k(x)) \quad \text{for} \quad k = 1(1)\infty,$$

or, after truncating the series, by

$$b_k = \langle F(x), \omega_k(x)\rangle/\langle \omega_k(x), \omega_k(x)\rangle \quad \text{for} \quad k = 1(1)n-1$$
$$\tfrac{1}{2}b_0 = \langle F(x), \omega_0\rangle/\langle \omega_0, \omega_0\rangle.$$

The inner products $\langle F(x), \omega_k(x)\rangle$ are usually not so easily calculated as in the preceding case because they contain some definite integrals. In order to obtain the coefficients b_k it is therefore further assumed that the orthogonal polynomials $\omega_k(x)$ satisfy the linear differential equation

$$\tilde{W}(x)\omega_k(x)+B_k\frac{\mathrm{d}}{\mathrm{d}x}[(1-x^2)\tilde{W}(x)\omega_k'(x)]=0 \qquad (7.7.3)$$

with constant real numbers $B_k \in \mathbb{R}$. (Equations of this type are valid for Legendre, Tschebyscheff, Jacobi, Hermite, and Laguerre polynomials.) Integrating by parts we find that

$$\begin{aligned}(F(x), \omega_k(x)) &= \int_{-1}^{1} \tilde{W}(x)\left[F(-1)+\int_{-1}^{x} W(t)f(t)\,\mathrm{d}t\right]\omega_k(x)\,\mathrm{d}x\\ &= -\left[F(-1)+\int_{-1}^{x} W(t)f(t)\,\mathrm{d}t\right]B_k(1-x^2)\tilde{W}(x)\omega_k'(x)\Big|_{x=-1}^{x=1}\\ &\quad +B_k\int_{-1}^{1} W(x)f(x)(1-x^2)\tilde{W}(x)\omega_k'(x)\,\mathrm{d}x\end{aligned}$$

and, hence, the weight function $\tilde{W}(x)$ can be replaced by $\hat{W}(x)\equiv(1-x^2)\tilde{W}(x)$. The orthogonal polynomials $\hat{\omega}_k(x)$ which are appropriate to $\hat{W}(x)$ are the polynomials $\omega_{k+1}'(x)$, since for $(f*g)\equiv\int_{-1}^{1}\hat{W}(x)f(x)g(x)\,\mathrm{d}x$ we have that

$$\begin{aligned}(\omega_k, \omega_m) &= \int_{-1}^{1} \tilde{W}(x)\omega_k(x)\omega_m(x)\,\mathrm{d}x = -B_k(1-x^2)\tilde{W}(x)\omega_k'(x)\omega_m(x)\Big|_{-1}^{1}\\ &\quad +B_k\int_{-1}^{1} \tilde{W}(x)\omega_k'(x)\,\omega_m'(x)(1-x^2)\,\mathrm{d}x\\ &= B_k(\omega_k' * \omega_m') = \delta_{k,m}\,\|\omega_k'\|^2\,B_k.\end{aligned}$$

The orthogonality $\omega_k \perp \mathbb{P}_{k-1}$ with respect to $\tilde{W}(x)$ causes $\omega_k' \perp \mathbb{P}_{k-2}$ with respect to $\hat{W}(x)$ by virtue of the differential equation (7.7.3). Let the n zeros of ω_{n+1}' be denoted by $\hat{x}_i$, $i=1(1)n$ and define the discrete inner product $\langle f*g\rangle\equiv\sum_{i=1}^{n}\hat{A}_i f(\hat{x}_i)g(\hat{x}_i)$, where

$$\hat{A}_i=\int_{-1}^{1}\hat{W}(x)[\omega_{n+1}'(x)/[(x-\hat{x}_i)\omega_{n+1}''(\hat{x}_i)]]^2\,\mathrm{d}x>0, \qquad i=1(1)n,$$

are the weights of the corresponding Gauss-type quadrature formula $Q_nf=\sum_{i=1}^{n}\hat{A}_i f(\hat{x}_i)$, which has degree $2n-1$.

Note that there exist other possibilities for choosing the inner products

$\langle\cdot,\cdot\rangle$ and $\langle\cdot*\cdot\rangle$. It can happen, that there is yet another set of nodes y_j, z_j and weights C_j, D_j such that, for example,

$$\langle\omega_i(x),\omega_k(x)\rangle\equiv\sum_{j=1}^{n}C_j\omega_i(y_j)\omega_k(y_j)=c_n\delta_{ik}$$

or

$$\langle\omega_i'(x)*\omega_k'(x)\rangle\equiv\sum_{j=1}^{n}D_j\omega_i'(z_j)\omega_k'(z_j)=d_n\delta_{ik}.$$

We write the derivative of the Fourier series expansion of $F'(x)$ for $|x|\leqslant 1$ as

$$F'(x)=\sum_{k=1}^{\infty}{}' b_k\omega_k'(x) \tag{7.7.4}$$

and hence obtain the coefficients b_k as

$$b_k=\frac{(F'(x)*\omega_k'(x))}{(\omega_k'(x)*\omega_k'(x))}=\frac{(W(x)f(x)*\omega_k'(x))}{(\omega_k'(x)*\omega_k'(x))},\quad\text{for}\quad k=1(1)\infty$$

or, on truncating (7.7.4), as

$$\begin{aligned}b_k&=\frac{\langle W(x)f(x)*\omega_k'(x)\rangle}{\langle\omega_k'(x)*\omega_k'(x)\rangle},\quad\text{for}\quad k=1(1)n\\&=\frac{1}{\langle\omega_k'(x)*\omega_k'(x)\rangle}\sum_{i=1}^{n-1}\hat{A}_iW(\hat{x}_i)f(\hat{x}_i)\omega_k'(\hat{x}_i).\end{aligned}$$

Thus we are able to compute the coefficients of the truncated series expansion $f_{n-1}(x)=\sum_{k=1}^{n}{}' b_k\omega_k'(x)$ without any analytic integration. The above two techniques based upon the expansion of f and F can be characterised as follows.

SCHEME 1

$$f(x)=\sum_{k=0}^{\infty}{}' a_k\omega_k(x)\xrightarrow{\text{integration}}F(x)=F(-1)+\int_{-1}^{x}f\,dt=F(-1)+\sum_{k=0}^{\infty}{}' a_k\int_{-1}^{x}\omega_k\,dt$$

best approximation by truncation ↓ (from $f(x)$ to $f_{n-1}(x)$)

finite not best approximation ↕ (between $F(x)$ and $F_{n-1}(x)$)

$$f_{n-1}(x)\xrightarrow[\text{integration}]{}F_{n-1}(x)$$

SCHEME 2

$$f(x)=\sum_{k=0}^{\infty}{}' b_k\omega_k'(x) \xleftarrow{\text{differentiation}} F(x)=F(-1)+\sum_{k=0}^{\infty}{}' b_k\omega_k(x)$$

not best approximation by truncation ↕ (left) best approximation by truncation ↓ (right)

$$f_{n-1}(x) \xleftarrow[\text{differentiation}]{} F_{n-1}(x)$$

The truncated Fourier series expansions of f giving f_n produces a best approximation of f in the L_2-norm so that

$$\|f-f_n\|^2=(f-f_n, f-f_n)=\int_{-1}^{1} \tilde{W}(x) \sum_{k=n+1}^{\infty} a_k^2\omega_k^2(x)\,dx = \sum_{k=n+1}^{\infty} a_k^2$$

where ω_k are *normalised*. If $p_n \in \mathbb{P}_n$ is another approximation of f, then $p_n(x)$ has an expansion of the form

$$p_n(x)=\sum_{k=0}^{n}{}' c_k\omega_k(x)$$

so that

$$\begin{aligned}\|f-p_n\|^2&=(f-p_n, f-p_n)\\ &=\int_{-1}^{1} \tilde{W}(x)\left(\sum_{k=0}^{n} (a_k-c_k)\omega_k(x)+\sum_{i=n+1}^{\infty} a_k\omega_k(x)\right)^2 dx\\ &=\sum_{k=0}^{n}{}' (a_k-c_k)^2+\sum_{k=n+1}^{\infty} a_k^2 \geqslant \sum_{k=n+1}^{\infty} a_k^2=\|f-f_n\|^2.\end{aligned}$$

Hence $|E_nf|=|If-Q_nf|\leqslant|I(f-f_n)|\leqslant\|I\|\,\|f-f_n\|$, for the case of upper limit $x=1$ and the error estimate

$$|E_nf|\leqslant\left(\int_{-1}^{1} W(x)\,dx \cdot \sum_{k=n+1}^{\infty} a_k^2\right)^{\frac{1}{2}}$$

is obtained. The same arguments are valid for the expansion of $F(x)$ and F_n respectively. Hence, convergence in L_2-norm of f_n to f and F_n to F follows from the convergence of the Fourier series expansions. If the upper limit $x=1$ is fixed, we have quadrature procedures of the normal type. In general, for integrals of this type two schemes lead to quadrature formulae which are defined respectively as

$$Q_nf = \sum_{k=0}^{n-1}{}' a_k \int_{-1}^{1} W(x)\omega_k(x)\,\mathrm{d}x = \sum_{k=0}^{n-1}{}' \frac{\langle f(x), \omega_k(x)\rangle}{\langle \omega_k(x), \omega_k(x)\rangle} \int_{-1}^{1} W(x)\omega_k(x)\,\mathrm{d}x$$

$$= \sum_{i=1}^{n} \left[\sum_{k=0}^{n-1}{}' \frac{A_i\omega_k(x_i)}{\langle \omega_k(x), \omega_k(x)\rangle} \int_{-1}^{1} W(x)\omega_k(x)\,\mathrm{d}x \right] f(x_i) \qquad (7.7.5)$$

and

$$Q_{n-1}f = \sum_{k=0}^{n-1}{}' b_k\omega_k(1) = \sum_{k=0}^{n-1}{}' \frac{\langle W(x)f(x) * \omega_k'(x)\rangle}{\langle \omega_k'(x) * \omega_k'(x)\rangle} \omega_k(1)$$

$$= \sum_{i=1}^{n-1} \left[\sum_{k=0}^{n-1}{}' \frac{\hat{A}_i W(\hat{x}_i)\omega_k'(\hat{x}_i)}{\langle \omega_k'(x) * \omega_k'(x)\rangle} \omega_k(1) \right] f(\hat{x}_i) \qquad (7.7.6)$$

which are based on different sets of nodes and which have polynomial accuracy. It is not generally possible to say whether or not they are positive, and hence proof of convergence is not straightforward except in the L_2-norm.

7.7.2 Applications for Jacobi polynomials

As an example for the above theory we treat the Jacobi polynomials $P_n^{\alpha,\beta}(x)$ and, for simplicity, omit the superscripts α and β. The weight function $\tilde{W}(x)$ in this case has the form

$$\tilde{W}(x) = (1-x)^\alpha (1+x)^\beta, \qquad \alpha, \beta > -1.$$

The polynomial $P_n(x)$ is normalised by making $P_n(1) = \binom{n+\alpha}{\alpha}$ so that

$$(P_k, P_k) = \frac{2^{\alpha+\beta+1}\Gamma(k+\alpha+1)\Gamma(k+\beta+1)}{(2k+\alpha+\beta+1)k!\,\Gamma(k+\alpha+\beta+1)} = \langle P_k, P_k\rangle, \qquad k = 0(1)n-1.$$

In the first scheme, the Fourier coefficients of the truncated series (7.7.2)

$$a_k = \frac{\langle f(x), P_k(x)\rangle}{\langle P_k(x), P_k(x)\rangle} = \frac{\langle f(x), P_k(x)\rangle}{(P_k(x), P_k(x))}$$

$$= \frac{(2k+\alpha+\beta+1)k!\,\Gamma(k+\alpha+\beta+1)}{2^{\alpha+\beta+1}\Gamma(k+\alpha+1)\Gamma(k+\beta+1)} \sum_{i=1}^{n} A_i P_k(x_i) f(x_i),$$

and a sufficient condition for the explicit quadrature formula $Q_n f$ to be positive is that

$$\sum_{k=0}^{n-1}{}' \left[P_k(x_i)\int_{-1}^{1} W(x)P_k(x)\,dx\right] \geq 0$$

(see formula (7.7.5))

with $A_i > 0$ and $\langle P_k(x), P_k(x)\rangle > 0$. The second scheme is governed by the linear differential equation for the Jacobi polynomials (see Problem P.5.8)

$$\tilde{W}(x)P_k(x) + \frac{1}{k(k+\alpha+\beta+1)}\frac{d}{dx}[(1-x^2)\tilde{W}(x)P'_k(x)] = 0. \quad (7.7.7)$$

Hence $\hat{W}(x) = (1-x^2)\tilde{W}(x)$, $B_k = \dfrac{1}{k(k+\alpha+\beta+1)}$ and

$$\begin{aligned}\langle P'_k * P'_k\rangle &= k(k+\alpha+\beta+1)(P_k, P_k)\\ &= \frac{k(k+\alpha+\beta+1)2^{\alpha+\beta+1}\Gamma(k+\alpha+1)(\Gamma(k+\beta+1)}{(2k+\alpha+\beta+1)k!\,\Gamma(k+\alpha+\beta+1)}.\end{aligned}$$

The Fourier coefficients b_k of the truncated series are

$$\begin{aligned}b_k &= \frac{\langle W(x)f(x) * P'_k(x)\rangle}{\langle P'_k(x) * P'_k(x)\rangle}\\ &= \frac{(2k+\alpha+\beta+1)k!\,\Gamma(k+\alpha+\beta+1)}{k(k+\alpha+\beta+1)2^{\alpha+\beta+1}\Gamma(k+\alpha+1)\Gamma(k+\beta+1)}\sum_{i=1}^{n-1}\hat{A}_i W(\hat{x}_i)P'_k(\hat{x}_i)f(\hat{x}_i).\end{aligned}$$

The sufficient conditions for the positivity of $Q_{n-1}f$, i.e.

$$\sum_{k=0}^{n-1}{}' P'_k(\hat{x}_i)P_k(1) \geq 0 \qquad \text{(see formula (7.7.6))}$$

are that $\hat{A}_i > 0$, $W(\hat{x}_i) > 0$, and $\langle \omega'_k(x) * \omega'_k(x)\rangle > 0$.

7.7.3 Applications for Tschebyscheff polynomials

The results of Section 7.7.2 can be specialised to Tschebyscheff polynomials by taking $\alpha = \beta = -\frac{1}{2}$. In this case $\omega_n(x) = P_n^{-\frac{1}{2},-\frac{1}{2}}(x) = T_n(x)$ are the Tschebyscheff polynomials of the first kind with respect to $\tilde{W}(x) = 1/(1-x^2)^{\frac{1}{2}}$. Because of the factor $1-x^2$ in (7.7.7) $\hat{W}(x) = (1-x^2)^{\frac{1}{2}}$ and, hence, $\omega'_n(x) = P_{n-1}^{\frac{1}{2},\frac{1}{2}}(x) = U_{n-1}(x)$ are Tschebyscheff polynomials of the second kind. Thus

$$\langle T_k, T_k\rangle = \int_{-1}^{1} (1-x^2)^{-\frac{1}{2}} T_k^2(x)\,dx = \int_0^{\pi} \cos^2(kt)\,dt = \begin{cases} \pi/2 & \text{for } k \neq 0 \\ \pi & \text{for } k = 0 \end{cases} \tag{7.7.8}$$

and the Fourier coefficients are

$$\begin{aligned} a_k &= \frac{2}{\pi}\langle f, T_k\rangle = \frac{2}{\pi}\sum_{i=1}^{n} f(x_i)T_k(x_i) \\ &= \frac{2}{\pi}\sum_{i=1}^{n} f\left(\cos\frac{2i-1}{2n}\pi\right)T_k\left(\cos\frac{2i-1}{2n}\pi\right), \qquad k = 0(1)n, \end{aligned}$$

with respect to the zeros of $T_k(x) = \cos(k \arccos x)$. Since

$$T_k\left(\cos\frac{2i-1}{2n}\pi\right) = \cos\left(k \arccos\left(\cos\frac{2i-1}{2n}\pi\right)\right) = \cos\left(\frac{2i-1}{2n}k\pi\right),$$

the coefficients a_k become

$$a_k = \frac{2}{\pi}\sum_{i=1}^{n} \cos\left(\frac{2i-1}{2n}k\pi\right) f\left(\cos\frac{2i-1}{2n}\pi\right).$$

Equation (7.7.8), however, is not the only possible choice for an inner product with the discrete orthogonality property of $T_k(x)$. By replacing $x = \cos t$ we have that $T_k(\cos t) = T_k(x) = \cos(kt)$ and the discrete inner product

$$\langle f \mid g\rangle \equiv \sum_{i=0}^{n}{}'' f\left(\frac{\pi i}{n}\right) g\left(\frac{\pi i}{n}\right)$$

with respect to t can be defined which, for the $T_k(x)$'s, is

$$\langle T_k \mid T_m\rangle = \sum_{i=0}^{n}{}'' \cos\left(k\frac{\pi i}{n}\right)\cos\left(m\frac{\pi i}{n}\right) = \begin{cases} 0 & \text{for } k \neq m, \quad 0 < k, m < n \\ n & \text{for } k = m = 0, n \\ n/2 & \text{for } 0 < k = m < n \end{cases}$$

(see Problem P.5.5). This discrete inner product governs the Clenshaw–Curtis quadrature scheme in Section 7.7.4.

The corresponding coefficients a_k of the truncated Fourier series are given by $a_k = (2/n)\langle f(\cos t) | T_k(\cos t)\rangle$, $k = 0(1)n$ which, on substitution of $T_k(\cos \pi i/n) = \cos(k \arccos(\cos \pi i/n)) = \cos(k\pi i/n)$ becomes $a_k = 2/n \sum_{i=0}^{n}{}'' \cos(k\pi i/n) f(\cos(\pi i/n))$ and, thus, the truncated Fourier series is

$f_n(x) = \sum_{k=0}^{n}{}' a_k T_k(x)$. From the differential equation (7.7.7) specified to the Tschebyscheff polynomials

$$T_k(x) + \frac{1}{k^2}(1-x^2)^{\frac{1}{2}} \frac{\mathrm{d}}{\mathrm{d}x}[(1-x^2)^{\frac{1}{2}} T_k'(x)] = 0$$

we see that $B_k = 1/k^2$ and $\hat{W} = (1-x^2)^{\frac{1}{2}}$ and, hence, that

$$b_k = \frac{2}{\pi k^2}\langle f * T_k' \rangle = \frac{2}{\pi k^2} \sum_{i=1}^{n-1} \hat{A}_i T_k'(\hat{x}_i) f(\hat{x}_i).$$

Using the property of Tschebyscheff polynomials that

$$T_k'(x) = kU_{k-1}(x) = k\frac{\sin(k \arccos x)}{(1-x^2)^{\frac{1}{2}}} = k\frac{\sin kt}{\sin t},$$

with $x = \cos t$, we obtain the zeros to be $t_i = i\pi/n$, $i = 1(1)n-1$ or, equivalently, $x_i = \cos(\pi i/n)$, $i = 1(1)n-1$. Hence, since

$$(T_k', T_k') = k^2(T_k, T_k) = \tfrac{1}{2}\pi k^2, \qquad k = 1(1)\ldots,$$

we obtain that the coefficients of the truncated Fourier series are

$$b_k = \frac{2}{\pi k^2}\langle f(x) * T_k'(x) \rangle = \frac{2}{\pi k} \sum_{i=1}^{n-1} \hat{A}_i U_{k-1}\left(\cos\frac{\pi i}{n}\right) f\left(\cos\frac{\pi i}{n}\right)$$

and, therefore, that

$$F_{n-1}(x) = F(-1) + \sum_{k=1}^{n-1}{}' b_k T_k'(x).$$

An alternative choice of a discrete inner product is

$$\langle f \circ g \rangle \equiv \sum_{k=1}^{n} f\left(\frac{k\pi}{n+1}\right) g\left(\frac{k\pi}{n+1}\right)$$

which corresponds to the composite rectangle rule. In this case the functions $g_m(t) = \sin mt$, $m = 1(1)n$ are orthogonal with respect to $\langle \cdot \circ \cdot \rangle$. The new function $G_n(t)$ is defined as

$$G_n(\cos t) \equiv F_n(\cos t) - F(-1) = \sum_{k=1}^{n} kb_k \frac{\sin kt}{\sin t},$$

and the coefficients b_k of

$$\sin t \cdot G_n(\cos t) = \sum_{k=1}^{n} kb_k \sin kt$$

can then be obtained by using

$$\langle \sin t \cdot G_n(\cos t) \circ \sin mt \rangle = \sum_{k=1}^{n} k b_k \langle \sin kt \circ \sin mt \rangle = m b_m (n+1)/2 \qquad (7.7.9)$$

The above equation follows (see Problem P.7.4) from the fact that

$$\langle \sin lt \circ \sin mt \rangle = \sum_{k=1}^{n} \sin \frac{\pi lk}{n+1} \sin \frac{\pi mk}{n+1}$$
$$= \begin{cases} 0 & \text{for} \quad l \neq m, \quad 0 \leq l, \quad m \leq n+1 \\ \dfrac{n+1}{2} & \text{for} \quad l = m, \quad 0 \leq l, \quad m \leq n+1. \end{cases}$$

The coefficients b_k are found in the usual way from (7.7.9) as

$$b_m = \frac{2}{m(n+1)} \sum_{k=1}^{n} \sin \frac{k\pi}{n+1} \sin \frac{k\pi m}{n+1} G_n\left(\cos \frac{k\pi}{n+1}\right).$$

The Clenshaw–Curtis scheme has previously been mentioned and is described in Section 7.7.4. The discrete inner product $\langle \cdot \circ \cdot \rangle$ is used in Filippi's modified Clenshaw–Curtis scheme, described in Section 7.7.5.

7.7.4 The Clenshaw–Curtis quadrature scheme (CC scheme)

The basic algorithm

The Clenshaw–Curtis quadrature scheme is structured in the manner of the first scheme in Section 7.7.1. It is based upon the Tschebyscheff polynomials $T_k(x)$ of the first kind, together with the discrete inner product $\langle \cdot \mid \cdot \rangle$ of Section 7.7.3. The integrand $f(x)$ is thus approximated by its series expansion in terms of Tschebyscheff polynomials so that $f(x) = \sum_{k=0}^{\infty}{}' a_k T_k(x)$. This series converges for every f with bounded variation in $[-1, 1]$ and hence *a fortiori* for every $f \in C^0[-1, 1]$. On integration, we have that

$$F(x) = F(-1) + \int_{-1}^{x} f(t)\,dt = F(-1) + \sum_{k=0}^{\infty}{}' a_k \int_{-1}^{x} T_k(t)\,dt.$$

The function $F(x)$ also possesses a convergent series expansion in terms of Tschebyscheff polynomials of the form

$$F(x) = F(-1) + \sum_{k=0}^{\infty}{}' \alpha_k T_k(x).$$

The coefficients a_k and α_k are connected by the relationship

$$2k\alpha_k = a_{k-1} - a_{k+1}, \qquad k = 1(1)\ldots$$

which is seen as follows. By definition, $F'(x) = f(x)$ and, hence,

$$\sum_{k=1}^{\infty} \alpha_k T_k'(x) = \sum_{k=0}^{\infty}{}' a_k T_k(x) \quad \text{(note that } T_0' \equiv 0\text{).} \tag{7.7.10}$$

Recalling that $T_k' = kU_{k-1}$, while $x = \cos t$, then

$$U_k(x) = \frac{\sin(k+1)t}{\sin t} = \frac{\sin kt \cos t + \cos kt \sin t}{\sin t} = xU_{k-1}(x) + T_k(x)$$

with the analogous relationship that $U_{k-2}(x) = xU_{k-1}(x) - T_k(x)$ from which it follows that

$$T_k(x) = \tfrac{1}{2}(U_k(x) - U_{k-2}(x)) = \frac{1}{2}\left[\frac{T_{k+1}'}{k+1} - \frac{T_{k-1}'}{k-1}\right].$$

We now replace $T_k(x)$ in the series expansion (7.7.10) so that

$$\begin{aligned}\sum_{k=1}^{\infty} \alpha_k T_k' &= \tfrac{1}{2}a_0 + a_1 T_1 + \frac{1}{2}\sum_{k=2}^{\infty} a_k \left[\frac{T_{k+1}'}{k+1} - \frac{T_{k-1}'}{k-1}\right] \\ &= \tfrac{1}{2}a_0 + a_1 T_1 + \frac{1}{2}\sum_{k=3}^{\infty} \frac{1}{k} a_{k-1} T_k' - \frac{1}{2}\sum_{k=1}^{\infty} \frac{1}{k} a_{k+1} T_k'.\end{aligned}$$

Comparison of the coefficients of T_k' shows that

$$\alpha_k = \frac{1}{2k}(a_{k-1} - a_{k+1}) \quad \text{for} \quad k = 3(1)\ldots$$

and, further, since $T_0 = 1$, $T_1 = x$, $T_2 = 2x^2 - 1$ we have that

$$\alpha_1 T_1' + \alpha_2 T_2' = \tfrac{1}{2}a_0 T_1' + \tfrac{1}{4}a_1 T_2' - \tfrac{1}{2}a_2 T_1' - \tfrac{1}{4}a_3 T_2',$$

and hence $\alpha_1 = \frac{1}{2}(a_0 - a_2)$, $\alpha_2 = \frac{1}{4}(a_1 - a_3)$. Thus $\alpha_k = (1/2^k)(a_{k-1} - a_{k+1})$ for $k = 1(1)\ldots$. The coefficient α_0 is finally determined from the initial condition

$$0 = F(-1) = \sum_{k=0}^{\infty}{}' \alpha_k T_k(-1) = \sum_{k=0}^{\infty}{}' \alpha_k (-1)^k,$$

so that

$$\alpha_0 = 2[\alpha_1 - \alpha_2 + \alpha_3 - + \ldots] = 2\sum_{k=1}^{\infty} (-1)^{k+1}\alpha_k \tag{7.7.11}$$

and we are able to approximate $F(x)=F(-1)+\int_{-1}^{x} f(t)\,dt$ for every $x\in[-1,1]$ by evaluating the series (7.7.11). In the case $x=1$ we have that $F(1)=F(-1)+\int_{-1}^{1} f(t)\,dt=\sum_{k=0}^{\infty}{}' \alpha_k T_k(1)=\sum_{k=0}^{\infty}{}' \alpha_k=2\sum_{k=0}^{\infty}\alpha_{2k+1}$ where the last result is obtained by substituting for α_0 in the series (7.7.11). Hence

$$\int_{-1}^{1} f(x)\,dx=2\sum_{k=0}^{\infty}\alpha_{2k+1}.$$

In numerical computations we cannot evaluate infinitely many terms. We therefore truncate the series and use the discrete inner product $\langle\cdot\mid\cdot\rangle$ of Section 7.7.3 in order to compute the coefficients. Thus, in the truncated series $f_n=\sum_{k=0}^{n}{}' a_k T_k(x)$ the coefficients are

$$a_k=\frac{2}{n}\langle f\mid T_k\rangle=\frac{2}{n}\sum_{r=0}^{n}{}''\cos\frac{\pi kr}{n}\,f\!\left(\cos\frac{\pi r}{n}\right),\qquad k=0(1)n,$$

which produces a quadrature formula for $x=1$ by substituting for a_k.

Some properties of the CC scheme

As the explicit form of the quadrature formula has not yet been given, this is now done. On integration of the truncated series (7.7.10) we have that

$$\int_{-1}^{1} f(x)\,dx \doteq Q_n f=\sum_{k=0}^{n}{}' a_k\int_{-1}^{1} T_k(x)\,dx$$

in which the T_{2k+1}, $k=1(1)\ldots$ are odd functions, so that their integrals vanish. For every even index it is known (see Problem P.5.7) that

$$\int_{-1}^{1} T_{2k}(x)\,dx=\frac{-2}{4k^2-1},\qquad k=0(1)\ldots.$$

The final quadrature is

$$Q_n f=a_0-2\sum_{k=1}^{[n/2]}\frac{1}{4k^2-1}\,a_{2k}=\frac{2}{n}\sum_{r=0}^{n}{}'' f\!\left(\cos\frac{\pi r}{n}\right)\left[1-2\sum_{k=1}^{[n/2]}\frac{1}{4k^2-1}\cos\frac{2\pi rk}{n}\right]$$

with weights

$$A_r\equiv\frac{2}{n}\left[1-2\sum_{k=1}^{[n/2]}\frac{1}{4k^2-1}\cos\frac{2\pi rk}{n}\right].$$

It is found that the A_r are positive, since

$$\left|\sum_{k=1}^{[n/2]} \frac{1}{4k^2-1} \cos \frac{2\pi rk}{n}\right| \leqslant \sum_{k=1}^{[n/2]} \frac{1}{4k^2-1} < \sum_{k=1}^{\infty} \frac{1}{4k^2-1} = \tfrac{1}{2}$$

(cf. Ryshik and Gradstein p. 8, [17]). Conversely,

$$\sum_{k=1}^{[n/2]} \frac{1}{4k^2-1} \cos \frac{2\pi rk}{n} \geqslant -\sum_{k=1}^{[n/2]} \frac{1}{4k^2-1} > -\sum_{k=1}^{\infty} \frac{1}{4k^2-1} = -\tfrac{1}{2}$$

thus providing the inclusion $0 < A_r < 4/n$. Since every polynomial has a finite and unique expansion in terms of Tschebyscheff polynomials for n large enough, the Clenshaw–Curtis scheme integrates any polynomial exactly. Hence, we have convergence for every $f \in C^0[-1, 1]$ and the following theorem is valid:

Theorem 7.7.1 *The Clenshaw–Curtis scheme is a positive quadrature procedure which converges for every $f \in C^0[-1, 1]$.*

Numerical performance

Assume, first, that $\{f_n\}$ *is a rapidly-convergent sequence.* An arbitrary value of n is then chosen and tested to see whether this value produces an approximation of sufficient accuracy. We therefore compare the three coefficients α_n, α_{n-1}, and α_{n-2}. Clenshaw and Curtis [3] propose that the value of n be accepted, provided that α_n, $k\alpha_{n-1}$, $k^2\alpha_{n-2}$ are each negligible for $k \in (0, 1)$. From experience, they propose the value $k = 1/8$. If α_n, $k\alpha_{n-1}$ and $k^2\alpha_{n-1}$ are not all negligible, we replace n by $2n$, calculate α_{2n}, $k\alpha_{2n-1}$, $k^2\alpha_{2n-2}$ proceeding in the same way. In going from n to $2n$ we again use the values of $f(x)$ calculated previously. These are the alternate members of the set $f(x_k) = f \cos(k\pi/2n)$, $k = 0(1)2n$. It is normal to start the computation with $n = 4$ and then to proceed to $n = 8$, $n = 16$ etc. with an upper bound N on n which is specified by the user. If the check fails even for $n = N$, we assume that $\{f_n\}$ *is a slowly-convergent sequence.* In this situation, suppose that f possesses a convergent infinite series expansion

$$f(x) = \sum_{k=0}^{\infty}{}' A_k T_k(x). \tag{7.7.12}$$

Then, as before,

$$a_m = \frac{2}{n} \langle f \mid T_m \rangle = \frac{2}{n} \sum_{k=0}^{n}{}'' \cos \frac{\pi mk}{n} \sum_{i=0}^{\infty}{}' A_i \cos \frac{\pi ik}{n} \tag{7.7.13}$$

which, on account of the discrete orthogonality of $\cos r\pi x$, $r=0(1)\ldots$, can be written as

$$a_m = A_m + A_{2n+m} + A_{2n-m} + A_{4n+m} + A_{4n-m} + \ldots$$

$$= \sum_{k=0}^{\infty} (A_{2kn+m} + A_{2kn-m}) \quad \text{with} \quad A_m = A_{-m}. \tag{7.7.14}$$

Hence

$$If = A_0 - 2\sum_{k=1}^{\infty} \frac{1}{4k^2-1} A_{2k}$$

and

$$E_{2n}f = If - Q_{2n}f = A_0 - a_0 + 2\sum_{k=1}^{n} \frac{1}{4k^2-1} a_{2k} - 2\sum_{k=1}^{\infty} \frac{1}{4k^2-1} A_{2k}.$$

Substituting for (7.7.13) the coefficients a_i by (7.7.14), this becomes

$$E_{2n}f = -\sum_{k=1}^{\infty} 2A_{2kn} + 2\sum_{k=1}^{n} \frac{1}{4k^2-1} \sum_{j=1}^{\infty} (A_{2jn+2k} + A_{2jn-2k})$$
$$-2\sum_{k=n+1}^{\infty} \frac{1}{4k^2-1} A_{2k}$$

which can be truncated to give

$$E_{2n}f \doteq -2A_{2n} + 2\sum_{k=1}^{[n/2]-1} \frac{1}{4k^2-1} (A_{2n+2k} + A_{2n-2k})$$
$$-2\sum_{k=1}^{[n/2]-1} \frac{1}{(2n+2k)^2-1} A_{2n+2k},$$

in which only the coefficients $A_{n+2}, \ldots A_{3n-2}$ appear. It follows from the convergence of (7.7.12) that we can bound A_m by

$$|A_m| \leqslant \frac{1}{m} K_n \quad \text{for} \quad m > n \tag{7.7.15}$$

where K_n is independent of m. Assume that the bound K_n is taken as small as possible. Then

$$|A_{2n+2k} + A_{2n-2k}| \leqslant \tfrac{1}{2} K_n \left(\frac{1}{n+k} + \frac{1}{n-k} \right) = \frac{n}{n^2-k^2} K_n$$

and, since $k<[n/2]$, we have that

$$|A_{2n+2k}+A_{2n-2k}|\leqslant\frac{1}{n}K_n\sum_{m=0}^{\infty}\left(\frac{k}{n}\right)^{2m}<\frac{1}{n}K_n\left[1+\left(\frac{k}{n}\right)^2\sum_{m=0}^{\infty}(\tfrac{1}{4})^m\right]$$
$$=\frac{1}{n}\left(1+\frac{4k^2}{3n^2}\right)K_n.$$

Finally, amalgamation of these results gives that

$$|E_{2n}f|<\frac{1}{n}\left\{1+2\sum_{k=1}^{[n/2]-1}\left[\frac{1}{4k^2-1}\left(1+\frac{4k^2}{3n^2}\right)+\frac{1}{(2n+2k)^2-1}\right]\right\}K_n$$
$$<\frac{1}{n}\left\{1+2\sum_{k=1}^{\infty}\frac{1}{4k^2-1}+\frac{2}{3n^2}\sum_{k=1}^{[n/2]-1}\left[\frac{4k^2}{4k^2-1}+\frac{3(n+k)^2}{4(n+k)^2-1}\right]\right\}K_n$$
$$<\frac{1}{n}\left\{2+\frac{2}{3n^2}2\left(\frac{n}{2}+\sum_{k=1}^{\infty}\frac{1}{4k^2-1}\right)\right\}K_n=\frac{2}{n}\left(1+\frac{1}{3n}+\frac{1}{3n^2}\right)K_n$$

where, for large n, the terms $1/3n$ and $1/3n^2$ can be neglected. The error is thus estimated by $|E_{2n}f|\approx(2/n)K_n$.

Equation (7.7.11) shows that $K_n\geqslant m\,|A_m|>n\,|A_m|$ for $m>n$ so that $|E_{2n}f|\approx 2\,|A_m|$ for $m>n$ where the A_m can be expressed in terms of the coefficients a_i which, in turn, can be represented in terms of α_i.

***Example* 7.7.1** Approximate the integral $\int_{-1}^{1}\frac{dx}{1+x^2}=2\text{ arc tg}1=$ 1.570 796 327. For $n=4$ we evaluate the integrand at $x_i=\cos(\pi i/4)$, $i=0(1)4$ and obtain from $a_k=\frac{2}{n}\sum_{i=0}^{n}{}''\frac{\cos(\pi ik/n)}{1+\cos^2(\pi i/n)}$ for $k=0(1)n$ and $a_{n+1}=0$ first that

$a_4=8.333\bar{3}(-2)$ and, hence, that $\alpha_5=8.333\bar{3}(-3)$.

These coefficients are clearly not small enough so that we proceed to $n=8$ and find that

$a_8=2.45098(-3)$ and, hence, that $\alpha_9=1.36166(-4)$

which, again, is not acceptable. Doubling the number of points once more we have that for $n=16$

$a_{16}=2.1238(-6)$ and, hence, that $\alpha_{17}=6.2(-8)$

$a_{14}=-6.373(-6)$ $\quad\frac{1}{8}\alpha_{15}=-3.5(-8)$

$a_{12}=3.6106(-5)$ $\quad\frac{1}{64}\alpha_{13}=2.6(-8)$

These coefficients are sufficiently small so that $2\sum_{i=0}^{8}\alpha_{2i+1}=1.570796320$

results. The error is $-7.0(-9)$ and arises solely on account of round off.

TABLE 7.7.1

i	a_i	α_i
17		6.2(−8)
16	2.124(−6)	
15		−2.83(−7)
14	−6.373(−6)	
13		1.634(−6)
12	3.6106(−5)	
11		−1.1199(−5)
10	−2.10266(−4)	
9		7.9764(−5)
8	1.225490(−3)	
7		−5.97726(−4)
6	−7.142675(−3)	
5		4.8773236(−2)
4	4.1630560(−2)	
3		−4.7378541(−2)
2	−2.42640687(−1)	
1		8.28427125(−1)
0	1.414213563	

Comparison with classical Gauss quadrature formulae gives

$n=5$ error 3.75(−4)

$n=9$ error 3.28(−7)

$n=17$ round-off errors only.

In general it seems that the Gauss quadrature formulae are more accurate for a fixed number of nodes, but the CC procedure gives an additional reliable test for finding the appropriate number of nodes. ■

Program

We present a program which evaluates the integral $F(x)$ of a function $f(x)$ such that $F(x)=\int_a^x f(t)\,dt$, $x\in[a, b]$ by the CC scheme.

```
      SUBROUTINE CCQ(F, A, B, N, C, M, XTAB, QTAB)
C
C
C BY THIS PROGRAM THE INTEGRAL OF F(X) IS
C TABULATED FOR THE INTERVAL (A, B) USING THE
C CC QUADRATURE SCHEME
```

```
C
C  INPUT:
C   F: INTEGRAND
C   A: LEFT LIMIT OF INTEGRATION
C   B: RIGHT LIMIT OF INTEGRATION
C   N: NUMBER OF NODES
C   C: ARRAY OF LENGTH N
C   M: LENGTH OF THE TABLE
C   XTAB: ARRAY OF LENGTH M FOR THOSE POINTS
C   AT WHICH THE INTEGRATED FUNCTION
C   MUST BE TABULATED
C  OUTPUT:
C   QTAB: ARRAY OF LENGTH M FOR THE VALUES
C   OF THE TABULATED FUNCTION
C
C
C
      EXTERNAL F
      DIMENSION C(N), XTAB(M), QTAB(M)
      DATA PI/3.141592653589793/
      BPA2=(B+A)/2
      IF(N.EQ.1) GOTO 1Ø
      BMA2=(B-A)/2
      NM=N-1
      PIDNM=PI/NM
      DO 1 I=1, N
    1 C(I)=Ø
      X=Ø
      DO 3 I=1, N
      Y=F(BMA2×COS (X)+BPA2)/NM
      IF(I.EQ.1.OR.I.EQ.N) Y=Y/2
      Z=Ø
      DO 2 J=1, N
      C(J)=C(J)+Y×COS (Z)
    2 Z=Z+X
    3 X=X+PIDNM
      C(N)=C(N)/2
      IF(N.EQ.2) GOTO 6
      J=NM-1
      DO 5 I=1, J
    5 C(I)=(C(I)-C(I+2))/I
    6 C(NM)=C(NM)/NM
```

```
      C(N)=C(N)/N
      CØ=C(N)
      DO 7 I=1, NM
    7 CØ=C(N-I)-CØ
      DO 9 J=1, M
      T1=(XTAB(J)-BPA2)/BMA2
      Y=CØ+T1×C(1)
      TØ=1
      X=T1+T1
      DO 8 I=2, N
      T2=X×T1-TØ
      Y=Y+C(I)×T2
      TØ=T1
    8 T1=T2
    9 QTAB(J)=Y×BMA2
      RETURN
   10 DO 11 J=1, M
   11 QTAB(J)=F(BPA2)×(XTAB(J)-A)
      RETURN
      END
```

7.7.5 Filippi's modified CC scheme (FCC scheme)

The FCC scheme

In this scheme we expand the function $F(x)$

$$F(x) \equiv F(-1)+\int_{-1}^{x} f(t)\,\mathrm{d}t, \qquad |x| \leqslant 1,$$

as a series in terms of Tschebyscheff polynomials $T_k(x)$. Thus

$$F(x)=\sum_{l=0}^{\infty}{}' \, b_k T_k(x) \tag{7.7.16}$$

and the coefficients b_k are determined from

$$b_k=\frac{2}{\pi}\int_{-1}^{1}\frac{F(x)T_k(x)}{(1-x^2)^{\frac{1}{2}}}\,\mathrm{d}x, \qquad k=0(1)\ldots.$$

Using the differential equation (see Section 7.7.3)

$$\frac{1}{(1-x^2)^{\frac{1}{2}}}\,T_k(x)=\frac{-1}{k^2}\frac{\mathrm{d}}{\mathrm{d}x}[(1-x^2)^{\frac{1}{2}}T_k'(x)], \qquad k \geqslant 1$$

and integrating by parts, we obtain $(F'(x)=f(x))$

$$b_k = -\frac{2}{\pi} F(x) \frac{1}{k^2} (1-x^2)^{\frac{1}{2}} T_k'(x)\Big|_{-1}^{1} + \frac{2}{\pi k^2} \int_{-1}^{1} f(x) T_k'(x)(1-x^2)^{\frac{1}{2}}\, dx$$

$$= \frac{2}{\pi k^2} \int_{-1}^{1} (1-x^2)^{\frac{1}{2}} f(x) T_k'(x)\, dx, \qquad k=1(1)\ldots$$

and

$$b_0 = \frac{2}{\pi} \int_{-1}^{1} F(x) \frac{dx}{(1-x^2)^{\frac{1}{2}}} = \frac{2}{\pi} \int_{-1}^{1} f(x) \arccos x\, dx.$$

The polynomials $T_k'(x) = kU_{k-1}(x)$ are Tschebyscheff polynomials of the second kind and hence are orthogonal, i.e. $\int_{-1}^{1} (1-x^2)^{\frac{1}{2}} T_k'(x) T_m'(x)\, dx = (\pi/2) m^2 \delta_{km}$. The series (7.7.16) is truncated to give $F_n(x) = \sum_{k=0}^{n}{}' b_k T_k(x)$ (where $F_n(x)$ approximates $F(x)$) and hence a best approximation

$$\|F(x) - F_n(x)\|^2 = \left\| \sum_{k=n+1}^{\infty} b_k T_k(x) \right\|^2 = \sum_{k=n+1}^{\infty} \frac{\pi}{2} b_k^2$$

where $\|f\|^2 = \int_{-1}^{1} \frac{1}{(1-x^2)^{\frac{1}{2}}} f^2(x)\, dx$. Clearly, $F_n'(x)$ is not a best approximation to $f(x)$ in the L_2-norm; it is in the sense that

$$\int_{-1}^{1} (f(x) - F_n'(x))^2 (1-x^2)^{\frac{1}{2}}\, dx = \min.$$

If we now expand $f(x)$ in terms of $T_k'(x)$ and truncate, then $f_n(x) = \sum_{k=1}^{n} \beta_k T_k'(x) \in \mathbb{P}_{n-1}$ or, on integration, $F_n(x) = \sum_{k=0}^{n}{}' \beta_k T_k(x) \in \mathbb{P}_n$ where β_0 is determined by $0 = F_n(-1) = \sum_{k=0}^{n}{}' \beta_k T_k(-1) = \sum_{k=0}^{n}{}' \beta_k (-1)^k$ so that

$$\beta_0 = 2F_n(-1) - 2 \sum_{k=1}^{n} \beta_k (-1)^k.$$

The coefficients β_k of the truncated series $f_n(x)$ are determined by using the discrete inner product (see Section 7.7.3) $\langle f \circ g \rangle \equiv \sum_{k=1}^{n} f\left(\frac{k\pi}{n+1}\right) g\left(\frac{k\pi}{n+1}\right)$. Since $x = \cos t$ we find that for $m > 0$

$$\beta_m = \frac{2}{m(n+1)} \langle \sin t\, f_n(\cos t) \circ \sin mt \rangle$$

$$= \frac{2}{m(n+1)} \sum_{k=1}^{n} \sin \frac{k\pi}{n+1} \sin \frac{k\pi m}{n+1} f\left(\cos \frac{k\pi}{n+1}\right).$$

The values b_k are substituted in the expansion of $F(x)$ with coefficients β_k to produce the truncated series $F_n(x)=\frac{1}{2}\beta_0+\sum_{k=1}^{n}\beta_k T_k(x)$ where β_0 is determined by the condition $F(-1)=F_n(-1)$. This yields

$$\beta_0=2[F(-1)+\beta_1-\beta_2+\beta_3-+\ldots+(-1)^{n+1}\beta_n].$$

Hence $F_n(x)=F(-1)+\sum_{k=1}^{n}\beta_k[T_k(x)+(-1)^{k+1}]$ and, for the case $x=1$,

$$F(1)=F(-1)+\int_{-1}^{1}f(x)\,dx \doteq F(-1)+\sum_{k=1}^{n}\beta_k[1-(-1)^k]=F_n(1)$$

since $T_k(1)=1$ or $F(1)-F(-1)=\int_{-1}^{1}f(x)\,dx \doteq 2\sum_{k=1}^{[(n+1)/2]}\beta_{2k-1}\equiv Q_n f.$
This is again a quadrature formula of the familiar type. If x is not specified to be a fixed number, we have that the quadrature weights depend on x and hence the approximation can be evaluated for any value $x\in[-1,1]$. In particular, for varying x the approximating function can be plotted or listed.

Properties of the FCC scheme

The FCC scheme is also a positive convergent quadrature procedure. This can be shown by applying the scheme for $x=1$ so that

$$\begin{aligned}Q_n f&=2\sum_{i=1}^{[(n+1)/2]}\beta_{2i-1}\\&=\frac{4}{n+1}\left\{\sum_{k=1}^{n}\sin\frac{\pi k}{n+1}\sum_{i=1}^{[(n+1)/2]}\frac{1}{2i-1}\sin\frac{k\pi(2i-1)}{n+1}\right\}f\left(\cos\frac{k\pi}{n+1}\right).\end{aligned}$$

The weights in this case are

$$A_k=\frac{4}{n+1}\sin\frac{k\pi}{n+1}\sum_{i=1}^{[(n+1)/2]}\frac{1}{2i-1}\sin\frac{(2i-1)k\pi}{n+1},\quad k=1(1)n$$

and are always positive (cf. Filippi [8]). For sufficiently large n this can be seen from the formula (cf. Ryshik and Gradstein [17], p. 38)

$$\sum_{i=1}^{\infty}\frac{\sin(2i-1)x}{2i-1}=\frac{\pi}{4}\quad\text{for}\quad 0<x<\pi \tag{7.7.17}$$

by replacing x by $k\pi/(n+1)$.

Since the indefinite integral of a polynomial is again a polynomial and the Tschebyscheff series expansion of every polynomial is finite and unique, we have polynomial accuracy of the FCC scheme and hence convergence of $Q_n f$:

Theorem 7.7.2 *Filippi's modified Clenshaw–Curtis scheme for* $[-1, 1]$ *is a positive quadrature procedure which converges for every* $f \in C^0[-1, 1]$.

Taking into account formula (7.7.17) we have asymptotically for sufficiently large n that $A_k = \frac{\pi}{n+1} \sin \frac{k\pi}{n+1} = \frac{\pi}{n+1}(1-x_k^2)^{\frac{1}{2}}$, $k = 1(1)n$, since the nodes x_k are given by $x_k = \cos\left(\frac{\pi k}{n+1}\right)$. The nodes and weights are thus closely related. We have (note that the nodes x_k are now decreasing with increasing k) that

$$\begin{aligned}\tfrac{1}{2}(x_{k-1} - x_{k+1}) &= \frac{1}{2}\left(\cos \frac{k-1}{n+1}\pi - \cos \frac{k+1}{n+1}\pi\right) = \sin \frac{k\pi}{n+1} \sin \frac{\pi}{n+1}\\ &\approx \frac{\pi}{n+1} \sin \frac{k\pi}{n+1} = A_k.\end{aligned}$$

In general terms every weight asymptotically equals the length of the interval $[x_k, x_{k+1}] \approx [x_{k-1}, x_k]$ and $Q_n f$ is asymptotically a Riemann sum.

Numerical performance

The numerical performance of the FCC scheme is similar to that of the CC scheme. In both cases the inner products have to be calculated. This process is facilitated by the following algorithm: Let $p(t) = c_1 \sin t + c_2 \sin 2t + \ldots + c_n \sin nt$, and let us assume that the value $p(T)$ for some T is desired. (The coefficients are specified by the scheme.) We then compute the subsequent numbers $z_n, z_{n-1}, \ldots, z_1$, starting with $z_{n+2} = z_{n+1} = 0$, by means of the relationship $z_k = 2z_{k+1} \cos T - z_{k+2} + c_k$. Once the z_k are known, the coefficients c_k of $p(T)$ can be replaced by $z_k + z_{k+2} - 2z_{k+1} \cos T$ so that

$$p(T) = \sum_{k=1}^{n} (z_k - 2z_{k+1} \cos T + z_{k+2}) \sin kT.$$

On rearranging terms and taking into account that $z_{n+1} = z_{n+2} = 0$, we have that

$$\begin{aligned}p(T) &= z_1 \sin T + \sum_{k=2}^{n} [\sin kT - 2 \cos T \sin (k-1)T + \sin (k-2)T] z_k\\ &= z_1 \sin T,\end{aligned}$$

since the terms in square brackets vanish. Thus, we obtain $p(T)=z_1 \sin T$ and hence

$$\beta_m=\frac{2\sin T}{m(n+1)}z_1, \qquad \beta_0=2[F(-1)+\beta_1-+\ldots+(-1)^{n+1}\beta_n].$$

Note that $z_1=z_1(m)$ depends on m.

Example 7.7.2 We wish to compute for $x=0$ and for $x=1$:

$$F(x)=\int_{-1}^{x}\frac{du}{1+(1+u)^2}=\operatorname{arc\,tg}(1+x), \qquad x\in[-1,1], \qquad F(-1)=0.$$

Taking $n=8$ we find that

$$\beta_m=\frac{2}{9m}z_1(m), \quad \text{with} \quad c_k=\sin\frac{k\pi}{9}f\left(\cos\frac{k\pi}{9}\right),$$

$$\beta_0=2[\beta_1-\beta_2+-\ldots-\beta_8].$$

The values of c_k, $k=1(1)8$ can thus be computed giving the eight values $z_k=z_k(m)$, $k=8(-1)1$ and $m=1(1)8$. The coefficients β_i, $i=1(1)8$ and β_0 can be calculated. The values of these are as in Table 7.7.2. ■

The great advantage of the CC and FCC schemes is that they allow $F_n(x)\doteq F(x)$ to be tabulated. On the other hand, if $F(x)$ is needed only for a very few values of x, then the total amount of work involved seems large in comparison with, for example, the Gauss quadrature. However, both schemes are reliable since the decreasing of $|\beta_k|$ is a very good criterion for the accuracy to be expected in the final result. Last but not

TABLE 7.7.2

k	β_k	
0	$2\cdot 0.666\,223\,8775$	$F_8(x)=\frac{1}{2}\beta_0+\sum_{k=1}^{8}\beta_k T_k(x)$
1	0.544 039 2829	
2	−0.116 342 0140	$F_8(0)=0.785\,419\,1705$
3	0.011 460 8465	arc tg 1 = 0.785 398 1635
4	0.003 288 2456	
		error = 0.000 021 0070
5	−0.001 948 0998	
6	0.000 431 0062	$F_8(1)=1.107\,123\,022$
7	0.000 009 4813	arc tg 2 = 1.107 148 718
8	−0.000 039 6043	
		error = −0.000 025 696

least the convenience for many technical applications, that the function $F_n(x)$ can be plotted automatically, should be mentioned.

7.8 Remarks, problems, references

7.8.1 Remarks

R.7.1: The preassigned nodes in Section 7.1 are such that f' is not interpolated at these nodes. If this additional interpolation is desired, we take $H_{n+k}(f, x)$ with k preassigned nodes. The weights of the preassigned derivatives are usually not all positive. If only ± 1 are preassigned the derivatives $f'(\pm 1)$ can be replaced by a finite-difference formula involving values of f. This possibility was treated by Filippi [9] (cf. Section 7.4.3).

R.7.2: Formulae of the type

$$Q_{n+2}f = \sum_{i=1}^{n} A_i f(x_i) + B_1(f(1) - f(-1))$$

are sometimes considered. If these are applied as composite rules, the boundary nodes from adjacent intervals cancel. Formulae of this type seem only to be possible by pseudo-interpolation; i.e. the Lagrange interpolating polynomial for the preassigned nodes has the pseudo-interpolating property

$$L_2(f, 1) = f(1), \qquad L_2(f, -1) = -f(-1),$$

(see Engels [4]). Such formulae were introduced by Ralston [16] and Filippi [9] (cf. Section 7.4.3).

R.7.3: Though Wilf-type quadratures with preassigned nodes have not yet been researched, such formulae do exist. The basis functions are $h_i(x) = x - x_i$, and in each case the nodal function is $\omega_n(x) = \prod_{i=1}^{n} (x - x_i)/(1 - \varepsilon^2 x x_i)$. If $k_i(x) = x - z_i$ and $\Omega_m = \prod_{i=1}^{m} k_i(x)$, then there exist positive formulae having preassigned nodes $z_i \notin (-1, 1)$. Clearly, $\varepsilon < 1$ must be sufficiently small (cf. Section 7.5).

R.7.4: The conditions

$$\int_{-1}^{1} \omega_n(x)\lambda_i(x)\,dx = 0 \quad \text{or equivalently} \quad \int_{-1}^{1} \prod_{j=1}^{n} \left(\frac{x - x_j}{1 - \varepsilon^2 x x_j}\right)^2 \frac{dx}{x - x_i} = 0,$$

$$i = 1(1)n,$$

for the nodes of minimal quadratures establish a nonlinear system of equations for determining the nodes. The first extended table of minimal quadratures for $\varepsilon = 1$ was calculated by solving this system (see Engels and Eckhardt [7]). The system itself may be written in the form

$$\int_{-1}^{1} \frac{p_n(x) q_i(x)}{(\varepsilon^2 x)^{2n} p_n^2(1/\varepsilon^2 x)}\,dx = 0, \qquad i = 1(1)n$$

with $p_n(x)=(x-x_1)\dots(x-x_n)$ and $q_i(x)=p_n(x)/(x-x_i)$ which, because of the linearity of polynomials with respect to their coefficients, is equivalent to

$$\int_{-1}^{1}\frac{p_n(x)x^k}{(\varepsilon^2x)^{2n}p_n^2(1/\varepsilon^2x)}\,dx=0,\qquad k=0(1)n-1.$$

Hence $p_n(x)$ is an implicitly-determined orthogonal polynomial; such polynomials have been treated in Chapter 5 (cf. Section 7.5).

R.7.5: Stancu and Stroud [18] proposed multiple preassigned nodes. This means that derivatives of the integrand, in addition to integrand values, are preassigned. This was done independently by Filippi [9] (cf. Section 7.4.3).

7.8.2 Problems

P.7.1: For $f(x)=\exp(x)$ compare the quadrature results of a Gaussian quadrature formula $Q_{n+1}f$ and a Lobatto quadrature formula $Q_{n+2}f$ with small n (cf. Sections 7.3, 7.4).

P.7.2: Construct a Filippi formula Q_9f with four exterior nodes (cf. Section 7.4.3).

P.7.3: Show that the Hermitian operator $H_n(f, x)$, formula (7.5.1) has the following properties:

(i) $H_n(1/(1-\varepsilon^2xx_k), x)=1/(1-\varepsilon^2xx_k)$, $k=1(1)n$,
(ii) $H_n(\lambda_i(x), x)=\lambda_i(x)$, $i=1(1)n$.
(Hint: expand $\lambda_i(x)$ into partial fractions.)

P.7.4: Show that for the discrete inner product (Section 7.7)

$$\langle f\circ g\rangle\equiv\sum_{k=1}^{n}f\left(\frac{k}{n+1}\right)g\left(\frac{k}{n+1}\right)$$

the system of functions $\sin mt$ is orthogonal, i.e. that

$$\langle\sin lt\circ\sin mt\rangle=\tfrac{1}{2}(n+1)\delta_{lm}.$$

7.8.3 References

1. Bouzitat, J. (1949). Sur l'intégration numérique approchée par la méthode de Gauss généralisée et sur une extension de cette méthode. *C.R. Acad. Sci. Paris*, **229**, 1201–1203.
2. Christoffel, E. B. (1858). Über die Gauss'sche Quadratur und eine Verallgemeinerung derselben. *J. reine angew. Math.* (*Crelle's J.*), **55,** 61–82.
3. Clenshaw, C. W. and Curtis, A. R. (1960). A method for numerical integration on an automatic computer. *Num. Math.*, **2,** 197–205.
4. Engels, H. (1972). Über allgemeine Gauss'sche Quadraturen. *Computing*, **10,** 83–95.
5. Engels, H. (1972). Ein numerisches Differentiationsverfahren mit ableitungsfreien Fehlerschranken. *ZAMM*, **52,** 145–154.

6. Engels, H. (1974). Positive Interpolationsquadraturen mit teilweise vorgeschriebenen Stützstellen. *Computing*, **13,** 121–141.
7. Engels, H. and Eckhardt, U. (1974). Symmetrische Wilf'sche Quadraturformeln. Nuclear Research Centre Jülich. Report Jül–1094–Ma.
8. Filippi, S. (1964). Angenäherte Approximation einer Stammfunktion—eine Modifikation des Verfahrens von Clenshaw und Curtis. *Num. Math.*, **6,** 320–328.
9. Filippi, S. (1964). "Neue Gauss Typ-Quadraturformeln." Habilitations-Thesis, Aachen.
10. Gauss, C. F. (1866). Methodus nova integralium valores per approximationem inveniendi. [Latin] Collected papers, Göttingen, Vol. 3, 163–196. Translated into German by A. Kowalewski (see historical remarks, Section 1.3).
11. Jacobi, C G. J. (1825). Über Gauss' neue Methode, die Werte der Integrale näherungsweise zu finden. *J. reine angew. Math.* (*Crelle's J.*), **1,** 301–308.
12. Kronrod, A. S. (1965). "Nodes and Weights of Quadrature Formulas." Consultants Bureau, New York.
13. Markoff, A. (1885). Sur la méthode de Gauss pour le calcul approché des intégrales. *Math. Ann.*, **25,** 427–432.
14. Piessens, R., and Branders, M. (1972). Optimal addition of abszissas to quadrature formulas of Gauss and Lobatto type. Report TW 10 Applied Mathematics Division of the kath. Univ. Leuven. (See also (1974), *Math. Comput.*, **28,** 135–139.)
15. Radau, R. (1880). Étude sur les formules d'approximation qui servent à calculer la valeur numérique d'une intégrale défine. *J. Math. pure appl.*, **6,** 283–336.
16. Ralston, A. (1959). A family of quadratures which achieve high accuracy in composite rules. *J. Assoc. Comput. Mach.*, **6,** 384–394.
17. Ryshik, I. M. and Gradstein, I. S. (1957). "Tables of series, products and integrals." VEB Berlin.
18. Stancu, D. D. and Stroud, A. H. (1963). Quadrature formulas with simple Gaussian nodes and multiple fixed nodes. *MTAC*, **17,** 384–394.
19. Szegö, G. (1935). Über gewisse orthogonale Polynome, die zu einer oszillierenden Belegungsfunktion gehören. *Math. Ann.*, **110,** 501–513.
20. Wilf, H. S. (1961). Exactness conditions in numerical quadrature. *Num. Math.*, **6,** 315–319.

8

Non-interpolatory Quadratures

In the previous two chapters, quadrature and cubature procedures derived from interpolatory schemes have been discussed. There are, however, several well-known useful quadrature and cubature schemes which are not produced in this way. Examples of these are *Romberg's quadrature*, based upon extrapolation techniques for the error; *adaptive quadrature methods*, useful when no properties of the integrand are known; *Sard's optimal quadratures* obtained by minimising the L_2-norm of the Peano kernels of the error functional; and *Euler–Maclaurin quadrature* which no doubt has great theoretical importance but less practical value.

Romberg's quadrature will be discussed in detail here, because it is in common use. In addition, we shall look for an adaptive method, although detailed discussion of methods of this type is beyond the scope of this book. In this discussion we mention briefly methods of proceeding if the integrand is oscillating rapidly, a problem which arises, for example, in numerical performance of Fourier transforms.

8.1 Richardson extrapolation

8.1.1 General background

In his comprehensive paper in 1927, Richardson [20] proposed his famous extrapolation method (he was, however, not the first person to have this idea). As an example of this, suppose that we are given a problem with exact solution T; for example, $T = \int_{-1}^{1} f(x)\,dx$. This problem is solved approximately using discretisation methods. To this end we

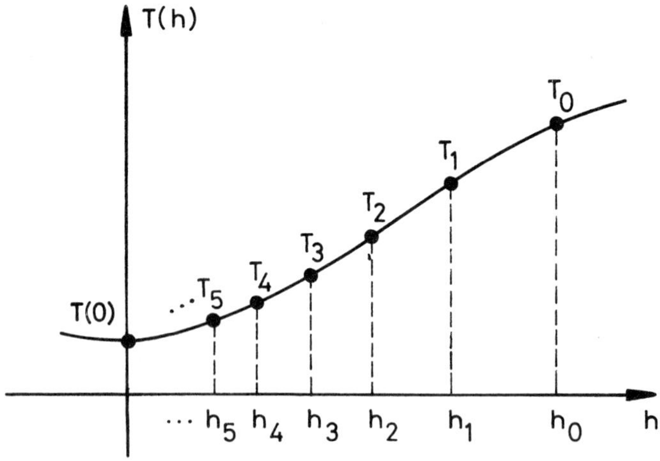

FIG. 8.1.1. Extrapolation of $T(h)$ at $h = 0$.

introduce a *discretisation parameter* or a *step-size h*. The solution of the discretised problem, $T(h)$, depends on h, (see Fig. 8.1.1), for example by the composite trapezoidal rule,

$$T(h) = h\left[\tfrac{1}{2}f(-1) + \sum_{i=1}^{n-1} f(-1+ih) + \tfrac{1}{2}f(1)\right], \qquad h = 2/n.$$

Assume that $\lim_{h \to 0} T(h) = T(0) = T$ is valid, which means that $T(h)$ is continuous in the interval $[0, H]$ with $H > 0$ and that the solution of the discrete problem converges to the solution of the original problem. Assume, further, that there exists an asymptotic expansion in powers of h such that: $T(h) = C_0 + C_1 h + C_2 h^2 + \ldots + C_m h^m + O(h^{m+1})$. On truncating this expansion and omitting the $O(h^{m+1})$ terms, the expansion becomes a $\hat{T}(h) \in \mathbb{P}_m$ in h having the value $C_0 = T(0) = T$, the solution, at $h = 0$. It is therefore desirable either to determine or to eliminate $C_0, C_1, \ldots, C_m$.

We remember that every polynomial of degree m is completely determined by prescribing values at $m+1$ separate points. Hence, if we know $m+1$ values $T_i \equiv T(h_i)$, $i = 0(1)m$ at different points h_i, $i = 0(1)m$ we are able to construct the unique polynomial $\hat{T}(h)$, for example by means of Lagrange interpolation, and then C_0 would approximate T up to $O(h^{m+1})$. Once this is done, the question of convergence of the Lagrange interpolation arises. It was seen in Chapter 4 that this type of interpolation does not necessarily converge at every point as the number of nodes tends to ∞. We therefore seek convergence conditions for

Lagrange interpolation at $h=0$ as $m\to\infty$. Clearly, there may be other extrapolation formulae more appropriate than Lagrange interpolation. For example, if there is no asymptotic expansion in integer powers of h then Lagrange interpolation should not be used. Alternative methods are those due to Bulirsch and Stoer [6] using rational interpolation, and to Engels [8], [9] using spline interpolation.

8.1.2 The algorithm for polynomial extrapolation

We use here only Lagrange polynomial interpolation. In order to perform the extrapolation explicitly this interpolation may be undertaken by the algorithm of Aitken and Neville, which we now describe. The linear interpolant between the two discrete values T_i of $T(h)$, $T_i \equiv T(h_i$, is

$$L_1(T,h)_i \equiv \frac{h_{i+1}-h}{h_{i+1}-h_i}T_i + \frac{h-h_i}{h_{i+1}-h_i}T_{i+1} = \frac{\begin{vmatrix} T_i & h_i-h \\ T_{i+1} & h_{i+1}-h \end{vmatrix}}{\begin{vmatrix} 1 & h_i \\ 1 & h_{i+1} \end{vmatrix}}, \qquad i=0(1)n-1.$$

Repeating this process with $L_1(T,h)_i$ and $L_1(T,h)_{i+1}$ we have that

$$L_{12}(T,h)_i \equiv \frac{\begin{vmatrix} L_1(T,h)_i & h_i-h \\ L_1(T,h)_{i+1} & h_{i+2}-h \end{vmatrix}}{\begin{vmatrix} 1 & h_i \\ 1 & h_{i+2} \end{vmatrix}}, \qquad i=0(1)n-2,$$

which is a second-degree polynomial in h which interpolates $T(h)$ at the points h_i, h_{i+1} and h_{i+2}. This can be verified by a straightforward calculation. More generally, using induction, we obtain the Lagrange interpolating polynomial $L_n(T,h)$ with nodes $h_i,\ldots,h_{i+n}$ in the form

$$L_{12\ldots n}(T,h)_i = \frac{\begin{vmatrix} L_{12\ldots(n-1)}(T,h)_i & h_i-h \\ L_{12\ldots(n-1)}(T,h)_{i+1} & h_{i+n}-h \end{vmatrix}}{\begin{vmatrix} 1 & h_i \\ 1 & h_{i+n} \end{vmatrix}}, \qquad i=0.$$

We require the result only for $h=0$ and so have the algorithm

$$L_1(T,0)_i = \frac{T_i h_{i+1} - T_{i+1}h_i}{h_{i+1}-h_i} \equiv T_1^{(i)}, \qquad i=0(1)\ldots$$

$$L_{12}(T,0)_i = \frac{L_1(T,0)_i h_{i+2} - L_1(T,0)_{i+1}h_i}{h_{i+2}-h_i}$$

$$= \frac{T_1^{(i)}h_{i+2} - T_1^{(i+1)}h_i}{h_{i+2}-h_i} \equiv T_2^{(i)}, \qquad i=0(1)\ldots$$

$$L_{12\ldots n}(T,0)_i=\frac{L_{12\ldots(n-1)}(T,0)_i h_{i+n}-L_{12\ldots(n-1)}(T,0)_{i+1}h_i}{h_{i+n}-h_i}$$
$$=\frac{T_{n-1}^{(i)}h_{i+n}-T_{n-1}^{(i+1)}h_i}{h_{i+n}-h_i}\equiv T_n^{(i)}, \qquad i=0(1)\ldots.$$

The values $T_k^{(i)}$ are usually arranged in a triangular T-scheme which is useful for applications. The scheme usually indicates the rate of convergence, (see Table 8.1.1).

TABLE 8.1.1.

h_i	$T_i\equiv T_0^{(i)}$	$T_1^{(i)}$	$T_2^{(i)}$	$T_3^{(i)}$	$T_4^{(i)}$	...
h_0	$T_0^{(0)}$					
		$T_1^{(0)}$				
h_1	$T_0^{(1)}$		$T_2^{(0)}$			
		$T_1^{(1)}$		$T_3^{(0)}$		
h_2	$T_0^{(2)}$		$T_2^{(1)}$		$T_4^{(0)}$	...
		$T_1^{(2)}$		$T_3^{(1)}$	....	
h_3	$T_0^{(3)}$		$T_2^{(2)}$	....		
		$T_1^{(3)}$	....			
h_4	$T_0^{(4)}$	....				
...	...					

There is often an asymptotic expansion of $T(h)$ which is an *even* function, $T(h)=T(-h)$, since it contains only even powers of h. In this case the additional values $T_{-i}=T(-h_i)=T(h_i)=T_i$ are known and may be taken into account. The resulting interpolating polynomial must also be an even function and hence only even powers of h appear in $L_n\,(T,h)$. The same result is produced if *a priori* we replace h by h^2 and extrapolate with respect to the $n+1$ values $T_0,\ldots,T_n$. The algorithm is then changed into

$$\begin{aligned}T_1^{(i)}&=\frac{T_ih_{i+1}^2-T_{i+1}h_i^2}{h_{i+1}^2-h_i^2}=\frac{T_0^{(i)}h_{i+1}^2-T_0^{(i+1)}h_i^2}{h_{i+1}^2-h_i^2}\\ &\ldots\\ T_n^{(i)}&=\frac{T_{n-1}^{(i)}h_{i+n}^2-T_{n-1}^{(i+1)}h_i^2}{h_{i+n}^2-h_i^2}, \qquad i=0(1)\ldots, n=1(1)\ldots.\end{aligned} \tag{8.1.1}$$

8.1.3 Romberg's scheme

An important example of the extrapolation scheme is Romberg's scheme which evaluates $\int_a^b f(x)\,dx$ by means of the trapezoidal rule with step-sizes $\{h_i\}=\{(b-a)/2^i\}$. Then the algorithm (8.1.1) is simplified

$$T_n^{(i)}=\frac{4^nT_{n-1}^{(i+1)}-T_{n-1}^{(i)}}{4^n-1}, \qquad i=0(1)\ldots, n=1(1)\ldots \tag{8.1.2}$$

and for the approximation of $\overset{*}{I}f$ (i.e. $a=0$, $b=1$) we have that

$$\left.\begin{aligned}
T_0^{(0)} &= \tfrac{1}{2}[f(0)+f(1)] \\
T_0^{(1)} &= \tfrac{1}{4}[f(0)+2f(\tfrac{1}{2})+f(1)] = \tfrac{1}{2}T_0^{(0)} + \tfrac{1}{2}M_1 f \\
T_0^{(2)} &= \tfrac{1}{8}[f(0)+2f(\tfrac{1}{4})+2f(\tfrac{1}{2})+2f(\tfrac{3}{4})+f(1)] = \tfrac{1}{2}T_0^{(1)} + \tfrac{1}{2}M_2 f \\
&\text{etc.} \\
T_0^{(k)} &= \frac{1}{2^{k+1}}\left[f(0)+2\sum_{i=1}^{2^k-1} f\left(\frac{i}{2^k}\right)+f(1)\right] = \tfrac{1}{2}T_0^{(k-1)} + \tfrac{1}{2}M_{2^{k-1}} f
\end{aligned}\right\} \qquad (8.1.3)$$

where $M_{2^k}f = \dfrac{1}{2^k}\sum_{i=1}^{2^k} f\left(\dfrac{2i-1}{2^{k+1}}\right)$ is the midpoint rule. Once values of $f(x)$ have been computed, they are used in all subsequent steps. Equation (8.1.2) and appropriate substitution for (8.1.3) give, successively,

$$\begin{aligned}
T_1^{(0)} &= \tfrac{1}{3}[4T_0^{(1)} - T_0^{(0)}] = \tfrac{1}{6}[f(0)+4f(\tfrac{1}{2})+f(1)] \\
T_1^{(1)} &= \tfrac{1}{12}[f(0)+4f(\tfrac{1}{4})+2f(\tfrac{1}{2})+4f(\tfrac{3}{4})+f(1)] \\
T_1^{(2)} &= \tfrac{1}{24}[f(0)+4f(\tfrac{1}{8})+2f(\tfrac{1}{4})+\ldots+f(1)] \\
&\ldots\ldots
\end{aligned}$$

which are composite Simpson rules, i.e. composite three-point Newton–Cotes formulae. Further,

$$T_2^{(0)} = \tfrac{1}{15}(16T_1^{(1)} - T_1^{(0)}) = \tfrac{1}{90}[7f(0)+32f(\tfrac{1}{4})+12f(\tfrac{1}{2})+32f(\tfrac{3}{4})+7f(1)]$$

which is, again, a Newton–Cotes formula (cf. Section 6.2) and

$$T_2^{(1)} = \tfrac{1}{180}[7f(0)+32f(\tfrac{1}{8})+12f(\tfrac{1}{4})+32f(\tfrac{3}{8})+12f(\tfrac{1}{2})+\ldots+7f(1)].$$

The formula $T_3^{(0)} = \frac{1}{63}(64T_2^{(1)} - T_2^{(0)})$, as well as $T_3^{(1)}$, $T_3^{(2)}, \ldots,$ are no longer Newton–Cotes formulae and hence no longer interpolatory quadrature formulae. This is clear since the nine-point formula $T_3^{(0)}$ has degree 7, while the nine-point Newton–Cotes formula has degree 9. The scheme consisting of these special quadrature formulae is known as the *Romberg quadrature* scheme, although Romberg was not the first person to find this scheme (cf. Section 1.3). The numerical values are again arranged in a T-scheme.

Example 8.1.1 We calculate approximately $\int_0^2 \sin x/x \, dx = \int_0^1 \sin 2x/x \, dx = 1.605\,412\,976$ (Table 8.1.2). The correct figures in each entry of the table have been underlined. The approximations $T_j^{(i)}$ improve with increasing j, particularly proceeding down the leading diagonal. ■

TABLE 8.1.2.

i	h_i	$T_0^{(i)}$	$T_1^{(i)}$	$T_2^{(i)}$	$T_3^{(i)}$	$T_4^{(i)}$
0	1	1.454 648 714				
			1.606 844 218			
1	1/2	1.568 795 342		1.605 407 119		
			1.605 496 938		1.605 412 986	
2	1/4	1.596 321 539		1.605 412 893		1.605 412 976
			1.605 418 146		1.605 412 976	
3	1/8	1.603 143 994		1.605 412 975		
			1.605 413 299			
4	1/16	1.604 845 973				

8.1.4 Program

A subroutine for implementing the Romberg algorithm can easily be written. Since the algorithm works like an iterative procedure, a stopping criterion has to be defined. This is done by comparing successive values $T_k^{(i)}$ and $T_{k+1}^{(i-1)}$, or by reaching a prescribed upper bound for k. The computation is performed as follows. Assume that $T_0^{(k)}, T_1^{(k-1)}, \ldots, T_k^{(0)}$ have been computed and that $|T_k^{(0)} - T_{k-1}^{(1)}|$ is too large. The set of values $T_0^{(k+1)}, T_1^{(k)}, \ldots, T_{k+1}^{(0)}$ with h_{k+1} is then computed and accuracy is checked, so that the procedure stops or continues in the same way. The following FORTRAN subroutine implements this algorithm.

```
1ØØ   SUBROUTINE RQ(F, A, B, EPS, T, NMIN, NMAX, N, Q)
11ØC  THIS PROGRAM CALCULATES THE INTEGRAL OF
12ØC  F(X) BETWEEN THE LIMITS A AND B USING ROM-
13ØC  BERG QUADRATURE
14ØC  IF THE ABSOLUTE VALUE OF THE DIFFERENCE
15ØC  OF TWO SUCCESSIVE VALUES OF THE T-SCHEME
16ØC  HAPPENS TO BE SMALLER THAN EPS, THEN
17ØC  THE PROGRAM STOPS. AT LEAST NMIN AND AT
18ØC  MOST NMAX ROWS OF THE T-SCHEME ARE
19ØC  CALCULATED
2ØØC  T: ARRAY OF LENGTH NMAX
21ØC  OUTPUT PARAMETERS:
22ØC  N: NUMBER OF EXECUTED ROWS OF THE
23ØC  T-SCHEME
24ØC  Q: RESULT, IF N=Ø, THEN THE ERROR BOUND EPS
25ØC  WAS NOT REACHED.
```

```
      DIMENSION T(N)
      EXTERNAL F
      D=B-A
      K=1
      N=Ø
      T(1)=(F(A)+F(B))/2
    1 N=N+1
      IF(N.GE.NMAX)GOTO 4
      T(N+1)=Ø
      K=K+K
      DO 2 I=1, K, 2
    2 T(N+1)=T(N+1)+F(I×D/K+A)
      T(N+1)=T(N+1)/K+T(N)/2
      J=N
      G=1
      DO 3 I=1, N
      G=4×G
      T(J)=(G×T(J+1)-T(J))/(G-1)
    3 J=J-1
      Q=T(1)×D
      IF(ABS((T(1)-T(2))×D)-ABS(EPS)) 5, 5, 1
    4 N=Ø
    6 RETURN
    5 IF(NMIN-N) 6, 6, 1
      END
```

The number NMIN is introduced in order to avoid trivial mistakes. If, for example, $\sin^2(4\pi x)$ is integrated in $[0, 1]$ by this subroutine, then for NMIN $= 0$ we would obtain the approximate value 0 instead of $\frac{1}{2}$.

8.2 Convergence of the extrapolation

In pictorial form, the T-scheme and the underlying extrapolation process can be represented as:

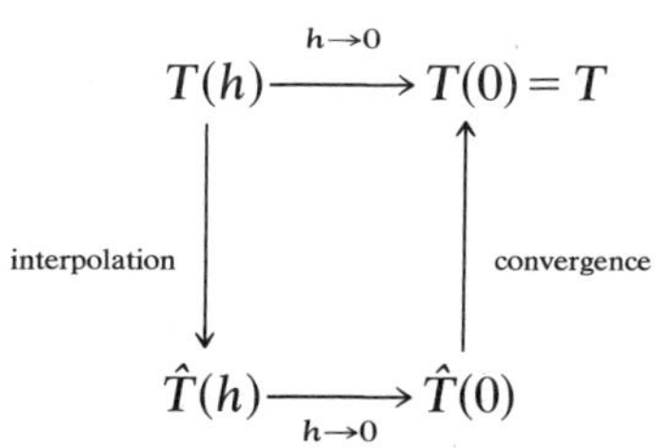

Clearly, we cannot always expect that $\lim_{n\to\infty} T_n^{(i)} = \lim_{h\to 0} \hat{T}(h) = T(0)$, since Lagrange polynomial interpolation does not converge for every continuous function (cf. Theorem 4.2.6). Since polynomial interpolation is unique, the algorithm of Aitken and Neville is precisely Lagrangian polynomial interpolation. Roughly speaking, we presume that the nodes cluster at $h=0$ in order to obtain convergence at $h=0$, for example the nodes h_i tend to zero in the manner of the elements of a geometric sequence. We have the following:

Theorem 8.2.1 *If $h_i \geqslant a h_{i+1}$ with $a>1$ for every $i \in \mathbb{N}_0$ and $T(h)$ is continuous for $h \in [0, H]$, $H>0$, then the polynomial extrapolation converges, i.e.* $\lim_{n\to\infty} T_n^{(i)} = T(0)$.

Proof It suffices to verify that the provisions of the Banach–Steinhaus Theorem 4.3.2 are fulfilled. This necessitates showing that:

(i) $\lim_{n\to\infty} T_n^{(i)} = T(0)$ for every polynomial,

(ii) the norm $\|L_{n+1}\| \equiv \sum_{k=0}^{n} |\lambda_k(0)| = \sum_{k=0}^{n} \left| \frac{\omega_{n+1}(0)}{\omega'_{n+1}(h_k)h_k} \right|$ with $\omega_{n+1}(h) = \prod_{k=0}^{n} (h-h_k)$ is bounded for every n.

The first condition (i) is satisfied trivially since $L_{n+1}(f, x) = f$ for $n \geqslant m$ if $f \in \mathbb{P}_m$. In order to simplify the proof of (ii) we start the T-scheme of Section 8.1.2 with T_0 and use the Lagrange interpolating polynomial $L_{n+1}(T(h), h) = \sum_{k=0}^{n} \lambda_k(h) T_k$. Then, from Chapter 3,

$$\|L_{n+1}\| = \sum_{k=0}^{n} |\lambda_k(0)|. \tag{8.2.1}$$

Since $|\omega_{n+1}(0)| = \prod_{k=0}^{n} h_k$, and $\omega'_{n+1}(h) = \sum_{j=0}^{n} \prod_{\substack{k=0\\ \neq j}}^{n} (h-h_k)$ or, equivalently,

$$|\omega'_{n+1}(h_\nu)| = \prod_{\substack{k=0\\ \neq \nu}}^{n} |h_\nu - h_k|,$$

we have that

$$|\lambda_k(0)| = \left| \frac{\omega_{n+1}(0)}{\omega'_{n+1}(h_k)h_k} \right| = \left| \frac{\prod_{j=0}^{n} h_j}{\prod_{\substack{j=0\\ \neq k}}^{n} (h_j - h_k) h_k} \right| = \prod_{\substack{j=0\\ \neq k}}^{n} \frac{h_j}{|h_j - h_k|} = \prod_{\substack{j=0\\ \neq k}}^{n} \frac{1}{|1 - h_k/h_j|},$$

and hence $\|L_{n+1}\| = \sum_{k=0}^{n} \left[1 \Big/ \prod_{\substack{j=0\\ \neq k}}^{n} \left|1 - \frac{h_k}{h_j}\right|\right]$. Since $h_i \geqslant a h_{i+1}$ we conclude successively that $h_i \geqslant a h_{i+1} \geqslant a^2 h_{i+2} \geqslant \ldots \geqslant a^r h_{i+r}$ and therefore that

$$\frac{h_i}{h_{i+r}} \geqslant a^r \quad \text{and} \quad \frac{h_{i+r}}{h_i} \leqslant \frac{1}{a^r}$$

and hence

$$1 \Big/ \left|1 - \frac{h_{i+r}}{h_i}\right| = 1 \Big/ \left(1 - \frac{h_{i+r}}{h_i}\right) \leqslant 1 \Big/ \left(1 - \frac{1}{a^r}\right) = \frac{a^r}{a^r - 1}$$

and

$$1 \Big/ \left|1 - \frac{h_i}{h_{i+r}}\right| = 1 \Big/ \left(\frac{h_i}{h_{i+r}} - 1\right) \leqslant \frac{1}{a^r - 1}.$$

The product in the expression for $\|L_{n+1}\|$ is split into two parts so that

$$\sum_{k=0}^{n} |\lambda_k(0)| = \sum_{k=0}^{n} \left(\prod_{j=0}^{k-1} 1 \Big/ \left|1 - \frac{h_k}{h_j}\right|\right)\left(\prod_{j=k+1}^{n} 1 \Big/ \left|1 - \frac{h_k}{h_j}\right|\right)$$

from which the above inequalities immediately give that

$$\sum_{k=0}^{n} |\lambda_k(0)| \leqslant \sum_{k=0}^{n} \left(\prod_{j=0}^{k-1} \frac{a^{k-j}}{a^{k-j} - 1}\right)\left(\prod_{j=k+1}^{n} \frac{1}{a^{j-k} - 1}\right)$$
$$= \sum_{k=0}^{n} \frac{a^k \ldots a}{(a^k - 1) \ldots (a-1) \cdot (a-1) \ldots (a^{n-k} - 1)} \equiv \sum_{k=0}^{n} A_k.$$

We now have to show that the A_k are bounded, and for this we consider $A_k, A_{k+1}, \ldots, A_{k+\nu}$, $k+\nu \leqslant n$. By definition

$$A_{k+\nu} = \frac{a^{k+\nu} \ldots a^{k+1} \cdot a^k \ldots a}{(a^{k+\nu} - 1) \ldots (a^{k+1} - 1)(a^k - 1) \ldots (a-1) \cdot (a-1) \ldots (a^{n-k-\nu} - 1)}$$
$$= \frac{a^{k+\nu} \ldots a^{k+1}}{(a^{k+\nu} - 1) \ldots (a^{k+1} - 1)} \left(\frac{a^k \ldots a}{(a^k - 1) \ldots (a-1)(a-1) \ldots (a^{n-k} - 1)}\right)$$
$$\times (a^{n-k-\nu+1} - 1) \ldots (a^{n-k} - 1)$$
$$= \frac{a^{k+\nu} \ldots a^{k+1}(a^{n-k-\nu+1} - 1) \ldots (a^{n-k} - 1)}{(a^{k+\nu} - 1) \ldots (a^{k+1} - 1)} A_k$$
$$= \frac{a^{n+1} - a^{k+\nu}}{a^{k+\nu} - 1} \cdots \frac{a^{n+1} - a^{k+1}}{a^{k+1} - 1} A_k \,. \tag{8.2.2}$$

In (8.2.2) there are ν factors multiplying A_k and the last of these possesses both the largest numerator (since $a^{k+\nu} > a^{k+\nu-j}$, $j > 0$) and the smallest denominator. Thus, we can write

$$A_{k+\nu} \leqslant \left(\frac{a^{n+1} - a^{k+1}}{a^{k+1} - 1}\right)^{\nu} A_k.$$

In a similar manner we find that

$$A_{k-\nu} \leqslant \left(\frac{a^k - 1}{a^{n+1} - a^k}\right)^{\nu} A_k. \tag{8.2.3}$$

In each case there is equality only for $\nu = 1$. Let us now define k to be the index for which $A_k = \max_i A_i$. In this case $A_{k+1} < A_k$ since for $A_{k+1} = A_k$ we have $(a^{n+1} - a^{k+1})/(a^{k+1} - 1) = 1$ or, equivalently, that $a^{n+1} - 2a^{k+1} + 1 = 0$. This condition is generally not satisfied. Hence, we may assume that $A_{k+1} < A_k$ and so see immediately from (8.2.3) that $0 < q_n \equiv (a^{n+1} - a^{k+1})/(a^{k+1} - 1) < 1$. Analogously, $A_{k-1} < A_k$ and hence $0 < p_n \equiv (a^k - 1)/(a^{n+1} - a^k) < 1$. The following four possible cases must be considered:

(i) $\overline{\lim_{n\to\infty}}\, q_n < 1$, $\overline{\lim_{n\to\infty}}\, p_n < 1$,

(ii) $\overline{\lim_{n\to\infty}}\, q_n = 1$, $\overline{\lim_{n\to\infty}}\, p_n < 1$,

(iii) $\overline{\lim_{n\to\infty}}\, q_n = \overline{\lim_{n\to\infty}}\, p_n = 1$, and

(iv) $\overline{\lim_{n\to\infty}}\, q_n < 1$, $\overline{\lim_{n\to\infty}}\, p_n = 1$.

Considering first case (i) we have, from (8.2.1), that

$$\begin{aligned} \|L_{n+1}\| &\leqslant \sum_{j=0}^{n} A_j = \sum_{j=0}^{k-1} A_j + A_k + \sum_{j=k+1}^{n} A_j \\ &< p_n^k A_k + \ldots + p_n A_k + A_k + q_n A_k + \ldots + q_n^{n-k} A_k \\ &< A_k \sum_{\nu=0}^{\infty} (p_n^\nu + q_n^\nu) = \left(\frac{1}{1-p} + \frac{1}{1-q}\right) A_k \end{aligned} \tag{8.2.4}$$

where $p = \sup p_n$ and $q = \sup q_n$. This last expression is bounded if A_k is bounded.

In case (ii) $\overline{\lim_{n\to\infty}}\, q_n = 1$ means that

$$\overline{\lim_{n\to\infty}} \left(\frac{a^{n+1} - a^{k+1}}{a^{k+1} - 1}\right) = \overline{\lim_{n\to\infty}} \left(\frac{a^{k+1}}{a^{k+1} - 1} (a^{n-k} - 1)\right) = 1.$$

But $a^{k+1}/(a^{k+1}-1)\geqslant 1$ and this tends to 1 as k tends to ∞. Further, for fixed finite $k<n$ we have that $a^{n-k}-1>1$ for all $n\in\mathbb{N}$ and if k and n both tend to ∞ then $\overline{\lim\limits_{n\to\infty}}\,(a^{n-k}-1)=1$, i.e. $\overline{\lim\limits_{n\to\infty}}\,a^{n-k}=2$. Thus, in the limit n and k can differ only by a finite number $[\varkappa]\in\mathbb{N}$ defined by $a^{\varkappa}=2$ (for example $\varkappa=1$ if $a=2$). Hence, we have only a *finite* number of terms A_ν with $\nu>k$. From the definition of p_n we have that

$$p_n=\frac{a^k-1}{a^{n+1}-a^k}=\frac{1-a^{-k}}{a^{n-k+1}-1}$$

and, using the result $\overline{\lim\limits_{k,n\to\infty}}\,a^{n-k}=2$, we find that $\overline{\lim\limits_{n\to\infty}}\,p_n=\dfrac{1}{2a-1}<1$ for $a>1$. We thus obtain from (8.2.4) the bound

$$\begin{aligned}\|L_{n+1}\|&<p_n^{n-\varkappa}A_k+\ldots+A_k+\varkappa A_k=(p^{n-\varkappa}+\ldots p+1+\varkappa)A_k\\&<A_k\left(\varkappa+\frac{1}{1-p}\right),\qquad p\equiv\sup_n p_n,\end{aligned}$$

which is always finite if A_k is bounded.

Case (iii): The discussion of case (ii) demonstrates that this case cannot occur.

Case (iv): For this situation we have by the definition of p_n that $\overline{\lim\limits_{n\to\infty}}\,[(a^k-1)/(a^{n+1}-a^k)]=1$ which, again, can only be true if $\overline{\lim\limits_{n\to\infty}}\,a^{n-k}=2/a$ (see case (ii)). Hence, again, $k\to\infty$ as $n\to\infty$ and they differ by at most a finite number. Thus the sequence $\{A_{k+1},\ldots,A_n\}$ contains only a finite number of terms, say $\varkappa$, which are all related to A_k. We also need the limit of q_n as $n\to\infty$. By definition

$$q_n=\frac{a^{n+1}-a^{k+1}}{a^{k+1}-1}=\frac{a^{n-k}-1}{1-a^{-k-1}}\to\frac{2-a}{a}\quad\text{as}\quad k,n\to\infty. \qquad (8.2.5)$$

Notice the special importance of $a=2$ for which we must have, from (8.2.5), that $k=n$. For $a\geqslant 2$ the term A_n of the sequence $\{A_k\}$ is the largest. For $1<a\leqslant 2$ we have $q_n<1$ but at the same time we may have $\lim\limits_{n\to\infty}p_n=1$. If we consider the coefficients $A_{k-\nu}$, we find that

$$\begin{aligned}A_{k-\nu}&=A_k\frac{(a^k-1)\ldots(a^{k-\nu+1}-1)}{(a^{n+1}-a^k)\ldots(a^{n+1}-a^{k-\nu+1})}\\&<A_k/[(a^{n-k+1}-1)\ldots(a^{n-k+\nu}-1)]\\&=A_k\Big/\prod_{i=1}^{\nu}(a^{n-k+\nu}-1)],\end{aligned}$$

and, after at most a finite number of steps, say ρ, we have that $a^{n-k+\nu}>2$ if $\nu>\rho$ in which case $\overline{\lim_{n\to\infty}}\,(a^{n-k+\nu}-1)^{-1}<1$. Hence, the sequence $\{A_1,\ldots,A_{k-1}\}$ is finite and has the common factor A_k. Collecting the above results for the A_k we have that

$$\|L_{n+1}\|<\sum_{j=0}^{k-\rho-1}A_j+\sum_{j=k-\rho}^{k-1}A_j+A_k+\sum_{j=k+1}^{n}A_j$$

$$<(p_n^{k-\rho}\tau_n+\ldots+p_n\tau_n+\rho+1+\varkappa)A_k<\left(\frac{\tau}{1-p}+\varkappa+\rho+1\right)A_k$$

where $\tau_n\equiv\prod_{i=1}^{\rho}[1/(a^{n-k+1}-1)]$, $p\equiv\sup_n p_n$ and $\tau\equiv\sup_n \tau_n$. Hence $\|L_{n+1}\|$ is always bounded provided that A_k is bounded. The final step of the proof is therefore to show that A_k is bounded. We recall that

$$A_k=\frac{a^k\ldots a}{(a^k-1)\ldots(a-1)}\frac{1}{(a-1)\ldots(a^{n-k}-1)}\equiv P_1\cdot P_2.$$

The product

$$p_1\equiv\prod_{j=1}^{k}\frac{a^j}{a^j-1}=\prod_{j=1}^{k}\left(1+\frac{1}{a^j-1}\right)<\prod_{j=1}^{\infty}\left(1+\frac{1}{a^j-1}\right)$$

converges if the series $S=\sum_{j=1}^{\infty}1/(a^j-1)$ converges. However, this is true since $a>1$. The product P_2 is bounded for $a>1$ since for every $a>1$ there exists an integer $k(a)$ such that for $k>k(a)$ we have $a^k>2$ and hence $(a^k-1)^{-1}\leqslant 1$.

Hence A_k and therefore $\|L_{n+1}\|$ is bounded. The Theorem of Banach and Steinhaus 4.3.2 thus applies, and Theorem 8.2.1 is proved. ■

Theorem 8.2.1 states that the extrapolation process always converges if the step-sizes h_i decrease in geometric progression. From the point of view of numerical computation this result, although restrictive, is fortunate. This can be seen in the case where $h_{i+1}=\frac{1}{2}h_i$, i.e. the h_i is successively halved, for then the values T_i appear in T_{i+1} so that much repetitive function evaluation is avoided. However, although proving convergence of the extrapolation, the theorem gives no information about the quadrature procedure which is thus generated. For $h_i=H/2^i$ we obtain the Romberg quadrature procedure and we shall see now that this special quadrature has very important properties.

8.3 Properties of Romberg's quadrature

The Romberg quadrature has the following important striking properties: it is positive and it converges for every $f \in C^0[0, 1]$.

8.3.1 Positivity and convergence of Romberg's quadrature

We have already seen that for an odd number of nodes the single formulae of Romberg's quadrature coincide with Newton–Cotes formulae when these have only positive weights. We now prove that the Romberg formulae are all positive.

Theorem 8.3.1 *Romberg's quadrature procedure is positive.*

Proof We recall that (see p. 273 for $M_k f$)

$$T_0^{(k)} = \tfrac{1}{2}[T_0^{(k-1)} - M_{2^{k-1}} f], \qquad k = 1(1) \ldots.$$

Replacing $T_0^{(k-1)}$ successively and writing $M_k \equiv M_{2^k} f$ we have that

$$\begin{aligned} T_0^{(k)} &= \tfrac{1}{2}[\tfrac{1}{2} T_0^{(k-2)} + \tfrac{1}{2} M_{k-2} + M_{k-1}] \\ &= \ldots. \\ &= \frac{1}{2^k} [T_0^{(0)} + M_0 + 2M_1 + \ldots + 2^{k-1} M_{k-1}] = \frac{1}{2^k} \left[T_0^{(0)} + \sum_{i=0}^{k-1} 2^i M_i \right] \end{aligned}$$

which is clearly positive. Straightforward calculations give the formula

$$\begin{aligned} T_1^{(k-1)} &= \tfrac{1}{3}[4T_0^{(k)} - T_0^{(k-1)}] = \frac{2^{1-k}}{3} \left[T_0^{(0)} + \sum_{i=0}^{k-2} 2^i M_i + 2^k M_{k-1} \right] \\ &\equiv \frac{2^{1-k}}{3} \left[a_0^{(1)} \left(T_0^{(0)} + \sum_{i=0}^{k-2} 2^i M_i \right) + 2^{k-1} a_1^{(1)} M_{k-1} \right], \end{aligned}$$

with $a_0^{(1)} = 1$ and $a_1^{(1)} = 2$,

$$\begin{aligned} T_2^{(k-2)} &= \tfrac{1}{15}[16 T_1^{(k-1)} - T_1^{(k-2)}] \\ &= \frac{2^{2-k}}{15 \cdot 3} \left[7 a_0^{(1)} \left(T_0^{(0)} + \sum_{i=0}^{k-3} 2^i M_i \right) + (8 a_0^{(1)} - a_1^{(1)}) 2^{k-2} M_{k-2} \right. \\ &\qquad \left. + 2^{k+2} a_1^{(1)} M_{k-1} \right] \\ &\equiv \frac{2^{2-k}}{15 \cdot 3} \left[a_0^{(2)} \left(T_0^{(0)} + \sum_{i=0}^{k-3} 2^i M_i \right) + a_1^{(2)} 2^{k-2} M_{k-2} + a_2^{(2)} 2^{k-2} M_{k-1} \right], \end{aligned}$$

with $a_0^{(2)} = 7$, $a_1^{(2)} = 8a_0^{(1)} - a_1^{(1)} = 6$, $a_2^{(2)} = 16 a_1^{(1)} = 32$ and

$$\begin{aligned} T_3^{(k-3)} &= \tfrac{1}{63}[64 T_2^{(k-2)} - T_2^{(k-3)}] \\ &= \frac{2^{3-k}}{63 \cdot 15 \cdot 3} \left[a_0^{(3)} \left(T_0^{(0)} + \sum_{i=0}^{k-4} 2^i M_i \right) + 2^{k-3} \sum_{i=1}^{3} a_i^{(3)} M_{k+i-4} \right] \end{aligned}$$

with $a_0^{(3)}=(2^5-1)a_0^{(2)}=217$, $a_1^{(3)}=2^5a_0^{(2)}-a_1^{(2)}=218$, $a_2^{(3)}=2^6a_1^{(2)}-a_2^{(2)}=352$, $a_3^{(3)}=2^6a_2^{(2)}=2048$ (see Problem P.8.3). These lead to the pattern which can be verified by induction, that

$$T_m^{(k-m)}=\frac{2^{m-k}}{(4^m-1)\ldots(4-1)}\left[a_0^{(m)}\left(T_0^{(0)}+\sum_{i=0}^{k-m-1}2^iM_i\right)\right.$$
$$\left.+2^{k-m}\sum_{i=1}^{m}a_i^{(m)}M_{k+i-m-1}\right]$$

where

$$a_0^{(0)}=1,$$

and

$$a_0^{(m)}=(2^{2m-1}-1)a_0^{(m-1)}$$
$$a_1^{(m)}=2^{2m-1}a_0^{(m-1)}-a_1^{(m-1)}$$
$$a_i^{(m)}=2^{2m}a_{i-1}^{(m-1)}-a_i^{(m-1)},\qquad i=2(1)m-1$$
$$a_m^{(m)}=2^{2m}a_{m-1}^{(m-1)}.$$

For $m=4$, for example, we have that $a_0^{(4)}=27\,559$, $a_1^{(4)}=27\,558$, $a_2^{(4)}=55\,456$, $a_3^{(4)}=88\,064$, and $a_4^{(4)}=524\,288$. The coefficients $a_0^{(m)}$, $a_1^{(m)}$, $a_m^{(m)}$ are now calculated and the other $a_i^{(m)}$ can be estimated. Thus, we have

$$a_0^{(m)}=(2^{2m-1}-1)(2^{2m-3}-1)\ldots(2-1)$$
$$a_1^{(m)}=a_0^{(m)}+(-1)^{m+1}$$
$$a_m^{(m)}=2^{2m}\cdot2^{2m-2}\ldots2^4\cdot2$$

and the inclusion

$$3\cdot2^{2m-2}a_{i-1}^{(m-1)}<a_i^{(m)}<2^{2m}a_{i-1}^{(m-1)}\quad\text{for}\quad i=2(1)m-1,\qquad m\geqslant3,$$

which again can be easily shown by induction. Hence, the coefficients $a_i^{(m)}$ are always positive and, since $T_m^{(k-m)}$, $k=m(1)\ldots$ consists only of $T_0^{(0)}$ and midpoint sums, we have demonstrated the positivity of Romberg's scheme. ■

Having the positivity of Romberg's quadrature procedure we need to show its *convergence* for every polynomial because then its convergence for every $f\in C^0\,[0,1]$ follows from the Banach–Steinhaus Theorem 4.3.2. Let $f\in\mathbb{P}_n$. Then, on construction, $T(h)=C_0+C_1h+\ldots+C_nh^n\in\mathbb{P}_n$, but depends on h rather than x. Hence we need at most the $n+1$ values $T_0^{(k)}$ of $T(h)$ in order to achieve the identity $T(h)=\hat{T}(h)$. Thus, the scheme has polynomial accuracy and we have:

Theorem 8.3.2 *Romberg's quadrature scheme converges for every* $f\in C^0[0,1]$.

8.3.2 The error of the Romberg quadrature formulae

In order to obtain the error of the Romberg quadrature formulae we start with the general error expression for $T_0^{(k)}$ and then perform the extrapolation process. The order of the derivative is raised through integration by parts. We shall see that the Peano kernels of $T_i^{(k)}$ always have constant sign in $[0, 1]$ and hence a mean-value theorem applies and allows a simple error representation. By way of illustration, an example for the low-order formulae is given.

From Peano's kernel Theorem 3.1.1 we have that

$$\int_a^b f(x)\,\mathrm{d}x-\tfrac{1}{2}[f(a)+f(b)]=\int_a^b \tfrac{1}{2}(a-x)(b-x)f''(x)\,\mathrm{d}x,$$

which leads immediately to

$$\overset{*}{I}f-T_0^{(0)}=-\int_0^1 \tfrac{1}{2}x(1-x)f''(x)\,\mathrm{d}x\equiv\int_0^1 K_{2,1}(x)f''(x)\,\mathrm{d}x,$$

where the Peano kernel $K_{2,1}(x)$ satisfies $K_{2,1}(x)=K_{2,1}(1-x)$. Halving the interval $[0, 1]$ we obtain

$$\begin{aligned}\overset{*}{I}f-T_0^{(1)}&=-\int_0^{\frac{1}{2}} \tfrac{1}{2}x(\tfrac{1}{2}-x)f''(x)\,\mathrm{d}x+\int_{\frac{1}{2}}^1 \tfrac{1}{2}(\tfrac{1}{2}-x)(1-x)f''(x)\,\mathrm{d}x\\&\equiv\int_0^1 K_{3,1}(x)f''(x)\,\mathrm{d}x,\end{aligned}$$

where the Peano kernel $K_{3,1}(x)$ now has the properties

$$\begin{aligned}&K_{3,1}(x)=K_{3,1}(1-x),\\&K_{3,1}(x)=\tfrac{1}{4}K_{2,1}(2x)\quad\text{for}\quad x\in[0,\tfrac{1}{2}],\\&K_{3,1}(x)=\tfrac{1}{4}K_{2,1}(2x-1)\quad\text{for}\quad x\in[\tfrac{1}{2},1].\end{aligned}$$

Extrapolation in the usual Richardson manner gives

$$\begin{aligned}&\overset{*}{I}f-\tfrac{1}{3}[4T_0^{(1)}-T_0^{(0)}]=\tfrac{1}{3}\int_0^1 [4K_{3,1}(x)-K_{2,1}(x)]f''(x)\,\mathrm{d}x\\&=\tfrac{1}{3}\int_0^{\frac{1}{2}}[K_{2,1}(2x)-K_{2,1}(x)]f''(x)\,\mathrm{d}x+\tfrac{1}{3}\int_{\frac{1}{2}}^1[K_{2,1}(2x-1)-K_{2,1}(x)]f''(x)\,\mathrm{d}x\\&\equiv\int_0^1 \tilde{K}_{3,1}(x)f''(x)\,\mathrm{d}x\quad\text{(the factorials are included here)},\end{aligned}$$

where

$$\begin{aligned}\tilde{K}_{3,1}(x)&=\begin{cases}\tfrac{1}{3}[K_{2,1}(2x)-K_{2,1}(x)] & \text{for}\quad x\in[0,\tfrac{1}{2}]\\ \tilde{K}_{3,1}(1-x) & \text{for}\quad x\in[\tfrac{1}{2},0]\end{cases}\\&=\begin{cases}-\tfrac{1}{6}x(1-3x) & \text{for}\quad x\in[0,\tfrac{1}{2}]\\ \tfrac{1}{6}(1-x)(2-3x) & \text{for}\quad x\in[\tfrac{1}{2},0]\end{cases}.\end{aligned}$$

On integrating by parts twice we obtain for $x \in [0, \frac{1}{2}]$

$$K_{3,2}(x) \equiv \int_0^x \tilde{K}_{3,1}(t)\,dt = -\int_0^x \tfrac{1}{6}t(1-3t)\,dt = -\tfrac{1}{6}(\tfrac{1}{2}-x)x^2$$

and

$$K_{3,3}(x) \equiv \int_0^x K_{3,2}(t)\,dt = -\int_0^x \tfrac{1}{6}t^2(\tfrac{1}{2}-t)\,dt = -\tfrac{1}{72}x^3(2-3x).$$

Since $K_{3,2}(\frac{1}{2}) = 0$ we have that

$$K_{3,2}(1-x) = -K_{3,2}(x) \quad \text{and} \quad K_{3,3}(1-x) = K_{3,3}(x), \qquad x \in [0, \tfrac{1}{2}]$$

and thus $K_{3,2}(x)$ and $K_{3,3}(x)$ are defined for $[0, 1]$. All these kernel functions have zeros at $x=0$ and at $x=1$, so that two further integrations by parts give

$$\int_0^1 \tilde{K}_{3,1}(x) f''(x)\,dx = \int_0^1 K_{3,2}(x) f'''(x)\,dx = \int_0^1 K_{3,3}(x) f^{(4)}(x)\,dx,$$

where, of course, $K_{3,3}(x)$ is the Peano kernel of Simpson's rule (cf. Problem P.3,1). We note, in addition, that

$$K_{3,3}(1) = \int_0^1 K_{3,2}(x)\,dx = 0, \quad \text{and} \quad K_{3,3}(x) \leq 0 \text{ in } [0, 1],$$

$$\tilde{K}_{3,1}(0) = \tilde{K}_{3,1}(1) = 0, \qquad K_{3,3}(x) = K_{3,3}(1-x),$$

$K_{3,3}(x)$ is monotone in $[0, \frac{1}{2}]$ and in $[\frac{1}{2}, 1]$. In the same way we find that for $T_0^{(k)}$, $k > 1$

$$\overset{*}{I}f - T_0^{(k)} = \sum_{i=0}^{2^k-1} \int_{2^{-k}\cdot i}^{2^{-k}\cdot(i+1)} \tfrac{1}{2}\left(\frac{i}{2^k} - x\right)\left(\frac{i+1}{2^k} - x\right) f''(x)\,dx \equiv \int_0^1 K_{2^k+1,1}(x) f''(x)\,dx$$

with

$$K_{2^k+1,1}(x) = \tfrac{1}{2}\left(\frac{i}{2^k} - x\right)\left(\frac{i+1}{2^k} - x\right) \quad \text{for} \quad x \in [2^{-k}i, 2^{-k}(i+1)], \quad i = 0(1)2^k - 1,$$
$$= 4^{-k} K_{2,1}(2^k x - i) = K_{2^k+1,1}(1-x).$$

Hence, $K_{2^k+1,1}(x)$ is a periodic function with period 2^{-k} and in every period interval is a smaller version of $K_{2,1}(x)$; in addition, $K_{2^k+1,1}(x) = K_{2^k+1,1}(1-x)$. An analogous kernel is obtained if we replace k by $k+1$. Extrapolation gives

$$\overset{*}{I}f - \tfrac{1}{3}[4T_0^{(k+1)} - T_0^{(k)}] = \tfrac{1}{3}\int_0^1 [4K_{2^{k+1}+1,1}(x) - K_{2^k+1,1}(x)] f''(x)\,dx$$
$$\equiv \int_0^1 \tilde{K}_{2^{k+1}+1,1}(x) f''(x)\,dx.$$

The kernel $\tilde{K}_{2^{k+1}+1,1}(x)$ is also periodic with period 2^{-k} and is produced by piecing together smaller versions of $\tilde{K}_{3,1}(x)$. Thus

$$K_{2^{k+1}+1,2}(x)\equiv\int_0^x \tilde{K}_{2^{k+1}+1,1}(t)\,\mathrm{d}t \quad \text{and} \quad K_{2^{k+1}+1,3}(x)\equiv\int_0^x K_{2^{k+1}+1,2}(t)\,\mathrm{d}t$$

are periodic with period 2^{-k} and are made up of smaller versions of $K_{3,2}(x)$ and $K_{3,3}(x)$. Hence, as before,

$$\overset{*}{I}f-T_1^{(k)}=\int_0^1 K_{2^{k+1}+1,1}(x)f''(x)\,\mathrm{d}x=\int_0^1 K_{2^{k+1}+1,2}(x)f'''(x)\,\mathrm{d}x$$

$$=\int_0^1 K_{2^{k+1}+1,3}(x)f^{(4)}(x)\,\mathrm{d}x.$$

The second extrapolation must now be performed. Again, we consider the simple case $T_2^{(0)}=[16T_1^{(1)}-T_1^{(0)}]/15$, since, on constructing $T_2^{(k)}$, we again produce periodic kernels consisting of smaller versions of those belonging to $T_2^{(0)}$. We have

$$K_{3,3}(x)=\begin{cases}-x^3(2-3x)/72 & \text{for} \quad x\in[0,\tfrac{1}{2}]\\ K_{3,3}(1-x) & \text{for} \quad x\in[\tfrac{1}{2},1]\end{cases}$$

and hence

$$K_{5,3}(x)=\tfrac{1}{16}K_{3,3}(2x)-K_{3,3}(x), \qquad K_{5,3}(x)=K_{5,3}(1-x) \quad \text{for} \quad x\in[0,\tfrac{1}{2}].$$

Thus for $x\in[0,\frac{1}{2}]$ we have the kernel

$$\tilde{K}_{5,3}(x)\equiv\tfrac{1}{15}[16K_{5,3}(x)-K_{3,3}(x)]$$
$$=\tfrac{1}{15}\begin{cases}K_{3,3}(2x)-K_{3,3}(x) & \text{for} \quad x\in[0,\tfrac{1}{2}]\\ K_{3,3}(2x-1)-K_{3,3}(x) & \text{for} \quad x\in[\tfrac{1}{2},1].\end{cases}$$

The kernel $K_{3,3}(x)$ is monotonically decreasing in $[0,\frac{1}{2}]$ so that $K_{3,3}(2x)-K_{3,3}(x)\leqslant 0$ in $[0,\frac{1}{4}]$. For $x\in[\frac{1}{4},\frac{1}{2}]$ we have $K_{3,3}(2x)=K_{3,3}(1-2x)$, and hence for $x\in[\frac{1}{4},\frac{1}{2}]$

$$K_{3,3}(2x)-K_{3,3}(x)\begin{cases}\leqslant 0\\ \geqslant 0\end{cases} \text{for} \begin{cases}1-2x\leqslant x\\ 1-2x\leqslant x\end{cases} \text{or} \begin{cases}x\geqslant 1/3\\ x\leqslant 1/3\end{cases}.$$

Further, $K_{3,3}(x)$ is monotonically increasing for $x\in[\frac{1}{3},\frac{1}{2}]$ and since

$$\int_0^{\frac{1}{2}} K_{3,3}(2x)\,\mathrm{d}x=\tfrac{1}{2}\int_0^1 K_{3,3}(x)\,\mathrm{d}x=\tfrac{1}{2}\left[2\int_0^{\frac{1}{2}} K_{3,3}(x)\,\mathrm{d}x\right],$$

we have that

$$\int_0^{\frac{1}{2}} \tilde{K}_{5,3}(x)\,dx = \tfrac{1}{15}\left[\int_0^{\frac{1}{2}} K_{3,3}(2x)\,dx - \int_0^{\frac{1}{2}} K_{3,3}(x)\,dx\right] = 0.$$

Hence, $K_{5,4}(x) \equiv \int_0^x \tilde{K}_{5,3}(t)\,dt$ satisfies $K_{5,4}(0) = K_{5,4}(\frac{1}{2}) = 0$ and, in addition, $K_{5,4}(1-x) = -K_{5,4}(x)$. Let $K_{5,5}(x) \equiv \int_0^x K_{5,4}(t)\,dt$, then $K_{5,5}(0) = K_{5,5}(\frac{1}{2}) = 0$, and $K_{5,5}(1-x) = K_{5,5}(x)$ for $x \in [0, \frac{1}{2}]$. We integrate by parts twice again to obtain

$$\int_0^1 K_{5,3}(x) f^{(4)}(x)\,dx = \int_0^1 K_{5,4}(x) f^{(5)}(x)\,dx = \int_0^1 K_{5,5}(x) f^{(6)}(x)\,dx.$$

The essential properties of the kernels which allow simple integration by parts are maintained if this process is continued, as may be verified by induction. The essential details of the process are now given, although the above notation for the kernels is changed for convenience. Since the first index is uniquely determined by the second, we omit this and write K_{2j+1} instead of $K_{2^{k+i}+1,2j+1}$.

Lemma 8.3.1 (Bauer) If $K_1(x) = -\frac{1}{2}x(1-x)$, then the recursively-defined kernels

$$\tilde{K}_{2j-1}(x) = \frac{1}{4^j - 1}\begin{cases} K_{2j-1}(2x) - K_{2j-1}(x) & \text{for } x \in [0, \frac{1}{2}] \\ K_{2j-1}(2x-1) - K_{2j-1}(x) & \text{for } x \in [\frac{1}{2}, 1] \end{cases}, \qquad j = 1(1)\ldots$$

$$K_{2j}(x) \equiv \int_0^x \tilde{K}_{2j-1}(t)\,dt, \qquad K_{2j+1}(x) \equiv \int_0^x K_{2j}(t)\,dt$$

have the following properties:

(i) $K_j(1-x) = (-1)^{j+1} K_j(x)$,
(ii) $K_j(x) \leq 0$ in $[0, \frac{1}{2}]$,
(iii) $K_j(0) = K_j(1) = 0$, $K_{2j}(\frac{1}{2}) = 0$, $\qquad j = 1(1)\ldots$
(iv) $K_{2j-1}(x)$ decreases monotonically in $[0, \frac{1}{2}]$,

and $\overset{*}{I}f - T_j^{(k)} = \int_0^1 K_{2j-1}(x) f^{(2j)}(x)\,dx$ where $K_{2j-1}(x)$ has constant sign in $[0, 1]$. ■

Hence we may apply a mean-value theorem and obtain:

Theorem 8.3.3 *The error of $T_j^{(k)}$ is given by*

$$\overset{*}{I}f - T_j^{(k)} = f^{(2j)}(t) \int_0^1 K_{2j-1}(x)\,dx, \qquad t \in (0, 1).$$

For $k \neq 0$ the Peano kernel is a periodic function with period 2^{-k} formed as previously by piecing together smaller versions of the kernel with $k=0$. By calculating the integral of the Peano kernel exactly, we now obtain the error coefficient. This may be done by using properties of the Bernoulli polynomials which were introduced in Section 3.5.3. From the definition given there we obtain, by differentiation with respect to x, that

$$\frac{t^2 e^{tx}}{e^t-1}=\sum_{r=0}^{\infty} B_r'(x)\frac{t^r}{r!}$$

and, on comparing with the defining equation in Section 3.5.3, we have that $B_r'(x)=rB_{r-1}(x)$, $r=1(1)\ldots.$ This equation is integrated to give $B_r(x)=B_r(0)+r\int_0^x B_{r-1}(t)\,dt$, where $B_r=B_r(0)$ are the Bernoulli numbers. The polynomials $B_r(x)-B_r$ have essentially the same behaviour as the above kernels. Moreover, the Bernoulli polynomials have the following properties:

$$\int_0^1 B_{r-1}(x)\,dx=\frac{1}{r}(B_r(1)-B_r)=0 \quad \text{for} \quad r>1,$$

since $B_r(1)=B_r$ for $r>1$,

$$B_r(1-x)=(-1)^r B_r(x).$$

The first Bernoulli polynomials are

$$\begin{aligned}
B_0(x)&=1\\
B_1(x)&=x-1/2\\
B_2(x)&=x^2-x+1/6\\
B_3(x)&=x^3-3/2x^2+1/2x\\
B_4(x)&=x^4-2x^3+x^2-1/30\\
B_5(x)&=x^5-5/2x^4+5/3x^3-1/6x\\
B_6(x)&=x^6-3x^5+5/2x^4-1/2x^2+1/42\\
B_7(x)&=x^7-7/2x^6+7/2x^5-7/6x^3+1/6x.\\
B_8(x)&=x^8-4x^7+14/3x^6-7/3x^4+2/3x^2-1/30
\end{aligned}$$

so that

$$K_1(x)=(B_2(x)-B_2)/2 \quad \text{and} \quad \int_0^1 K_1(x)\,dx=-B_2/2=-1/12,$$

$$\tilde{K}_1(x)=\frac{1}{3!}(B_2(2x)-B_2(x))=\tfrac{1}{6}x(3x-1),$$

$$K_2(x)=\int_0^x \tilde{K}_1(t)\,dt=\tfrac{1}{6}x^2(x-\tfrac{1}{2})=\tfrac{1}{36}[(B_3(2x)-B_3)-2(B_3(x)-B_3)]$$

$$K_3(x)=\int_0^x K_2(t)\,dt=\tfrac{1}{72}x^3(3x-2)=\tfrac{1}{288}[(B_4(2x)-B_4)-4(B_4(x)-B_4)]$$

hence

$$\int_0^1 K_3(x)\,dx = \tfrac{1}{288}(\tfrac{1}{2} - B_4 + 4B_4) = \tfrac{1}{720}.$$

The notation is simplified by writing

$$b_k^{(0)}(x) \equiv (B_k(x) - B_k)/k! \qquad b_{2k+j}^{(2n)}(x) \equiv \frac{1}{2^{jn}}\,[b_{2k+j}^{(2n-2)}(2x) - 2^{jn} b_{2k+j}^{(2n-2)}(x)],$$

for $j = 0, 1, 2$, $k = 1(1)\ldots$, and $n = 1(1)\ldots$. Then

$$\begin{aligned}
&K_1(x) = b_2^{(0)}(x),\\
&\tilde{K}_1(x) = \tfrac{1}{3}[b_2^{(0)}(2x) - b_2^{(0)}(x)] = \tfrac{1}{3}b_2^{(2)}(x),\\
&K_2(x) = \tfrac{1}{3}\cdot\tfrac{1}{2}[b_3^{(0)}(2x) - 2b_3^{(0)}(x)] = \tfrac{1}{3}b_3^{(2)}(x),\\
&K_3(x) = \tfrac{1}{3}\cdot\tfrac{1}{4}[b_4^{(0)}(2x) - 4b_4^{(0)}(x)] = \tfrac{1}{3}b_4^{(2)}(x),\\
&\tilde{K}_3(x) = \tfrac{1}{3}\cdot\tfrac{1}{15}[b_4^{(2)}(2x) - b_4^{(2)}(x)] = \tfrac{1}{3}\cdot\tfrac{1}{15}b_4^{(4)}(x),\\
&K_4(x) = \tfrac{1}{3}\cdot\tfrac{1}{15}\cdot\tfrac{1}{4}[b_5^{(2)}(2x) - 4b_5^{(2)}(x)] = \tfrac{1}{3}\cdot\tfrac{1}{15}b_5^{(4)}(x),\\
&K_5(x) = \tfrac{1}{3}\cdot\tfrac{1}{15}\cdot\tfrac{1}{16}[b_6^{(2)}(2x) - 16b_6^{(2)}(x)] = \tfrac{1}{3}\cdot\tfrac{1}{15}b_6^{(4)}(x),\\
&\text{etc.}
\end{aligned}$$

It is easily seen, by induction, that this representation of the kernels is valid. If we do not start with $T_0^{(0)}$ but, instead, with $T_0^{(k)}$ then the corresponding kernel is a periodic function with period 2^{-k} and $K_1(x) = b_2^{(0)}(2^k x)/4^k$ in $[0, 2^{-k}]$. From then on everything proceeds as previously. For example, we have that

$$\int_0^1 K_1(x)\,dx = \int_0^1 K_{2^k-1,1}(x)\,dx = 2^k\int_0^{2^{-k}} \frac{1}{4^k}\, b_2^{(0)}(2^k x)\,dx = \frac{4^{-k}}{2}\,B_2,$$

and are able to calculate the error coefficient

$$\int_0^1 K_{2i+1}(x)\,dx = \prod_{j=1}^{i}(4^j - 1)^{-1}\int_0^1 b_{2i+2}^{(2i)}(x)\,dx$$

since the integrals of the Bernoulli polynomials vanish over $[0, 1]$. We have ($j = 2$, $k = i$, $n = i$) by successive substitution that

$$\int_0^1 b_{2i+2}^{(2i)}(x)\,dx = \frac{1-2^{2i}}{2^{2i}}\cdot\frac{1-2^{2i-2}}{2^{2i-2}}\cdot\ldots\cdot\frac{1-4}{4}\int_0^1 b_{2i+2}^{(0)}(x)\,dx$$

and that $\int_0^1 b_{2i+2}^{(0)}(x)\,dx = -B_{2i+2}/(2i+2)!$ giving finally

$$\int_0^1 b_{2i+2}^{(2i)}(x)\,dx = (-1)^{i+1}\left[\prod_{j=1}^{i}(4^j - 1)\right]B_{2i+2}\frac{1}{(2i+2)!\cdot 2^{i(i+1)}}$$

which corresponds to $T_i^{(0)}$. For $T_i^{(k)}$ we obtain an additional factor $4^{-(i+1)k}$ and hence have the following theorem:

Theorem 8.3.4 (Bauer) *The Romberg quadrature formulae $T_i^{(k)}$ have the error representation*

$$\overset{*}{I}f - T_i^{(k)} = (-1)^{i+1} \frac{1}{(2i+2)!} \frac{1}{4^{(i+1)k}} \frac{1}{2^{i(i+1)}} B_{2i+2} f^{(2i+2)}(t),$$

where $t \in (0, 1)$.

The Bernoulli numbers B_{2i+2} have the asymptotic expansion

$$B_{2i+2} \sim 2(-1)^i (2i+2)!/(2\pi)^{2i+2}, \quad \text{for large } i, \tag{8.3.1}$$

and hence:

Corollary 8.3.1 (Bauer) For large values of i the Romberg quadrature formulae $T_i^{(k)}$ have the asymptotic error representation

$$\overset{*}{I}f - T_i^{(k)} \sim \frac{-2}{2^{(i+1)(i+2+2k)} \pi^{2i+2}} f^{(2i+2)}(t), \qquad t \in (0, 1). \quad \blacksquare$$

Some absolute values of the error coefficient

$$\sigma_{i,k} = \frac{1}{(2i+2)!} \frac{1}{4^{(i+1)k}} \frac{1}{2^{i(i+1)}} B_{2i+2}$$

are tabulated in order to illustrate the rate of convergence to 0 (Table 8.3.1).

TABLE 8.3.1. Table of $\sigma_{i,k}$.

$k \backslash i$	0	2	4	6	8	10
0	8.33(−2)	5.17(−7)	1.99(−14)	3.04(−24)	1.82(−36)	4.24(−51)
2	5.21(−3)	1.26(−10)	1.90(−20)	1.13(−32)	2.65(−47)	2.41(−64)
4	3.26(−4)	3.08(−14)	1.81(−26)	4.22(−41)	3.85(−58)	1.37(−77)

8.3.3 Superlinear convergence of the Romberg quadrature formulae

If the integrand $f(x)$ is an analytic function within a domain containing the interval of integration, we can prove an unusual convergence result.

First, we define:

Definition 8.3.1 A series $\sum_{k=0}^{\infty} a_k$ is said to *converge superlinearly* if it converges asymptotically faster than any geometric series. ■

In general terms, 'superlinear convergence' means that the quotient a_{n+1}/a_n of the series tends to zero as $n \to \infty$. For example,

$$\sum_{n=1}^{\infty} a^n/n, \qquad \sum_{n=0}^{\infty} a^n/2^n, \qquad \sum_{n=0}^{\infty} \frac{a^n}{n!}, \qquad |a|<1 \text{ constant},$$

are convergent series, but only the third converges superlinearly.

Let C be a closed contour containing $[0, 1]$ and contained within the domain of analyticity of $f(x)$. Further, let d be the distance between C and the points of $[0, 1]$. Then, from Section 3.5.1, we have the bound

$$|f^{(2i+2)}(x)| \leq \frac{(2i+2)!}{d^{2i+3}} \frac{|C|M}{2\pi}.$$

and, from Theorem 8.3.4, we obtain

$$|\overset{*}{I}f - T_i^{(k)}| \leq \frac{|C|M}{2\pi d} \left(\frac{1}{d^2 2^{2k+i}}\right)^{i+1} |B_{2i+2}|.$$

Introducing the asymptotic expansion (8.3.1) of the Bernoulli numbers together with Stirling's formula for the factorials, we obtain the asymptotic expression

$$|B_{2i+2}| \sim 4(\pi(i+1))^{\frac{1}{2}} \left(\frac{i+1}{\pi e}\right)^{2i+2},$$

and hence

$$|\overset{*}{I}f - T_i^{(k)}| \sim \frac{|C|M}{2\pi d} \left(\frac{i+1}{\pi e d 2^{2k+i/2}}\right)^{2i+2} \equiv D_i,$$

which leads to

$$D_{i+1}/D_i \sim \left(\frac{i+2}{\pi e d \cdot 2^{i+k+1}}\right)^2 \left[\left(1+\frac{1}{i+1}\right)^{i+1}\right]^2 \doteq \left(\frac{i+2}{\pi d \cdot 2^{k+i+1}}\right)^2 \to 0$$

as $i \to \infty$. In particular, for fixed k, increasing i has the effect of moving along the diagonal lines of the triangular T-scheme thus producing superlinear convergence there.

Theorem 8.3.5 (Bauer, Rutishauser and Stiefel) *If $f(z)$ is analytic in a domain R containing the closed contour C which again contains the interval* $[0, 1]$, *then any diagonal of the T-table (i.e. the sequence $T_i^{(k)}$ with fixed k and moving i) converges superlinearly.*

8.4 Romberg cubature

Since the application of Richardson extrapolation assumes only the existence of a discretisation $T(h) = C_0 + C_1 h + C_2 h^2 + \ldots$, we are able to apply this technique also to cubature problems if we succeed in finding such a $T(h)$. We restrict our attention to the successive halving of h. Hence, the domain Ω must be subdivisible (cf. Sections 2.6.1 and 4.4). Now, we choose an elementary cubature formula in order to obtain approximate values $T_0^{(k)}$ for $I_\Omega f$. As examples we shall treat the square, the cube, the N-cube and the triangle. We choose elementary cubature formulae for these domains such that the non-vanishing elements of their expansion (see Section 2.2) correspond to the non-vanishing terms of the expansion of $I_\Omega f$. Then the resulting $T(h)$ always has an expansion in even powers of h, if f possesses sufficiently many derivatives. The above familiar extrapolation algorithm then applies. A remarkable difference between quadrature and cubature arises: the cubature formulae which are obtained in this way are not all positive, even if we apply the simplest known cubature formulae of degree 1 as elementary formulae for the calculation of $T_0^{(k)}$.

8.4.1 Romberg cubature for the square

We assume that Ω is the square $0 \leq |x|, |y| \leq 1$ and let

$$T_0^{(0)} \equiv C_4 f = f(\pm 1, \pm 1) \equiv f(1, 1) + f(1, -1) + f(-1, 1) + f(-1, -1).$$

We shall use this compact notation in the following to simplify the formulae. In addition, we define $f(\pm 1, 0) \equiv f(1, 0) + f(-1, 0)$.

Halving the step-size once in both dimensions we obtain

$$\begin{aligned} T_0^{(1)} = C_9 f &\equiv \tfrac{1}{4}[f(\pm 1, \pm 1) + 2(f(\pm 1, 0) + f(0, \pm 1)) + 4f(0, 0)] \\ &= \tfrac{1}{4} T_0^{(0)} + \tfrac{1}{2}[f(\pm 1, 0) + f(0, \pm 1) + 2f(0, 0)]. \end{aligned}$$

and, for a second time,

$$\begin{aligned} T_0^{(2)} = C_{25} f \equiv \tfrac{1}{16}\{&f(\pm 1, \pm 1) + 2[f(\pm 1, 0) + f(0, \pm 1) + f(\pm 1, \pm\tfrac{1}{2}) + f(\pm\tfrac{1}{2}, \pm 1)] \\ &+ 4[f(0, 0) + f(\pm\tfrac{1}{2}, \pm\tfrac{1}{2}) + f(0, \pm\tfrac{1}{2}) + f(\pm\tfrac{1}{2}, 0)]\} \\ = \tfrac{1}{4} T_0^{(1)} + \tfrac{1}{2}\{&f(\pm 1, \pm\tfrac{1}{2}) + f(\pm\tfrac{1}{2}, \pm 1) + 2[f(0, \pm\tfrac{1}{2}) + f(\pm\tfrac{1}{2}, 0) + f(\pm\tfrac{1}{2}, \pm\tfrac{1}{2})]\} \end{aligned}$$

$\ldots$.

$$T_0^{(k)} = C_{(2^k+1)^2} f \equiv \frac{1}{4^k} \left\{ f(\pm 1, \pm 1) + 2 \sum_{i=1}^{2^k-1} \left[f\left(\pm 1, -1 + \frac{i}{2^{k-1}}\right) + f\left(-1 + \frac{i}{2^{k-1}}, \pm 1\right) \right] \right.$$
$$\left. + 4 \sum_{i=1}^{2^k-1} \sum_{j=1}^{2^k-1} f\left(-1 + \frac{i}{2^{k-1}}, -1 + \frac{j}{2^{k-1}}\right) \right\}$$
$$= \tfrac{1}{4} T_0^{(k-1)} + \frac{1}{2^{2k-1}} \left\{ \sum_{i=1}^{2^{k-1}-1} \left[f\left(\pm 1, -1 + \frac{2i+1}{2^{k-1}}\right) + f\left(-1 + \frac{2i+1}{2^{k-1}}, \pm 1\right) \right] \right.$$
$$\left. + 2 \sum_{i=1}^{2^{k-1}-1} \sum_{j=1}^{2^{k-1}-1} f\left(-1 + \frac{2i+1}{2^{k-1}}, -1 + \frac{2j+1}{2^{k-1}}\right) \right\}.$$

On extrapolation we obtain

$$T_1^{(0)} = \tfrac{1}{3}(4T_0^{(1)} - T_0^{(0)}) = \tfrac{2}{3}[2f(0,0) + f(\pm 1, 0) + f(0, \pm 1)] \equiv C_5 f,$$
$$T_1^{(1)} = \tfrac{1}{3}(4T_0^{(2)} - T_0^{(1)}) = \tfrac{1}{6}\{2[f(\pm\tfrac{1}{2}, 0) + f(0, \pm\tfrac{1}{2}) + f(\pm\tfrac{1}{2}, \pm\tfrac{1}{2})]$$
$$+ f(\pm\tfrac{1}{2}, \pm 1) + f(\pm 1, \pm\tfrac{1}{2})\} \equiv C_{16} f.$$
$$T_2^{(0)} = \tfrac{1}{15}(16T_1^{(1)} - T_1^{(0)}) = \tfrac{2}{45}\{8[f(\pm\tfrac{1}{2}, 0) + f(0, \pm\tfrac{1}{2}) + f(\pm\tfrac{1}{2}, \pm\tfrac{1}{2})]$$
$$+ 4[f(\pm 1, \pm\tfrac{1}{2}) + f(\pm\tfrac{1}{2}, \pm 1)] - 2f(0,0) - f(\pm 1, 0) - f(0, \pm 1)\} \equiv C_{25} f$$

etc.

Here a cubature formula with negative weights appears for the first time. It is worth mentioning that $T_1^{(0)}$ has only 5 nodes, while $T_0^{(1)}$ has 9 and that $T_1^{(1)}$ has 16 nodes only while $T_0^{(2)}$ has 25. $T_2^{(0)}$ also has 25 nodes. Fortunately, from the first extrapolation some weights are zero and hence the corresponding nodes disappear. However, these same nodes re-enter in $T_2^{(0)}$ with negative weights! The appearance of negative weights is unfortunate and can cause an accumulation of rounding errors in critical examples. If $\|C_n\| \equiv \Sigma |A_i|$ becomes unbounded because of the negative weights, we cannot have convergence for every $f \in C^0(\Omega)$.

We now analyse the errors. Using Sard's kernel Theorem 3.2.2 we have the error representation

$$I_\Omega f - T_0^{(0)} = \int_0^1 -\tfrac{1}{2}u(1-u) f_{2,0}(u, 0)\, du + \int_0^1 -\tfrac{1}{2}v(1-v) f_{0,2}(0, v)\, dv$$
$$+ \lim_{\varepsilon \to 0} \int_\varepsilon^1 \int_\varepsilon^1 (uv - u - v + \tfrac{3}{4}) f_{1,1}(u, v)\, dv\, dv$$

which becomes (using Bernoulli polynomials and dropping the lim)

$$= \tfrac{1}{2} \int_0^1 (B_2(u) - B_2) f_{2,0}(u, 0)\, du + \tfrac{1}{2} \int_0^1 (B_2(v) - B_2 f_{0,2}(0, v)\, dv$$
$$+ \tfrac{1}{2} \int_0^1 \int_0^1 (2B_1(u)B_1(v) - B_1(u) - B_1(v)) f_{1,1}(u, v)\, du\, dv \quad (8.4.1)$$

since the Sard kernel of $T_0^{(0)}$ is obtained as

$$
\begin{aligned}
K_{1,1}(u,v) &= \int_0^1\int_0^1 (x-u)_+^0(y-v)_+^0\,dx\,dy \\
&\quad -\tfrac{1}{4}[(0-u)_+^0(0-v)_+^0+(1-u)_+^0(0-v)_+^0 \\
&\quad +(0-u)_+^0(1-v)_+^0+(1-u)_+^0(1-v)_+^0] \\
&= (1-u)(1-v)-\begin{cases} 1 & \text{for} \quad u=v=0 \\ \tfrac{3}{4} & \text{for} \quad u=0,\ \nu\neq 0 \quad \text{or} \quad u\neq 0,\ v=0 \\ \tfrac{1}{4} & \text{for} \quad u\neq 0 \quad \text{and} \quad v\neq 0. \end{cases}
\end{aligned}
$$

The discontinuities on the boundary of Ω do not affect the double integral of $K_{1,1}(u,v)f_{1,1}(u,v)$. Note that $K_{1,1}$ changes sign in Ω. Let us define by

$$
\begin{aligned}
&b_1(x)\equiv B_1(x),\quad b_k(x)\equiv(B_k(x)-B_k)/k! \quad \text{for} \quad k>1 \quad \text{and} \quad x\in[0,1) \\
&\qquad b_k(x+1)=b_k(x) \quad \text{for} \quad k\geqslant 1 \quad \text{and} \quad x\in[0,1),
\end{aligned}
$$

the periodically-continued Bernoulli polynomials. We then have analogously, that replacing u by x and v by y

$$
\begin{aligned}
I_\Omega f-T_0^{(k)} &= 4^{-k}\int_0^1 b_2(2^kx)f_{2,0}(x,0)\,dx+4^{-k}\int_0^1 b_2(2^ky)f_{0,2}(0,y)\,dy \\
&\quad +4^{-k}\int_0^1\int_0^1 b_1(2^kx)b_1(2^ky)f_{1,1}(x,y)\,dx\,dy \\
&\quad -2^{-k-1}\int_0^1\int_0^1 (b_1(2^kx)+b_1(2^ky))f_{1,1}(x,y)\,dx\,dy \\
&= 4^{-k}\int_0^1 b_2(2^kx)f_{2,0}(x,0)\,dx+4^{-k}\int_0^1 b_2(2^ky)f_{0,2}(0,y)\,dy \\
&\quad +4^{-k}\int_0^1\int_0^1 [(2^kx-1)(2^ky-1)-\tfrac{1}{4}]f_{1,1}(x,y)\,dx\,dy.
\end{aligned}
$$

Note that $4^{-k}=h_k^2$ and hence $T(h)=C_0+C_2h^2+\ldots$. This error representation does not reflect well the product-type property of Ω and $T_0^{(k)}$. If we apply tensor-product techniques (see Section 2.5.7 and 2.6.2) we obtain a product-type error representation. For this, see Remark R.8.11 and Problem P.8.4. On systematic integration twice by parts followed by extrapolation, such as in the one-dimensional case, we obtain an error representation of $I_\Omega f-T_i^{(k)}$ and, in addition, we see that $T(h)$ possesses an expansion in powers of h^2 if $f(x,y)$ has sufficiently many derivatives.

8.4.2 Romberg cubature for the N-cube

The previous formulae for the square are now extended to the N-cube. In order to illustrate the manner of generalisation we first write down the formulae for $N=3$. If $\Omega=\{(x_1, x_2, x_3) \mid, \ |x_1|, |x_2|, |x_3| \leqslant 1\}$ the elementary cubature formula is the eight-point formula

$$T_0^{(0)}=f(\pm 1, \pm 1, \pm 1) \equiv f(1,1,1)+f(1,1,-1)+f(1,-1,1)+f(1,-1,-1)$$
$$+f(-1,1,1)+f(-1,1,-1)+f(-1,-1,1)+f(-1,-1,-1)$$

and

$$\begin{aligned}
T_0^{(1)}&=\tfrac{1}{8}\{f(\pm 1, \pm 1, \pm 1)+4[f(\pm 1,0,0)+f(0, \pm 1,0)+f(0,0, \pm 1)] \\
&\quad +2[f(\pm 1, \pm 1,0)+f(\pm 1,0, \pm 1)+f(0, \pm 1, \pm 1)]+8 f(0,0,0)\} \\
&=\tfrac{1}{8} T_0^{(0)}+\tfrac{1}{4}\{f(\pm 1, \pm 1,0)+f(\pm 1,0, \pm 1)+f(0, \pm 1, \pm 1) \\
&\quad +2[f(\pm 1,0,0)+f(0, \pm 1,0)+f(0,0, \pm 1)]+4 f(0,0,0)\} \\
T_0^{(2)}&=\tfrac{1}{8} T_0^{(1)}+\tfrac{1}{32}\{f(\pm 1, \pm 1, \pm \tfrac{1}{2})+f(\pm 1, \pm \tfrac{1}{2}, \pm 1)+f(\pm \tfrac{1}{2}, \pm 1, \pm 1) \\
&\quad +2[f(\pm 1, \pm \tfrac{1}{2}, \pm \tfrac{1}{2})+f(\pm \tfrac{1}{2}, \pm 1, \pm \tfrac{1}{2})+f(\pm \tfrac{1}{2}, \pm \tfrac{1}{2}, \pm 1) \\
&\quad +f(\pm 1, \pm \tfrac{1}{2}, 0)+f(\pm 1,0, \pm \tfrac{1}{2})+f(\pm \tfrac{1}{2}, \pm 1,0)+f(\pm \tfrac{1}{2}, 0, \pm 1) \\
&\quad +f(0, \pm 1, \pm \tfrac{1}{2})+f(0, \pm \tfrac{1}{2}, \pm 1)] \\
&\quad +4[f(\pm \tfrac{1}{2}, \pm \tfrac{1}{2}, 0)+f(\pm \tfrac{1}{2}, 0, \pm \tfrac{1}{2}),+f(0, \pm \tfrac{1}{2}, \pm \tfrac{1}{2}) \\
&\quad +f(0,0, \pm \tfrac{1}{2})+f(0, \pm \tfrac{1}{2}, 0)+f(\pm \tfrac{1}{2}, 0,0)+f(\pm \tfrac{1}{2}, \pm \tfrac{1}{2}, \pm \tfrac{1}{2})]\},
\end{aligned}$$

with, more generally,

$$\begin{aligned}
T_0^{(k)}&=2^{-3k}\Big\{f(\pm 1, \pm 1, \pm 1)+2 \sum_{i=1}^{2^k-1}[f(\pm 1, \pm 1,-1+2^{k-1} i) \\
&\quad +f(\pm 1,-1+2^{k-1} i, \pm 1)+f(-1+2^{k-1} i, \pm 1, \pm 1)] \\
&\quad +4 \sum_{i=1}^{2^k-1} \sum_{j=1}^{2^k-1}[f(\pm 1,-1+2^{k-1} i,-1+2^{k-1} j) \\
&\quad +f(-1+2^{k-1} i, \pm 1.-1+2^{k-1} j)+f(-1+2^{k-1} i,-1+2^{k-1} j, \pm 1)] \\
&\quad +8 \sum_{i=1}^{2^k-1} \sum_{j=1}^{2^k-1} \sum_{l=1}^{2^k-1} f(-1+2^{k-1} i,-1+2^{k-1} j,-1+2^{k-1} l)\Big\} \\
&=\tfrac{1}{8} T_0^{(k-1)} \\
&\quad +2^{1-3k}\left\{\sum_{i=0}^{2^{k-1}-1}\left[f\left(\pm 1, \pm 1,-1+\frac{2 i+1}{2^{k-1}}\right)+f\left(\pm 1,-1+\frac{2 i+1}{2^{k-1}} \pm 1\right)\right.\right. \\
&\quad \left.+f\left(-1+\frac{2 i+1}{2^{k-1}}, \pm 1, \pm 1\right)\right]
\end{aligned}$$

$$+2\sum_{i=0}^{2^{k-1}-1}\sum_{j=0}^{2^{k-1}-1}\left[f\left(\pm 1,-1+\frac{2i+1}{2^{k-1}},-1+\frac{2j+1}{2^{k-1}}\right)\right.$$

$$+f\left(-1+\frac{2i+1}{2^{k-1}},\pm 1,-1+\frac{2j+1}{2^{k-1}}\right)$$

$$\left.+f\left(-1+\frac{2i+1}{2^{k-1}},-1+\frac{2j+1}{2^{k-1}},\pm 1\right)\right]$$

$$\left.+4\sum_{i=0}^{2^{k-1}-1}\sum_{j=0}^{2^{k-1}-1}\sum_{l=1}^{2^{k-1}-1}f\left(-1+\frac{2i+1}{2^{k-1}},-1+\frac{2j+1}{2^{k-1}},-1+\frac{2l+1}{2^{k-1}}\right)\right\}.$$

From these formulae we obtain, by extrapolation, that

$$\begin{aligned}T_1^{(0)}&=\tfrac{1}{3}(4T_0^{(1)}-T_0^{(0)})\\&=\tfrac{1}{6}\{-f(\pm 1,\pm 1,\pm 1)+2[f(\pm 1,\pm 1,0)+f(\pm 1,0,\pm 1)+f(0,\pm 1,\pm 1)]\\&\quad+4[f(\pm 1,0,0)+f(0,\pm 1,0)+f(0,0,\pm 1)]+8f(0,0,0)\},\quad\text{etc.}\end{aligned}$$

Some indication of the formulae $T_0^{(k)}$ for the N-dimensional case is given. Let $\Omega=\{(x_1,x_2,\ldots x_N)\,|\,0\leqslant|x_1|,|x_2|,\ldots,|x_N|\leqslant 1\}$, and the elementary cubature formula for $T_0^{(0)}$ be $T_0^{(0)}=f(\pm 1,\pm 1,\ldots,\pm 1)$. From this we obtain, recursively, that

$$T_0^{(k)}=2^{-N}T_0^{(k-1)}$$

$$+2^{1-Nk}\left\{\sum_{i_1=0}^{2^{k-1}-1}Pf\left(-1+\frac{2i_1+1}{2^{k-1}},\pm 1,\ldots,\pm 1\right)\right.$$

$$+2\sum_{i_1=0}^{2^{k-1}-1}\sum_{i_2=0}^{2^{k-1}-1}Pf\left(-1+\frac{2i_1+1}{2^{k-1}},-1+\frac{2i_2+1}{2^{k-1}},\pm 1,\pm 1,\ldots,\pm 1\right)$$

$$+\ldots$$

$$+2^{N-2}\sum_{i_1=0}^{2^{k-1}-1}\sum_{i_2=0}^{2^{k-1}-1}\cdots\sum_{i_{N-1}=0}^{2^{k-1}-1}Pf\left(-1+\frac{2i_1+1}{2^{k-1}},\ldots,-1+\frac{2i_{N-1}+1}{2^{k-1}},\pm 1\right)$$

$$+2^{N-1}\sum_{i_1=0}^{2^{k-1}-1}\sum_{i_2=0}^{2^{k-1}-1}\cdots\sum_{i_N=0}^{2^{k-1}-1}$$

$$\left.\times f\left(-1+\frac{2i_1+1}{2^{k-1}},-1+\frac{2i_2+1}{2^{k-1}},\ldots,-1+\frac{2i_N+1}{2^{k-1}}\right)\right\}$$

where P means that we have to sum over all permutations of the N components of $f(x_1,\ldots x_N)$ (such as $Pf(1,i,j)=f(1,i,j)+f(i,1,j)+f(i,j,1)$).

8.4.3 Romberg cubature for the triangle

If we choose an elementary cubature formula $T_0^{(0)}$ such as

$$T_0^{(0)} = \tfrac{1}{6}[f(0,0)+f(1,0)+f(0,1)] \doteq I_\Omega f,\ \Omega = \{(x,y)\,|\,0 \leq x+y \leq 1,\ x, y \geq 0\},$$

which takes into account the symmetries of the triangle, then the expansion of this formula corresponds to that of $I_\Omega f$. On successive halving in both dimensions we obtain formulae $T_0^{(k)}$ which are symmetric with respect to the triangle. Hence, $T(h)$ can be expressed in powers of h as $T(h) = C_0 + C_2h^2 + C_3h^3 + \ldots$ and an extrapolation process can be applied and leads to a Romberg cubature procedure for triangular domains. A strict proof of the existence of an expansion of $T(h)$ in powers of h is due to the application of Sard's kernel theorem as was pointed out in Section 8.4.1 for the square. We obtain the following formulae:

$$\begin{aligned}
T_0^{(1)} &= \tfrac{1}{24}\{f(0,0)+f(0,1)+f(1,0)+3[f(0,\tfrac12)+f(\tfrac12,0)+f(\tfrac12,\tfrac12)]\} \equiv C_6 f\\
&= \tfrac14 T_0^{(0)} + \tfrac18[f(0,\tfrac12)+f(\tfrac12,0)+f(\tfrac12,\tfrac12)]\\
T_0^{(2)} &= \tfrac14 T_0^{(1)} + \tfrac{1}{32}\{f(0,\tfrac14)+f(0,\tfrac34)+f(\tfrac14,0)+f(\tfrac34,0)+f(\tfrac14,\tfrac34)+f(\tfrac34,\tfrac14)\\
&\quad +2[f(\tfrac14,\tfrac14)+f(\tfrac14,\tfrac12)+f(\tfrac12,\tfrac14)]\}\quad \text{etc.}
\end{aligned}$$

$$\begin{aligned}
T_0^{(k)} &= \tfrac14 T_0^{(k-1)} + \frac{1}{2^{2k+1}}\Bigg\{2\sum_{i=0}^{2^{k-1}-1}\sum_{j=0}^{2^{k-1}-1}{}' f\left(\frac{2i+1}{2^{k-1}}, \frac{2j+1}{2^{k-1}}\right)\\
&\quad + \sum_{i=0}^{2^{k-1}-1}{}' \left[f\left(0, \frac{2i+1}{2^{k-1}}\right) + f\left(\frac{2i+1}{2^{k-1}}, 0\right) + f\left(\frac{2i+1}{2^{k-1}}, 1-\frac{2i+1}{2^{k-1}}\right)\right]\Bigg\}.
\end{aligned}$$

On extrapolation,

$$\begin{aligned}
T_1^{(0)} &= \tfrac13(4T_0^{(1)} - T_0^{(0)}) = \tfrac16[f(0,\tfrac12)+f(\tfrac12,0)+f(\tfrac12,\tfrac12)] \equiv C_3 f\\
T_1^{(1)} &= \tfrac13(4T_0^{(2)} - T_0^{(1)}) = \tfrac{1}{24}\{f(0,\tfrac14)+f(0,\tfrac34)+f(\tfrac14,0)+f(\tfrac34,0)+f(\tfrac14,\tfrac34)\\
&\quad +f(\tfrac34,\tfrac14)+2[f(\tfrac14,\tfrac14)+f(\tfrac14,\tfrac12)+f(\tfrac12,\tfrac14)]\} \equiv C_9 f\\
T_2^{(0)} &= \tfrac17(8T_1^{(1)} - T_1^{(0)}) = \tfrac{1}{42}\{2[f(0,\tfrac14)+f(0,\tfrac34)+f(\tfrac14,0)+f(\tfrac34,0)\\
&\quad +f(\tfrac14,\tfrac34)+f(\tfrac34,\tfrac14)]+4[f(\tfrac14,\tfrac14)+f(\tfrac12,\tfrac14)+f(\tfrac14,\tfrac12)]\\
&\quad -[f(0,\tfrac12)+f(\tfrac12,0)+f(\tfrac12,\tfrac12)]\} = C_{12} f
\end{aligned}$$

etc.

in which negative weights appear for the first time; note that the vertex nodes do not appear in $T_i^{(k)}$ and that they have the weight zero for $i \geq 1$.

The error representation of $T_0^{(0)}$ is obtained by applying Sard's kernel

Theorem 3.2.2 and we find that the kernel is

$$K_2(u)=\int_0^1\int_0^{1-x}(x-u)_+\,dy\,dx-\tfrac{1}{3}[2(0-u)_++(1-u)_+]$$
$$=\tfrac{1}{6}(1-u)(u^2-2u-1),$$
$$K_2(v)=\tfrac{1}{6}(1-v)(v^2-2v-1),$$

$$\begin{aligned}K_{1,1}(u,v)&=\tfrac{1}{2}\int_0^1\int_0^{1-x}(x-u)_+^0(y-v)_+^0\,dy\,dx\\&\quad+\tfrac{1}{2}\int_0^1\int_0^{1-y}(x-u)_+^0(1-v)_+^0\,dx\,dy\\&\quad-\tfrac{1}{3}[(0-u)_+^0(0-v)_+^0+(0-u)_+^0(1-v)_+^0+(1-u)_+^0(0-v)_+^0]\\&=\begin{cases}-\tfrac{1}{2} & u=v=0\\ -v/2-\tfrac{1}{4}(1-v)^2-\tfrac{5}{12} & u=0,\quad v\neq 0\\ -u/2-\tfrac{1}{4}(1-u)^2-\tfrac{5}{12} & v=0,\quad u\neq 0\\ \tfrac{1}{4}(1-u)^2+\tfrac{1}{4}(1-v)^2-\tfrac{1}{2}(1-u)-\tfrac{1}{2}(1-v)-\tfrac{1}{3}, & u\neq 0,\quad v\neq 0.\end{cases}\end{aligned}$$

Hence

$$I_\Omega f-T_0^{(0)}=\int_0^1 K_2(u)f_{2,0}(u,o)\,du+\int_0^1 K_2(v)f_{0,2}(0,v)\,dv$$
$$+\lim_{\varepsilon\to 0}\int_\varepsilon^1\int_\varepsilon^{1-u}K_{1,1}(u,v)f_{1,1}(u,v)\,dv\,du$$

since the discontinuities of $K_{1,1}(u, v)$ on the boundaries of Ω do not affect the value of the double integral. In a similar manner as for the square we obtain the error representation of $T_0^{(k)}$ by subsequent halving of h in both dimensions and continuing the kernel of $I_\Omega f-T_0^{(0)}$ as a periodic function in both dimensions.

8.5 Numerical integration of rapidly-oscillating functions

The numerical integration of rapidly-oscillating functions presents serious difficulties if we automatically apply the usual quadrature procedures to such functions. The difficulties usually arise from a mixture of slow convergence and numerical instability. A function is said to be rapidly oscillating if it possesses many maxima and minima in the interval of integration, for example, more than 20 extreme values. The functions

$\sin nx$, $x \in [0, 3]$ and $L_n(x)$ (the n-th Legendre polynomial) are rapidly-oscillating functions if n is large enough, say $n > 20$. Clearly, the product of a smooth function and a rapidly-oscillating function is again a rapidly-oscillating function, for example, $e^x \sin nx$, $x \in [0, 3]$, $n > 20$, or $x^{\frac{1}{2}} L_n(x)$, $x \in [0, 1]$, $n > 20$.

If we apply one of the familiar quadrature procedures, such as Gauss quadrature or Romberg quadrature, to a rapidly-oscillating function, we obtain during the calculation many positive and negative values which have to be added together. The sum of the positive values and the absolute value of the sum of the negative values are often nearly equal so that, on addition, the values essentially cancel thus causing significant loss of accuracy in the final result. It is difficult to prevent this in this situation. There is also a further difficulty which is closely connected with the oscillating behaviour of the integrand. The quadrature error is usually expressed in terms of high-order derivatives. Now, for rapidly-oscillating functions the absolute values of the derivatives often increase steadily as the degree of differentiation increases: we have the bound n^k for the k-th derivative of $\sin(nx)$ and the bound $k!\binom{n}{k}$ for the k-th derivative of the n-th Legendre polynomial $L_n(x) = x^n + \ldots$. Such unfortunate growth of the derivatives naturally causes slow convergence. There exist several possibilities for weakening or even circumventing the difficulties in integrating rapidly-oscillating functions; these will now be discussed.

8.5.1 Numerical integration between the zeros

Assume that for the rapidly-oscillating functions we know the zeros $0 \leq t_1 < t_2 < \ldots < t_n \leq 1$, at which the integrand changes sign. We then split the integral such that

$$\int_0^1 f(x)\,dx = \sum_{i=0}^{n} \int_{t_i}^{t_{i+1}} f(x)\,dx, \qquad t_0 = 0,\ t_{n+1} = 1.$$

In every interval $[t_i, t_{i+1}]$ we have the standard situation unless $f(x)$ is again a rapidly-oscillating function. This situation occurs if $f(x)$ does not change sign in $[t_i, t_{i+1}]$ while $f'(x)$ does. If this happens, we choose a function $g(x)$ (which can be simply and analytically integrated) so that $f(x) - g(x)$ changes sign for every oscillation. Having done this we treat the interval $[t_i, t_{i+1}]$ in the same way as the interval $[0, 1]$ before. Thus, the original situation arises with increased n and we now integrate numerically in every subinterval $[t_i, t_{i+1}]$. Since, usually, $f(t_i) = f(t_{i+1}) = 0$, the most economic quadrature formulae are those of Lobatto (see Section 7.4.3) for which the boundary points are nodes. Then the function values in these nodes need not be calculated. Assume first that we have the Lobatto quadrature formula

$$If \doteq Q_{m+2}f = A_0 f(0) + \sum_{i=1}^{m} A_i f(x_i) + A_{m+1} f(1)$$

and, further, that $f(t_i)=0$ for $i=1(1)n$. Then

$$\int_0^1 f(x)\,dx \doteq B_0 f(0) + \sum_{i=1}^{m}\sum_{j=1}^{n} A_i(t_{j+1}-t_j) f(t_j + x_i(t_{j+1}-t_j)) + B_{n+1} f(1)$$

where

$$B_0 = \begin{cases} A_0 t_1 & \text{if } t_1 \neq 0 \\ 0 & \text{if } t_1 = 0 \end{cases}, \qquad B_{n+1} = \begin{cases} A_{n+1}(1-t_n) & \text{if } t_n \neq 1 \\ 0 & \text{if } t_n = 1 \end{cases}.$$

We therefore have a total of $mn+2$ function evaluations whereas, in order to apply a Gaussian quadrature formula $Q_{n+1}f$ (having the same degree), we require a total of $mn+n$ function evaluations.

8.5.2 The use of weight functions

There are special cases where the integrand can be split into factors such that one is rapidly oscillating and the other is a smooth function. This situation arises, for example, in the calculation of Fourier transforms (cf. Section 1.2.1). We can then consider this rapidly-oscillating part as a weight function $W(x)$. There sometimes exist Gaussian or other quadrature formulae with positive weights with respect to $W(x)$, although the weight function $W(x)$ changes sign in [0, 1]. As we have seen in Section 7.4.6, this happens if $W(x)$ is the n-th Legendre polynomial $L_n(x)$ which changes sign n times in [0, 1]. The corresponding quadrature formulae are those of Kronrod–Piessens–Szegö. Using a simple trick, we circumvent the property of sign-changing of $W(x)$ as follows: $W_0 \equiv \sup|W(x)|$ for $0 \leqslant x \leqslant 1$ then $W(x)+W_0$ does not change sign in [0, 1]. Hence

$$\int_0^1 W(x)f(x)\,dx = \int_0^1 [W(x)+W_0]f(x)\,dx - W_0\int_0^1 f(x)\,dx$$

and there exist Gaussian quadrature formulae with respect to the weight function $W(x)+W_0$. We can construct such formulae and apply these to the first integral. The second integral is calculated as usual. Clearly, the price of this method is very high. Instead of n function evaluations for one integral we have $2n$ for the two integrals since, in general, the nodes are different. In addition, we have to set up the quadrature formulae with respect to the weight function $W(x)+W_0$.

8.5.3 Analytic integration of the oscillating part

Assume, as above, that the integrand is the product of a smooth function and a rapidly-oscillating multiplier which can be integrated analytically. Then it is reasonable to approximate the smooth multiplier by suitable functions so that the resulting product with the oscillating part can again be integrated analytically. For example, suppose that the smooth factor $f(x)$ in $\int_0^1 W(x)f(x)\,dx$ is approximated by functions $g_1(x), \ldots, g_n(x)$ such that the integrals $b_i = \int_0^1 W(x)g_i(x)\,dx$ can be easily evaluated or approximated. Assume, further, that the functions $g_i(x)$, $i=1(1)n$, are chosen to make $f(x) \approx \sum_{i=1}^{n} a_i g_i(x)$ a good approximation. Then

$$\int_{-1}^{1} W(x)f(x)\,dx \doteq \sum_{i=1}^{n} a_i \int_{-1}^{1} g_i(x)W(x)\,dx \equiv \sum_{i=1}^{n} a_i b_i$$

is also a good approximation. Frequently, the $g_i(x)$ are taken to be splines with local support. We consider the case where $W(x) = \sin n\pi x$ and the $g_i(x)$ are continuous splines of second order. The resulting quadrature method is usually called Filon's method (see [10]).

Example 8.5.1 Assume that we have to calculate an approximation to $\int_0^1 f(x) \sin n\pi x$, $n = 1(1)\ldots$ where $f(x)$ is a smooth function. We subdivide the interval $[0, 1]$ in $2m$ equidistant subintervals of length $1/2m$ and denote these by $[x_i, x_{i+1}] = [x_i, x_i + h] = [x_i, x_i + 1/2m]$, $i = 1(1)2m-1$. For any adjacent pair of subintervals $[x_i - h, x_i]$ and $[x_i, x_i + h]$ we replace $f(x)$ by a parabola passing through the three function values $f(x_i - h) \equiv f_{i-1}$, $f(x_i) \equiv f_i$, and $f(x_i + h) \equiv f_{i+1}$, $i = 1(2)2m-1$, so that $f(x) \approx a_i + b_i x + c_i x^2$, $x \in [x_i - h, x_i + h]$, $i = 1(2)2m-1$.

The interpolation requirements give three conditions for the unknown coefficients so that a_i, b_i, c_i are

$$a_i = \frac{1}{2h^2}[x_i x_{i-1} f_{i+1} - 2x_{i-1}x_{i+1} f_i + x_i x_{i+1} f_{i-1}]$$

$$b_i = \frac{-1}{2h^2}[(x_i + x_{i-1})f_{i+1} - 4x_i f_i + (x_i + x_{i+1})f_{i-1}]$$

$$c_i = \frac{1}{2h^2}[f_{i+1} - 2f_i + f_{i-1}].$$

These approximations are now integrated to give

$$\int_{x_i-h}^{x_i+h} (a_i + b_i x + c_i x^2) \sin n\pi x\,dx \equiv a_i J_{0,i} + b_i J_{1,i} + c_i J_{2,i}, \qquad i = 1(2)2m-1.$$

in which the $J_{k,i}$, $k=0, 1, 2$ are the integrals $J_{k,i} \equiv \int_{x_i-h}^{x_i+h} x^k \sin n\pi x \, dx$, with $n\pi = m\theta$, having the values

$$J_{0,i} = \frac{2}{m\theta} \sin\theta \sin i\theta$$

$$J_{1,i} = \frac{i}{m} J_{0,i} + \frac{2}{m^2\theta^2} (\sin\theta - \theta\cos\theta)\cos i\theta$$

$$J_{2,i} = \frac{1}{m^2\theta^2} (2-\theta^2(1-i^2))J_{0,i} + \frac{2i}{m} J_{1,i} + \frac{4}{m^3\theta^2} \sin i\theta \cos\theta$$

Hence

$$\int_0^1 f(x) \sin n\pi x \, dx \doteq \sum_{i=1(2)}^{2m-1} (a_i J_{0,i} + b_i J_{1,i} + c_i J_{2,i}).$$

The integrals $J_{k,i}$ are sometimes approximated by finite sums, as was done by Filon (cf. for example [10]). ■

8.6 Automatic integration–Adaptive methods

An automatic integration procedure generates the nodes and weights without further interference. It also improves the numerical result successively until a certain accuracy is achieved or some stopping parameter is exceeded. In the latter case automatic integration procedures commonly used behave in one of several ways. Some output an error message while many of them accept the final calculated approximation to be correct and output this (with the result that many errors pass undetected). It is better to stop the procedure not by printing the final approximate value without comment but at least to add some unsuppressible failure warning.

An example of an automatic integration procedure is the Romberg quadrature. Another type of automatic integration procedure are the *adaptive procedures*, the philosophy of which can be determined as follows.

(i) Divide the interval of integration into a certain number of (equidistant) subintervals D_i, $i=1(1)n$. The number n is specified either by the programmer or by the user.
(ii) Approximate the integral I_1 over the first subinterval D_1 with a suitable quadrature formula $Q_k f$ which is usually a low-degree formula. $Q_k f$ is an unchangeable component of the program.
(iii) Test whether the calculated value of I_1 is sufficiently accurate, or "too" accurate. This may be done, for example, by using one or more Richardson steps. Assume that we are given an error bound

e for the total error. Then we have to require e/n to be the error bound for the intervals D_i.

(iv) If the test shows that $|\int_{D_1} f(x)\,dx - Q_k f| < e/n$ then go to the next interval D_2 and do the same. If $|\int_{D_1} f(x)\,dx - Q_k f| \ll e/n$ (where "$\ll$" has to be specified in a practicable way) then enlarge (for example double) D_1 and start again the approximate calculation of I_1. If $|\int_{D_1} f(x)\,dx - I_1| > e/n$, then choose a smaller D_1 (for example by halving the original D_1) and start again. Thus, a first interval D_1 is obtained which has a "good" length. This means that the error is neither too small nor too large in comparison with the desired accuracy. (If the error is larger than e/n then we cannot guarantee that the total error is smaller than e while, if the error is too small, excessive unnecessary calculation is done.)

(v) Proceed to the second interval D_2. Usually the next interval has nearly the same "good" length as the previous one. Hence there is now less work to be done than for the first interval in order to obtain a "good" length D_2. Continue in this way until the interval of integration is exhausted.

(vi) Assume that $\bar{n}$ subintervals are in fact needed to exhaust the interval of integration. We then have the error bound $\bar{n}e/n$ for the total error and not the desired e. This is not a disadvantage if $\bar{n} \leq n$. Otherwise the calculation has to be repeated with smaller e (say e is replaced by $e/100$). The output should contain a true error bound as well as the approximate value of the integral.

If $\bar{n}$ exceeds n only slightly, then normally the result is still reliable since the bound $\bar{n}e/n$ is in most cases overpessimistic. The bound $\sqrt{\bar{n}}e/n$ is in general more realistic. However, if $\bar{n} \gg n$, for example if $\bar{n} = n^3$, then the total error bound is $\bar{n}e/n = n^2 e$, which may considerably exceed the limits of tolerance, especially if n is large. In this case e can be replaced, for example by $e/(\bar{n}/n) = ne/\bar{n}$ and then — hopefully — repetition of the procedure gives the desired accuracy.

If we specify $n = 20$ for example, and finally obtain $\bar{n} = 10\,000$, then $500e$ is a rigorous bound for the total error whereas $\sqrt{\bar{n}}e/n = 50e$ is a more realistic estimate. This simple example shows that the difference between n and $\bar{n}$ is usually not a very sensitive parameter.

The six above requirements should be satisfied by every adaptive and likewise by every automatic integration procedure in order to have reliable procedures. The possible number of different realisations of the six requirements, and hence of adaptive procedures, is so large that here we merely list the following references: [1], [4], [7], [11], [14–19], [21], [22].

8.7 Remarks, problems, references

8.7.1 Remarks

R.8.1: If we do not start with the trapezoidal rule in order to construct Romberg's scheme but, for example, with the midpoint rule instead, we again have an asymptotic expansion $T(h) = C_0 + h^2C_2 + h^4C_4 + \ldots$, if $f(x)$ has sufficiently many derivatives. The resulting quadrature scheme is not positive. We have, for example, $T_0^{(0)} = 2f(0)$, $T_0^{(1)} = f(\frac{1}{2}) + f(-\frac{1}{2})$ and hence, $T_1^{(0)} = \frac{1}{3}\{[4f(\frac{1}{2}) + f(-\frac{1}{2})] - 2f(0)\}$, which is not a positive quadrature formula (cf. Section 8.1.3).

R.8.2: In Section 2.5.1 we introduced a very general form of Lagrange interpolation. A corresponding version of a general Neville–Aitken algorithm is presented in [9]. This may also be used for extrapolation but a proof of convergence is much more difficult. Laurent [13] also presents a very general extrapolation process, which is not based upon *polynomial* Lagrange interpolation and for which he proves convergence. Stoer [24] performs rational extrapolation using an interpolation algorithm which is very similar to that of Aitken and Neville. The properties of the resulting quadrature scheme were considered by Bulirsch and Stoer [6].

R.8.3: A good survey on papers and material concerned with Richardson extrapolation, especially in the context of Romberg quadrature, is given by Joyce [12], who reviews the literature up to 1969 (cf. Section 8.1.2).

R.8.4: It is an open question whether Romberg's quadrature procedure can be shown to be an interpolatory quadrature procedure of Gauss-type or, better, of Lobatto-type (since all formulae contain as nodes the limits of integration). This can be conjectured since Romberg's procedure has all significant properties of Lobatto quadratures: positivity and convergence for every $f \in C^0[-1, 1]$. Further, trivially the nodes are always real, distinct and in $[-1, 1]$ (cf. Section 8.3).

R.8.5: Instead of the rectangle rule we can start the extrapolation, for example, with the Gaussian two-point rule $Q_2f = f(-1/\sqrt{3}) + f(1/\sqrt{3})$. Then

$$T(h) = \frac{h}{2}\sum_{i=0}^{n-1}\left[f\left(-1 + \frac{2i+1}{2}h - \frac{1}{2\sqrt{3}}h\right) + f\left(-1 + \frac{2i+1}{2}h + \frac{1}{2\sqrt{3}}h\right)\right] \doteq If, \qquad h = 2/n$$

possesses an expansion $T(h) = C_0 + C_4h^4 + C_6h^6 + \ldots$, if $f(x)$ has sufficiently many derivatives. The first step of extrapolation which eliminates the term C_2h^2 cannot improve the result, since this term is not present in $T(h)$. We therefore omit the column $T_1^{(k)}$ and immediately calculate $T_2^{(k)} = \frac{1}{15}(16T_0^{(k+1)} - T_0^{(k)})$ (cf. Section 8.1.3). Analogously, let Q_mf have degree of precision $2m-1$, $m > 0$. Then

$$T(h) = \frac{h}{2}\sum_{i=0}^{n-1}\sum_{j=1}^{m} A_jf\left(-1 + \frac{2i+1}{2}h - \frac{h}{2}x_i\right), \qquad h = 2/n$$

possesses the expansion $T(h) = C_0 + C_{2m}h^{2m} + C_{2m+2}h^{2m+2} + \ldots$, if $f(x)$ possesses sufficiently many derivatives and if Q_nf is symmetric. Hence, the elimination of the terms C_2h^2, $C_4h^4, \ldots, C_{2m-2}h^{2m-2}$ which is executed in the columns $T_1^{(k)}, T_2^{(k)}, \ldots, T_{m-1}^{(k)}$ cannot improve the result since these terms are not contained in $T(h)$. In this case we start by calculating the column $T_m^{(k)}$ immediately by means of $T_m^{(k)} = (4^m T_0^{(k+1)} - T_0^{(k)})/(4^m - 1)$. This application is most important for improving such approximations, which are obtained from Gaussian quadrature formulae Q_nf since then $n = m$.

R.8.6: If the integrand $f(x)$ has a singularity (such as $\sqrt{1+x}$ in $[-1, 1]$), then of course no expansion of $T(h)$ in integer powers of h exists. In such cases the extrapolation algorithm does not improve the result efficiently, although convergence does take place; it is, however, very slow (cf. Section 8.1).

R.8.7: If the basic quadrature formula for which an extrapolation process is intended happens not to be symmetric, then provided that $f(x)$ possesses sufficiently many derivatives we do not have an expansion of $T(h)$ in powers of h^2 but, instead, of h (cf. Section 8.1).

R.8.8: If we do not know whether there exists an expansion of $T(h)$ in powers of h, then Romberg's quadrature scheme assures only the convergence, which may be extremely slow. In this case we can concentrate on reducing the absolute value of the error coefficient instead of raising the degree of precision. This is to be performed, for example, by the use of interpolating splines instead of Lagrange interpolation. This is done in [9], and the resulting quadrature procedures are shown to be positive and convergent for every $f \in C^0\,[-1, 1]$. They have constant degree of precison (cf. Section 8.1).

R.8.9: The Romberg procedure does not take into account weight functions as do interpolatory quadratures. This is due to the fact that every given weight function is fixed for the whole interval of integration. Therefore, on subdividing this interval, the weight function changes from one subinterval to the next. The construction of the corresponding elementary quadrature formulae for the calculation of $T_0^{(k)}$ is tedious.

R.8.10: We do not need to halve the interval of integration successively in order to apply Richardson's extrapolation process. In the method of halving, the number of function evaluations grows rapidly. This growth has been overcome by Rutishauser [23], who interpolates $T(h)$ at additional points between any two values of the sequence $\{h/2^i,\ i = 0(1)\ldots\}$. His sequence is $\{h, h/2, h/3, h/4, h/6, \ldots\}$ or, more generally, $\left\{h,\ h/2^i,\ \frac{h}{3}\Big/2^i,\ i = 1(1)\ldots\right\}$. The corresponding extrapolation algorithm is more complicated than (8.1.1) and consists of two stages which are alternately applied to calculate the columns of the T-scheme. For further details the reader is referred to [23].

R.8.11: If Ω is the square $0 \leq x, y \leq 1$, then it is a product-type region and the cubature formulae $T_0^{(k)}$ are product formulae. Let $\bar{T}_n^x f(x, y)$ denote the trapezoidal rule with n nodes applied to $f(x, y)$ with respect only to x. Then $\bar{T}_n^x f(x, y)$ is a function of y and we can apply $\bar{T}_n$ with respect to y writing $T_0^{(k)} = \bar{T}_{2^k+1}^x \bar{T}_{2^k+1}^y f(x, y)$. If we apply this product operator to a monomial $x^m y^n$ we already know, from the one-dimensional case, that $\bar{T}_k^x \bar{T}_k^y x^m y^n = \bar{T}_k^x x^m \bar{T}_k^y y^n$ is a polynomial in even powers of $h = 2^{-k}$, since this is true for $\bar{T}_k^x x^m$ and for $\bar{T}_k^y y^n$. Hence, $T_0^{(k)}$ possesses an expansion in powers of h^2 if $f(x, y)$ possesses a sufficiently large number of partial derivatives. We apply the tensor-product technique in order to obtain another error expansion. For one dimension we had

$$\overset{*}{I}f - T_0^{(k)} = \overset{*}{I}f - \bar{T}^x_{2^k+1}f(x) = 4^{-k}\int_0^1 b_2(2^k x)f''(x)\,dx$$

and hence, for two dimensions

$$\int_0^1\int_0^1 f(x, y)\,dx\,dy = \overset{*}{I}(x)\overset{*}{I}(y)f - \bar{T}^x_{2^k+1}\bar{T}^y_{2^k+1}$$

$$= \overset{*}{I}(x)[\overset{*}{I}(y)f - \bar{T}^y_{2^k+1}f] + \bar{T}^y_{2^k+1}[\overset{*}{I}(x)f - \bar{T}^x_{2^k+1}f]$$

$$= 4^{-k}\int_0^1\int_0^1 b_2(2^k y)f_{0,2}(x, y)\,dx\,dy$$

$$+4^{-k}\int_0^1\int_0^1 b_2(2^k x)f_{2,0}(x, y)\,dx\,dy$$

$$+4^{-2k}\int_0^1\int_0^1 b_2(2^k x)b_2(2^k y)f_{2,2}(x, y)\,dx\,dy.$$

We integrate by parts twice in order to obtain partial derivatives of the same order. This is only necessary for the third integral in order to obtain the total order 2 or, for the first and second integrals, in order to obtain the total order 4. In the first case we obtain

$$I_\Omega f - T_0^{(k)} = 4^{-k}\int_0^1\int_0^1 [b_2(2^k x)f_{2,0} + b_2(2^k y)f_{0,2} - b_1(2^k x)b_1(2^k y)]f_{1,1}\,dx\,dy.$$

8.7.2 Problems

P.8.1: Calculate the quadrature formulae corresponding to the $T_3^{(k)}$ of Romberg's quadrature scheme (cf. Section 8.1.3).

P.8.2: Compute approximate values for $\int_0^{10}\exp(x^2)\,dx$ using Romberg's scheme and compare with Gaussian approximations up to 80 nodes (cf. Section 8.1.3).

P.8.3: Show that in Romberg's scheme the expansion

$$T_3^{(k)} = \frac{2^{-k}}{63\cdot 15\cdot 3}\left[a_0^{(3)}\left(T_0^{(0)} + \sum_{i=0}^{k-1} 2^i M_i\right) + 2^k\sum_{i=1}^{3} a_i^{(3)} M_{k+i-1}\right]$$

with $a_0^{(3)} = 217$, $a_1^{(3)} = 218$, $a_2^{(3)} = 352$, and $a_3^{(3)} = 2048$ is valid (cf. Section 8.3.1).

P.8.4: Show that the Sard kernel error representation (8.4.1) and the tensor product representation in Remark R.8.11 are equal. Hints: Start with the tensor product representation of Remark R.8.11, reduce the order of the partial derivatives $f_{2,0}(x, y)$ and $f_{0,2}(x, y)$ on integration by parts, replace $f_{1,0}(x, 1) = f(x, 0) + \int_0^1 f_{1,1}(x, y)\,dy$, and $f_{0,1}(1, y)$ correspondingly. Finally, raise the order of the derivatives in the one-dimensional integrals by integration by parts (cf. Section 8.4.1).

P.8.5: Look for an elementary cubature formula for the domain $\Omega = \{(x, y) \mid, |x|, |y| \leq 1\}$ such that the Romberg cubature formulae $T_2^{(k)}$ are positive (cf. Section 8.4.1).

8.7.3 References

1. Amble, O. (1969). An autoadaptive method for numerical integration. *Norske Vid. Selsk. Forh. (Trondheim)*, **42,** 43–50.
2. Bauer, F. L. (1961). La méthode d'intégration numérique de Romberg. *Centre Belg. Rech. Mathém. Coll. Anal. Numér. Mons*, 119–129.
3. Bauer, F. L., Rutishauser, H. and Stiefel, E. (1963). New aspects in numerical quadrature. *Proc. Sympos. Appl. Mathem.*, **15,** 199–218.
4. de Boor, C. (1971). On writing an automatic integration algorithm, *In* "Mathematical Software," (J. R. Rice, Ed.). Academic Press, New York.
5. Bulirsch, R. (1964). Bemerkungen zur Romberg Integration. *Num. Math.*, **6,** 1–13.
6. Bulirsch, R. and Stoer, J. (1967). Numerical quadrature by extrapolation. *Num. Math.*, **9,** 271–278.
7. Cranley, R. and Patterson, T. N. L. (1971). On the automatic numerical evaluation of definite integrals. *Comput. J.*, **14,** 189–198.
8. Engels, H. (1971). "Richardson-Extrapolation mit Hilfe kubischer Splines." Report Jül–791-Ma. Nuclear Research Centre Jülich.
9. Engels, H. (1971). "Über einige allgemeine lineare Interpolationsoperatoren und ihre Anwendung auf Quadratur und Richardson-Extrapolation." Report Jül–831–Ma. Nuclear Research Centre Jülich,
10. Filon, L. N. G. (1928/29). On a quadrature formula for trigonometric integrals. *Proc. Roy. Soc. Edin.* **49,** 38–47.
11. O'Hara, H. and Smith, F. J. (1969). The evaluation of definite integrals by interval subdivision. *Computer J.*, **12,** 179–182.
12. Joyce, D. C. (1970). "A Survey of Extrapolation Processes in Numerical Analysis." Technical Report No. 8, University of Newcastle upon Tyne.
13. Laurent, P. J. (1964). Formules de quadrature approchée sur domaines rectangulaires convergentes pour toute fonction intégrable Riemann. *C.R. Acad. Sci. Paris*, **258,** 798–801.
14. Lyness, J. N. (1969). The effect of inadequate convergence criteria in automatic routines. *Computer J.*, **12,** 279–281.
15. Lyness, J. N. (1969). Notes on the adaptive Simpson quadrature routine. *J. ACM*, **16,** 483–495.
16. Miller, E. K. (1970). A variable interval width quadrature technique based on Romberg's method. *J. Comput. Phys.*, **5,** 265–279.
17. Rice, J. R. (1973). "A Metalgorithm for Adaptive Quadrature." CSD–TR 89 Purdue University, Lafayette, Indianna.
18. Rice, J. R. (1973). An educational adaptive quadrature algorithm. *SIGNUM Newsletter*, **8,** 27–41,

19. Rice, J. R. (1973). "Parallel Algorithms for Adaptive Quadrature—Convergence." CSD–TR 104 Purdue University, Lafayette, Indianna.
20. Richardson, L. F. (1927). The deferred approach to the limit. *Phil. Transact. Roy. Soc. London* (Series A), **226,** 299–342.
21. Robinson, I. G. (1971). Adaptive Gaussian integration. *Austral. Comput. J.*, **3,** 126–129.
22. Russell, D. L. "Numerical Integration by Midpoint Procedure with Preferential Interval Placement." SHARE File No. 0704–1017 and 107.
23. Rutishauser, H. (1963). Ausdehnung des Romberg'schen Prinzips. *Num. Math.*, **5,** 48–54.
24. Stoer, J. (1961). Über zwei Algorithmen zur Interpolation mit rationalen Funktionen. *Num. Math.*, **3,** 285–305.

See also: Patterson, T. N. L., on p. 417.

9

Auxiliary Material

We present some auxiliary material which facilitates the application of numerical quadrature and cubature, and should help those who are doing research on this subject. This includes bibliographical material, lists of functions for testing quadrature subroutines, published programs, and tables of nodes and weights.

9.1 Bibliographic material

Most papers concerned with numerical quadrature and cubature up to 1910 are reviewed in Runge, C. and Willers, F. A., "Enzyklopädie der mathematischen Wissenschaften," Vol. 2, Analysis 3.1, p. 50 ff. No comparable review of subsequent material exists; perhaps because the field has expanded too much. Nevertheless, some recent useful collections of papers do exist in such books as those by Davis and Rabinowitz, and by Stroud.

Many papers concerned mainly with spline functions and spline analysis, as far as these are involved with quadrature and cubature, have been collected by van Roij and Schurer and published in: Böhmer, K., Meinardus, G. and Schempp, W. "Spline-Funktionen." pp. 315–415. Bibliographisches Institut Mannheim/Wien/Zürich. The bibliography of Stroud should also be mentioned. This was published in 1961 in *Mathematics of Computation*, **15,** 52–80.

Many papers concerning theory and application of Richardson extrapolation, especially if applied to quadrature problems, are carefully listed and reviewed by Joyce, D. C. (1970). A survey of extrapolation processes in numerical analysis. *Tech. Report,* Series 8.

Papers which are concerned with automatic integration are collected in: De Doncker, E. and Piessens, R. (1976). A bibliography on automatic integration. *J. Comput. App. Math.*, **2,** 273–280. See also Dixon, V.

(1974) in "Software for Numerical Mathematics" (ed. D. J. Evans), Academic Press, New York, pp. 105–137.

Finally, we make reference to the report by Engels, H. (1972). "Eine Zusammenstellung von Arbeiten über Numerische Quadratur and Kubatur". *Report, Jül-Bibl.*, **16** Nuclear Res. Centre Jülich, which lists about 1400 titles published between 1910 and 1970. Review sources, as far as they were available, have been added there.

9.2 Testing quadrature subroutines

There exists a vast number of quadrature programs and subroutines which are usually reliable for the problems for which they were constructed, but which do not necessarily perform so well for completely different problems. It is therefore desirable to have a broad collection of test functions for which a subroutine under consideration can be checked. To date there exist few tables of such test functions. We list those of which we are aware and amalgamate the results which they contain in Table 9.2.1.

TABLE 9.2.1.

Integrand		Interval	Value of the integral
	k		
$P_k(x)=\sum_{j=0}^{k-1} x^{k-j-1}(j+1)(-1)^j$	1	[0, 1)	1
	2		−3/2
	3		7/3
for $k=1(1)20$	4		−35/12
	5		37/10
	6		−259/60
	7		533/105
	8		−1599/280
	9		1627/252
	10		−17897/2520
	11		18107/2310
	12		−235391/27720
	13		237371/25740
	14		−237371/24024
	15		95549/9009
	16		−1624333/144144
	17		1632341/136136
	18		−31014479/2450448
	19		155685007/11639628
	20		−155685007/11085360

TABLE 9.2.1. (*Continued*)

Integrand		Interval	Value of the integral
$(10x-1)(10x-1.1)(10x-1.2)(10x-1.3)$		[0, 1]	1627879/1500
$\frac{1}{x^k}$, $k=3(1)5$	k 3 4 5	[0.01, 1.1]	 $5000-1/2.42$ $(1.0(06)-1/1.331)/3$ $2.5(07)-1/5.8564$
$\frac{1}{1+x}$		[0, 1]	$\ln 2 = 0.69314718\ldots$
$\frac{1}{1+x^2}$		[0, 1]	$\frac{\pi}{4} = 0.7853981635\ldots$
$\frac{1}{1+x^4}$		[0, 1]	$\left(\ln\left(\frac{2+\sqrt{2}}{2-\sqrt{2}}\right)+\pi\right)\Big/\sqrt{32}$ $= 0.8669729873\ldots$
$\frac{1}{x^4+x^2+1}$		[−1, 1]	$\frac{\ln 9}{4}+\frac{\pi}{\sqrt{12}} = 1.456205826\ldots$
$\frac{1}{x^2+1.005}$		[−1, 1]	$\frac{2}{\sqrt{1.005}}\text{arc tg}\frac{1}{\sqrt{1.005}}$ $= 1.564396444\ldots$
$\frac{50}{\dot{\pi}(2500x^2+1)}$, $\dot{\pi}=3.14159$		[0, 10]	$\frac{\text{arc tg } 500}{\dot{\pi}} = 0.4993633811\ldots$
$\frac{1}{x^2+10^{-k}}$, $k=2(1)6$		[−1, 1]	$2\cdot 10^{k/2}\,\text{arc tg}\,(10^{k/2})$
$\frac{1}{1+5x^2}$		[0, 1]	$\frac{\text{arc tg}\sqrt{5}}{\sqrt{5}} = 0.5144128002\ldots$
$\frac{1}{1+10x^2}$		[0, 1]	$\frac{\text{arc tg}\sqrt{10}}{\sqrt{10}} = 0.3998760026\ldots$
$\frac{1}{1-0.5x^2}$		[0, 1]	$\ln\left(\frac{1+\sqrt{0.5}}{1-\sqrt{0.5}}\right)\Big/2\sqrt{0.5}$ $= 1.246450479\ldots$

TABLE 9.2.1. (*Continued*)

Integrand			Interval	Value of the integral
$\frac{1}{1-0.98x^2}$			[0, 1]	$\ln\left(\frac{1+\sqrt{0.98}}{1-\sqrt{0.98}}\right)/(2\sqrt{0.98})$ $=2.670965332\ldots$
$\frac{1}{1-0.998x^2}$			[0, 1]	$\ln\left(\frac{1+\sqrt{0.998}}{1-\sqrt{0.998}}\right)/(2\sqrt{0.998})$ $=3.803756264\ldots$
$\frac{2^r}{1+2^r(x-\beta)^2}$	β	r		
		2	[0, 1]	π
$r=2(1)9,\quad \beta=\frac{1}{2},\frac{1}{3}$	$\frac{1}{2}$	>2		$2^{(r+2)/2}\operatorname{arc\,tg}(2^{(r-2)/2})$
	$\frac{1}{3}$	$\geqslant 2$		$2^{r/2}(\operatorname{arc\,tg}(\frac{2}{3}2^{r/2})+\operatorname{arc\,tg}(\frac{1}{3}2^{r/2}))$
$\sqrt{x^3}$			[0, 1]	0.4
$\sqrt{x^5}$			[0, 1]	2/7
$\sqrt[2^k]{x},\quad k=1(1)4$			[0, 1]	$2^k/(1+2^k)$
		k		
$\sqrt{\|x^2-1/4\|^k},\quad k=1(2)5$		1		$0.4647425054\ldots$
		3	[0, 1]	$0.1488716212\ldots$
		5		$0.06551476837\ldots$
$1/\sqrt{\|x\|}$			[−9, 100]	26
$\sqrt[5]{x}$			[0, 1]	5/6
$\sqrt[10]{x}$			[0, 1]	10/11
$\sqrt{\|x+0.5\|}$			[−1, 1]	$\sqrt{1.5}+\sqrt{0.5}/3=1.460447131\ldots$
$\|x-\beta\|^\alpha,\quad \alpha=0.1(0.1)0.5,\quad \beta=0,\frac{1}{2},\frac{1}{3}$			[0, 1]	$\frac{\beta^{\alpha+1}+(1-\beta)^{\alpha+1}}{\alpha+1}$
$\|x-\beta\|^\alpha\ln\|x-\beta\|$ $\alpha=0.1(0.1)0.5$ $\alpha=-0.55(-0.05)-0.75$ $\beta=0,\frac{1}{2},\frac{1}{3}$			[0, 1]	$\frac{\beta^{\alpha+1}}{(\alpha+1)^2}[(\alpha+1)\ln\beta-1]$ $+\frac{(1-\beta)^{\alpha+1}}{(\alpha+1)^2}[(\alpha+1)\ln(1-\beta)-1]$

TABLE 9.2.1 (*Continued*)

Integrand	Interval	Value of the integral
e^x	[0, 1]	$e-1=1.718281828\ldots$
$\dfrac{1}{e^x+1}$	[0, 1]	$\ln\dfrac{2e}{e+1}=0.379885493\ldots$
$\dfrac{x}{e^x+1}$	[0, 1]	$0.1705573501\ldots$
$\sqrt{50}e^{-50\dot{\pi}x^2}$ $\dot{\pi}=3.14159$	[0, 10]	$\dfrac{1}{2}\sqrt{\dfrac{\pi}{\dot{\pi}}}\operatorname{erf}(50\sqrt{2\dot{\pi}})\approx 1$
$25e^{-25x}$	[0, 10]	$1-e^{-250}=1.0000000000\ldots$
$e^{-x}-e^{-10x}$	[0, 1]	$0.9+\dfrac{e^{-10}}{10}-\dfrac{1}{e}=0.5321250988\ldots$
$\dfrac{2}{\sqrt{\pi}}(e^{-9x^2}+e^{-1024(x-1/4)^2})$	[0, 1]	$\frac{1}{16}+\frac{1}{3}\operatorname{erf}(3)=0.3958259695\ldots$
$\ln x$	[1, 10]	$10\ln 10-9=14.02585093\ldots$
$\ln x\cdot\sin x$	$[10^{-4}, 7]$	$\ln 10^{-4}\cos 10^{-4}-\ln 7\cos 7$ $+Ci(10^{-4})-Ci(7)$ $=-1.967546396\ldots$
$\dfrac{2^x x}{2^x-1}$	[0, 1]	$1.711857375\ldots$
$\sin(\pi x)$	[0, 1]	$2/\pi=0.6366197722\ldots$
$\cos x$	[0, 1]	$\sin 1=0.841470984\ldots$
$\dfrac{2}{2+\sin(10\pi x)}$	[0, 1]	$2/\sqrt{3}=1.154700538\ldots$
$x\cdot\sin(30x)\cdot\cos x$	$[0, 2\pi]$	$-60\pi/899=-0.2096724796\ldots$
$x\cdot\sin(30x)\cdot\cos(50x)$	$[0, 2\pi]$	$3\pi/80=0.1178097245\ldots$
$\dfrac{2\pi x\sin(30x)}{(4\pi^2-x^2)^{\frac{1}{2}}}$	$[0, 2\pi]$	$\pi^2 J_1(60\pi)$

TABLE 9.2.1. (*Continued*)

Integrand	k	m	Interval	Value of the Integral	
$0.92 \cosh x - \cos x$			$[-1, 1]$	$0.92(e - 1/e) - 2 \sin 1$ $= -0.479428228\ldots$	
$\sin (100\pi x)$			$[0, 1]$	0	
$\frac{1}{x} \sin \frac{1}{x}$			$[\frac{1}{20}, \frac{1}{3}]$	$Si(20) - Si(3)$	
$\tan (\pi x/2) \sin (m\pi x)$ $m = 1(1)4$		m	$[0, 1]$		
		1		1	
		2		-1	
		3		1	
		4		-1	
$\frac{\sin (m\pi x)}{\sin (\pi x)}$, $m = 2(1)4$		2		0	
		3	$[0, 1]$	1	
		4		0	
$x^k \sin (m\pi x)$ $m = 5(5)25$ $k = 0(1)3$	k	m	$[0, 1]$		
	0	5		$2/5\pi$	
		10		0	
		15		$2/15\pi$	
		20		0	
		25		$2/25\pi$	
	1	5		$1/5\pi$	
		10		$-1/10\pi$	
		15		$1/15\pi$	
		20		$-1/20\pi$	
		25		$1/25\pi$	
	2	5		$1/m\pi - 4/(m\pi)^3$,	$m = 5, 15, 25$
		10		$-(1/m\pi)$	$m = 10, 20$
		15			
		20			
		25			
	3	5		$1/m\pi - 6/(m\pi)^3$	$m = 5, 15, 25$
		10		$6/(m\pi)^3 - 1/m\pi$	$m = 10, 20$
		15			
		15			
		20			
		25			

TABLE 9.2.1. (*Continued*)

Integrand		Interval	Value of the integral
$\operatorname{sech}^2(10x-2)+\operatorname{sech}^4(100x-40)$ $+\operatorname{sech}^6(1000x-600)$		$[0, 1]$	$(\operatorname{tgh}8+\operatorname{tgh}2)/10+2/150$ $+2/1875=0.2108027355006\ldots$
$[mx]x^{k-1}$, $[x]$ step function	k	$[0, 1]$	
	0		$\sum_{i=1}^{m-1} i \ln\left(\frac{i+1}{i}\right)$
$m=5(5)25$	1		$\frac{m-1}{2}$
$k=0(1)2$	2		$\frac{4m^2-3m-1}{12m}$
$x+i$ for $i/3 \leq x<(i+1)/3$, $i=0(1)2$		$[0, 1]$	1.5
$\begin{cases}0 & 0.49<x<0.5\\ 1000(x-x^2) & \text{otherwise}\end{cases}$		$[0, 1]$	$1000(0.49^2/2-0.49^3/3+1/12)$ $=164.167$
$\begin{cases}1/(x+2) & 0 \leq x<\mathrm{e}-2\\ 0 & \mathrm{e}-2 \leq x \leq 1\end{cases}$		$[0, 1]$	$1-\ln 2=0.3068528195\ldots.$
$\begin{cases}0 & 0 \leq x<0.3\\ 1 & 0.3 \leq x \leq 1\end{cases}$		$[0, 1]$	0.7
$\begin{cases}0 & x=0\\ 1/\sqrt{x} & 0<x \leq 1\end{cases}$		$[0, 1]$	2.0
$\begin{cases}1 & x=0\\ x/(\mathrm{e}^x-1) & 0<x \leq 1\end{cases}$		$[0, 1]$	$0.777504635\ldots$
$\begin{cases}100 & x=0\\ \dfrac{\sin(100\dot{\pi}x)}{x} & 0<x \leq 1, \quad \dot{\pi}=3.14159\end{cases}$		$[0, 1]$	$\frac{1}{\dot{\pi}}(Si(100\dot{\pi})-Si(10\dot{\pi}))$
$\begin{cases}50 & x=0\\ \dfrac{\sin^2(50\dot{\pi}x)}{50(\dot{\pi}x)^2} & 0<x \leq 1, \quad \dot{\pi}=3.14159\end{cases}$		$[0, 1]$	$\frac{1}{\dot{\pi}} Si(100\dot{\pi})+\frac{\cos(100\dot{\pi})-1}{100(\dot{\pi})^2}$
$\begin{cases}0 & 0 \leq x<10^{-15}\\ \ln x & 10^{-15} \leq x \leq 1\end{cases}$		$[0, 1]$	$10^{-15}-1+15 \cdot 10^{-15} \ln 10 \approx -1$
$\begin{cases}\mathrm{e}^x & -1 \leq x<0\\ \mathrm{e}^{1-x} & 0 \leq x \leq 2\end{cases}$		$[-1, 2]$	$1+\mathrm{e}-2/\mathrm{e}=2.982522946\ldots$
$\begin{cases}\mathrm{e}^{10x} & -1 \leq x<0.5\\ \mathrm{e}^{10(1-x)} & 0.5 \leq x \leq 1.5\end{cases}$		$[-1, 1.5]$	$\frac{2\mathrm{e}^5+\mathrm{e}^{-5}-\mathrm{e}^{-10}}{10}=29.68330105\ldots$

Casaletto, J., Picket, M. and Rice, J. (1960). A comparison of some numerical integration programs. *SIGNUM Newsletter,* **4,** 30–40.

Gentleman, W. M. (1972). Implementing Clenshaw–Curtis quadrature I. *Comm. ACM,* **15,** 337–342.

Kahaner, D. K. (1971). Comparison of numerical quadrature formulas., *In* "Mathematical Software" (J. Rice, Ed.). pp. 229–259. Academic Press, London and New York.

Roothart, C. J. and Fiolet, H. (1972). "Quadrature Procedures." Mathematisch Centrum–Rekenafdeling–Leuven. Rep. MR137, and collected in: Piessens, R. and de Doncker. E. (1975). "Testing a New One-dimensional Quadrature (Preliminary Results)" Report TW24, Leuven.

9.3 Some published programs

The number of published programs for numerical quadrature and cubature is very large. There exist simple programs which are straightforward, using only one quadrature formula, as well as extended subroutine packages which give a choice between several formulae taking into account special properties of the integrand. Such properties are: the integrand function is (i) only tabulated, (ii) periodic, (iii) peaked, (iv) a rational function, (v) singular but integrable, (vi) integrable only in the Cauchy sense, (vii) only continuous, (viii) only piecewise continuous, (ix) polynomial-like.

Without claiming completeness we give references to some published programs which are written in a high-level programming language such as ALGOL, FORTRAN, PL/1 and FORMAC.

9.3.1 Romberg quadrature and modifications

Program presented in Chapter 8. (FORTRAN)

Antes, H. (1972). *Computing,* **9,** 51. (Cubature in 4 dimensions, FORTRAN)

Bauer, F. L. (1961). *Comm. ACM,* **4,** 255. (ALGOL); (1962). *Comm. ACM,* **5,** 168 and 881. (ALGOL); (1964). *Comm. ACM,* **7,** 420. (ALGOL)

de Boor, C. (1971). *In* "Mathematical Software" (J. Rice, Ed.). pp. 413–449. Academic Press, New York. (FORTRAN)

Burlirsch, R. (1964). Num. Math., **6,** 15. (ALGOL)

Burlirsch, R. and Stoer, J. (1964). *Num. Math.,* **6,** 413–427. (ALGOL); (1967). *Num. Math.,* **9,** 274. (ALGOL)

Davis, P. and Rabinowitz, P. book, pp. 375 and 376. (FORTRAN)
Fairweather, G. (1969). *Comm. ACM,* **12,** 324. (FORTRAN); (1970). *Comm. ACM,* **13,** 263, 374 and 449. (FORTRAN)
Gram, C. (1964). *BIT,* **4,** 54–60 and 118–120. (ALGOL)
Håvie, T. (1966). *BIT,* **6,** 24–30. (ALGOL)
IBM Scientific Subroutine Packages. (PL/1 and FORTRAN)
Kubik, R. N. (1965). *Comm. ACM,* **8,** 381. (ALGOL)
Laurent, J. P. (1967). *Procéd. ALGOL en Analyse Numér.* (*Paris*), pp. 239–240 (ALGOL) pp. 241–242. (ALGOL)
Stroud, A. H. book, p. 374. (N-cube, $2 \leqslant N \leqslant 5$, FORTRAN)

9.3.2 Adaptive and automatic integration programs

de Boor, C. (1971). *In* "Mathematical Software" (J. Rice, Ed.) pp. 201–209 and 417–449. Academic Press, New York. (FORTRAN)
Davis, P. and Rabinowitz, P., book p. 380. (FORTRAN)
Engels, H. (1969). "elektronische datenverarbeitung," p. 264. (FORTRAN)
Gallagher, L. J. (1967). *Comm. ACM,* **10,** 373–374. (ALGOL)
Kahaner, D. (1971). *In* "Mathematical Software" (J. Rice, Ed.) pp. 229–259. Academic Press, New York. (FORTRAN)
McKeeman, W. M. (1962). *Comm. ACM,* **5,** 604. (ALGOL)
(1963). *Comm. ACM,* **6,** 167. (ALGOL)
(1965). *Comm. ACM,* **8,** 171. (ALGOL)
Melendez, J. and Kahaner, D. (1971) *In* "Mathematical Software" (J. Rice, Ed.) p. 239. Academic Press, New York. (FORTRAN)
Oliver, J. (1972). *Comput. J.,* **15,** 141–147. (ALGOL)
Patterson, T. N. L. (1973). *Comm. ACM,* **16,** 694–699. (FORTRAN); see also Section 9.3.3
Piessens, R. (1973). *Angewandte Informatik,* pp. 399–401. (FORTRAN)
Piessens, R. (1975). Report TW 25, Leuven. (FORTRAN)
Piessens, R. and de Doncker, E. (1975). Report TW 24, Leuven. (FORTRAN)
Piessens, R., Mertens, I. and Branders, M. (1974). *Angewandte Informatik* pp. 65–68. (FORTRAN)
Robinson, I. G. (1971). *Austral. Comput. J.,* **3,** 126–129.

9.3.3 Gaussian quadrature programs

Davis, P. and Rabinowitz, P. book, p.363. (FORTRAN)
Gautschi, W. (1968). *Comm. ACM,* **11,** 432–436. (ALGOL)

(1969). *Comm. ACM,* **12,** 280–281. (ALGOL)
(1970). *Comm. ACM,* **13,** 512. (ALGOL)
Golub, G. H. and Welsch, J. H. (1969). *Math. Comput.*, **23,** 221–230. (ALGOL)
IBM Scientific Subroutine Packages. (FORTRAN and PL/1)

Computation of nodes and weights

Davis, P. and Rabinowitz, P. book, p.369. (FORTRAN)
Stroud, A. H. and Secrest, D. book.

Computation of weights of interpolatory quadrature rules

Engels, H. and Meuer, H. W. (1970). Report Jül-662-MA. (Nuclear Research Centre, Jülich. (PL/1-FORMAC)
Gustafson, S. A. (1971). *Comm. ACM,* **14,** 807. (ALGOL)

Optimal addition of nodes to Gaussian and Lobatto quadrature formulae: Calculation of nodes and weights via FORTRAN-program

Patterson, T. N. L. (1968). *Math. Comput.*, **22,** 842–856.
Piessens, R. and Branders, M. (1972). Report TW 10, Leuven.

Tschebyscheff quadrature

Hopgood, F. R. A. and Litherland, G. (1966). *Comm. ACM,* **9,** 270 and 434. (ALGOL)
(1967). *Comm. ACM,* **10,** 294 and 666. (ALGOL)

Wilf quadrature

Engels, H. and Eckhardt, U. (1977). *Computing,* **16.**

9.3.4 Clenshaw–Curtis quadrature

Davis, P. and Rabinowitz, P. book, p.384. (FORTRAN)
Gentleman, W. M. (1972). *Comm, ACM,* **15,** 353–355. (FORTRAN)
(1972). *Comm. ACM,* **16,** 490. (FORTRAN)
Oliver, J. (1972). *Comput. J.*, **15,** 141–147. (ALGOL)
Program presented in Chapter 7. (FORTRAN)

9.3.5 Rapidly oscillating integrand

Chase, S. M. and Fosdick, L. D. (1969). *Comm. ACM,* **12,** 457. (FORTRAN)
(1970). *Comm. ACM,* **13,** 263. (FORTRAN)

Davis, P., and Rabinowitz, P. book, p. 269. (FORTRAN)
Einarsson, B. (1972). *Comm. ACM,* **15,** 47–48. (FORTRAN)
(1974). *Comm. ACM,* **17,** 324. (FORTRAN)
Linz, P. (1972). *Comm. ACM,* **15,** 350–360. (FORTRAN)
Piessens, R. and Poleunis, F. (1971). *BIT,* **11,** 317–327. (FORTRAN)

Inversion of Laplace-Transforms

Piessens, R. (1973). *Comm. ACM,* **16,** 486–487. (FORTRAN)
Piessens, R. and Haegemans, A. (1974). *Computing,* **13,** 184.

9.3.6 Cubature programs (dimension N)

Antes, H. (1972). *Computing,* **9,** 45–52. (FORTRAN, $N=4$, Romberg)
Boland, W. R. (1972). Product-type formulae. *Comm. ACM,* **15,** 1070–1072. (FORTRAN: Trapezoidal, Simpson, Gauss two-point and Gauss three-point)
Davis, P. and Rabinowitz, P., book pp. 366, 371, 372 and 387. (FORTRAN)
Engels, H. (1967). "elektronische datenverarbeitung." pp. 345–350. (ALGOL, N-cube, $N=4$).
Freeman, Jr. R. D. (1961). *Comm. ACM,* **4,** 106. (ALGOL)
(1963). *Comm. ACM,* **6,** 69. (ALGOL)
(1968). *Comm. ACM,* **11,** 826. (ALGOL)
Gallagher, L. J. (1973). *Comm. ACM,* **16,** 49–50. (ALGOL, Monte Carlo method)
Good, I. J. and Gaskins, R. A. (1971). *Num. Math.,* **16,** 356–358. (FORTRAN, N-dimensional polygonal domain or N-simplex)
McKeeman, W. M. (1962). *Comm. ACM,* **5,** 604–605. (ALGOL)
(1963). *Comm. ACM,* **6,** 443–444. (ALGOL)
(1964). *Comm. ACM,* **7,** 296. (ALGOL)
Olynik, F. (1964). *Comm. ACM,* **7,** 348–349. (ALGOL)
(1970). *Comm. ACM,* **13,** 512. (ALGOL)
Stroud, A. H., book, pp. 347–348. ($N=2, 3$, iterated integrals, FORTRAN):
p. 350. ($N=2$, square, FORTRAN)
pp. 351, 352. ($N=3$ sphere, FORTRAN)
p. 353. (N-sphere for $N=2(1)8$, FORTRAN)
p. 359. ($N=1$, circumference of circle, FORTRAN)
p. 360. ($N=2$, surface of a 3-sphere, FORTRAN)
p. 362. ($N=3$, tori, FORTRAN)

p. 366. ($N=2(1)8$, N-simplex, FORTRAN)
p. 374. (N-cube, $N=2(1)5$, Romberg, FORTRAN)
Verjus-Chein, (1967). *PAAN* (procéd. ALGOL en anal. numér.) *Paris*, pp. 243–244. (Romberg, ALGOL)
Wilson, M. W. (1969). Program for the construction of positive *cubature formulae*. *Math. Comput.*, **23,** 253–258.

9.4 A list of tables

It is impossible to list here all the existing tables of nodes and weights of quadrature and cubature formulae. We present only a representative selection of such tables which will hopefully be sufficient for most practical purposes. The following abbreviations are chosen:

I = interval; Ω = domain of integration; N = dimension of Ω; n = number of nodes and weights; W = weight function; DFig = number of decimal figures of nodes or weights.

9.4.1 Quadrature formulae

Stroud and Don Secrest, book, 30 DFig.

W	I	n
1	$[-1, 1]$	2(1)63; 64(4)92; 96(8)168; 256; 384; 512
$(1-x^2)^\alpha$	$[-1, 1]$	2(1)20; $\alpha=-0.5;\ 0.5;\ 1;\ 1.5$
$(1+x)^\beta$	$[-1, 1]$	2(1)30; $\beta=1$ 2(1)20; $\beta=2;\ 3;\ 4$
$\|x\|^\alpha$	$[-1, 1]$	2(1)20; $\alpha=1(1)4$
$\exp(-x^2)$	$(-\infty, \infty)$	2(1)63; 64(4)92; 96(8)136
$\exp(-x)$	$(0, \infty)$	2(1)31; 32(4)68
$\|x\|^\alpha \exp(-x^2)$	$(-\infty, \infty)$	2(1)20; $\alpha=1;2;3$
$\|x\|^\alpha \exp(-\|x\|)$	$(-\infty, \infty)$	2(1)20; $\alpha=1;2;3$
$\ln 1/x=-\ln x$	$[0, 1]$	2(1)16
$\exp(x)/x$	$(c+i\infty, c-i\infty)$	2(1)24
1	$[-1, 1]$	2(1)18; 19(4)47; -1 preassigned, Radau formulae
1	$[-1, 1]$	2(1)31; 32(4)96; ± 1 preassigned, Lobatto formulae
1	$[-1, 1]$	2(2)16; $f^{(k)}(x)$ preassigned at $x=0$, $k=1;2;3$
$\exp(-x^2)$	$(-\infty, \infty)$	2(2)16; $f^{(k)}(x)$ preassigned at $x=0$, $k=1;2;3$

Abramowitz, M. and Stegun, I. (1965). "Handbook of Mathematical Functions." Dover Publications, New York.

W	I	n
1	$[-1, 1]$	2(1)10; 12; 16(4)24; 32(8)48; 64; 80; 96 15DFig 20DFig
1	$[-1, 1]$	3(1)10; 10DFig; ± 1 preassigned; Lobatto formulae
x^k, $k=0(1)5$	$[0, 1]$	1(1)8; 10DFig
$\exp(-x)$	$[0, \infty)$	2(1)10; 12; 15; 12DFig
$\exp(-x^2)$	$(-\infty, \infty)$	2(1)10; 12; 16; 20; 15DFig

Further: Coefficients of Filon quadrature.

The following tables usually contain only one type of formula.

Aizensthat, V. S., Krylov, V. I. and Metelskii, A. S. (1962). Izdat. Akad. Nauk BSSR, Minsk.

• Laplace Transforms and integrals with $W = x^s \exp(-x)$, $I = [0, \infty)$ for $s = -0.9(0.02)0$; $0.55(0.05)3.0$; $-\frac{3}{4}$; $-\frac{1}{4}$; $m + k/3$, $m = -1(1)2$, $k = 1; 2$ $n = 1(1)15$, 8DFig

Berger, B. S. Tables of zeros and weights of Gauss–Laguerre quadrature to 23–24DFig for $n = 100$; 150; 200(100)900, and Tables of zeros and weights for Gauss–Hermite quadrature to 26–27DFig for $n = 200(200)10000$; 2000. Department of Mechanical Engineering, University of Maryland.

Engels, H. (1969). Zur iterativen Behandlung der Interpolationsquadratur. "elektronische datenverarbeitung." p. 263. $W = (1-x^2)^2$; $n = 1(1)13$; 15DFig.

Engels, H. and Eckhardt, U. (1974). "Symmetrische Wilf'sche Quadraturformeln." Report Jül–1094–MA. Nuclear Research Centre Jülich. $n = 2(1)32$, 25DFig.

Engels, H. and Mangold, R. (1971). "Über eine Klasse Wilf'scher Quadraturformeln." Report Jül–789–MA. Nuclear Research Centre Jülich. $I = [-1, 1]$, $n = 1(1)10$, open Wilf formulae for preassigned equidistant nodes and for the nodes

$$x_i = 1 - \left(\frac{n+1-i}{n+1}\right)^{i+1}, \qquad i = 1(1)n, \quad \text{for} \quad I = [0, 1]$$

$$x_i = \pm\left(1 - \left(\frac{n-i}{n}\right)^{i+1}\right), \qquad i = 1(1)n, \quad \text{for} \quad I = [-1, 1].$$

Engels, H. and Meuer, H. W. (1970). "Die Entwicklung von Interpolationsquadratur- und Differentiationsformeln zu gegebenen

Stützstellen mit PL/I–FORMAC." Report Jül–662–MA. Nuclear Research Centre Jülich.

$W = x^\alpha \ln 1/x$, $I = [0, 1]$, $n = 2(1)10$,
rational weights, $\alpha = -0.5(0.5)3.0$

$W = x^\alpha e^{-x}$, $I = [0, \infty)$, $n = 2(1)10$,
rational weights, $\alpha = -0.5(0.5)3.0$

Engels, H. and Meuer, H. W. (1970). "Quadraturformeln mit äquidistanten Abszissen zur Berechnung von Integralen der Form

$$\int_0^1 x^A (1-x)^B f(x)\,dx$$

für spezielle Werte von A und B." Report Jül–648–MA. Nuclear Research Centre Jülich.
Closed quadrature formulae with preassigned equidistant nodes for $I = [0, 1]$ with respect to the Jacobi weight function for $n = 2(1)10$ and for $A = -0.9(0.1)1.0$; $B = A(0.1)1.0$; weights: rational numbers. The Newton–Cotes formulae are contained for $A = B = 0$.

Glonti, E. N. (1971). Tables of roots and quadrature coefficients of Jacobi polynomials. [Russian] *In* "Mathematical Tables." (A. Ditkin, Ed.). Computer Centre of the Academy of Science, USSR.
$W = x^{q-1}(1-x)^{p-q}$, $I = [0, 1]$, 15DFig,
$n = 2(1)15$, $q = 0.1(0.1)1$; $p = 2q - 1(0.1)q + 1$.

Kronrod, A. S. (1965). "Nodes and Weights of Quadrature Formulas." Consultants Bureau, New York. $n = 1(1)40$, $I = [0, 1]$ 20DFig and 20 octal figures. (Extended formulae: $n = 3(2)81$)

Kruglikova, L. (1964). "Tables for the Numerical Fourier Transform." [Russian] Nauka i Tehnika, Minsk. Formulae exact for special rational functions.

Krylov, V. I., Lugin, V. V. and Janovič, L. A., (1963). "Tables for the Numerical Integration of Functions with Power Singularities." Akad. Nauk BSSR, Minsk,

$W = x^\alpha (1-x)^\beta$, $\alpha, \beta = -0.9(0.1)3.0$, $\alpha < \beta$, $n = 1(1)8$, 8DFig.

Krylov, V. I. and Pal'cev, A. A. (1967). "Tables for the Numerical Integration of Functions with Logarithmic and Power Singularities

$$\int_0^1 x^\alpha \ln \frac{e}{x} f(x)\,dx, \qquad \int_0^1 x^\beta \ln \frac{e}{x} \ln \frac{e}{1-x} f(x)\,dx,$$

$$\int_0^1 \ln \frac{1}{x}\,dx, \qquad \int_0^\infty x^\beta e^{-x} \ln \left(1+\frac{1}{x}\right) f(x)\,dx."$$

[Russian] Nauka i Tehnika, Minsk.

$n = 1(1)10, \quad \alpha = -0.9(0.01)0.0(0.1)5, \quad \beta = 0(1)5, \quad$ 15DFig.

Krylov, V. I. and Skoblia, N. S. (1974). "Handbook of Laplace Transforms." Israel. Progr. Scient. Transl., Jerusalem.

Formulae: $\frac{1}{2\pi i}\int_{c-i\infty}^{c+i\infty} e^x x^{-s} f(x)\,dx \doteq \sum_{k=1}^{n} A_k f(x_k)$

exact for $f = x^{-j}, \quad j = 0(1)2n-1$

$s = 1(1)5, \quad n = 1(1)15,$

$s = 0.01(0.01)3.00, \quad s \neq 1; 2; 3, \quad n = 1(1)10.$ 7–8DFig

Krylov, V. I. and Sul'gina, L. T. (1966). "Handbook for Numerical Integration." Nauka, Moskau.

Some topics: bibliography: 46 pages;
various tables for numerical integration: 166 pages;
Fourier, Laplace, Mellin transforms;
integration of periodic functions;
indefinite integration and its stability;
multiple integration.

Krylov, V. I. and Vorob'eva, A. A. (1971). "Tables for Calculating Integrals of Functions which contain Power Singularities

$$\int_0^1 x^\alpha f(x)\,dx, \qquad \int_0^1 x^\alpha (1-x)^\alpha f(x)\,dx."$$

[Russian] Nauka i Tehnika, Minsk.

$\alpha = -\frac{2}{3}(\frac{1}{3})\frac{8}{3}, \quad -\frac{3}{4}(\frac{1}{2})\frac{11}{4}, \quad -0.9(0.1)3.0, \quad n = 1(1)15,$ 20DFig.

$f(x) = (1+x)^{-s-i}, \quad i = 0(1)2n-1,$

for $s = m+\frac{1}{4}, \ m+\frac{1}{3}, \ m+\frac{1}{2}, \ m+\frac{2}{3}, \ m+\frac{3}{4}, \ m = 1; 2; 3,$ and $s = 4, \ n = 1(1)8,$ 10DFig.

Miller, J. C. P. (1960). "Quadrature of Equally-spaced Function Values." Techn Report 167, MRC, Madison, Wisconsin. Very extended collection of formulae with preassigned equally-spaced nodes.

Piessens, R. (1969). "Gaussian Quadrature Formulas for the Numerical Integration of Bromwich's Integral and the Inversion of the Laplace Transform." Report TW 1, Leuven.

Piessens, R. (1969). "Integration Formulas of Interpolatory Type for the Inversion of the Laplace Transform." Report TW 2, Leuven.

Piessens, R. (1969). "Tables for the Inversion of the Laplace Transform." Report TW 3, Leuven.

Piessens, R. and Branders, M. (1972). "Optimal Addition of Abscissas to Quadrature Formulae of Gauss and Lobatto Type." Report TW 10, Leuven. (See also (1974) *Math. Comput.*, **28,** 135–139.)
$n = 3(1)10(10)100$; 200 (extended formulae: $n = 7(2)21(20)201$, 401 $I = [-1, 1]$, 20DFig.

Shao, T. S., Chen, T. C. and Frank, R. M. (1964). "Tables of Zeros and Gaussian Weights of certain Associated Laguerre Polynomials and the Related Generalized Hermite Polynomials." Techn. Rep. 00.1100 IBM Data Systems Division, New York. (See also (1964) *Math. Comput.*, **18,** 598–616.)

$W = x^\alpha e^{-x}$, $\alpha = -0.5(0.5)10$, $n = 4, 8(8)32(16)64(32)128(64)256$, 25DFig
$W = |x|^\alpha \exp(-x^2)$, $\alpha = 0(1)10$, $n = 8(8)32(16)64(32)128(64)256$, 25DFig.

Skoblia, N. S. (1964). "Tables for the Numerical Inversion of Laplace Transforms." Nauka i Tehnika, Minsk.

$n = 1(1)10$, 7–8DFig.

Welsch, J. H. (1966). Abscissas and weights for Gregory quadrature. *Comm. ACM*, **9,** 271. (Special formulae with equidistant nodes.)

Welsch, J. H. (1967). Abscissas and weights for Romberg quadrature. *Comm. ACM*, **9,** 271, and *Comm. ACM*, **10,** 188.

9.4.2 Cubature formulae

Most of the known cubature formulae are given by Stroud, book. We present here a guide to the numerous formulae of that book and add some recent papers which are not contained there.

Degree	Number of nodes	Ω
1	$1;\ 2^N$	N-cube
2	$N+1;\ 2N+1$	
3	$2N;\ 2N+1;\ 2^N;\ 2^N+1;\ 3^N$	
5	$N^2+N+2;\ (N=4;5;6);\ 2N^2+1;\ 3N^2+3N+1;\ 2^N+2N$ $2^N+2N+1;\ 2^{N+1}+1;\ 2^N N+1;\ 2^N(N+1);\ 3^N$	
7	$(4N^3-6N^2+14N+3)/3;\ (4N^3+8N+3)/3$	
9	$4(N^4-5N^3+14N^2-7N+3)/3\ (N>4);\ (2N^4-4N^3+22N^2-8N+3)$	
11	$(4N^5-20N^4+140N^3-130N^2+96N+15)/15$	

Degree	Number of nodes	Ω
1	4; 9	2-cube
2	6	
3	4; 5; 9; 12	
5	7; 8; 9; 13; 24	
7	12; 13; 16; 25	
9	20; 21; 25	
11	25; 28	
13	37	
15	44; 48	
3	6; 8; 9; 13; 15; 19; 20	3-cube
5	13; 14; 21; 23; 25; 27; 29; 42	
7	27; 34	
5	31	
7	49	
3	$2N$; 2^N; 2^{N+1}; $2^N N$	shell of N-cube
2	$N+1$	N-sphere
3	$2N$ $(W=(\mathbf{x},\mathbf{x})^{\alpha/2})$; 2^N	
5	N^2+N+2; $2N^2+1$ $(W=(\mathbf{x},\mathbf{x})^{\alpha/2})$; 2^N+2N	
	$2^{N+1}-1$ $(W=(\mathbf{x},\mathbf{x})^{\alpha/2})$; $2^N(N+1)$	$\alpha>-N$
7	2^N+2N^2+1; $2^{N+1}+4N^2$; $(4N^3+8N+3)/3$	
9	$(2N^4-3N^3+22N^2-8N+3)/3$	
11	$(4N^5-20N^4+140N^3-130N^2+96N+15)/15$	
3	4	2-sphere
4	$6(W=(\mathbf{x},\mathbf{x})^{\alpha/2},\ \alpha>-2)$	
5	$7(W=(\mathbf{x},\mathbf{x})^{\alpha/2},\ \alpha>-2)$; 9	
7	12; 16	
9	19; 20; 21; 28	
11	28; 32	
13	37; 41	
15	84	
17	61	
3	6	3-sphere
5	$13\ (W=(\mathbf{x},\mathbf{x})^{\alpha/2},\ \alpha>-3)$; 21	
7	27; 32; 33; 64	
14	288	
5	25	4-sphere
7	49; 72	
3	$2N$	shell of N-sphere
7	$2MN$ (M specified by another formula)	

Degree	Number of nodes	Ω
3	$2N$; 2^N	surface of N-sphere
5	$2N^2$; 2^N+2N; $2^{N+1}-2$; $2^N N$	
7	2^N+2N^2 (one of a whole class); $2^N(N+1)$	
11	$2^N(N^2+N+1)$ $(N\geqslant 5)$	
3	12	surface of 3-sphere
5	12; 14; 18; 20; 30	
7	24; 26	
8	30	
9	32; 42; 50	
11	50; 56; 62	
14	72	
5	24	surface of 4-sphere
7	48	
11	240	
2	$N+1$	N-octahedron
3	$2N$; 2^N	
5	N^2+N+2; $4N^2-4N+1$; 2^N+2N; $2^{N+1}-1$	
5	13	3-octahedron
1	1; $N+1$	N-simplex
2	$N+1$; $(N+1)(N+2)/2$	
3	$N+2$; $2N+2$; $2N+3$; $\binom{N+2}{2}$; $(N+1)(N+4)/2$; $N(N+1)$; $(N^3+5N+12)/6$ $(N\geqslant 3)$; $(N^3+11N+12)/6$; $\binom{N+3}{3}$; $(N^3-N+3)/3$ $(N\geqslant 3, n\neq 5)$; $(N^3+2N+3)/3$ $(N\geqslant 3, N\neq 5)$	
4	$(N+4)!/(4!N!)$ $(N\geqslant 3)$	
5	N^2+3N+3; $(N+5)!/(5!N!)$ $(N\geqslant 4)$	
3	6	
5	7	2-simplex
7	16	
5	15	3-simplex
7	64	
3	16	5-simplex
3	$2N$; 2^N	N-space $W=e^{-(\mathbf{x},\mathbf{x})}$
5	N^2+N+2; $2N^2+1$; 2^N+2N; $2^{N+1}-1$; $2^N N+1$; $2^N(N+1)$	
7	2^N+2N^2+1; $2^{N+1}+4N^2$; $(4N^3+8N+3)/3$	
9	$(2N^4-4N^3+22N^2-8N+3)/3$	
11	$(4N^5-20N^4+140N^3-130N^2+96N+15)/15$	

Degree	Number of nodes	Ω
4	6	2-space
5	7; 9	$W=e^{-(\mathbf{x},\mathbf{x})}$
7	12; 16	
9	20	
11	28	
13	37	
15	44	
5	13	3-space
5	14; 15; 21	$W=e^{-(\mathbf{x},\mathbf{x})}$
7	27; 33	
14	288	
5	25	4-space
7	49	$W=e^{-(\mathbf{x},\mathbf{x})}$
3	$2N$; 2^N	N-space
5	$2N^2+1$; 2^N+2N $(N\neq 3)$; $2^{N+1}-1$; $2^N N+1$; $2^N(N+1)$	$W=e^{-\lvert\mathbf{x}\rvert}$
7	2^N+2N^2+1	
4	6	2-space
5	7	$W=e^{-\lvert\mathbf{x}\rvert}$
7	12	
9	20	
11	28	
13	37	
15	44	
5	13; 15; 21	3-space
7	27; 33	$W=e^{-\lvert\mathbf{x}\rvert}$
5	25	4-space
7	49	$W=e^{-\lvert\mathbf{x}\rvert}$
3	4; 7	Hexagon
5	7	$N=2$
3	4; 4	Ellipse
5	7	
7	12	$W=((x-c)^2+y^2)((x+c)^2+y^2)$, $(\pm c, 0)$ foci

We suppress a few more or less unnatural regions without practical importance.

For further formulae the reader is referred to Chapters 3 and 4 of the Stroud book and the references given there.

We present the corresponding table for the results of Haegemans who calculated many new formulae in her thesis (cf. Chapter 3); some of these formulae are listed there for the 2-sphere.

Degree	Number of nodes	Ω
7	12	2-cube
9	18; 19	
11	25; 26; 28	
9	18; 19	2-sphere
11	25; 26; 28	
13	41	
19	71; 76	
21	90; 99	
23	97; 108	
25	127	
31	172	
9	18; 19	2-space $W=e^{-(\mathbf{x},\mathbf{x})}$
11	25; 26; 28	
13	41	
19	71; 76	
21	90; 99	
23	97; 108	
25	127	
27	140	
31	172	
7	12	2-space $W=e^{-\lvert\mathbf{x}\rvert}$
9	19	
11	26; 28	
13	41	
15	44	
17	61	
19	71, 76	
21	90; 99	
23	108	
25	108	
27	140	
31	172	

Möller (cf. Chapter 5) obtained the minimal number of 17 nodes for a ninth-degree cubature formula for the square.

Author Index

Italic page numbers indicate that the name appears on that page and following pages.

Index of Key-words

Italic page numbers indicate that the key-word appears on that page and following pages.

Index of Programs and Tables in this Book

Page numbers are given in italic

Programs

Tables